INTRODUCTION TO OPERATIONAL AMPLIFIER THEORY AND APPLICATIONS

McGraw-Hill Series in Electrical Engineering

Consulting Editor

Stephen W. Director, *Carnegie-Mellon University*

Circuits and Systems
Communications and Signal Processing
Control Theory
Electronics and Electronic Circuits
Power and Energy
Electromagnetics
Computer Engineering
Introductory
Radar and Antennas
VLSI

Previous Consulting Editors

Ronald N. Bracewell, Colin Cherry, James F. Gibbons, Willis W. Harman, Hubert Heffner, Edward W. Herold, John G. Linvill, Simon Ramo, Ronald A. Rohrer, Anthony E. Siegman, Charles Susskind, Frederick E. Terman, John G. Truxal, Ernst Weber, and John R. Whinnery

Electronics and Electronic Circuits

Consulting Editor

Stephen W. Director, *Carnegie-Mellon University*

Colclaser and Diehl-Nagle: *Materials and Devices for Electrical Engineers and Physicians*
Fabricius: *Introduction to VLSI Design*
Ferendeci: *Physical Foundations of Solid State and Electron Devices*
Franco: *Design with Operational Amplifiers and Analog Integrated Circuits*
Grinich and Jackson: *Introduction to Integrated Circuits*
Hamilton and Howard: *Basic Integrated Circuits Engineering*
Hodges and Jackson: *Analysis and Design of Digital Integrated Circuits*
Long and Butner: *Gallium Arsenide Digital Integrated Circuit Design*
Millman and Grabel: *Microelectronics*
Millman and Halkias: *Integrated Electronics: Analog, Digital Circuits, and Systems*
Millman and Taub: *Pulse, Digital, and Switching Waveforms*
Paul: *Analysis of Linear Circuits*
Roulston: *Bipolar Semiconductor Devices*
Schilling and Belove: *Electronic Circuits: Discrete and Integrated*
Smith: *Modern Communication Circuits*
Sze: *VLSI Technology*
Taub: *Digital Circuits and Microprocessors*
Taub and Schilling: *Digital Integrated Electronics*
Wait, Huelsman, and Korn: *Introduction to Operational and Amplifier Theory Applications*
Yang: *Microelectronic Devices*
Zambuto: *Semiconductor Devices*

INTRODUCTION TO OPERATIONAL AMPLIFIER THEORY AND APPLICATIONS

Second Edition

John V. Wait

Department of Electrical and Computer Engineering
University of Arizona

Lawrence P. Huelsman

Department of Electrical and Computer Engineering
University of Arizona

Granino A. Korn

McGraw-Hill, Inc.

New York St. Louis San Francisco Auckland Bogotá
Caracas Lisbon London Madrid Mexico Milan
Montreal New Delhi Paris San Juan Singapore
Sydney Tokyo Toronto

This book was set in Times Roman by Publication Services.
The editors were Roger L. Howell and Jack Maisel;
the production supervisor was Richard A. Ausburn.
The cover was designed by Joseph Gillians.
R. R. Donnelley & Sons Company was printer and binder.

INTRODUCTION TO OPERATIONAL AMPLIFIER THEORY AND APPLICATIONS

2 3 4 5 6 7 8 9 0 DOC DOC 9 0 9 8 7 6 5 4 3 2 1

ISBN 0-07-067770-0

Library of Congress Cataloging-in-Publication Data

Wait, John V.
Introduction to operational amplifier theory and applications / John V. Wait, Lawrence P. Huelsman, Granino A. Korn. — 2nd ed.
p. cm.
Includes bibliographical references and index.
ISBN 0-07-067770-0
1. Operational amplifiers. 2. Electronic circuits. I. Huelsman, Lawrence P. II. Korn, Granino Arthur, (date). III. Title.
TK7871.58.O6W3 1992
621.39′5—dc20 91-121

To HAROLD N. and INEZ F. WAIT
ELIZABETH O. and DAVID L. HUELSMAN
ARTHUR and ELIZABETH P. KORN

CONTENTS

Preface xiii

1 Basic Operational Amplifier Circuits 1

1.1 General Characteristics of Operational Amplifiers 2
1.2 The Ideal Amplifier Model 4
1.3 The Summer-Inverter 7
1.4 The Inverting Summer-Integrator 12
1.5 The General Inverting Amplifier 24
1.6 The Noninverting Amplifier 28
1.7 The Differential Amplifier 32
1.8 The Combined Inverting and Noninverting Amplifier 35
1.9 Simple First-Order Transfer Functions 40
1.10 Miscellaneous Linear Applications 48
References and Bibliography 58
Problems 59

2 Performance Limitations: The Nonideal Amplifier 74

2.1 The General Nonideal Linear Model 75
2.2 Closed-Loop Gain, Output Impedance, and Input Impedance 79
2.3 Closed-Loop Stability; Bode Plot Methods; Compensation 95
2.4 Common-Mode Effects 103
2.5 DC Offsets and Temperature Drift 107
2.6 Amplifier Internal Noise Effects 114
2.7 Large-Signal Limitations 125
2.8 Output Power Boosting 129
2.9 Power Supplies, Grounding, and Shielding 131
2.10 Components: Resistors and Capacitors 135
2.11 Computer-Aided Design 137
References and Bibliography 141
Problems 141

3 Nonlinear Circuit Applications 150

3.1 General Nonlinear Design Approach 151
3.2 Feedback Limiters and Comparators 159
3.3 Additional Piecewise-Linear Functions 179
3.4 Logarithmic Operators 193
3.5 Multipliers, Dividers, and Applications 201
3.6 Waveform Generation 211
References and Bibliography 222
Problems 223

4 Active *RC* Filters 239

4.1 Filter Types 240
4.2 Normalized and Denormalized Filters 244
4.3 A Second-Order Low-Pass Filter—Sensitivity 247
4.4 Other Second-Order VCVS Filters 255
4.5 Second-Order Multiple-Feedback Filters 265
4.6 High-*Q* Filters 272
4.7 Cascade Realization of High-Order Filters—Tuning 276
4.8 Generalized Impedance Converter Filters 287
4.9 Leapfrog Filters 295
4.10 Switched Capacitor Filters 303
Bibliography 315

5 Electronic Switching Circuits 316

5.1 Introduction 316
5.2 Ideal Switches and Real Switching Circuits 317
5.3 Diode Switching Circuits 319
5.4 Bipolar-Transistor Switches 321
5.5 Junction Field-effect Transistor Switches 325
5.6 MOSFET and CMOS Integrated Circuits 328
5.7 Design Decisions and Improved Switches 332
5.8 Sample-Hold Circuits 333
5.9 Integrator Mode Control and Related Switching Circuits 337
References and Bibliography 343

6 Digital-to-Analog and Analog-to-Digital Converters 344

6.1 Digital Representation of Voltages; Simple Conversion Circuits 344
6.2 Practical Digital-to-Analog Converters; Voltage-Ladder Circuits 347
6.3 Current-Switching DACs 357
6.4 Static Error Specifications 360
6.5 DAC Dynamic Performance: Settling Time, Phase Shift, Spikes, and Glitches 362
6.6 Choice of a Digital-to-Analog Converter: Design Considerations 364
6.7 Analog-to-Digital Conversion 365
6.8 Parallel-Comparator and Cascade-Encoder ADCs 365

6.9 Ramp-Comparator Converters 366
6.10 Integrating Converters 368
6.11 Analog-to-Digital Converters Using Digital-to-Analog Feedback 370
6.12 ADC System Design Considerations 374
References and Bibliography 375
Problems 375

Index 377

PREFACE

During the 1960s transistorized solid-state amplifier technology evolved rapidly. Today such amplifiers in microminiature form have become basic building blocks that are used in a broad range of electronic circuit applications, e.g., amplification, filtering, nonlinear waveshaping, waveform generation, and switching. This text was developed to be used with a course in operational amplifier circuit design for senior and graduate students in electrical engineering; it covers techniques of analysis and design of electronic circuits, using operational amplifiers, resistors, capacitors, diodes, and other special components as needed.

In this second edition, coming over a decade later, we include the impact of integrated-circuit technology and computer-aided design (CAD) methods. We have also revised and augmented the problems sets.

This book does not discuss the internal design of operational amplifiers themselves. Considerable attention is paid, however, to the ways in which nonideal properties that arise from internal design limitations affect circuit performance. The text assumes the typical background of an electrical engineering senior, including some prior knowledge of Laplace transforms, frequency response operators, and Bode plots, as well as a basic feeling for solid-state device characteristics.

This book should also be quite useful to the practicing electrical engineer as a self-teaching aid and as a source of applications ideas.

Chapter 1 provides a thorough presentation of *basic linear operational amplifier circuit design*, using the *ideal operational amplifier model*. This chapter also reviews important active circuit analysis techniques in general by using many examples. The ideal model is carefully defined, with an eye toward a treatment of nonideal effects in Chap. 2. The following sections present detailed discussions of the major circuit configurations, including the summer-inverter, the inverting summer-integrator (including mode control), the generalized inverting amplifier with two-port networks, the noninverting amplifier, and the differential amplifier. The eighth section is a general discussion of combined inverting and noninverting summer-amplifiers, which is not readily available elsewhere. The ninth section describes important first-order filter circuits; we postpone a detailed treatment of active *RC* filter circuits until Chap. 4, however. The last section presents a few of the well-known miscellaneous applications, such as instrumentation amplifiers, reference sources, and voltage-current converters. The reader will also find that the more advanced exercises will suggest many more useful applications.

Chapter 2 discusses means for analyzing the effects of *amplifier performance limitations* (nonidealness) on circuit performance. A general nonideal small-signal linear model is presented first, and we discuss techniques based on this model that may be used to estimate closed-loop circuit gain and circuit output and input impedance. Next we present a discussion of Bode plot techniques for estimating closed-loop circuit stability, and the problem of compensation is examined. Common-mode effects are described in the fourth section. Dc offsets and temperature drift effects are described in the fifth section, and standard compensation and balancing techniques are also described. A description of random noise effects, based upon a power spectral density model, is presented in the sixth section; the reader may want to skip this part on a first reading, since some appreciation of the principles of random process theory is required to apply the model. Large-signal limitations are discussed in the seventh section, e.g., slewing rate, full-power bandwidth, settling time, overload recovery time, and maximum common-mode voltage. The last four sections are devoted to practical considerations in operational amplifier use. The eighth section discusses output power boosting techniques. The ninth section surveys power supply, and grounding and shielding considerations, while the tenth section discusses passive components. Finally, there is a discussion of CAD modelling methods with references.

Chapter 3 is concerned with the design of *nonlinear circuits*. This chapter presents a number of nonlinear circuit design methods that are hopefully of general value. Piecewise-linear diode and transistor models are reviewed and applied to some simple examples. The second section discusses a variety of limiter and comparator circuits and compares their merits. The use of hysteresis for noise-immune comparator design is included. General piecewise-linear function generation methods are presented in the third section. The fourth section is devoted to logarithmic operators to help the reader evaluate these types of nonlinear operators in a specific application. Conventional methods for multiplying, dividing, and squaring are described and compared in the fifth section, and typical applications are presented. Finally, the sixth section presents a collection of well-known waveform generation circuits. Again, the examples and exercises suggest additional applications.

The remaining chapters cover advanced applications. Chapter 4 discusses one of the more important applications of operational amplifiers, namely, *active RC filters*. In the first section, some general design and analysis techniques are introduced. A discussion of sensitivity is included to permit the reader to evaluate different filter realizations. The following two sections present several of the most important filter configurations. These include the Sallen and Key structures and the infinite-gain multiple-feedback configurations. The fourth section covers active *RC* filter realizations for the high-Q case. Both state-variable and

Tarmy-Ghausi realizations are presented. The effect of finite amplifier bandwidth on the realizations is discussed. In the fifth section, the use of cascade techniques to realize filters of higher than second order is covered. The sixth and seventh sections discuss alternative methods of active *RC* filter realization, including the use of gyrators or negative-impedance converters, and the recently proposed technique of using frequency-dependent negative resistors as produced by generalized impedance converters. In all cases, design tables are provided which enable the reader to easily realize specific filter configurations for several of the most common maximally flat magnitude (Butterworth), equal ripple (Chebychev), maximally flat delay (Thomson), and elliptic (Cauer) characteristics. Approximation tables for these characteristics are provided through the sixth order. In many cases alternative realizations are provided to make fabrication easier by specifying equal-valued *R* and *C* elements or to reduce sensitivity by using integer-valued gains. A discussion of tuning procedures for high-order realizations is also included. Many of the tables included in this chapter have never before appeared in the literature. Of special importance is the compilation of quadratic factors for a wide range of high-order filter characteristics.

Chapter 5 introduces up-to-date *electronic switching circuits*, including diode, transistor, JFET, MOSFET, and CMOS switches. Many examples show how to minimize circuit error due to leakage, on resistance, offsets, and switching spikes. Electronic multiplexer circuits, sample-holds, and intergrator-control switching are discussed together with error sources and modern design practice.

Chapter 6 deals with *digital-to-analog and analog-to digital converters*, both now available as relatively low-cost integrated circuits, indispensible in a designer's bag of tricks. Several sections discuss digital codes, negative voltages, and then voltage, inverted, and current-ladder circuits. The revised text goes on to describe the design of temperature-compensated CMOS and current-switching digital-to-analog converters and to discuss settling time and error sources. Flash, cascade, integrating, and successive-approximation analog-to-digital converters are introduced next. As in Chapter 5, tables provide up-to-date IC and module specifications, and separate sections list considerations for design decisions involving the choice of specific types of circuits.

There is clearly more material than one can thoroughly cover in a typical 3-hour one-semester course. This has been done intentionally, so that the book may be used for both an in-depth course on design fundamentals and a survey course on applications. We have been using class notes, from which this book originated, in a senior technical elective at the University of Arizona for four years, with considerable success. Our approach is to spend about 40 percent of the semester on the fundamental principles in Chaps. 1 and 2; we work many problems here. Then we use the material of Chap. 3 for about 20 percent of the course, with the primary goal of sharpening the student's ability to design and analyze nonlinear circuits using primarily piecewise-linear diode and transistor models. The remaining 40 percent of the course is devoted to surveying the applications described in Chaps. 4 to 6.

McGraw-Hill and the authors would like to thank the following reviewers for their many helpful comments and suggestions: Stan Burns, Iowa State University; Jerrold Krenz, University of Colorado at Boulder; A. A. Pandiscio, Harvard University; Edgar Sanchez-Sinencio, Texas A&M University; N. S. Smith, West Virginia University; and James Svoboda, Clarkson University.

John V. Wait
Lawrence P. Huelsman
Granino A. Korn

CHAPTER 1

BASIC OPERATIONAL AMPLIFIER CIRCUITS

During the 1960s the solid-state dc operational amplifier grew rapidly in importance to the electronic circuit designer. Today, such amplifiers, in microminiature form (Fig. 1.1), have become so *reliable, compact,* and *easy to use* that they are *basic electronic circuit building blocks* used in an extremely broad range of applications, including amplification, filtering, nonlinear waveshaping, wave generation, and switching. Today, instead of assembling an amplifier from dozens of components, the wise circuit designer frequently develops a circuit around a few standard, commercially available *prepackaged operational amplifiers*, which *comprise the active elements;* the rest of the circuit will be made up of resistors, capacitors, diodes, and other special components as needed.

The use of such standard active elements eliminates the need for detailed design of individual transistor stages. When they are properly used, the overall transfer characteristics of a circuit (gain, frequency response, etc.) can be precisely controlled by stable passive elements (e.g., resistors, capacitors, diodes). *Feedback techniques* are used to suppress any nonideal properties of the operational amplifier so that individual variations in particular amplifiers have a negligible effect on final circuit performance. Hence circuit designs based on the use of operational amplifiers usually have *highly predictable performance*.

When most of the active elements of a circuit are concentrated into the small space occupied by a modern solid-state operational amplifier, *miniaturization* of electronic circuits *is facilitated*. Moreover, today's operational amplifiers are *highly*

FIGURE 1.1
Packaging of integrated-circuit operational amplifiers. Metal can and epoxy-encapsulated units are shown, along with an internal view of an integrated circuit. (*Burr-Brown Rsch. Corp.*)

reliable, and circuit designers can expect that production copies of their final system will closely emulate the performance of the prototype and will require a minimum of initial debugging and little long-term maintenance (if they have chosen operational amplifiers supplied by a competent manufacturer).

1.1 GENERAL CHARACTERISTICS OF OPERATIONAL AMPLIFIERS

Today most solid-state operational amplifiers are fabricated in integrated-circuit form. In the 1970s discrete, i.e., cordwood, thin-film, and thick-film packages were more common; today most manufacturers primarily provide integrated-circuit (IC) units. These are compact and mechanically rugged. Moreover, today's IC manufacturing methods provide units with excellent performance. Indeed, for applications requiring typical performance, the IC operational amplifier is hardly more expensive than the other discrete components, for example, resistors, capacitors, diodes, and transistors, with which they are used. Often the operational amplifier includes connecting pins and is encapsulated in an epoxy or metal package (Fig. 1.1). The metal packages often provide enhanced heat-sink properties to improve heat dissipation. Both bipolar and field-effect transistors (FETs) are employed.

In this book we do not discuss the techniques for designing the operational amplifier itself; the reader is referred to Refs. 1 to 3 for information about the internal design of modern solid-state operational amplifiers. We are primarily interested here in *applications* of operational amplifiers to signal-processing tasks (see also Refs. 1, 3 to 6, and 12 to 17 as further sources of application ideas).

Figure 1.2 illustrates the typical environment in which an operational amplifier is used. Often several operational amplifiers in a given system are supplied from a single pair of matched positive and negative sources of regulated dc power (for example, ±15 V) via low-impedance dc buses. An internal ground lead is sometimes provided on the amplifier package, although typically *all signals* are merely *referenced* to the *power* supply common-ground potential (see Sec. 2.9 for details of power and grounding systems). Operational amplifiers usually have two input terminals (inverting and noninverting) and an output terminal. In addition to these main signal terminals, connector pins are sometimes provided for connection of frequency response compensation and dc offset balancing networks.

Figure 1.3 shows the operational amplifier symbol that is used throughout this text. In this symbol, only the principal signal terminals are illustrated; the

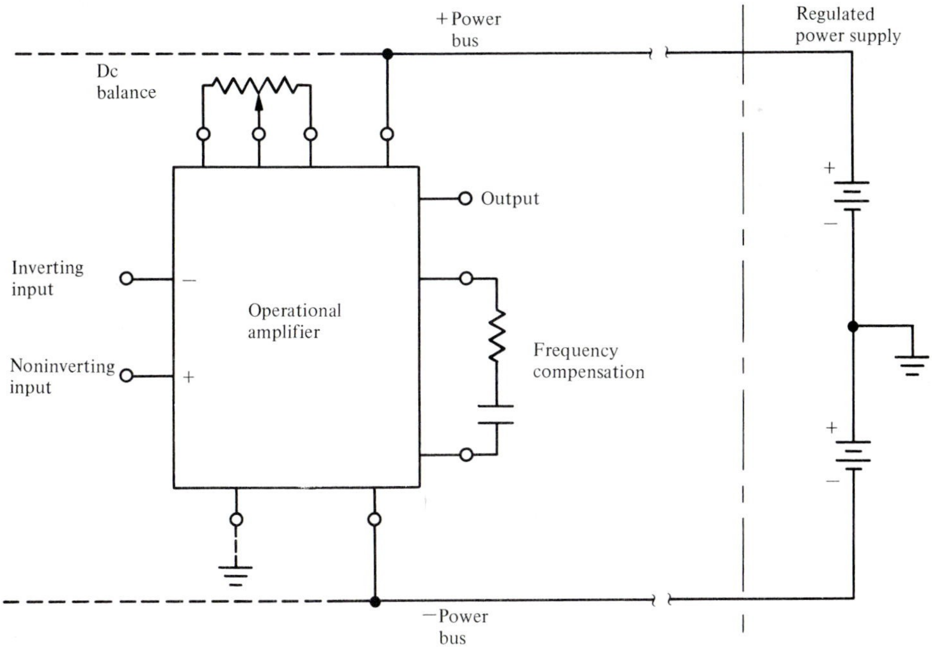

FIGURE 1.2
Typical operational amplifier environment. This diagram illustrates a typical manner in which power is supplied and external balancing and compensation networks are connected (not always required). Several amplifiers may be powered from a single pair of regulated power supplies.

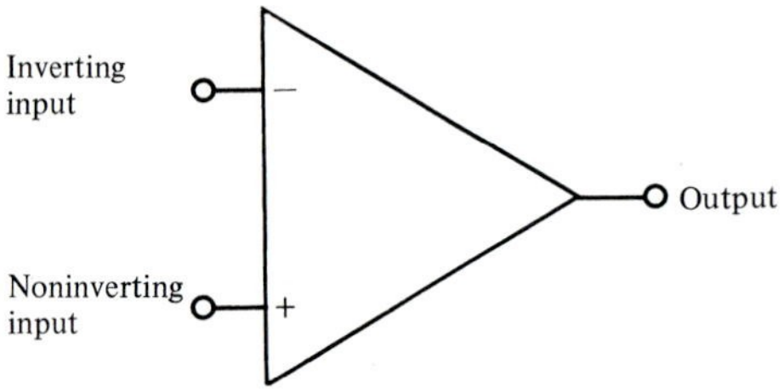

FIGURE 1.3
Conventional operational amplifier symbol; only active signal lines are shown, and all signals are referenced to ground.

other necessary connections to the amplifier (for power, etc.) are assumed to be made as specified by the amplifier manufacturer.

Table 1.1 lists the features normally required in an operational amplifier.

1.2 THE IDEAL AMPLIFIER MODEL

In this chapter, we present the so-called ideal operational amplifier model (Fig. 1.4*b* or *c*); in the next chapter, the important nonideal properties of actual amplifiers are described, and methods for estimating their effect on circuit performance are treated. Here, however, we neglect almost all the nonideal properties and describe an operational amplifier in terms of how we would like it to behave. First, we assume there are no internal dc offsets or nonlinearities associated with the amplifier. With this assumption, we can represent an operational amplifier fairly completely by the idealized linear model of Fig. 1.4*a*. Here we see that the oper-

TABLE 1.1
Features of operational amplifiers

1. *High,* relatively linear, *voltage gain,* down to and including direct current; open-loop gain at direct current may be 10^7 or greater.
2. There must be a *polarity inversion* between input and output. In some cases the noninverting input is not externally available, but it is preferable for a *differential input* to be provided, with both an *inverting* and *noninverting* terminal.
3. *Direct-current offsets should be minimized;* i.e., the input voltage (e_g in Fig. 1.4*c*) should be near zero when the output voltage is zero (good dc *balance*). *Temperature compensation* techniques (in some cases chopper stabilization systems) should be employed to provide long-term stability to this balance. Drifts due to power supply variations should also be minimized.
4. There should be *careful control of high-frequency response,* so that the amplifier will accommodate a large amount of negative feedback.
5. The *differential impedance* between the input terminals and the *common-mode impedance* between the terminals and ground should be *high*.
6. The amplifier *output impedance* should be *low*.
7. The amplifier *output stage should have the ability to deliver specified maximum current to or absorb it from* an output load over some nominal *bipolar* (±) *voltage range,* for example, ±10 V.
8. When a differential input is provided, there should be *good common-mode rejection;* i.e., the output depends only on the *difference* between the input voltages and is not dependent on the magnitude of either input voltage.

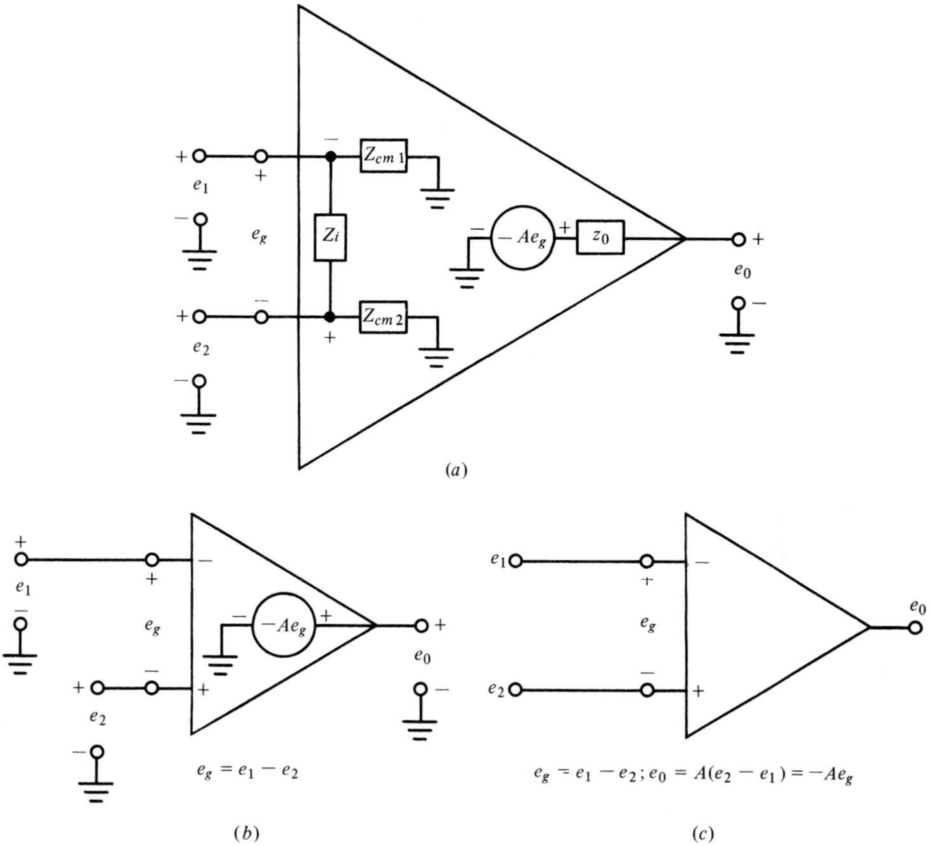

FIGURE 1.4
Operational amplifier circuit models. (*a*) Detailed linear model; (*b*) simplified model, neglecting internal impedances; (*c*) abbreviated notation for simplified model; voltages e_1, e_2, and e_0 are assumed to be referenced to ground.

ational amplifier is an active device. The Thévenin-equivalent circuit associated with the output terminal is modeled by an ideal controlled voltage source $-Ae_g$ in series with an equivalent output impedance Z_0. The input terminals are associated with a passive equivalent circuit representing the *common-mode impedances* Z_{cm1} and Z_{cm2} between each input and the common-signal ground and the *differential input impedance* Z_i between the input terminals.

In this chapter, *a yet simpler model suffices*, as depicted in Fig. 1.4*b* and *c*. In Fig. 1.4*b*, we have assumed that the internal impedances associated with the input terminals are negligibly large and that the internal output impedance is negligibly small. All that remains is the active voltage source $-Ae_g$ at the output terminal. Throughout the rest of this chapter, we adopt the more convenient notation of Fig. 1.4*c*, where the amplifier terminal voltages e_1, e_2, and e_0 are assumed to be referenced to a common ground.

In summary, we can describe the ideal operational amplifier model as follows:

1. The ideal operational amplifier is a *linear voltage-controlled voltage source* (VCVS), with

$$e_0 = A(e_2 - e_1) = -Ae_g \tag{1.1a}$$

$$e_g = e_1 - e_2 \tag{1.1b}$$

where e_g is the differential voltage between the amplifier input terminals.

2. The amplifier *open-loop voltage gain* A is assumed to be a *very large constant*, i.e.,

$$A \ggg 1 \qquad \text{essentially infinite} \tag{1.2}$$

3. Implicit in (1) is the assumption that the amplifier input terminals are essentially open-circuit control nodes, i.e., the impedance between terminals e_1 and e_2 is infinite, and the impedance between each input terminal and the ground is infinite. Thus the input terminal currents are zero.
4. Also implicit in (1) is the assumption that the amplifier output voltage is unaffected by external loads.

Note: We have made the choice of polarity indicated in Fig. 1.4 so that the amplifier *open-loop voltage gain* A can be treated as a large *positive* constant at low frequencies. The polarity relationship between the respective input terminals is as specified in (1.1); therefore with the sign marking on the Thévenin voltage generator in Fig. 1.4*a* or *b*, the value of the generator is $-Ae_g$. The reader is cautioned to study this thoroughly; it is easy to make mistakes in writing circuit equations. Remember that e_1 is associated with the inverting terminal and e_2 with the noninverting terminal.

As mentioned in Sec. 1.1, the operational amplifier is usually embedded in a circuit that provides *large amounts* of *negative feedback*, especially at low frequencies. The effect of the feedback is to make the amplifier behavior more ideal; i.e., dc offsets are suppressed, the amplifier characteristics are linearized, the amplifier is made less sensitive to external output loads, and the differential input voltage e_g is forced to be close to zero, so that input currents are indeed very low.

If circuit designers will accept the model in good faith, and use a few basic design practices to ensure that a reasonable amount of negative feedback is employed in the overall circuit design, they will normally find that the ideal model is quite valid for their initial design efforts.

Note: Even though the model described by (1.1), (1.2), and Fig. 1.4*b* is often adequate for analysis, it is frequently necessary to know the allowable deviations of the terminal voltages from ground, especially the maximum output voltage swing, which is typically ± 10 V. In special applications, the maximum allowable input swing for voltages e_1 and e_2 is also important. Also the available amplifier output current often must be considered.

Of course, the ideal model is not truly attainable; Table 1.1 hints at the myriad considerations one may have to include if the ideal model is not sufficiently complete to fully predict circuit performance. The reader may well ask how big the amplifier voltage gain A actually is. Indeed, this is a very important question. Typically, at low frequencies the gain may be very high, perhaps as large as 10^6 or 10^8. Of course, at higher frequencies, the voltage gain will be less and will include phase shift. In Chap. 2, we explore the ramifications of the variation of A with frequency in some detail, but in many simple applications, the assumptions of (1.1) and (1.2) suffice, at least to make a first try at a circuit design.

1.3 THE SUMMER-INVERTER

Operational amplifiers are normally *not operated open-loop*; some form of *feedback* is used to control the overall circuit transfer characteristics. If the amplifier were energized in an open-loop configuration (Fig. 1.5), the output voltage would saturate (amplitude-overload) with even a very small input signal because of the high open-loop gain; in fact, small internal dc offsets which are always present would normally be sufficient to cause overload without feedback.

Figure 1.6*a* shows a typical inverting operational amplifier configuration; the noninverting terminal is grounded for single-ended operation, either directly or through a drift-compensating resistor (see Sec. 2.5). The simplified diagram of Fig. 1.6*b*, where the noninverting terminal is assumed to be effectively at ground potential and all voltages are assumed to be referred to this same ground potential, is often used. Here also we use v's for the voltages on the external nodes of the complete circuit.

We make three basic assumptions in deriving the transfer function equations:

1. The amplifier output voltage v_0 is related to the *operational amplifier* input terminal voltage e_g (the summing junction) by

$$v_0 = -Ae_g$$

where A is termed the *operational amplifier open-loop gain* (assumed here to be a large positive constant). Typically $A > 10^4$. *Thus, as long as the amplifier output is within its rated range (for example, ±10 V), we will be able to assume* $e_g = 0$ *without serious error.*

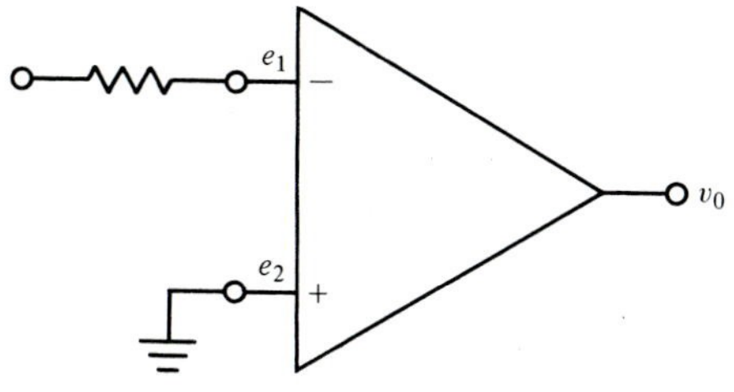

FIGURE 1.5
This circuit condition is normally to be avoided; operational amplifiers are not designed to run open-loop, without feedback.

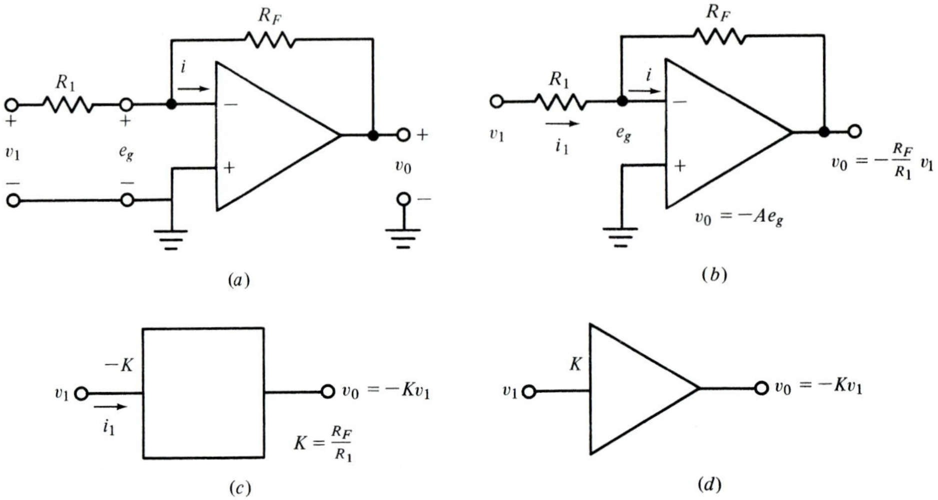

FIGURE 1.6
Simple resistive inverter-amplifier. (*a*) Complete circuit; (*b*) same circuit with all voltages referenced to ground; (*c*) shorthand or block-diagram symbol for complete circuit; (*d*) analog-computer symbol.

2. The output voltage v_O is not materially affected by loads applied to the amplifier output, including any feedback elements.
3. The amplifier input impedance is sufficiently high to permit us to assume that the amplifier input current i is approximately zero (since from the first assumption e_g is very small, the amplifier input impedance does not really have to be exceptionally high to ensure the validity of the final results).

In the simplified circuit of Fig. 1.6*b*, we can write a current-balance equation (nodal equation)* at node e_g:

$$\frac{v_o - e_g}{R_F} + \frac{v_1 - e_g}{R_1} = i \tag{1.3}$$

Now, if the circuit is working properly (no amplifier overload), we assume $e_g \cong 0$, $i \cong 0$, so that the above equation becomes

$$\frac{v_O}{R_F} + \frac{v_1}{R_1} \cong 0 \tag{1.4}$$

A general rule to remember in writing equations for amplifier performance is that *the sum of the currents entering the inverting terminal is approximately zero.* This, of course, is not strictly true, and we discuss more exact expressions later.

---

* For the reader whose background in undergraduate circuit theory is weak, Ref. 7 gives a modern presentation of circuit analysis techniques.

Thus, the transfer function for the circuit of Fig. 1.6 is simply

$$\frac{v_o}{v_1} = -\frac{R_F}{R_1} \tag{1.5}$$

We note that the gain is essentially controlled by the passive elements R_1 and R_F; *don't forget the minus sign!* Equation (1.5) is accurate to the extent that the above assumptions are valid. Resistance R_F is usually called the *feedback resistance*; R_1 is often called the *input resistance*.

Figure 1.6*c* shows a block-diagram symbol that is used in this book for the simple inverting amplifier; here we denote the voltage gain of the circuit by $-K$, that is,

$$\frac{v_0}{v_1} = -K \qquad K = \frac{R_F}{R_1} \tag{1.6}$$

When describing an inverting amplifier, we use the minus sign explicitly; later on we shall be describing noninverting amplifiers as well.

Figure 1.6*d* shows a conventional analog-computer symbol for an inverting amplifier; in analog-computer diagrams, the minus sign is assumed and not specified explicitly. To avoid confusion, we use the symbol of Fig. 1.6*c* here.

Note also that if $R_F = R_1 = R$, we have a *simple unity gain inverter* (Fig. 1.7), with

$$v_0 = -v_1 \qquad R_F = R_1 = R \tag{1.7}$$

Typical values for the resistances lie in the range from 100 to 10^6 Ω, and values of the ratio K are typically between 0.1 and 100. Note also that the input impedance of the circuit of Fig. 1.6 is

$$Z_{\text{in}} = \frac{v_1}{i_1} = R_1 \tag{1.8a}$$

since e_g is assumed to be zero. The output impedance is also predicted to be

$$z_{\text{out}} = 0 \tag{1.8b}$$

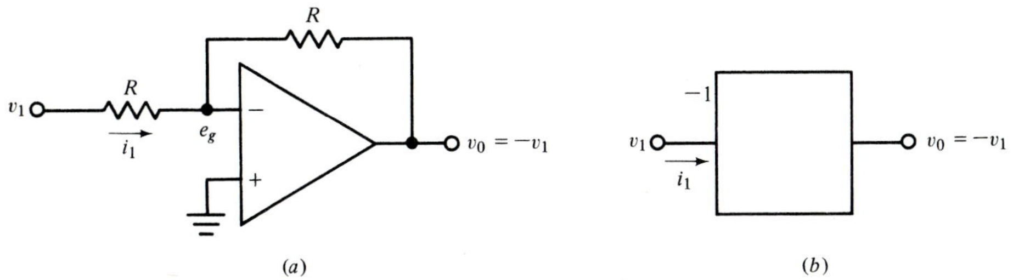

FIGURE 1.7
A simple unity gain inverter. (*a*) Detailed circuit; (*b*) block-diagram symbol.

since the ideal operational amplifier model itself has zero output impedance. In Chap. 2 we will see that these results are quite accurate in most cases, even for the nonideal case.

Example 1.1. In the circuit of Fig. 1.6, let

$$R_1 = 2000\ \Omega$$

and

$$R_F = 50{,}000\ \Omega$$

Then we have

$$K = \frac{50 \times 10^3}{2 \times 10^3} = 25$$

and

$$v_0 = -25 v_1$$

We might describe this circuit as either an *inverting amplifier* with a (voltage) gain of *magnitude 25* or, alternatively, an amplifier with a voltage gain of −25. Both descriptions are commonly used.

Because we have used the ideal model of Fig. 1.4*c* to represent the operational amplifier and have assumed that the input and feedback resistances are pure resistive elements, with no parasitic capacitance or inductance, (1.6) predicts that the circuit of Fig. 1.6 has a gain that is constant and independent of frequency. This, of course, is not strictly possible, and we will see later that there are methods for predicting the frequency-dependent behavior of the circuit. However, if the resistance values are kept fairly low (e.g., less than 100,000 Ω) and a relatively fast (broadband) amplifier is used, then (1.6) usually will be a useful description of the inverting amplifier for frequencies in the audio range or below (e.g., direct current to 10 kHz). Also we should note that (1.6) indicates that the only important thing is the *ratio* of R_F to R_1. This, too, will be essentially true as long as the individual resistance values fall within a reasonable range (for example, 100 to 1,000,000 Ω). Of course, if R_1 is made too low, the circuit of Fig. 1.6 may excessively load whatever system is providing its input signal. The reader again is reminded that (1.5) will hold only as long as the magnitude of v_0 is less than the specified maximum for the amplifier (typically ±10 V), and the magnitude of the total amplifier output current, including the current through R_F, does not exceed the rated value.

Figure 1.8*a* shows the circuit of a typical *summer-inverter*. If we write Kirchhoff's current law at node e_g, we have

$$\frac{v_1 - e_g}{R_1} + \frac{v_2 - e_g}{R_2} + \cdots + \frac{v_n - e_g}{R_n} + \frac{v_0 - e_g}{R_F} = i \tag{1.9}$$

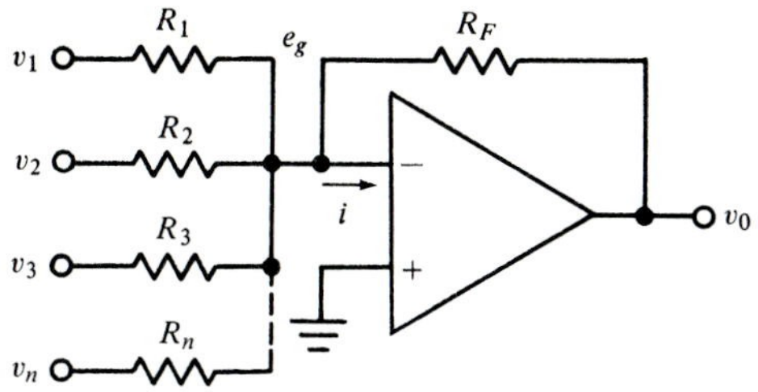

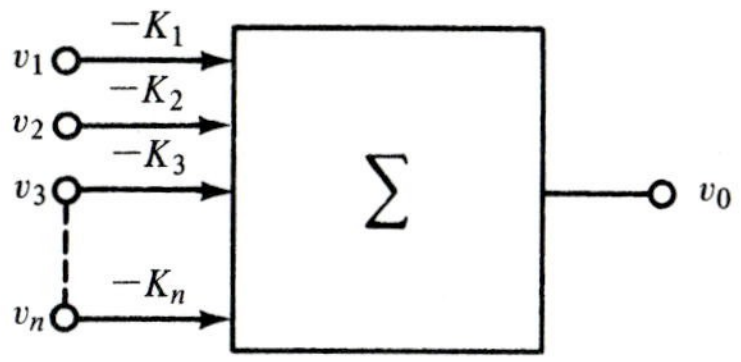

FIGURE 1.8
The summer-inverter circuit. (*a*) Complete circuit; (*b*) block-diagram symbol.

but since we assume $e_g \cong 0$ and $i \cong 0$, we can write

$$\frac{v_1}{R_1} + \frac{v_2}{R_2} + \cdots + \frac{v_n}{R_n} \cong \frac{-v_0}{R_F} \tag{1.10}$$

or

$$v_0 = -R_F\left(\frac{v_1}{R_1} + \frac{v_2}{R_2} + \cdots + \frac{v_n}{R_n}\right) \tag{1.11a}$$

$$= -(K_1v_1 + K_2v_2 + \cdots + K_nv_n) \qquad K_i = \frac{R_F}{R_i} \qquad i = 1, 2, \ldots, n \tag{1.11b}$$

Node e_g is often called the *summing junction*, and engineers often describe the results of the above analysis by saying that the summing junction current (labeled i in Fig. 1.8*a*) is essentially zero. In view of the result of (1.11), we can see why the circuit of Fig. 1.8 is called a *summer-inverter*. Each of the input signals is weighted by a gain constant K_i, a weighted sum is formed internally, and the output is equal to this weighted sum but inverted in polarity; Fig. 1.8*b* shows a simplified block-diagram symbol.

The summer-inverter circuit is generally very useful. In analog computers, it is used to sum computer variables precisely; in this application, fixed, precise resistor ratios are used to provide a standard set of gains.

Example 1.2. Figure 1.9 shows a two-input summer-inverter circuit. Assuming an ideal operational amplifier, we can predict the circuit performance from the resistor values by

$$v_0 = -\frac{R_F}{R_1}v_1 - \frac{R_F}{R_2}v_2$$

If $R_F = 20\ \text{k}\Omega$, $R_1 = 5\ \text{k}\Omega$, and $R_2 = 10\ \text{k}\Omega$, then

$$v_0 = -4v_1 - 2v_2$$

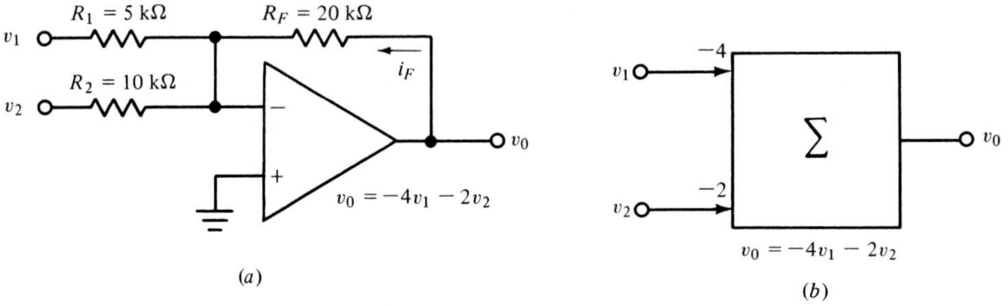

FIGURE 1.9
A two-input summer-inverter circuit (Example 1.2). (*a*) Complete circuit; (*b*) block-diagram symbol.

Suppose we also know that the output voltage of the amplifier should never exceed 10 V in magnitude, i.e.,

$$|v_O| \le 10\ \text{V}$$

Thus, in using the circuit, we must ensure that

$$|-4v_1 - 2v_2| \le 10 \tag{1.12}$$

If we have no prior information about the polarity of the individual input voltages, then to ensure that (1.12) is satisfied in the worst-case situation, we must ensure that

$$4|v_1| + 2|v_2| \le 10$$

(Note that it is not sufficient to limit v_1 to 2.5 V and v_2 to 5 V magnitude separately; why?)

Suppose also that we know our amplifier has a maximum rated output current of 5 mA; then R_F may require

$$|i_F| \le \frac{10\ \text{V}}{20{,}000\ \Omega} = 0.5\ \text{mA}$$

and thus only 4.5 mA is available to drive an external load. Here we have chosen $R_F = 20\ \text{k}\Omega$ so that stray leakage current and parasitic capacity effects will be negligible, but operational amplifier current demands will not be excessive.

In Chap. 2, we discuss the nonideal operational amplifier and how the ideal equations such as (1.11) are affected by such things as dc offset, finite bandwidth and gain, etc.

1.4 THE INVERTING SUMMER-INTEGRATOR

Figure 1.10 shows the essential features of the *inverting summer-integrator* circuit, commonly called an *integrator*. The principal difference between the integrator circuit of Fig. 1.10 and the summer-inverter circuit of Fig. 1.8 is that the *feedback*

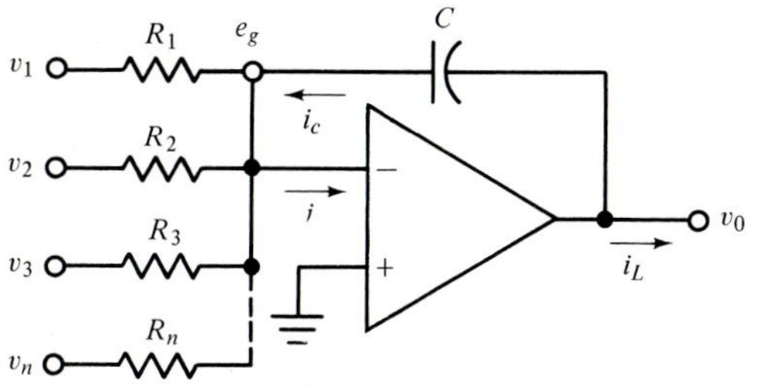

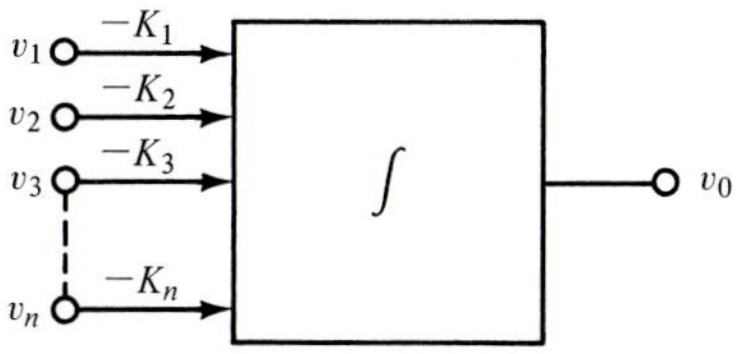

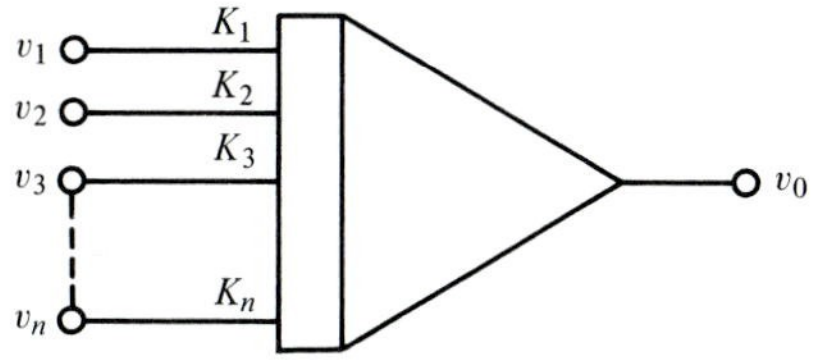

FIGURE 1.10
The summer-integrator circuit. (*a*) Complete circuit; (*b*) block-diagram symbol; (*c*) analog-computer symbol.

element is now a *capacitor* C. Again, as in Sec. 1.3, we assume an ideal amplifier with very high open-loop gain, and thus we can assume $e_g \cong 0$, $i \cong 0$. Kirchhoff's current equation at node e_g is

$$\frac{v_1 - e_g}{R_1} + \frac{v_2 - e_g}{R_2} + \cdots + \frac{v_n - e_g}{R_n} + i_c = i \tag{1.13}$$

but, of course, the terminal properties of a capacitance require that (again e_g is assumed to be very small)

$$i_c = C\left(\frac{dv_0}{dt} - \frac{de_g}{dt}\right) \cong C\,\frac{dv_0}{dt} \tag{1.14}$$

Substituting (1.14) in (1.13) and setting $e_g \cong 0$ and $i \cong 0$, we have

$$\frac{v_1}{R_1} + \frac{v_2}{R_2} + \cdots + \frac{v_n}{R_n} + C\,\frac{dv_0}{dt} = 0 \tag{1.15a}$$

or

$$C\,\frac{dv_0}{dt} = -\left(\frac{v_1}{R_1} + \frac{v_2}{R_2} + \cdots + \frac{v_n}{R_n}\right) \tag{1.15b}$$

and thus

$$\frac{dv_0}{dt} = -\frac{1}{C}\left(\frac{v_1}{R_1} + \frac{v_2}{R_2} + \cdots + \frac{v_n}{R_n}\right) \tag{1.16}$$

Integrating both sides of (1.16), we have

$$v_O(t) = -\frac{1}{C}\int_0^t \left(\frac{v_1}{R_1} + \frac{v_2}{R_2} + \cdots + \frac{v_n}{R_n}\right) dt + v_O(0) \qquad (1.17)$$

or

$$v_O(t) = -\int_0^t (K_1 v_1 + K_2 v_2 + \cdots + K_n v_n)\, dt + v_O(0) \qquad (1.18)$$

where

$$K_i = \frac{1}{CR_i} \quad \text{s}^{-1} \qquad i = 1, 2, \ldots, n \qquad (1.19)$$

and $v_O(0)$ is the initial voltage on the capacitance at $t = 0$ (more on initial-condition setting later).

Figure 1.10*b* shows the block-diagram symbol we intend to use in this book; again the polarity inversions are depicted explicitly. Figure 1.10*c* shows the symbol conventionally used in analog-computer block diagrams, where the polarity inversion is assumed.

Example 1.3. Consider the circuit of Fig. 1.11. We can use (1.18) and (1.19) to obtain

$$v_O = -\int_0^t (K_1 v_1 + K_2 v_2)\, dt + v_O(0)$$

with

$$K_1 = \frac{1}{R_1 C} = 1 \text{ s}^{-1}$$

and

$$K_2 = \frac{1}{R_2 C} = 5 \text{ s}^{-1}$$

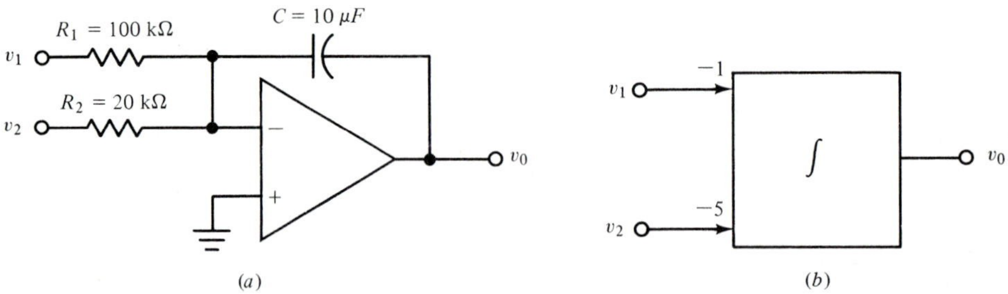

FIGURE 1.11
A two-input summer-integrator circuit (Example 1.3). (*a*) Complete circuit; (*b*) block-diagram symbol.

or

$$v_0 = -\int_0^t (v_1 + 5v_2)\,dt + v_0(0) \tag{1.20}$$

The factors K_i calculated by (1.19) are called the *integrator gain constants*. More precisely, we may say that each input to the integrator circuit of Fig. 1.10 has either a gain of $-K_i$ or an inverting gain of magnitude K_i. The gain constant K_i is obviously equal to the reciprocal of the time constant R_iC; thus K_i has units of s^{-1}. The concept of integrator gain is to some extent a shorthand way of specifying the (reciprocal) time constants associated with an integrator circuit.

We see also that one can achieve a particular integrator gain in many ways; i.e., it is the time constant that is important, not the individual values of R and C. In Example 1.3, we could also obtain the same circuit of performance, as predicted by (1.20), with

$$R_1 = 1{,}000{,}000\ \Omega \qquad R_2 = 200{,}000\ \Omega$$

and

$$C = 1\ \mu\text{F}$$

In actuality, there would be some difference in circuit performance, which we will be in a better position to predict after the presentation of Chap. 2. Of course, one does not have complete freedom in the choice of component values. Again, the input impedance at each input terminal of Fig. 1.10 is

$$Z_i = R_i \qquad i = 1, 2, \ldots, n \tag{1.21}$$

so that the circuit will excessively load its driving sources if the resistors are made too small. With regard to practical values of C, we should note that very small values cannot be easily obtained accurately (because of the stray capacity of wiring, etc.). On the other hand, very large values of capacity cannot be obtained inexpensively. Electrolytic capacitors do not permit bipolar potentials to be applied; they also have a large amount of dissipation, i.e., they have an appreciable leakage conductance. We will have more to say about component limitations and selection later (Sec. 2.11). For most applications, one should select low-loss capacitors (Mylar or polystyrene dielectric). Typical practical capacity values range from 0.001 to 10 μF. Resistor values should again range from perhaps 1000 to 1,000,000 Ω. Thus we see that a relatively wide range of integrator gains are possible.

A handy, relatively error-free way of calculating integrator gains is to work in megohms and microfarads, thereby obtaining a time constant directly in seconds, and thus the reciprocal of the time constant in seconds is the gain in s^{-1}.

Example 1.4. Given an integrator with

$$C = 1000\ \text{pF}$$

and

$$R = 10{,}000\ \Omega$$

we can find the gain as follows:

$$C = 10^{-3}\ \mu\text{F}$$
$$R = 10^{-2}\ \text{M}\Omega$$

Thus the time constant is

$$T = 10^{-2} \times 10^{-3} = 10^{-5}$$

and the integrator gain constant is

$$K = \frac{1}{T} = 100{,}000\ \text{s}^{-1}$$

Some additional practical considerations that should be mentioned are again output limitations. Of course, the amplifier output voltage should never be allowed to exceed its nominal rated magnitude; this requires some knowledge of the nature of the signals to be integrated. For example, one cannot integrate a signal with a nonzero average value indefinitely.

A more subtle point is that the *capacitor charging current must be considered in estimating the total output current required from the amplifier*. If the amplifier must supply a current of i_L to some external load, then we must ensure that

$$\left| C\frac{dv_0}{dt} + i_L \right| \le i_{\max} \tag{1.22}$$

where $i_{\max}$ is the rated amplifier output current.

From (1.15*b*) we see also that the capacitor current equals minus the sum of the input resistor currents; e.g., in Fig. 1.10

$$\left| i_L - \left(\frac{v_1}{R_1} + \frac{v_2}{R_2} + \cdots + \frac{v_n}{R_n} \right) \right| \le i_{\max} \tag{1.23}$$

Thus there are rate limitations on the integrator output voltage indicated by (1.22), which in turn dictate limitations on the instantaneous magnitude of the input signals (1.23).

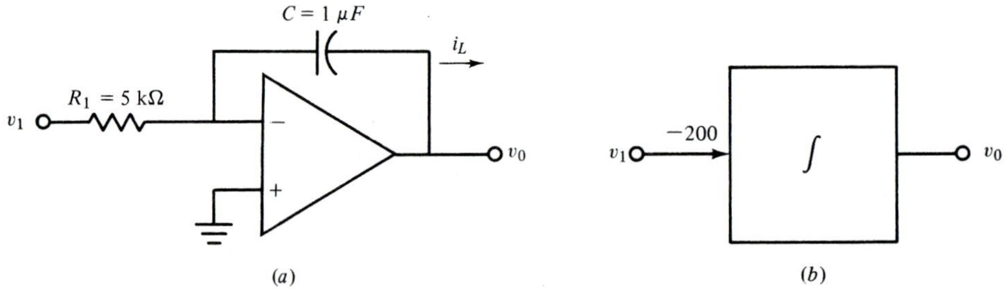

FIGURE 1.12
An inverting gain-of-−200 integrator (Example 1.5). (*a*) Complete circuit; (*b*) block-diagram symbol. (*continued on next page*)

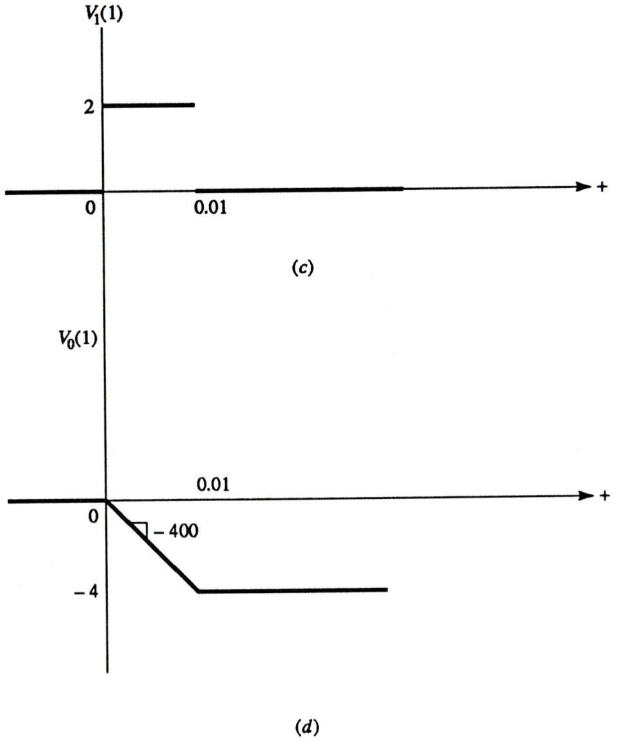

FIGURE 1.12 (continued)
(*c*) A pulse input; (*d*) the integrator output.

Example 1.5. Consider the integrator circuit of Fig. 1.12, with a gain of −200. Suppose the amplifier is rated so that

$$|v_0| \leq 10 \text{ V} \tag{1.24a}$$

$$|i_{max}| \leq 10 \text{ mA} \tag{1.24b}$$

Then from (1.22)

$$\left| C\frac{dv_0}{dt} + i_L \right| \leq 10 \text{ mA} \tag{1.25}$$

and from (1.23) this means that

$$\left| -\frac{v_1}{R_1} + i_L \right| \leq 10 \text{ mA} \tag{1.26}$$

If we want to reserve enough output current so that i_L can be as large as 5 mA, then

$$\left| \frac{v_1}{R_1} \right| \leq (10 - 5) \text{ mA} = 5 \times 10^{-3} \text{ A}$$

Thus with $R_1 = 5$ kΩ,

$$|2 \times 10^{-4} v_1| \leq 5 \times 10^{-3} \text{ A}$$

or

$$|v_1| \leq 25 \text{ V} \tag{1.27}$$

which is not a very strict requirement.

Suppose, however, that

$$v_1 = A\cos 2\pi ft = A\cos \omega t$$

that is, v_1 is a sinusoid of *peak* (not rms) value A, frequency f Hz, and $\omega = 2\pi f$ rad/s.

Assuming $v_O(0) = 0$, then

$$v_O(t) = -200\int_0^t A\cos \omega t\, dt = \frac{-200A}{\omega}\sin \omega t$$

Then to prevent a *voltage overload* at the output,

$$\left|\frac{200A}{\omega}\sin \omega t\right| < 10 \text{ V}$$

but $\sin\omega t \leq 1$, so that

$$\frac{200A}{\omega} \leq 10$$

or

$$\frac{A}{f} \leq \frac{2\pi}{20} = 0.314 \text{ V}\cdot\text{s} \tag{1.28}$$

Suppose we combine the results of (1.27) and (1.28). We find that if

$$v_1 = A\cos 2\pi ft$$

then for the circuit of Fig. 1.12, if 5 mA of output current is required, A is limited by

$$A \leq 0.314f \text{ V} \qquad \text{in the range } 0 < f \leq 79.6 \text{ Hz}$$

due to output voltage limitation (1.24*a*), and

$$A \leq 25 \text{ V} \qquad \text{in the range } f \geq 79.6 \text{ Hz}$$

due to amplifier output current limitation (1.24*b*).

We can gain more insight into the dynamic behavior of the integrator by frequency domain analysis. Taking the Laplace transform of (1.18), we obtain [with $v_O(0) = 0$]

$$v_O(s) = \frac{-1}{s}[K_1V_1(s) + K_2V_2(s) + \cdots + K_nV_n(s)] \tag{1.29}$$

Thus we see that the integrator transfer function (relating any input to the output) has a pole at $s = 0$; indeed, the operational amplifier integrator circuit provides a convenient way of implementing nearly ideal integration, with a transfer function having a pole *very close* to the origin of the s plane. Indeed, it is the ability of the operational amplifier circuit of Fig. 1.10*a* to implement the integration operation accurately that has made the electronic analog computer possible (see Ref. 3 for a good presentation of the role of operational amplifiers in analog computation).

It is interesting to examine the frequency response of the integrator. Let us represent sinusoidally varying quantities by a conventional phasor notation (Ref. 7, sec. 7.3); i.e., if a voltage has the form

$$v(t) = A\cos(\omega t + \theta)$$

we represent it by the notation

$$\mathcal{V} = A \angle \theta$$

where A is the magnitude and θ the angle (in radians) of a phasor $\mathcal{V}$.

We can then easily find the phasor operator relating input and output (sinusoidal) voltages by setting $s = j\omega$ in the Laplace transform transfer function. Thus for sinusoidal inputs, (1.29) becomes

$$\begin{aligned} \mathcal{V}_0 &= -\frac{1}{j\omega}(K_1\mathcal{V}_1 + K_2\mathcal{V}_2 + \cdots + K_n\mathcal{V}_n) \\ &= \frac{1}{\omega} \angle \frac{\pi}{2}(K_1\mathcal{V}_1 + K_2\mathcal{V}_2 + \cdots + K_n\mathcal{V}_n) \end{aligned} \tag{1.30}$$

Here we see that (in the steady state) the ideal integrator frequency response varies inversely with frequency, and the magnitude is infinite at direct current! Also, inputs are shifted $+\pi/2$ rad, or $+90°$.

Example 1.6. Consider again the circuit of Fig. 1.12 (an inverting integrator with a gain of 200). We can describe the integrator frequency response by

$$\frac{\mathcal{V}_0}{\mathcal{V}_1} = H(j\omega) = \frac{-200}{j\omega} = \frac{200}{\omega} \angle +\frac{\pi}{2}$$

Thus we arrive again at the conclusion of Example 1.5: If the circuit output is limited to 10 V peak and the input is a sinusoid of peak magnitude A, then

$$\frac{200A}{\omega} \leq 10$$

and with $\omega = 2\pi f$, we have

$$\frac{A}{f} \leq \frac{2\pi}{20} = 0.314 \text{ V}\cdot\text{s}$$

which is the same result as (1.28).

We can summarize the results of the last two examples as follows:

1. At low frequencies, the amplitude of the input signal to an integrator is in general limited by output voltage limitations. This is not too surprising, since an integrator with gain constant $-K$ has a sinusoidal frequency response

$$\frac{\mathcal{V}_0}{\mathcal{V}_1} = H(j\omega) = \frac{-K}{\omega} \angle -\frac{\pi}{2} = \frac{K}{\omega} \angle \frac{-3\pi}{2}$$

2. At higher frequencies, there may be an additional limitation on input signal magnitude resulting from the amplifier output current required to charge the integrating capacitor.

The above examples have been worked under the assumption that the input signals were sinusoids. For more complex waveforms, careful analysis must be made, considering the effects of output voltage and output current limitations separately, to determine whether a given signal can be accommodated.

Example 1.7. Again consider the integrator circuit of Figure 1.12, with a gain of -200. To reinforce our understanding of the physical significance of integrator operation, suppose the integrator input is that of Fig. 1.12*c*, i.e.,

$$v_1(t) = \begin{cases} 0 & t < 0 \\ 2 & 0 < t < 0.01 \\ 0 & 0.01 < t \end{cases}$$

Then the integrator output is

$$V_0(t) = -200 \int_0^t v_1(T)dT$$

or

$$V_0(t) = \begin{cases} 0 & t < 0 \\ -400t & 0 < t < 0.01 \\ -4 & 0.01 < t < \infty \end{cases}$$

See Fig. 1.12*d*.

Initial-Condition Setting and Mode Control

It is often necessary to establish an initial condition (voltage) on an integrator. Figure 1.13*a* shows one commonly used circuit which implements what is often called *three-mode control*. It requires two single-pole double-throw switches S_A and S_B and the initial-condition-setting network R_A and R_B. The *negative* of the desired initial-condition voltage is established at node v_X. Potentiometer P is used to permit establishment of an adjustable initial-condition value. The potentiometer is supplied with either +10 or −10V, depending on the desired initial condition. The operation of the circuit is best understood by analyzing it separately in three different modes (switch settings):

Position 1: RESET mode. In this state (Fig. 1.13*b*), we *acquire an initial condition*. Noting that node Y is forced by the amplifier feedback to be near ground potential, we can describe the circuit by

$$\frac{v_x}{R} + \frac{v_0}{R} + C\frac{dv_0}{dt} = i = 0 \tag{1.31}$$

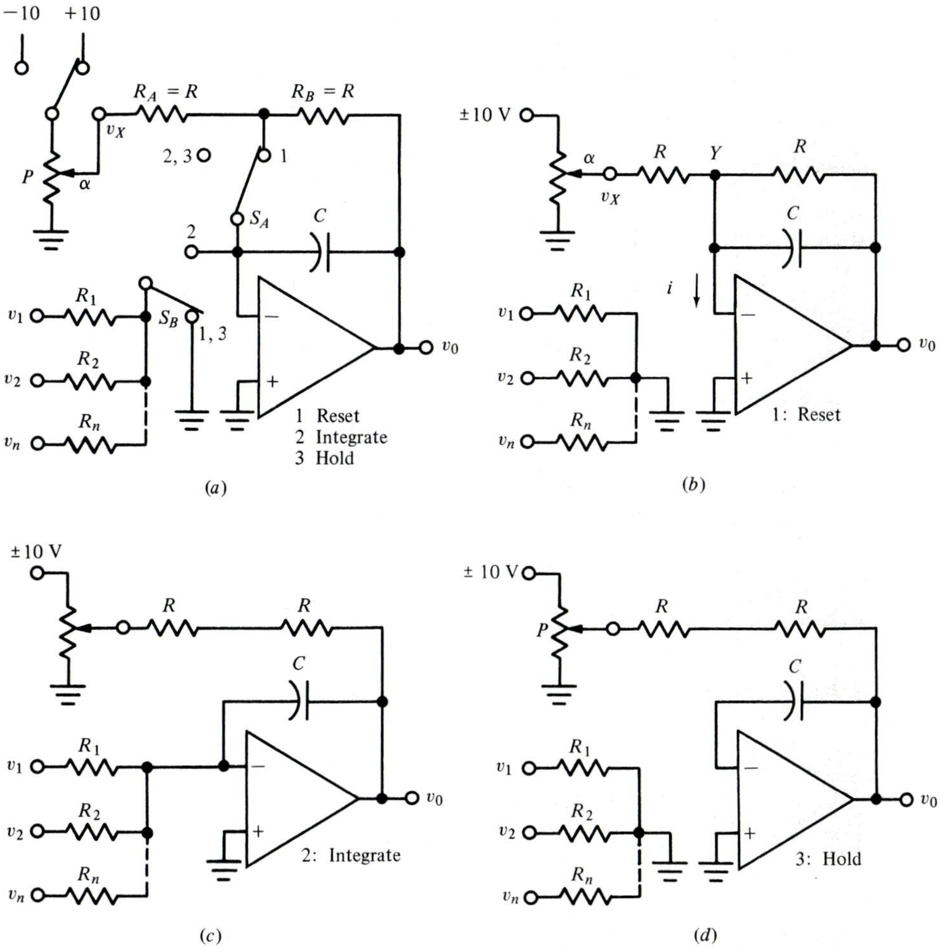

FIGURE 1.13
Three-mode integrator control circuit. (*a*) Complete circuit; (*b*) position 1, RESET or initial-condition mode; (*c*) position 2, INTEGRATE mode; (*d*) position 3, HOLD mode. (*continued on next page*)

which may be rewritten

$$\frac{dv_0}{dt} = -\frac{1}{RC}v_0 - \frac{1}{RC}v_x \tag{1.32}$$

We see that (1.32) is a simple time-invariant linear first-order ordinary differential equation. If we define α to be the loaded potentiometer ratio, i.e.,

$$v_x = \pm 10\alpha = \text{constant}$$

then v_0 will approach a steady-state value, which we can find from (1.32). Setting

$$\frac{dv_0}{dt} = 0$$

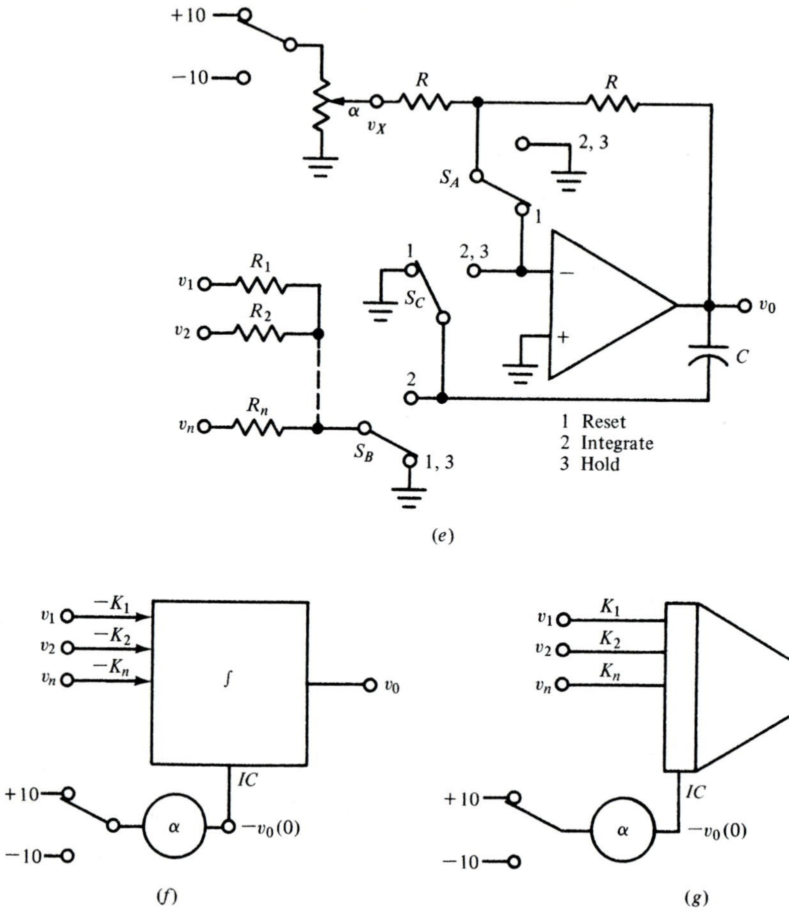

FIGURE 1.13 (continued)
(*e*) Three-switch alternative to *a* which provides faster reset; (*f*) integrator block-diagram symbol; (*g*) corresponding analog-computer symbol.

in (1.32), we obtain

$$v_{OSS} = -v_x \qquad \text{steady state} \tag{1.33}$$

Thus we see that the steady-state value of v_O is the *negative* of the initial-condition voltage established by the potentiometer (or any other source of voltage for v_x). This polarity inversion, while annoying, can often be accepted. We note also that (1.32) indicates that the basic time constant of the initial-condition-setting circuit is

$$\tau = RC \qquad \text{s}$$

Thus we can estimate the time required for initial-condition acquisition; e.g., 5 time constants is usually adequate.

Position 2: INTEGRATE mode. In this state (Fig. 1.13*c*) we have a conventional integrator circuit with inputs v_1 through v_n. The initial-condition-setting network is merely a low-current resistive load on the amplifier output, and the integrator equations derived previously apply for this mode. Now, however, if we go quickly from position 1 (RESET) to position 2 (INTEGRATE), the value of v_0 established in the RESET mode serves as the initial condition for the integration time interval.

Position 3: HOLD mode. The presence of the two switches S_A and S_B permits a third useful mode of operation (Fig. 1.13*d*), wherein the amplifier equation becomes

$$C \frac{dv_0}{dt} = 0 \tag{1.34}$$

or

$$v_0 = \text{constant} \tag{1.35}$$

In this mode, the circuit has *memory* of the value of v_0 present when the HOLD mode was established.

Example 1.8. Figure 1.14 shows a typical sequence of integrator modes; in this case, the initial-condition-setting circuit is supplied with -10 V dc. Therefore, the initial condition can be set anywhere between 0 and $+10$ V, depending on the setting of the potentiometer. Beginning at time T_2, a positive constant voltage E is integrated with a gain of -100, so that in the INTEGRATE mode, the output decreases linearly at the rate of $100E$ V/s. If the circuit is put into the HOLD mode at time T_3, the value of v_0 will be held, or stored.

We should mention that the ability of the integrator circuit of Fig. 1.13 to hold an output value in the HOLD mode is dependent on internal offsets in the amplifier (and possible switch leakage current, if the switches are solid-state devices); we have assumed here an ideal operational amplifier and switches. In practice, it may only be possible to hold the integrator output for a short time; more will be said about this in Chap. 2.

Figure 1.13*e* shows an alternative three-mode integrator control circuit which permits faster setting of the initial conditions. This circuit requires three single-pole double-throw switches. The operation is essentially the same as that of the circuit of Fig. 1.13*a* in HOLD and INTEGRATE, but differs in RESET, where the capacitor is now made the output load of the operational amplifier (see Prob. 1.9). Some operational amplifiers will not tolerate a large capacitive output load and cannot be used with this last circuit; it is preferred, however, where fast initial-condition setting is desirable. (See also Chap. 5 for a discussion of high-speed analog signal switching and the design of sample-hold circuits based on this circuit.)

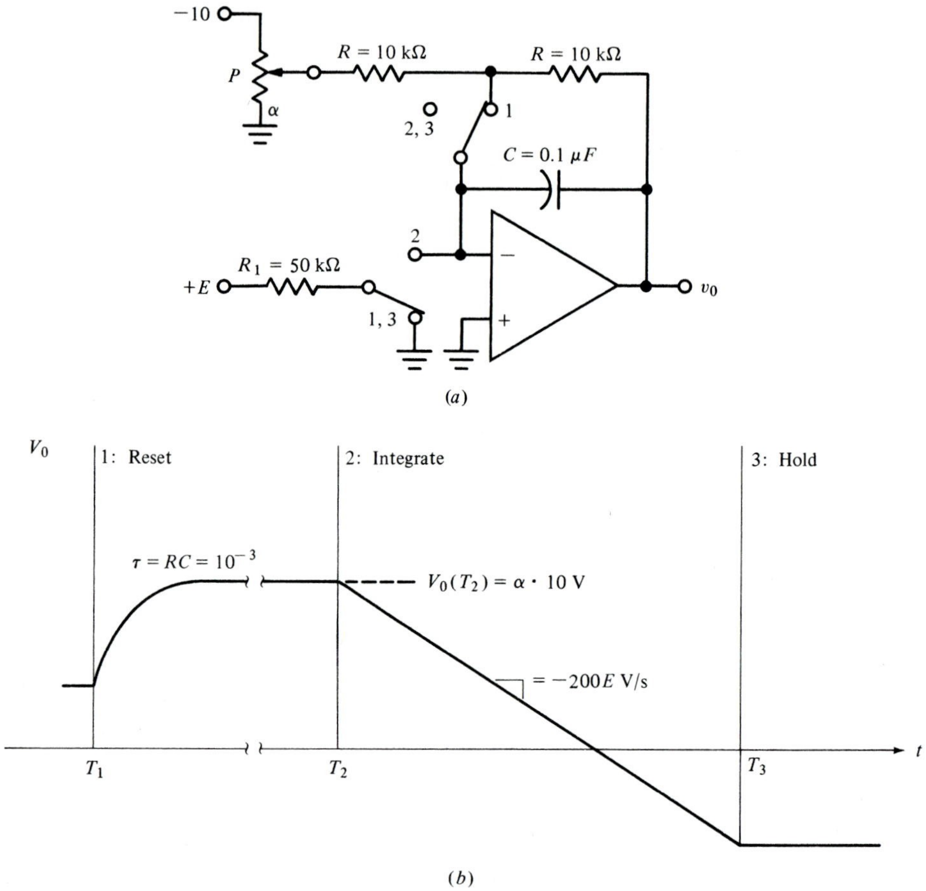

FIGURE 1.14
An inverting gain-of-−200 integrator with mode control (Example 1.7). (*a*) Circuit; (*b*) typical output waveform.

1.5 THE GENERAL INVERTING AMPLIFIER

Figure 1.15 shows a general (linear) inverting (and summing) operational amplifier circuit; here we allow the input summing networks and the feedback network to be generalized linear two-port (three-terminal) networks. In each network $Z_i(s)$ is defined to be the short-circuit transfer impedance. If the particular network is a two-terminal (one-port) network, then $Z_i(s)$ is merely the conventional terminal driving-point impedance. If the network is the more general three-terminal (two-port) network shown in Fig. 1.16, then we define the *short-circuit transfer impedance* as

$$\left| Z_{sc}(s) = \frac{V_1(s)}{I_2(s)} \right|_{V_2(s)=0} \tag{1.36}$$

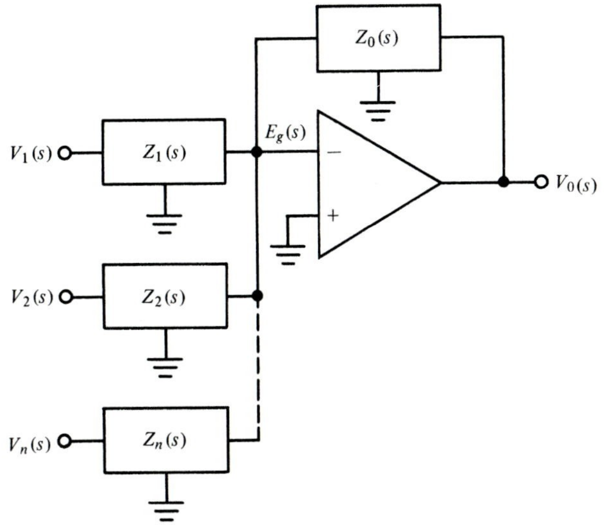

FIGURE 1.15
General inverting amplifier circuit with three-terminal (two-port) networks.

Note: We have not used the conventional network-theory definition, since our port currents are directed outward. However, our choice of direction eliminates a lot of minus signs in subsequent analyses: e.g., it would be unfortunate for a simple resistor to have a negative impedance.

Note that a tee network has the same value of Z_{sc} regardless of which terminal is used for input and which for output. In Fig. 1.17 the value of Z_{sc} can be shown to be

$$Z_{sc} = \frac{Z_1Z_2 + Z_2Z_3 + Z_3Z_1}{Z_2} = \left.\frac{V_1}{I_2}\right|_{V_2=0} = \left.\frac{V_2}{I_1}\right|_{V_1=0} \tag{1.37}$$

We again assume that the action of the amplifier, via the feedback network, is such that the summing-point voltage E_g is held at a virtual ground potential, that is, $E_g \cong 0$ and $I \cong 0$. With this assumption (using Laplace transform notation)

$$\frac{V_0(s)}{Z_0(s)} + \frac{V_1(s)}{Z_1(s)} + \frac{V_2(s)}{Z_2(s)} + \cdots + \frac{V_n}{Z_n(s)} = I(s) \cong 0 \tag{1.38}$$

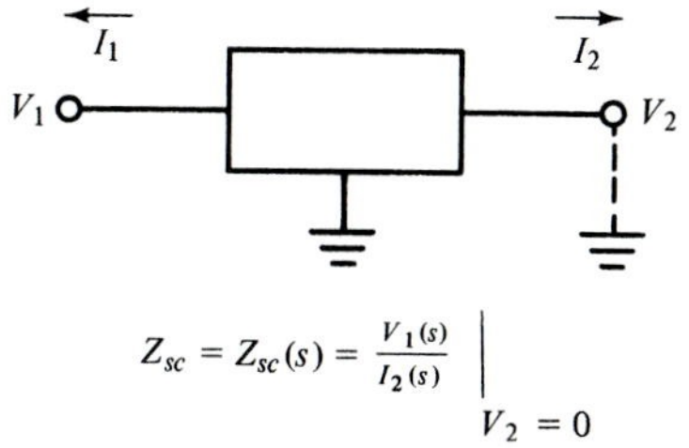

FIGURE 1.16
Definition of short-circuit transfer impedance for a three-terminal network.

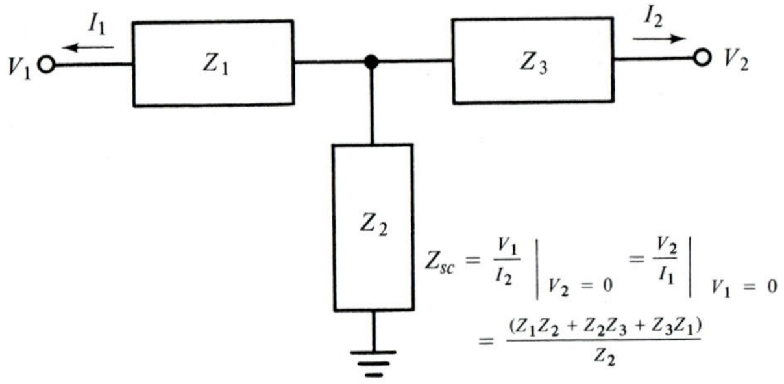

FIGURE 1.17
The short-circuit transfer impedance of a tee network.

or

$$V_0(s) = -Z_0(s)\left[\frac{V_1(s)}{Z_1(s)} + \frac{V_2(s)}{Z_2(s)} + \cdots + \frac{V_n(s)}{Z_n(2)}\right] \tag{1.39}$$

Example 1.9. Consider the single-input inverting amplifier circuit of Fig. 1.18. Suppose we want an inverting gain of 100. We also want to use fairly low resistance values throughout, so that the effects of stray wiring capacitances will be suppressed. We also want the circuit input impedance to be 10 kΩ. This latter requirement can be met by setting the input resistor $R_A = 10$ kΩ. From (1.39) we now see that the short-circuit feedback resistance should be

$$Z_0 = R_0 = 100 \times 10^4 = 10^6\,\Omega \tag{1.40}$$

The tee configuration consisting of R_1, R_2, and R_3 can be used to obtain a short-circuit transfer resistance that is considerably larger than any of the individual resistances. With $R_1 = R_3 = 10$ kΩ we can use (1.37) to find that

$$R_2 = 102.0408\,\Omega \simeq 102\,\Omega$$

will make the short-circuit feedback resistance 10^6. In practice, one might purchase three nominal 10-kΩ resistors and use an adjustable resistor for R_2, so that the voltage gain could be set to exactly -100.

We now have the capability to implement rather general transfer function operators with the circuit of Fig. 1.15. In Chap. 4 we will see even more general

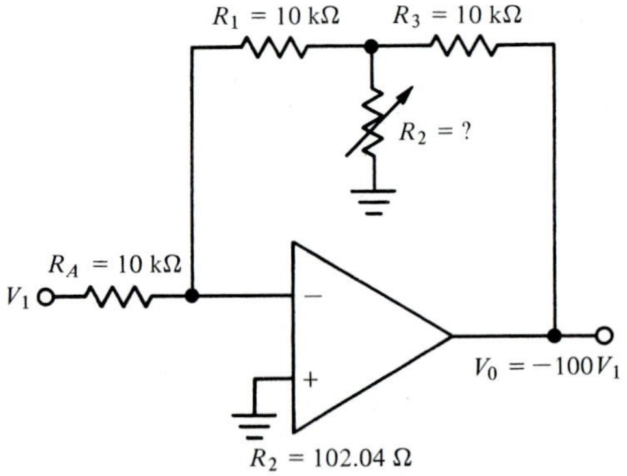

FIGURE 1.18
High-gain inverting amplifier using relatively low resistance values.

configurations specifically designed to implement transfer functions in an optimum fashion, but (1.39) is often sufficient if the designer can find the proper networks to use in the input and feedback paths to obtain a desired transfer function. Note again that it is the short-circuit transfer impedance that is important; Table A-1 of Ref. 3 provides a large collection of *RC* networks and describes their short-circuit transfer impedances. We discuss only a few important examples in this chapter, and we approach the problem of general transfer function synthesis more thoroughly in Chap. 4.

Example 1.10. Consider the circuit of Fig. 1.19*a*. We can analyze its performance, using (1.39), by treating the feedback and input networks separately. Clearly

$$Z_1(s) = R_1$$

and $Z_0(s)$ is the parallel combination of R_2 and C, that is,

$$Z_0(s) = \frac{R_2}{1 + R_2Cs}$$

Thus the transfer function of the circuit of Fig. 1.19*a* is

$$\frac{V_0(s)}{V_1(s)} = -\frac{R_2}{R_1}\frac{1}{1 + Ts} \qquad (1.41)$$

where

$$T = R_2C$$

Hence the circuit implements a *first-order low-pass filter* transfer function with

$$\text{Dc steady-state gain} = -\frac{R_2}{R_1} \qquad (1.42)$$

and the −3-dB cutoff frequency

$$\omega_{-3\,\text{dB}} = \frac{1}{T} = \frac{1}{R_2C} \quad \text{rad/s} \qquad (1.43)$$

This circuit is discussed further in Sec. 1.9.

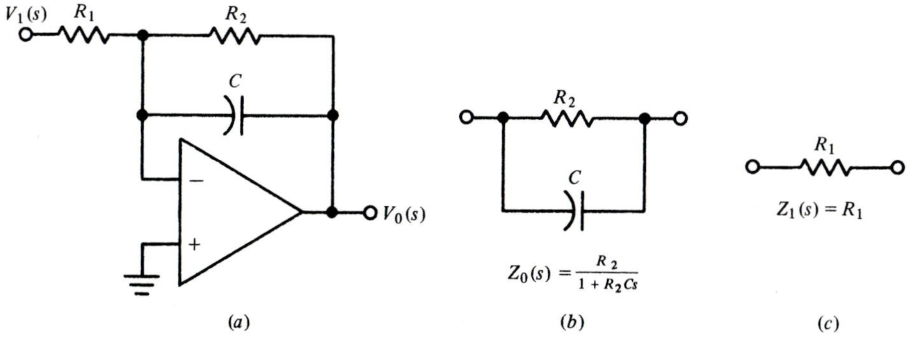

FIGURE 1.19
Simple inverting low-pass filter circuit. (*a*) Complete circuit; (*b*) feedback network; (*c*) input network.

1.6 THE NONINVERTING AMPLIFIER

So far we have always connected the noninverting input of the operational amplifier to ground. Most modern operational amplifiers, however, have this terminal available, as well as the inverting input; i.e., one usually deals with a differential input type of operational amplifier. If the noninverting input is available, then a broader variety of circuit applications are possible. In particular, one can easily implement the noninverting amplifier circuit of Fig. 1.20; here the + terminal is utilized along with the − terminal. If the amplifier has the ideal properties described in Sec. 1.3, we can assume that consequently $I_1(s)$ and $I_2(s)$ are zero and that

$$V_1(s) \cong E_n(s)$$

Thus

$$\frac{E_n(s)}{Z_1(s)} + \frac{E_n(s) - V_0(s)}{Z_2(s)} = -I_1(s) = 0 \tag{1.44}$$

or

$$\frac{V_0(s)}{V_1(s)} = \frac{Z_1(s) + Z_2(s)}{Z_1(s)} = 1 + \frac{Z_2(s)}{Z_1(s)} \tag{1.45}$$

Note that

$$\left|\frac{V_0(s)}{V_1(s)}\right| \geq 1 \tag{1.46}$$

We also see that there is no polarity inversion in this circuit, and the magnitude of the gain is always greater than unity. Above, (1.45) was derived for general two-terminal (one-port) impedances; some special cases merit attention. In Fig. 1.21*a*, we see the circuit of a simple noninverting amplifier circuit using only resistances; in this case

$$\frac{V_0(s)}{V_1(s)} = \frac{R_1 + R_2}{R_1} = 1 + \frac{R_2}{R_1} \geq 1 \tag{1.47}$$

The disarmingly simple circuit of Fig. 1.21*b* is that of a *unity gain follower*. Its

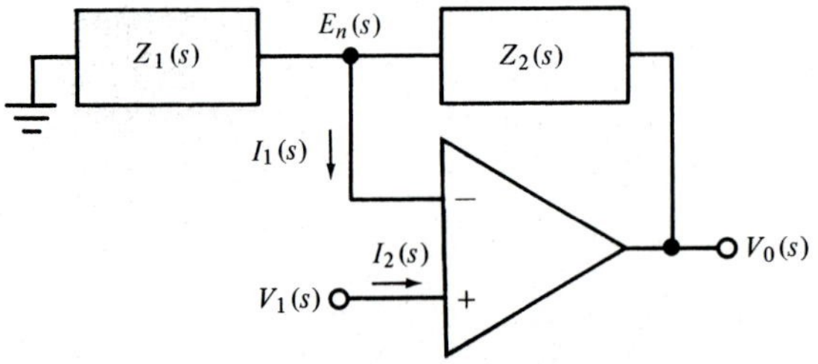

FIGURE 1.20
General noninverting amplifier circuit (two-terminal networks assumed).

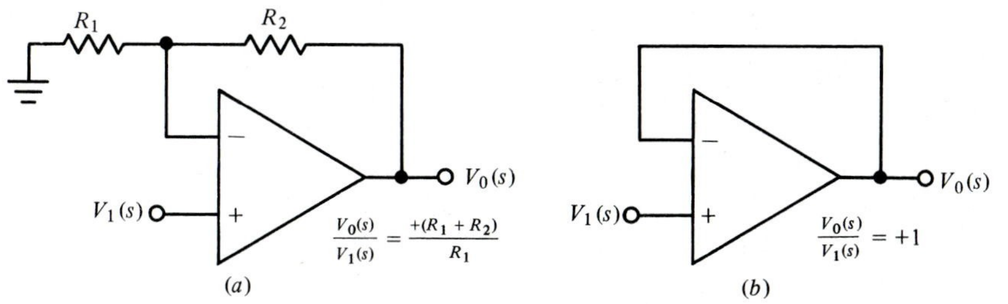

FIGURE 1.21
Noninverting amplifier circuit with resistive elements. (*a*) General circuit; (*b*) simple unity gain follower.

performance is predicted by setting $R_2 = 0$ and/or letting R_1 approach infinity. The voltage transfer function for the circuit of Fig. 1.21*b* is

$$\frac{V_0(s)}{V_1(s)} = 1 \tag{1.48}$$

Note that the noninverting circuit of Fig. 1.21 has a very high input impedance (for example, $10^{12}\ \Omega$ with an FET-input operational amplifier); on the other hand, the output impedance is very low (e.g., less than $0.1\ \Omega$). Thus the noninverting circuit usually does not load the signal source driving it and is therefore a suitable choice in many situations where a high Thévenin-equivalent source impedance is associated with some signal to be amplified. As mentioned earlier, the *inverting* amplifier circuit has an input impedance governed by the value of the input summing impedances associated with each input terminal (Fig. 1.6 or 1.15). Hence the simple unity gain follower of Fig. 1.21*b* serves a useful purpose as a buffer or unloading circuit to permit observing a high-impedance signal with some instrument that requires considerable driving current. In Chap. 2, when we discuss nonideal amplifiers, we will see how to estimate more precisely just how high the input impedance is and how low the output impedance is.

More complex noninverting arrangements are possible. For example, Fig. 1.22 shows a useful noninverting integrator circuit with the transfer function

$$V_0(s) = \frac{2}{sRC} V_1(s) \tag{1.49}$$

Readers should attempt to derive (1.49) for themselves (Prob. 1.16); in particular, it is important to note that resistor imbalance can yield a transfer function with undesirable properties (e.g., the circuit may be unstable!).

Example 1.11. Consider the instrumentation problem posed in Fig. 1.23. Here we have a signal source with a resistive internal impedance of 100,000 Ω. We want to amplify e_s by 100 (polarity not important) without distorting the signal level by loading the signal source by more than 1 percent. We want to keep all circuit resistances less than 10,000 Ω in order to reduce internal amplifier dc offsets (see

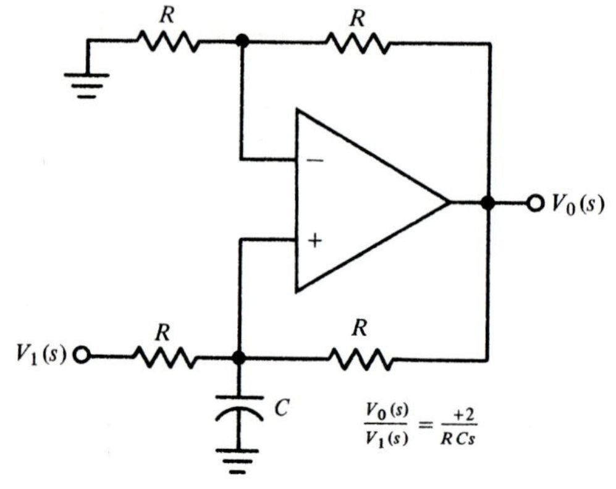

FIGURE 1.22
A noninverting integrator circuit (see also Prob. 1.16).

Sec. 2.5 for a discussion of this consideration). We can achieve the required gain of 100 (no polarity inversion), using the noninverting circuit of Fig. 1.23*a*, merely by selecting

$$R_2 = 99R_1$$

for example, $R_2 = 9900\ \Omega$ and $R_1 = 100\ \Omega$; assuming an ideal operational amplifier, the circuit of Fig. 1.23*a* would not load the signal source (we will see in Chap. 2 that some loading will occur, but probably not enough to be important in

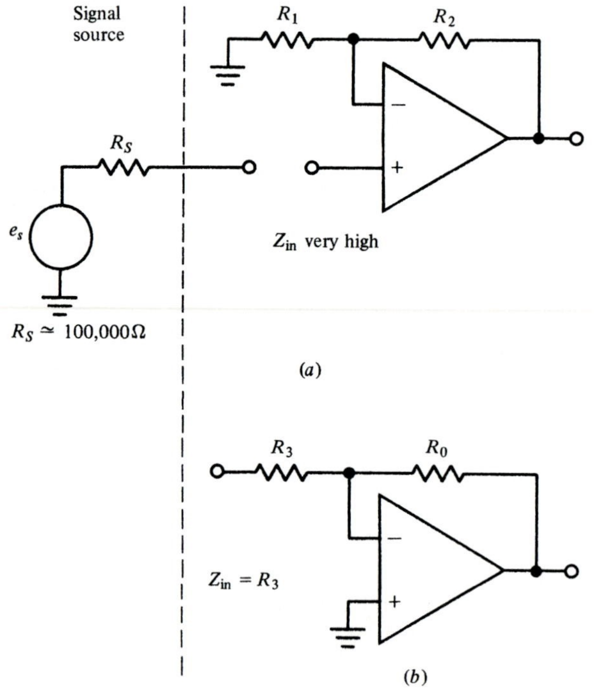

FIGURE 1.23
Circuit of Example 1.11.
(*a*) Noninverting amplifier;
(*b*) inverting amplifier.

this situation). Suppose the inverting circuit of Fig. 1.23*b* is selected. Then in order to keep the circuit input impedance sufficiently high, we are forced to make

$$R_0 \geq 100 \times 100{,}000 = 10 \text{ M}\Omega$$

and already we have violated our requirement that all resistors be less than 10,000 Ω in value.

Simple Kirchhoff summing may be employed with the noninverting amplifier circuit as well. In Fig. 1.24 we see a two-input noninverting summing amplifier circuit. Clearly, from (1.47),

$$V_0(s) = \left(1 + \frac{R_F}{R_S}\right) V_S(s) \tag{1.50}$$

Since the amplifier does not load node V_S,

$$V_S(s) = \frac{R_P}{R_1} V_1(s) + \frac{R_P}{R_2} V_2(s) \tag{1.51}$$

where R_P is the parallel combination of R_1 and R_2 [this last equation can be easily shown to be the Kirchhoff summing equation at node V_S, assuming $I(s) = 0$]. Thus, combining (1.50) and (1.51), we have

$$V_0(S) = \left(1 + \frac{R_F}{R_S}\right) [K_1 V_1(s) + K_2 V_2(s)] \tag{1.52a}$$

$$K_1 = \frac{R_P}{R_1} \qquad K_2 = \frac{R_P}{R_2} \tag{1.52b}$$

and

$$R_p = R_1 \parallel R_2 \tag{1.52c}$$

Note also that $K_1 \leq 1$ and $K_2 \leq 1$ and that (1.52) is derived by assuming zero source impedance for each input signal. Signal source impedances will affect the gain factors K_1 and K_2.

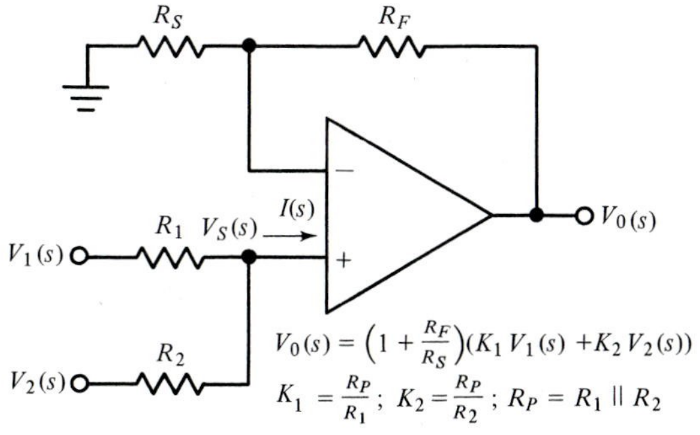

FIGURE 1.24
A simple two-input noninverting summing amplifier circuit.

1.7 THE DIFFERENTIAL AMPLIFIER

Figure 1.25 shows a special operational amplifier circuit commonly termed a *differential amplifier* or, more precisely, a differential-input single-ended-output amplifier. This circuit features a balanced (differential or push-pull) input and an unbalanced (single-ended) output. Assuming an ideal operational amplifier and ideal resistances, we can write time domain circuit equations

$$e_1\left(\frac{1}{R_1}+\frac{1}{R_0}\right)-\frac{v_0}{R_0}-\frac{v_1}{R_1}=0 \tag{1.53a}$$

and

$$e_2\left(\frac{1}{R_2}+\frac{1}{R_3}\right)-\frac{v_2}{R_2}=0 \tag{1.53b}$$

Combining (1.53*a*) and (1.53*b*), assuming $e_1 \cong e_2$, we have

$$v_0=\frac{(R_1+R_0)R_3}{(R_2+R_3)R_1}v_2-\frac{R_0}{R_1}v_1 \tag{1.54}$$

(The reader should verify the above.)

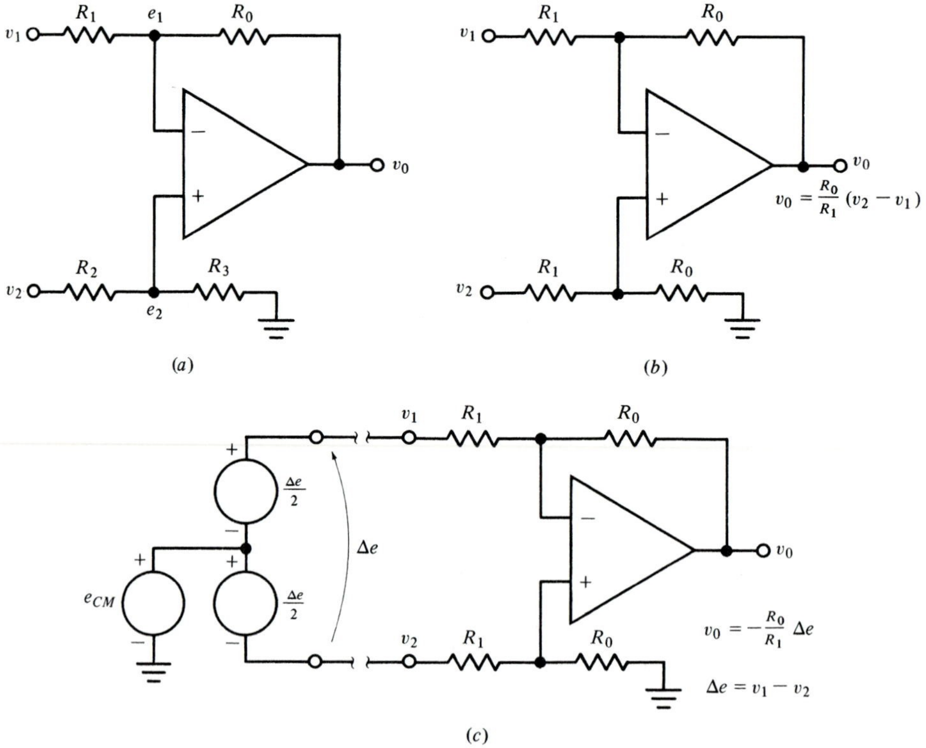

FIGURE 1.25
Single-output differential-input amplifier circuit. (*a*) General circuit; (*b*) circuit with balanced resistance ratios; (*c*) common-mode rejection.

Note that (1.54) has been derived under the assumption that we have an ideal operational amplifier. In Chap. 2, we will take nonideal behavior into account in more detail. In most situations, the special case of Fig. 1.25*b* is employed, that is, $R_3 = R_0$ and $R_2 = R_1$; in this case, (1.54) becomes

$$v_0 = \frac{R_0}{R_1}(v_2 - v_1) \tag{1.55}$$

Differential amplifiers such as Fig. 1.25*b* thus respond only to the *difference* between the input signals v_2 and v_1; they are said to have *good common-mode rejection* of noise induced in equal amounts in both signal sources. Figure 1.25*c* shows a typical situation where differential amplifiers are commonly employed. Here we are trying to observe, and usually also amplify, a small signal Δe, which may be generated in the presence of a relatively large common-mode signal e_{cm}. By using a differential amplifier, the common-mode signal is ideally rejected.

This type of situation is common in many instrumentation systems, where Δe originates from some low-level transducer; e_{cm} may be a dc offset in some push-pull signal source or some form of noise (for example, 60-Hz interference or some other crosstalk from a nearby electrical system)—in other words, an unwanted signal component which will hopefully be ignored by the amplifier.

In Fig. 1.25*c*, we assume

$$v_1 = e_{cm} + \frac{\Delta e}{2} \tag{1.56a}$$

$$v_2 = e_{cm} - \frac{\Delta e}{2} \tag{1.56b}$$

that is,

$$v_1 - v_2 = \Delta e \tag{1.57}$$

and thus (1.55) yields, for the case of Fig. 1.25*c*,

$$v_0 = -\frac{R_0}{R_1}\Delta e \tag{1.58}$$

independent of e_{cm}.

Note that the differential input resistance of the circuit of Fig. 1.25*b* is

$$R_{\text{in}} = 2R_1 \tag{1.59}$$

which creates a loading on the differential signal source that may attenuate it.

Example 1.12. Consider Fig. 1.26*a*; this depicts a low-level transducer whose signal is carried over a shielded cable with two floating conductors for some distance to a preamplifier. Induced noise (usually 60 Hz from the ac power lines in the vicinity of the transducer) is to be rejected, while the transducer signal itself is to be amplified by 10. The desired signal may be considered a differential signal in the presence of an unwanted common-mode noise signal. We have the further restriction that the transducer should be terminated in a total load resistance of 20 kΩ between

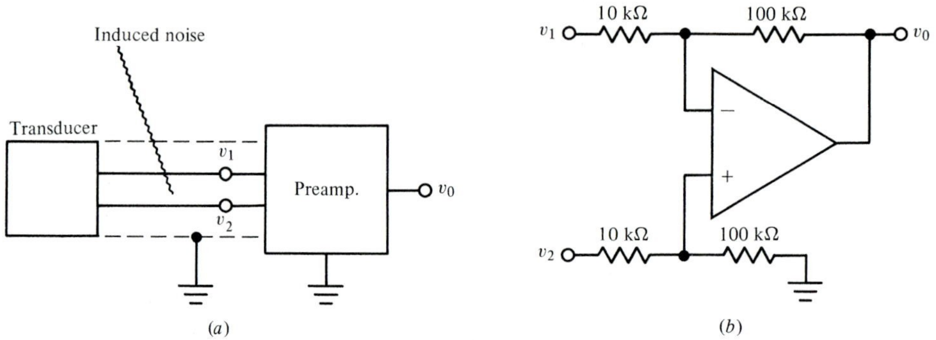

FIGURE 1.26
Circuit of Example 1.12. (*a*) Signal source with noise; (*b*) preamplifier circuit.

terminals. The differential amplifier circuit of Fig. 1.26*b* meets all the requirements stated above. It has a differential voltage gain of 10, supposedly a high common-mode rejection, if the amplifier and resistors are well balanced, and a total input impedance of 20 kΩ. The output signal is single-ended and may be subsequently processed by single-ended amplifiers for equalization and further amplification.

Note that if the common-mode voltage (e_{cm} in Fig. 1.25*c*) is fairly large, then the operational amplifier input terminals will have a similar voltage impressed upon them. Most operational amplifiers have a *maximum allowable common-mode input voltage* that is slightly less than the magnitude of the power supply voltages.

Figure 1.27 shows a *differential-input differential-output* amplifier circuit. This maintains a balanced (or push-pull) mode at both input and output. This circuit requires an operational amplifier with a differential output stage. Such amplifiers are less common than the unbalanced-output variety we have been discussing, but they are available from some suppliers. This configuration is mentioned here primarily for completeness; of course, such amplifiers are useful for

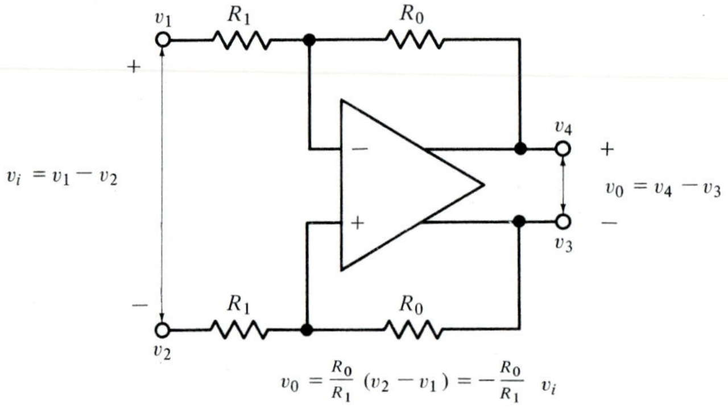

FIGURE 1.27
Differential-output differential-input amplifier circuit.

amplifying signals for delivery to balanced transmission lines, etc. The balanced circuit of Fig. 1.27 has a voltage gain described by

$$v_0 = v_4 - v_3 = -\frac{R_0}{R_1}(v_1 - v_2) = -\frac{R_0}{R_1}v_1 \tag{1.60}$$

where v_0 and v_1 are the differential output and input voltages, respectively.

1.8 THE COMBINED INVERTING AND NONINVERTING AMPLIFIER

Figure 1.28 shows a general amplifier circuit that permits summation with inversion and without. Such a general circuit provides considerable flexibility. Our method of analysis will be instructive when we come to similar circuits in Chap. 2. We assume that all the impedances are simple two-terminal driving-point impedances. We assume that there are m inverting inputs $V_{I1}(s), \ldots, V_{Im}(s)$ and n noninverting inputs $V_{N1}(s), \ldots, V_{Nn}(s)$, all obtained from ideal voltage sources. In addition, we have added two shunt impedances $Z_{IS}(s)$ and $Z_{NS}(s)$ for the sake of generality. These elements can be adjusted to affect the transfer function gains. The typical situation is for all the impedances to be resistances, but general fixed linear networks can also be used.

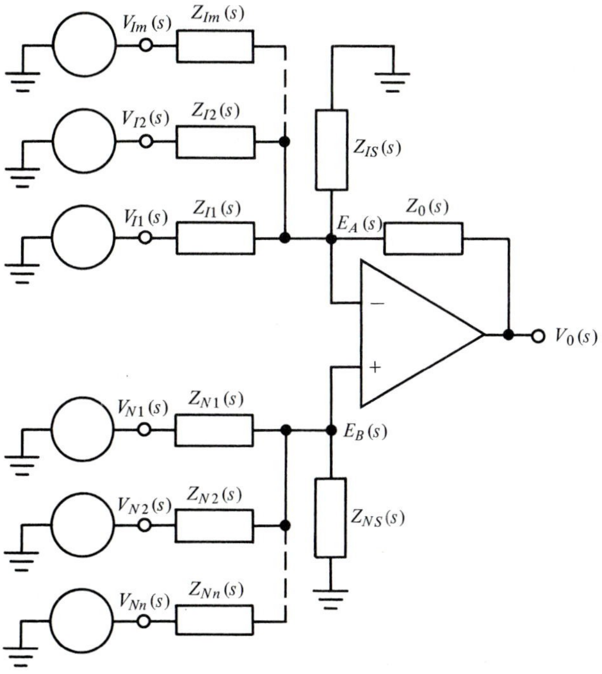

FIGURE 1.28
General inverting and noninverting summing amplifier.

The Kirchhoff nodal equations at nodes $E_A(s)$ and $E_B(s)$ are

$$E_A(s)\left[\frac{1}{Z_{IS}(s)}+\frac{1}{Z_{I1}(s)}+\cdots+\frac{1}{Z_{Im}(s)}+\frac{1}{Z_0(s)}\right]=\frac{V_0(s)}{Z_0(s)}+\sum_{i=1}^{m}\frac{V_{Ii}(s)}{Z_{Ii}(s)} \tag{1.61a}$$

and

$$E_B(s)\left[\frac{1}{Z_{NS}(s)}+\frac{1}{Z_{N1}(s)}+\cdots+\frac{1}{Z_{Nn}(s)}\right]=\sum_{j=1}^{n}\frac{V_{Nj}(s)}{Z_{Nj}(s)} \tag{1.61b}$$

Let us define

$$\begin{aligned}Z_A(s) &= Z_0(s)\,\|\,Z_{IS}(s)\,\|\,Z_{I1}(s)\,\|\cdots\|\,Z_{Im}(s)\\ &=\left[\frac{1}{Z_0(s)}+\frac{1}{Z_{IS}(s)}+\sum_{i=1}^{m}\frac{1}{Z_{Ii}(s)}\right]^{-1}\end{aligned} \tag{1.62a}$$

and

$$Z_B(s)=Z_{N1}(s)\,\|\,Z_{N2}(s)\,\|\cdots\|\,Z_{Nn}(s)\,\|\,Z_{NS}(s)=\left[\sum_{j=1}^{n}\frac{1}{Z_{Nj}(s)}+\frac{1}{Z_{NS}(s)}\right]^{-1} \tag{1.62b}$$

For an ideal operational amplifier,

$$E_A(s)\cong E_B(s) \tag{1.63}$$

Using (1.62) in (1.61), we obtain

$$\frac{E_A(s)}{Z_A(s)}=\frac{V_0(s)}{Z_0(s)}+\sum_{i=1}^{m}\frac{V_{Ii}(s)}{Z_{Ii}(s)} \tag{1.64a}$$

and

$$\frac{E_B(s)}{Z_B(s)}=\sum_{j=1}^{n}\frac{V_{Nj}(s)}{Z_{Nj}(s)} \tag{1.64b}$$

These last equations are generally of great importance to us in this text; they describe what we shall call *Kirchhoff voltage summing* at a node. Consider Fig. 1.29, which illustrates a general Kirchhoff summing network. We assume that there is no current flowing out of node V_k except via the branches indicated. (This is the situation for nodes E_A and E_B in Fig. 1.28, since we assume no current flows into either amplifier terminal.) A nodal analysis of Fig. 1.29 yields

$$V_k(s)\left[\frac{1}{Z_1(s)}+\frac{1}{Z_2(s)}+\cdots+\frac{1}{Z_n(s)}+\frac{1}{Z_s(s)}\right]=\frac{V_1(s)}{Z_1(s)}+\cdots+\frac{V_n(s)}{Z_n(s)} \tag{1.65}$$

If we define Z_p = parallel combination of all impedances at node V_k,

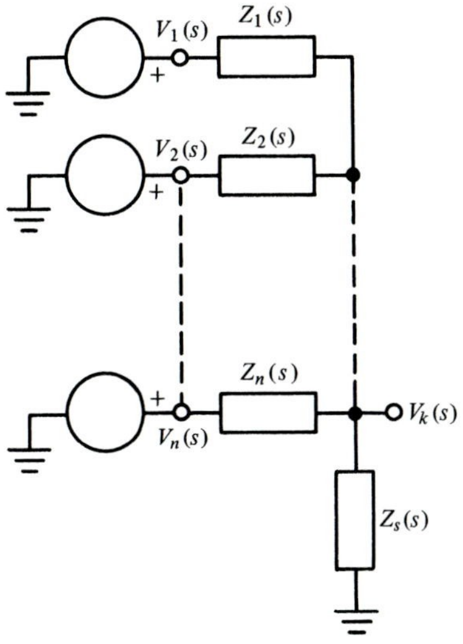

FIGURE 1.29
A general Kirchhoff voltage summing network.

$$Z_p = \left[\frac{1}{Z_1(s)} + \cdots + \frac{1}{Z_n(s)} + \frac{1}{Z_s(s)}\right]^{-1} \tag{1.66}$$

then (1.65) becomes

$$\frac{V_k(s)}{Z_p(s)} = \frac{V_1(s)}{Z_1(s)} + \cdots + \frac{V_n(s)}{Z_n(s)} \tag{1.67a}$$

or

$$V_k(s) = Z_p(s)\sum_{i=1}^{n} \frac{V_i(s)}{Z_i(s)} \tag{1.67b}$$

Thus in the circuit of Fig. 1.29 the voltages V_i, $i = 1, \ldots, n$, are summed at V_k, but with a weighting factor Z_p/Z_i. This is the general nature of such summing networks, and equations of the form of (1.67) will arise frequently in our work.

Employing (1.63) in (1.64a), we have

$$\frac{V_0(s)}{Z_0(s)} = -\sum_{i=1}^{m} \frac{V_{Ii}(s)}{Z_{Ii}(s)} + \frac{E_B(s)}{Z_A(s)} \tag{1.68}$$

Then combining (1.64b) and (1.68), we obtain

$$V_0(s) = Z_0(s)\left[\frac{Z_B(s)}{Z_A(s)}\sum_{j=1}^{n} \frac{V_{Nj}(s)}{Z_{Nj}(s)} - \sum_{i=1}^{m} \frac{V_{Ii}(s)}{Z_{Ii}(s)}\right] \tag{1.69}$$

This last equation permits us to design combined inverting and noninverting summing amplifiers. Relative values of the inverting gains can be selected by the proper choice of the $Z_{Ii}(s)$ and relative values of the noninverting gains by the $Z_{Nj}(s)$. Overall inverting and noninverting gains can then be adjusted by the proper choice of $Z_0(s)$, $Z_{IS}(s)$, and $Z_{NS}(s)$. Of course, the differential amplifier circuit of Fig. 1.25 is included in this general circuit, along with the more common inverting and noninverting circuits of Figs. 1.8 and 1.20.

Let us define $K_{Ii}(s)$ as the transfer function between $V_{Ii}(s)$ and $V_0(s)$ with all other inputs set to zero. *Warning!* To do this properly, it is conceptually necessary to set all other inputs to zero by replacing them with short circuits to ground, since all the development in this section has been predicated on the inputs $V_{Ii}(s)$ and $V_{Nj}(s)$ being ideal voltage sources. Thus in calculating the total shunt impedances at summing nodes, such as E_A and E_B in Fig. 1.28, the signal sources must be considered to have zero internal impedance. Let us also define $K_{Nj}(s)$ as the transfer function between $V_{Nj}(s)$ and $V_0(s)$. We note from (1.69) that

$$K_{Ii}(s) = -\frac{Z_0(s)}{Z_{Ii}(s)} \qquad i = 1, \ldots, m \tag{1.70a}$$

and

$$K_{Nj}(s) = +\frac{Z_B(s)}{Z_A(s)}\frac{Z_0(s)}{Z_{Nj}(s)} \qquad j = 1, \ldots, n \tag{1.70b}$$

These last equations facilitate the design process. Often we can pick $Z_{NS}(s)$ and/or $Z_{IS}(s)$ such that $Z_A(s) = Z_B(s)$, so that

$$K_{Nj}(s) = \frac{+Z_0(s)}{Z_{Nj}(s)} \qquad Z_A(s) = Z_B(s) \tag{1.71}$$

and the input summing impedances can all be designed individually. (Reference 9 provides some useful design guides and design examples for resistive summing amplifiers.)

Example 1.13. Consider the circuit of Fig. 1.30. Here we have a general summing amplifier with two inverting inputs and two noninverting inputs. Thus the general form of the amplifier gain equation is

$$V_0 = K_{I1}V_{I1} + K_{I2}V_{I2} + K_{N1}V_{N1} + K_{N2}V_{N2} \tag{1.72}$$

From (1.70),

$$K_{I1} = \frac{-R_0}{R_{I1}} \tag{1.73a}$$

$$K_{I2} = \frac{-R_0}{R_{I2}} \tag{1.73b}$$

$$K_{N1} = \frac{R_B}{R_A}\frac{R_0}{R_{N1}} \tag{1.73c}$$

$$K_{N2} = \frac{R_B}{R_A}\frac{R_0}{R_{N2}} \tag{1.73d}$$

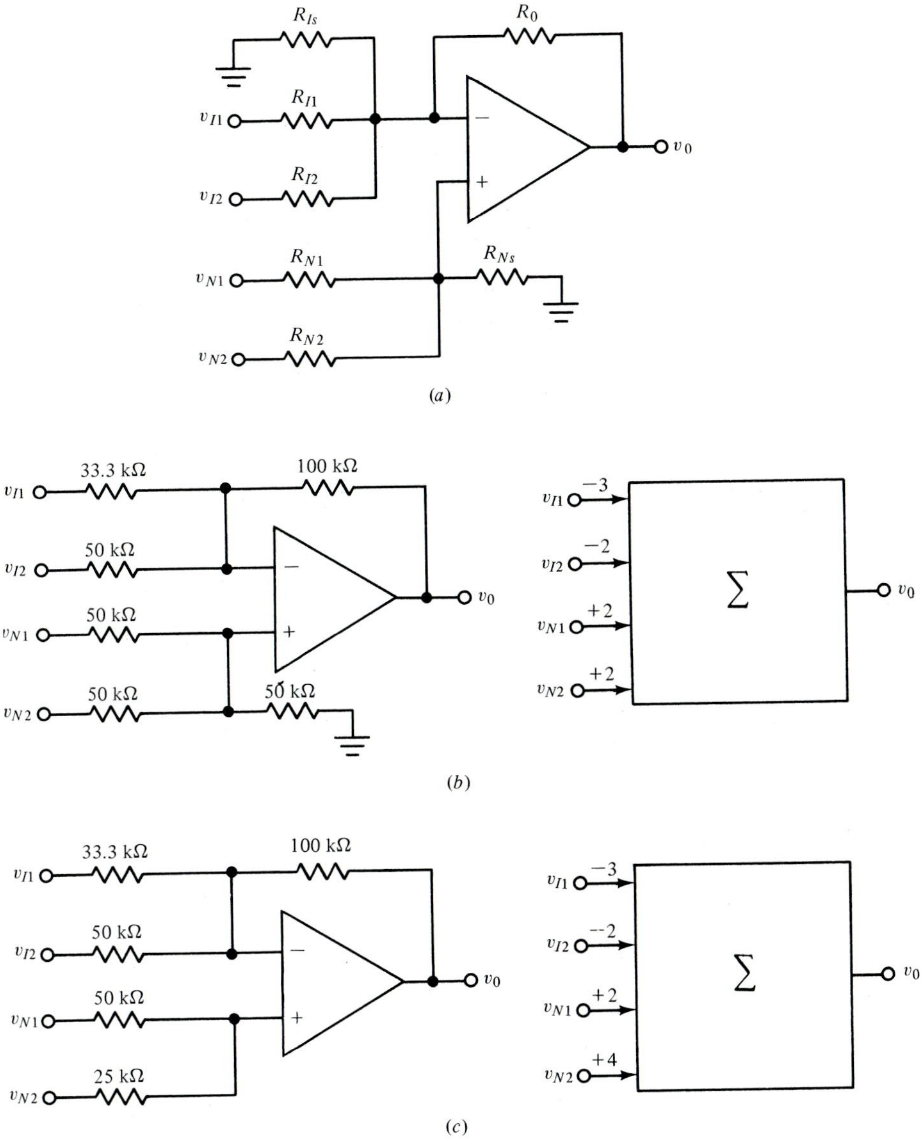

FIGURE 1.30
Circuit of Example 1.13. (*a*) General circuit with two inverting and two noninverting inputs; (*b*) circuit for gains of −3, −2, +2, and +2; (*c*) circuit for gains of −3, −2, +2, and +4.

where

$$R_A = R_0 \parallel R_{I1} \parallel R_{I2} \parallel R_{Is} \quad (1.74a)$$

and

$$R_B = R_{N1} \parallel R_{N2} \parallel R_{Ns} \quad (1.74b)$$

We can make $R_B = R_A$ by the proper choice of R_{Is} and R_{Ns} (usually one is made infinite, i.e., not used).

Suppose we want (Fig. 1.30*b*)

$$v_0 = -(3v_{I1} + 2v_{I2}) + 2v_{N1} + 2v_{N2} \quad (1.75)$$

Pick $R_0 = 100\ \text{k}\Omega$. Then from (1.73),

$$R_{I1} = \frac{100}{3}\ \text{k}\Omega \quad (1.76a)$$

$$R_{I2} = 50\ \text{k}\Omega \quad (1.76b)$$

$$R_{N1} = 50\ \text{k}\Omega \qquad \text{if } R_A = R_B \quad (1.76c)$$

and

$$R_{N2} = 50\ \text{k}\Omega \qquad \text{if } R_A = R_B \quad (1.76d)$$

From (1.74),

$$R_A = 16.67\ \text{k}\Omega \parallel R_{Is}$$

$$R_B = 25\ \text{k}\Omega \parallel R_{Ns}$$

If we let $R_{Is} = \infty$,

$$R_A = 16.67\ \text{k}\Omega \quad (1.77)$$

We can then make $R_B = R_A$ with

$$R_{Ns} = 50\ \text{k}\Omega \quad (1.78)$$

Figure 1.30*b* shows the final circuit.

Suppose instead we want

$$v_0 = -(3v_{I1} + 2v_{I2}) + 2v_{N1} + 4v_{N2}$$

First, this changes R_{N2} to 25 kΩ. We now have

$$R_A = 16.67\ \text{k}\Omega \parallel R_{Is} \quad (1.79a)$$

$$R_B = 16.67\ \text{k}\Omega \parallel R_{Ns} \quad (1.79b)$$

Now we need neither R_{Is} nor R_{Ns} to make $R_A = R_B$. Figure 1.30*c* shows the final circuit. Note that changing the gain for v_{N2} did not affect the values of the feedback resistor or the other summing resistors. Other examples will lead to a requirement for finite R_{IS} but not R_{NS}, etc. (see Prob. 1.22).

1.9 SIMPLE FIRST-ORDER TRANSFER FUNCTIONS

To illustrate the use of the basic linear amplifier design equations we have thus far developed, this section will present some basic configurations for implementing first-order (one-pole) linear transfer functions. Methods for designing higher-order transfer functions are presented in Chap. 4 (Active *RC* Filters). We have shown how to implement the basic linear operations of inversion (Fig. 1.7), multiplying by a constant (Figs. 1.6 and 1.21), summing (Figs. 1.8, 1.24, and 1.30), and integration (Figs. 1.10 and 1.22). Note that with a collection of these types of circuits, one can implement any linear transfer function by analog-computer programming techniques (see Ref. 3, pp. 10 to 16, and Ref. 8, Chap. 10). This approach yields designs with a relatively large number of components. We present here and in Chap. 4 methods of designing transfer function operators that require relatively few components and still provide accurate implementations.

First-Order Low-Pass Transfer Function

Consider the first-order low-pass transfer function

$$\frac{V_0(s)}{V_1(s)} = \frac{-H_0\omega_0}{s+\omega_0} = \frac{-H_0}{1+s/\omega_0} \tag{1.80}$$

As already suggested in Example 1.10, this transfer function can be implemented by the circuit of Fig. 1.19*a*, redrawn in Fig. 1.31*a*. We have already shown in Example 1.10 that we can use the approach of Sec. 1.5, and by means of (1.39), it follows that

$$\frac{V_0(s)}{V_1(s)} = \frac{-Z_0(s)}{Z_1(s)} \tag{1.81}$$

where

$$Z_0(s) = \frac{R_2}{1+R_2Cs} \tag{1.82a}$$

and

$$Z_1(s) = R_1 \tag{1.82b}$$

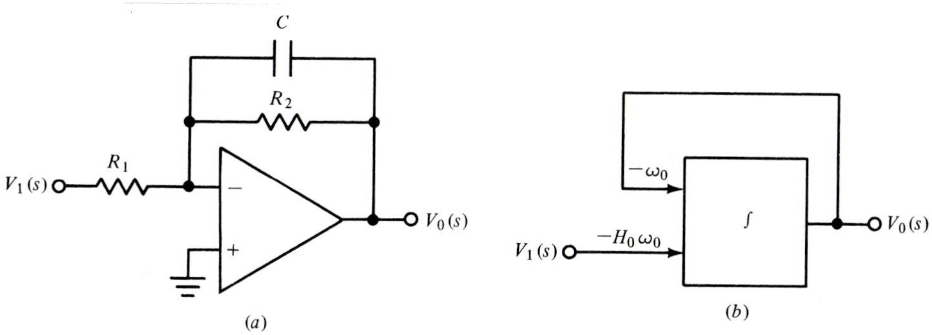

FIGURE 1.31
First-order low-pass filter circuit. (*a*) Circuit components; (*b*) circuit block diagram.

or

$$\frac{V_0(s)}{V_1(s)} = \frac{-R_2/R_1}{1 + R_2Cs} = \frac{-1/R_1C}{s + 1/R_2C} \tag{1.83}$$

Thus we have matched the transfer function of (1.80) with

$$H_0 = \frac{R_2}{R_1} = \text{dc steady-state gain} \tag{1.84a}$$

$$\omega_0 = \frac{1}{R_2C} = \text{3-dB cutoff frequency} \tag{1.84b}$$

As an alternative approach, let us write the time domain equation that corresponds to the frequency domain transfer function (1.80), that is,

$$\frac{dv_0}{dt} = -\omega_0 v_0(t) - H_0\omega_0 v_1(t) \tag{1.85}$$

Using the summing integrator circuit of Fig. 1.10, we implement (1.85) with the circuit of Fig. 1.31*b*. We see, however, that this is really merely a block-diagram sketch of the circuit of Fig. 1.31*a* with

$$\omega_0 = \frac{1}{R_2C} \tag{1.86a}$$

$$H_0\omega_0 = \frac{1}{R_1C} \tag{1.86b}$$

which agrees with (1.84).

Thus we have presented two related, but conceptually different, approaches to the implementation of (1.80), i.e.:

1. We can synthesize input and feedback *networks* for a circuit of the form of Fig. 1.15, 1.20, or 1.28, so that the overall transfer function is implemented (the approach of Fig. 1.31*a*).
2. Transfer functions may be implemented by *implementing their differential equation* with basic operations such as summing, inverting, integration, and amplification (the approach of Fig. 1.31*b*).

In this simple first-order example, we come up with the same design; in more complicated, higher-order cases, in general the resulting designs will differ in detail, depending on the approach used. The differential-equation-solving approach will generally lead to designs with more components. To see this better, let us consider a slightly different case.

Implementing a Pole-Zero Pair

Suppose we want to implement the transfer function

$$\frac{V_0(s)}{V_1(s)} = \frac{H_0(1 + sT_1)}{1 + sT_2} = \frac{H_0T_1(s + 1/T_1)}{T_2(s + 1/T_2)} \tag{1.87}$$

This transfer function has the properties

$$\left|\frac{V_0(s)}{V_1(s)}\right|_{s=0} = H_0 = \text{dc gain} \tag{1.88a}$$

$$\left|\frac{V_0(j\omega)}{V_1(j\omega)}\right|_{\omega=\infty} = \frac{H_0T_1}{T_2} = \text{high-frequency gain} \tag{1.88b}$$

with a pole at

$$s = -\frac{1}{T_2}$$

and zero at

$$s = -\frac{1}{T_1}$$

Such a transfer function is commonly encountered in a variety of equalizer and compensator circuits for communication and control systems.

We will again take two approaches to implementing such a transfer function. First, in Fig. 1.32, we have one-amplifier circuits which implement (1.87). These circuits may be analyzed by finding the short-circuit transfer impedances of the feedback networks. In Fig. 1.32*a*, we can find that (assuming the structure of Fig. 1.15)

$$Z_0(s) = R_1 + \frac{R_2}{1 + sR_2C} \tag{1.89}$$

Thus

$$\frac{V_0(s)}{V_1(s)} = \frac{-Z_0(s)}{Z_1(s)} = \frac{-[R_1 + R_2/(1 + sR_2C)]}{R_3}$$

$$= \frac{-(R_1 + R_2 + sR_1R_2C)}{R_3(1 + sR_2C)}$$

$$= H_0\frac{1 + sT_1}{1 + sT_2} \tag{1.90a}$$

with

$$T_1 = \frac{R_1}{R_1 + R_2}R_2C = \frac{R_1}{R_1 + R_2}T_2 \tag{1.90b}$$

$$T_2 = R_2C \tag{1.90c}$$

and

$$H_0 = \frac{-(R_1 + R_2)}{R_3} \tag{1.90d}$$

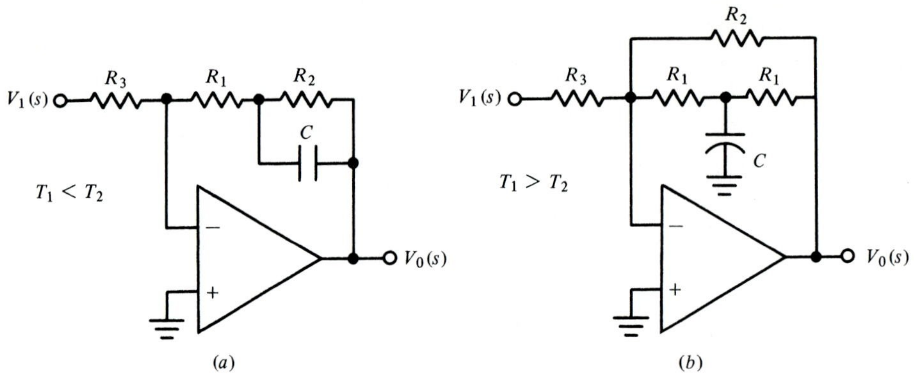

FIGURE 1.32
Two circuits for implementing one real zero and one real pole. (*a*) $T_1 < T_2$; (*b*) $T_1 > T_2$.

The pole is at

$$s = \frac{-1}{R_2C} \tag{1.90e}$$

and the zero is at

$$s = \frac{-R_1 + R_2}{R_1R_2C} \tag{1.90f}$$

The first -3-dB frequency is

$$\frac{1}{T_2} = \frac{1}{R_2C} \qquad \text{rad/s} \tag{1.90g}$$

Also we note that for Fig. 1.32*a*,

$$T_1 < T_2 \tag{1.90h}$$

This restriction can be overcome by the circuit of Fig. 1.32*b*. Here (see Prob. 1.28) we find that

$$\frac{V_0(s)}{V_1(s)} = H_0\frac{1 + sT_1}{1 + sT_2} \tag{1.91a}$$

with

$$T_1 = \frac{R_1C}{2} \tag{1.91b}$$

$$T_2 = \frac{2R_1}{2R_1 + R_2}\frac{R_1C}{2} \tag{1.91c}$$

and

$$H_0 = \frac{-2R_1R_2}{(2R_1 + R_2)R_3} \tag{1.91d}$$

For Fig. 1.32*b* we note that now

$$T_1 > T_2$$

Thus the two circuits in Fig. 1.32 together permit us to implement transfer functions of the form (1.87) for either $T_1 < T_2$ (Fig. 1.32*a*) or $T_1 > T_2$ (Fig. 1.32*b*).

Note that although we have implemented (1.87) with one amplifier, we have been limited to transfer functions which have a polarity inversion ($H_0 < 0$). Also the circuits of Fig. 1.32 present a constant input impedance over the useful frequency range; this might not be true of alternative approaches.

Figure 1.33 shows two alternative, but generally less desirable, circuits for implementing (1.87). Figure 1.33*a* looks acceptable at first glance, but the *circuit of Fig. 1.33a* should normally *not be used*. The main reason for this recommendation is not apparent from the presentation of this chapter. Briefly, the dc offsets in the circuit, including those in the amplifier, will be integrated, i.e., will cause a charge to build up on capacitor C, eventually causing an amplifier overload. The circuit of Fig. 1.33*a* might be safe to use, however, as part of a larger system, provided there is some external closed-loop path around the circuit to provide feedback cancellation of long-term (dc) drifts. The circuit of Fig. 1.33*b* also implements (1.87). It will perform properly; however, we note that the *input impedance* is highly capacitive, i.e.,

$$Z_{\text{in}} = R_1 \parallel C_1$$

Thus this last circuit (Fig. 1.33*b*) may provide a poor load for the signal source

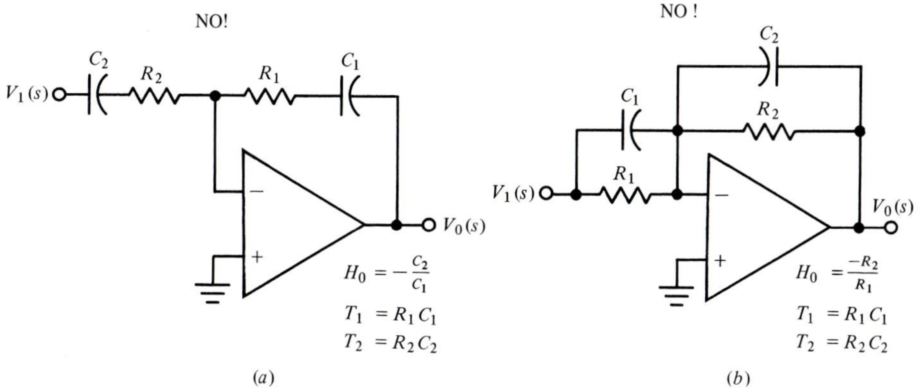

FIGURE 1.33
Alternative ways of implementing a pole-zero pair. (*a*) Avoid this (no dc feedback); (*b*) this works, but has capacitive input impedance.

driving it. In general, the circuits of Fig. 1.32 are preferred for implementing (1.87).

To illustrate the analog-computer or state-variable approach to implementing (1.87), let us write

$$\begin{aligned}\frac{V_0(s)}{V_1(s)} &= \frac{H_0(1+sT_1)}{1+sT_2} \\ &= \frac{T_1H_0}{T_2}\frac{s+1/T_1}{s+1/T_2} \\ &= \frac{T_1}{T_2}H_0 + \frac{(T_1-T_2)H_0}{T_2}\frac{-1/T_2}{s+1/T_2}\end{aligned} \tag{1.92}$$

or

$$V_0(s) = \frac{T_1H_0}{T_2}V_1(s) + \frac{(T_1-T_2)H_0}{T_2}V_2(s) \tag{1.93a}$$

where

$$V_2(s) = \frac{-1/T_2}{s+1/T_2}V_1(s) \tag{1.93b}$$

We recognize (1.93*b*) as being of the form of (1.80); i.e., we can implement it by the circuit of Fig. 1.31, noting that

$$\frac{dv_2(t)}{dt} = \frac{-1}{T_2}[v_2(t)+v_1(t)] \tag{1.94}$$

Thus (1.93) leads us to the block diagram of Fig. 1.34, which implies that *two amplifiers* are required (shown in dotted area); one implements (1.93*b*), and the other sums $V_2(s)$ and $V_1(s)$. Note that the polarity of the gains for the summing amplifier 2 depends on the relative values of T_1 and T_2 as well as the polarity of H_0. Note that H_0 may be either positive or negative, whereas the circuits

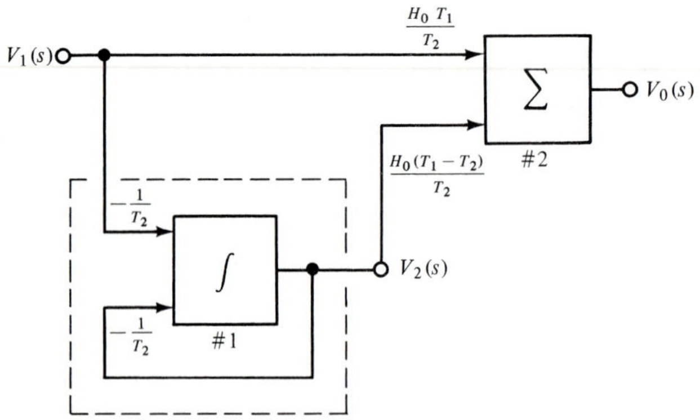

FIGURE 1.34
Two-amplifier circuit for implementing a pole-zero pair.

of Fig. 1.32 required $H_0 < 0$. The decomposition of (1.87) into (1.93*a*) and (1.93*b*) may seem somewhat arbitrary, since the relative gains might have been split up differently. We chose the form of (1.93) so that the voltage gain from $V_1(s)$ to $V_2(s)$ is always less than or equal to 1 at all frequencies, thereby avoiding an overload at the output of amplifier 1. Students with prior analog-computer experience will recognize that this last approach we have taken has *amplitude-scaling problems* associated with the actual design implementation. We will not pursue the amplitude scaling any further, but the reader is cautioned to remember this consideration in a general problem.

Example 1.14. Suppose we want the transfer function whose frequency response is sketched in Fig. 1.35*a*. From the corner or break frequencies we assume a transfer

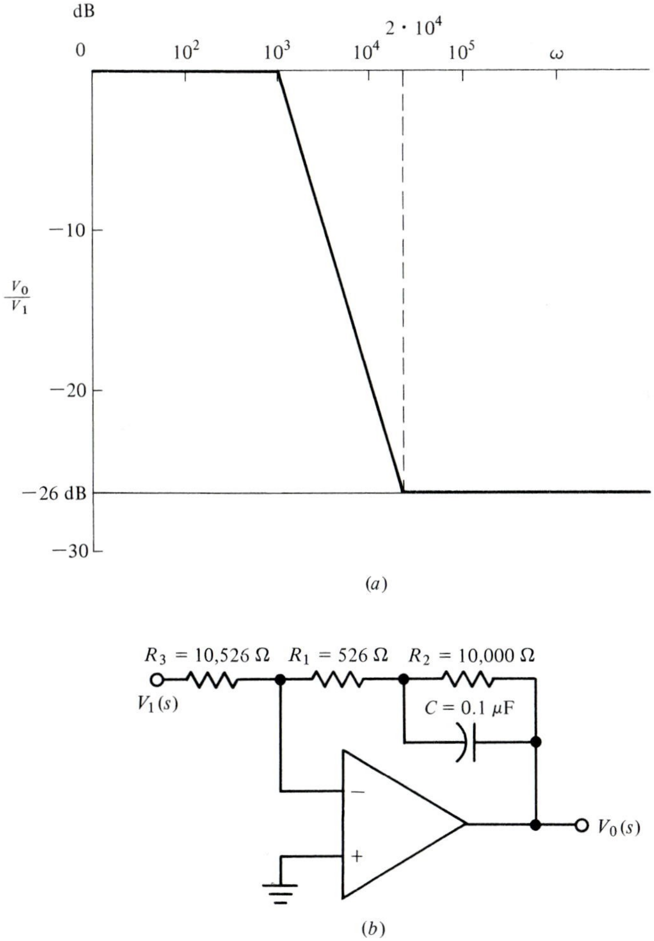

FIGURE 1.35
Circuit for Example 1.14. (*a*) Desired transfer function (gain magnitude only); (*b*) a one-amplifier implementation using (1.90).

function of the form of (1.87), with

$$T_1 = \frac{1}{2 \times 10^4} = 5 \times 10^5$$

$$T_2 = \frac{1}{10^3} = 10^{-3}$$

and

$$H_0 = \pm 1$$

Let us assume that a polarity inversion is acceptable. Then, using the circuit of Fig. 1.32*a* ($T_1 < T_2$), we obtain the design of Fig. 1.35*b*, where we have arbitrarily selected $R_2 = 10\ \text{k}\Omega$. Then from (1.90)

$$C = \frac{T_2}{R_2} = \frac{10^{-3}}{10^4}$$

$$= 0.1 \times 10^{-6} = 0.1\ \mu F$$

Since

$$\frac{R_1}{R_1 + R_2} = \frac{T_1}{T_2} = 0.05$$

then

$$R_1 = 526\ \Omega$$

and

$$R_3 = (R_1 + R_2) \times 1 = 10.526\ \text{k}\Omega$$

1.10 MISCELLANEOUS LINEAR APPLICATIONS

This section presents a variety of linear applications, discussed in terms of the ideal operational amplifier model, to broaden the reader's viewpoint about the potential range of operational amplifier applications. Since we are introducing these topics at this point, we cannot fully discuss the practical limitations imposed by the nonideal behavior of actual amplifiers. Later, after we have described methods for treating nonideal behavior (Chap. 2), the reader will be able to more fully appreciate some of the practical limitations associated with the performance of the circuits we describe here.

Differentiation, High-Pass Filtering, and DC Blocking

It is often desirable to implement the differentiator transfer function

$$\frac{V_0(s)}{V_1(s)} = H(s) = Ks \tag{1.95}$$

at least approximately. Please note that this transfer function is not physically realizable (there is one more zero than pole in the transfer function); however, we may be satisfied with an approximation, such as the one shown in Fig. 1.36*a*. That is, we might try to implement either

$$H(s) = \frac{Ks\omega_0}{s + \omega_0} \tag{1.96a}$$

which is a *high-pass filter* function, or, more realistically,

$$H(s) = \frac{Ks\omega_0\omega_1}{(s + \omega_0)(s + \omega_1)} \tag{1.96b}$$

Indeed, we never need the function of (1.95) in practice; rather, we want to perform the differentiation operation over some finite frequency range. The functions of (1.96) and Fig. 1.36 will block direct current, differentiate signals with spectral content well below $\omega = \omega_0$, and pass frequencies above ω_0 [and

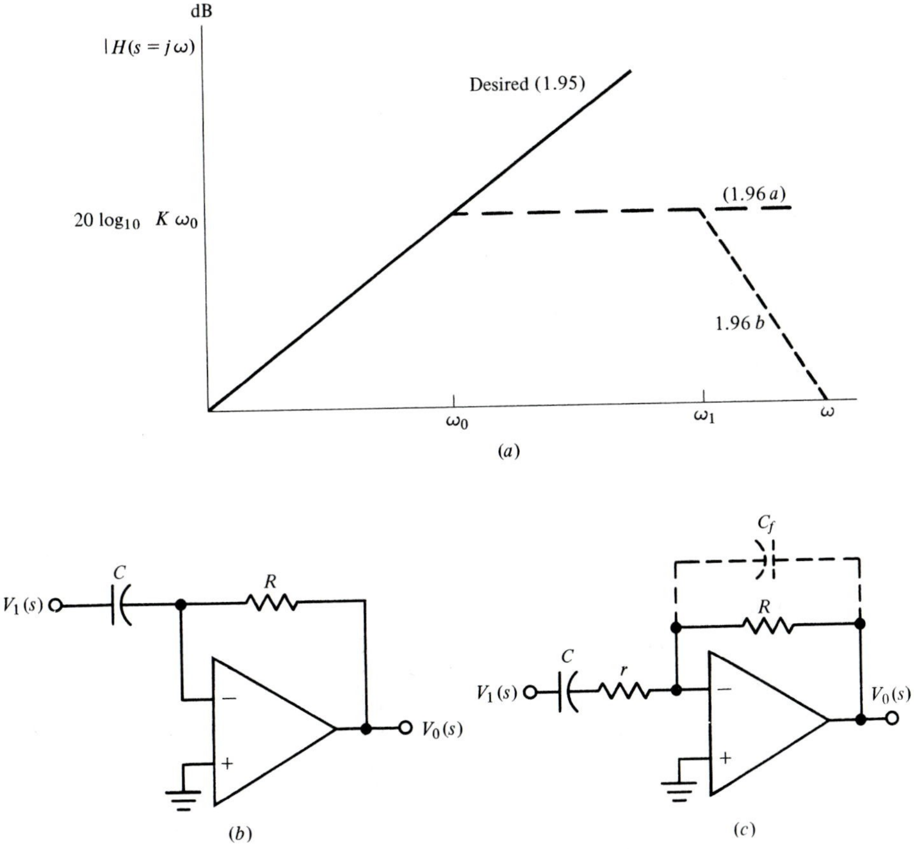

FIGURE 1.36
Differentiator and high-pass filter circuits. (*a*) Magnitude plots of ideal operators; (*b*) an ideal differentiator (*not recommended*); (*c*) a more practical circuit.

below ω_1 in the case of (1.96*b*)]. Differentiation of signals is a notoriously noisy operation, and a judicious choice of ω_0 is important. Note that, in actuality, the transfer function of (1.96*a*) is still not physically realizable, since we do not have amplifiers of infinite bandwidth and there are always some stray capacitances which cause the gain of an amplifier circuit to eventually drop off at high frequencies. The transfer function of (1.96*b*) is more realistic. Here the second corner frequency $\omega = \omega_1$ may either result directly from the internal bandwidth limitations of the operational amplifier or be purposely introduced by the circuit designer to further attenuate unwanted high-frequency noise.

Figure 1.36*b* shows the basic differentiator circuit that *ideally* ought to implement (1.95). If we had an ideal operational amplifier, then, for Fig. 1.36*b*,

$$H(s) = -RCs \tag{1.97}$$

i.e.,

$$K = -RC$$

(see Prob. 1.37). There are, however, serious difficulties with the circuit of Fig. 1.36*b*:

1. The input impedance of the circuit is a pure capacitance, possibly of fairly high value; this is a poor load for most signal sources.
2. Although we have not shown it yet, most typical operational amplifiers, because of their high-frequency behavior, will cause the actual circuit to be unstable; i.e., it will probably oscillate.
3. There is no control of the high-frequency response.

The *high-pass* circuit of Fig. 1.36*c* remedies all the above problems, if properly designed; the use of C_f is optional. With an ideal operational amplifier, but *without* C_f, we implement (1.96*a*), that is,

$$H(s) = \frac{-RCs}{1 + rCs} \tag{1.98a}$$

with

$$K = -RC \tag{1.98b}$$

$$\omega_0 = \frac{1}{rC} \tag{1.98c}$$

and

$$|H(s = j\infty)| = K\omega_0 = \frac{R}{r} \tag{1.98d}$$

The input impedance now has a resistive component, but is still highly capacitative. With C_f added to Fig. 1.36*c*, we obtain

$$H(s) = \frac{-RCs}{(1 + rCs)(1 + RC_f s)} \tag{1.99a}$$

i.e., we still have

$$K = -RC \tag{1.99b}$$

and

$$\omega_0 = \frac{1}{rC}$$

but now

$$\omega_1 = \frac{-1}{RC_f} \tag{1.99c}$$

and the high-frequency gain falls off 20 dB/decade above ω_1 rad/s.

A somewhat more expensive approach is taken in Fig. 1.37. This is a two-amplifier circuit which features a *constant input impedance*

$$Z_{\text{in}} = R_1 \parallel R_2 \tag{1.100}$$

Note that Fig. 1.37 is a *high-pass filter* circuit; without C_f, we have

$$\frac{V_0(s)}{V_1(s)} = \frac{-R_1Cs}{1+R_1Cs}\frac{R_3}{R_2} \tag{1.101a}$$

i.e.,

$$K = -R_1C\frac{R_3}{R_2} \tag{1.101b}$$

$$\omega_0 = \frac{1}{R_1C} \tag{1.101c}$$

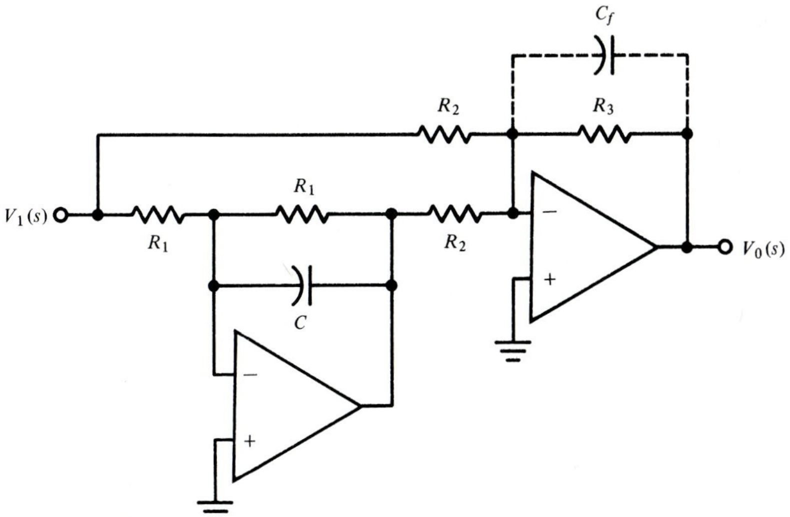

FIGURE 1.37
A two-amplifier high-pass filter circuit.

With C_f, we have

$$\frac{V_0(s)}{V_1(s)} = \frac{-R_1Cs}{(1+R_1Cs)(1+R_3C_fs)}\frac{R_3}{R_2} \tag{1.102a}$$

i.e.,

$$K = -R_1C\frac{R_3}{R_2} \tag{1.102b}$$

$$\omega_0 = \frac{1}{R_1C} \tag{1.102c}$$

$$\omega_1 = \frac{1}{R_3C_f} \tag{1.102d}$$

Note that Fig. 1.37 is really a special case of Fig. 1.34; that is, the numerator zero is now at the origin. Careful balancing of the resistors in the circuit is required to ensure precise location of the zero, or there will be some dc feedthrough. This circuit permits implementing good dc blocking and low-frequency differentiation.

Instrumentation Amplifiers

The differential amplifier circuit of Fig. 1.25*c* has at least two fairly serious disadvantages:

1. The input impedance is $2R_1$; in some cases this may cause serious loading of high-impedance low-level signal sources (e.g., in biological instrumentation).
2. It is not easy to adjust the overall gain.

The circuit of Fig. 1.38 eliminates these two limitations by using three operational amplifiers. Typically, the complete circuit is packaged in one module as a general-purpose *instrumentation amplifier* (see Prob. 1.38). The *input impedance* of the circuit is *very high*, particularly if operational amplifiers with FET input stages are used. The overall differential gain expression is

$$v_0 = -\left(1+\frac{2R_1}{R}\right)\frac{R_0}{R_2}(v_1 - v_2) \tag{1.103}$$

assuming well-balanced resistors throughout. A wide range of gains may be implemented merely by adjusting R.

Example 1.15. Suppose we want to design an instrumentation amplifier, using Fig. 1.38, so that the differential gain may be adjusted between 10 and 100; use $R_1 = R_2 = R_0 = 50\ \text{k}\Omega$. What values of R are required to get differential gains of 10 and 100?

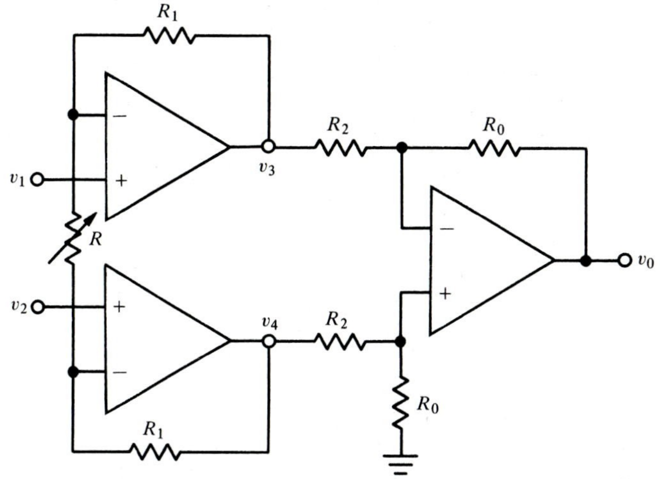

FIGURE 1.38
A three-amplifier differential-input instrumentation amplifier featuring high input impedance and easily adjustable gain.

Using (1.103),

$$v_0 = -\left(1 + \frac{10^5}{R}\right)(v_1 - v_2)$$

For a gain of 10:

$$R = \frac{10^5}{9} = 11.111\ \text{k}\Omega$$

For a gain of 100:

$$R = \frac{10^5}{99} = 1.0101\ \text{k}\Omega$$

Reference Voltage Sources and Voltage Regulators

Figure 1.39 presents circuits for providing a regulated voltage source, using an operational amplifier. In Fig. 1.39*a*, the amplifier is used in a simple noninverting circuit, which derives a reference voltage from the Zener diode D_1.

$$v_0 = \frac{R_2 + R_1}{R_1} v_z \tag{1.104}$$

In both circuits, v_s is a source of positive dc voltage; however, in Fig. 1.39*b*, improved immunity to variations in v_s is obtained by only using v_s to "start up" the circuit via R_3, R_4, and D_2; once the circuit is operating, the Zener diode D_1

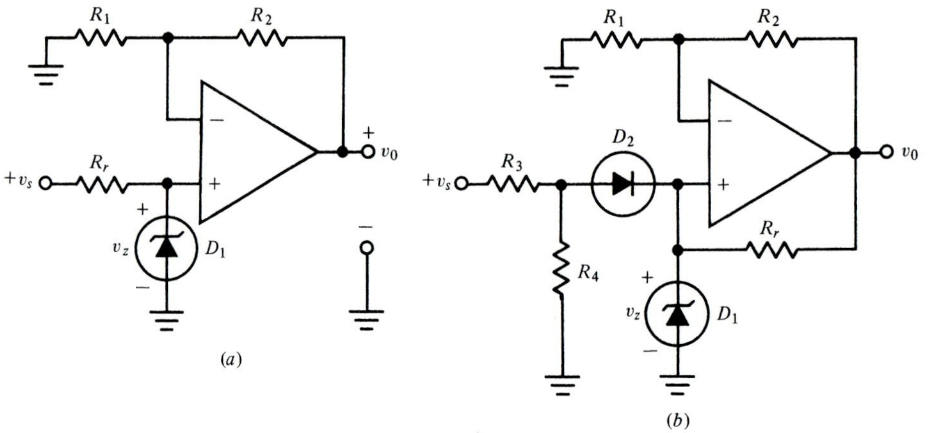

FIGURE 1.39
Voltage regulator circuits. (*a*) Simple noninverting regulator; (*b*) circuit with improved immunity to v_S.

obtains its "keep-alive" current from R_r, which is itself supplied by the well-regulated output voltage. Both of these circuits suffer at least two limitations:

1. The maximum output current obtainable is limited by the current-driving capability of the operational amplifier. The disadvantage can be alleviated by adding a booster output stage to the amplifier. Many manufacturers supply such a stage, or in many cases the output current may be boosted by using a transistor emitter follower (see also Fig. 1.40 and Sec. 2.8).

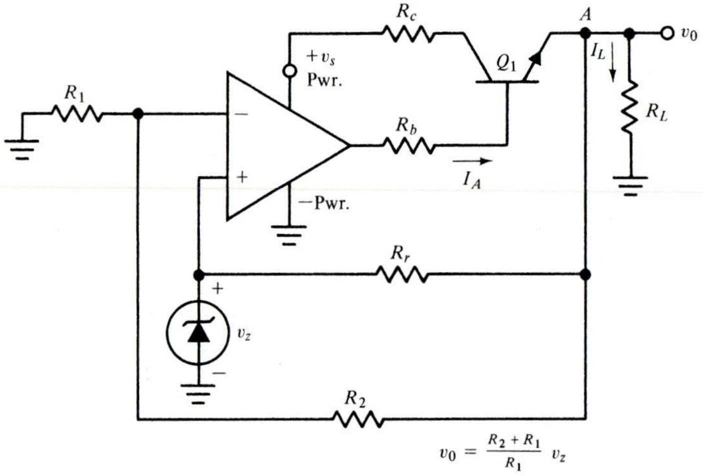

FIGURE 1.40
Current-boosted regulator requiring only one power supply potential.

2. The regulators of Fig. 1.39 require that the ± supply voltages be provided to the operational amplifier separately. This is often an undesirable nuisance.

The circuit of Fig. 1.40 eliminates both the above disadvantages and may be used to directly regulate the voltage supplied from power source v_s, without requiring any additional power voltages. Here is a special occasion where the operational amplifier is operated from one supply voltage (that is, v_s), since both input terminals are at a positive potential v_z. Transistor Q_1 is a series regulator which is actually an output booster for the operational amplifier. Resistor R_c should be chosen to protect Q_1 from excessive current overload. Base drive resistor R_b is used to protect the operational amplifier from excessive output current demand. Of course, Q_1 must be chosen with sufficient current gain (usually called β) to provide current I_L from amplifier current I_A.

Voltage-Current Converters

It is sometimes useful to be able to generate a current proportional to a voltage. Typical applications include driving a deflection coil for a magnetically deflected cathode-ray tube. For floating loads (i.e., both terminals may be connected into the driving circuit), the circuit of Fig. 1.41 is useful.

It is easy to show that

$$i_L = \frac{v_1}{R} \tag{1.105}$$

if $V_B < V_{\max}$ of the operational amplifier. Note also that the circuit presents a high impedance to the driving signal source. Of course, the operational amplifier must be capable of providing the requisite driving current and the total load voltage at point B. In some situations, it is necessary to drive a load which has one side already grounded externally. The circuit of Fig. 1.42, which requires two amplifiers, has a high input impedance and will drive a grounded load. Note also that the amplifier output swing must now be sufficient to provide the voltage drop associated with R_1 as well as the load. For Fig. 1.42, the current delivered to the load is (see Prob. 1.32)

$$i_L = \frac{-2v_1}{R_1} \tag{1.106a}$$

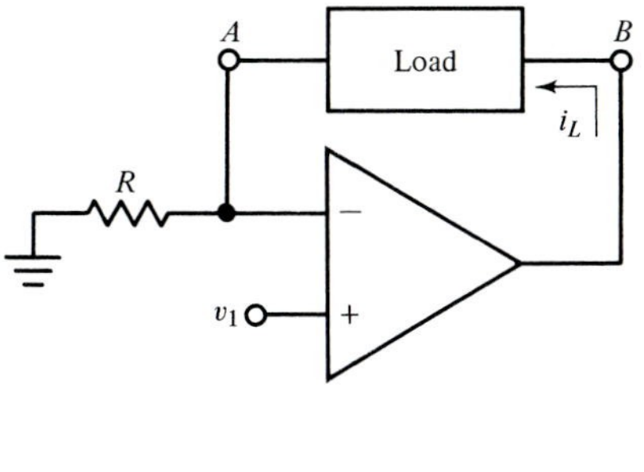

FIGURE 1.41
Simple voltage-controlled current source (floating load).

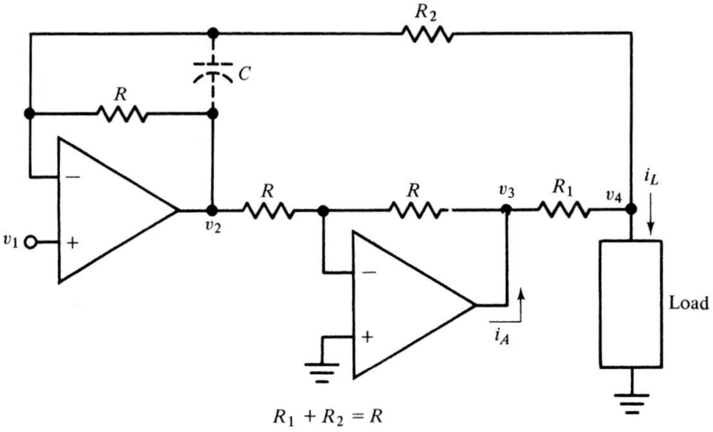

FIGURE 1.42
Voltage-controlled current source with high-input impedance and grounded load.

where it is required that

$$R_2 = R - R_1 \tag{1.106b}$$

This last circuit has the potential for presenting a negative output resistance (looking back into node v_4) and may cause oscillation if the load is highly inductive. A small stabilizing capacitor C may be required in some cases.

The Howland Current Source

This voltage-controlled voltage source (VCIS) is described in Refs. 16 and 17 and is shown in Fig. 1.43. Since we assume the operational amplifier is ideal, we can say

$$V_- = V_+ = V_x \tag{1.107}$$

Writing nodal equations, we have

$$\frac{V_2 - V_x}{nR_1} + \frac{V_0 - V_x}{nR_2} = 0 \tag{1.108}$$

multiplying by n gives

$$\frac{V_2 - V_x}{R_1} + \frac{V_0 - V_x}{R_2} = 0 \tag{1.109}$$

Also at the + terminal

$$\frac{V_1 - V_x}{R_1} + \frac{V_0 - V_x}{R_2} = I_L \tag{1.110}$$

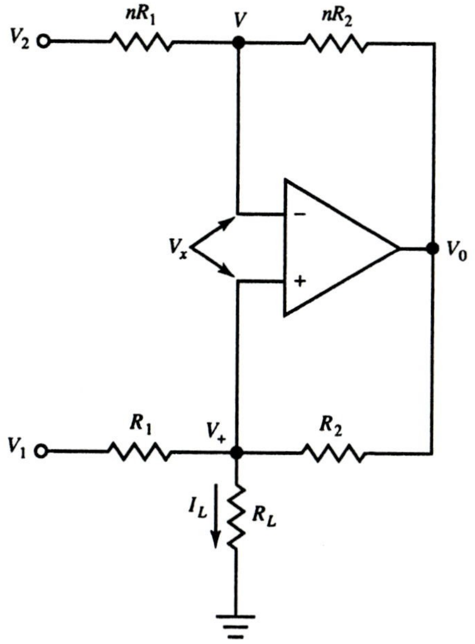

FIGURE 1.43
The Howland voltage-controlled current source. Note differential input and grounded load.

From (1.109)

$$\frac{V_0}{R_2} = V_x\left(\frac{1}{R_1} + \frac{1}{R_2}\right) - \frac{V_2}{R_1} \tag{1.111a}$$

From (1.110)

$$\frac{V_0}{R_2} = I_L + V_x\left(\frac{1}{R_1} + \frac{1}{R_2}\right) - \frac{V_1}{R_1} \tag{1.111b}$$

If we equate (1.109) and (1.110) and cancel like terms on either side, we obtain

$$I_L = \frac{1}{R_1}(V_1 - V_2) \tag{1.112}$$

which describes the VCIS characteristic of the Howland current source. Again note that we have a grounded load.

Single-Supply Operation

It is often convenient to operate from a *single-supply voltage*. Figure 1.44 shows a simple unity gain follower connection using a single power supply. Typically we have V_{in} and V_{out} restricted by

$$2 < V_{\text{in}} < V_s - 2$$

where V_s might be perhaps 30 V; in this case the usable range of the circuit would be from 2 to 28 V. Reference 15 discusses this technique further.

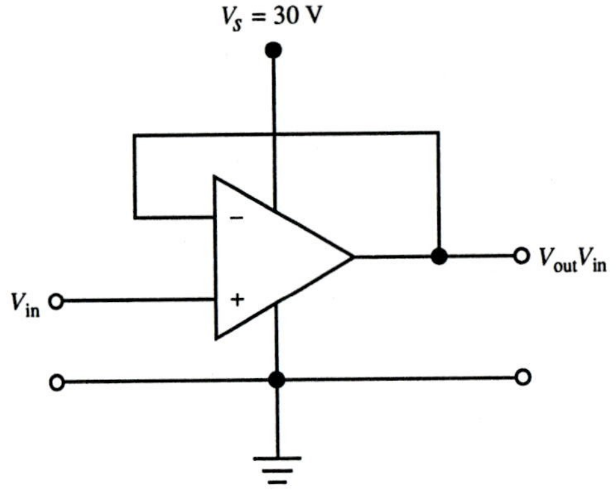

FIGURE 1.44
A single-supply unity gain follower.

Other Applications

There are many other applications of operational amplifiers to linear amplifying and signal-processing circuits. In Chap. 4 we discuss the general topic of active *RC* filter design. Chapter 6 of Ref. 1 also contains much useful information. Hopefully some of the problems at the end of this chapter will suggest other ways in which operational amplifiers may be used to precisely amplify, filter, or combine signals.

REFERENCES AND BIBLIOGRAPHY

1. Tobey, G. E., J. G. Graeme, and L. P. Huelsman: *Operational Amplifiers*, McGraw-Hill, New York, 1971.
2. Giacoletto, L. J.: *Differential Amplifiers*, Wiley, New York, 1970.
3. Korn, G. A., and T. M. Korn: *Electronic Analog and Hybrid Computers*, 2d ed., McGraw-Hill, New York, 1972.
4. Smith, John L.: *Modern Operational Circuit Design*, Wiley, New York, 1971.
5. *Electronics*: A biweekly journal for electronics engineers; in particular, see the Circuit Design features. McGraw-Hill, Inc.
6. *Electronic Design*: A biweekly journal for electronics engineers; in particular, see the articles in the Technology section. Hayden Publ. Co.
7. Huelsman, L. P.: *Basic Circuit Theory*, 3d ed., Prentice-Hall, Englewood Cliffs, N.J., 1990.
8. Hausner, A.: *Analog and Analog/Hybrid Computer Programming*, Prentice-Hall, Englewood Cliffs, N.J., 1971.
9. Kostanty, Raymond G.: "Doubling Op Amp Summing Power," *Electronics*, Feb. 14, 1972, pp. 73–75.
10. Moschytz, G. S.: "Inductorless Filters: A Survey," *IEEE Spectrum*, September 1970, pp. 63–75.
11. Shah, M. J.: "Stable Voltage Reference Uses Single Power Supply," *Electronics*, Mar. 13, 1972, p. 74.
12. Clayton, G. B.: *Operational Amplifiers*, Butterworth, London, 1971.
13. Graeme, J. G.: *Applications of Operational Amplifiers*, McGraw-Hill, New York, 1973.
14. Wojslaw, C. F., and E. A. Moustakas: *Operational Amplifiers*, McGraw-Hill, New York, 1986.
15. "Single-Supply Operation of Operational Amplifiers," AN-139, Burr-Brown Corp., Tucson, Ariz., March, 1986.

16. Scolof, S.: *Applications of Analog Integrated Circuits*, Prentice-Hall, Englewood Cliffs, N.J., 1985.
17. Franco, S.: *Design with Operational Amplifiers and Analog Integrated Circuits*, McGraw-Hill, New York, 1988.

PROBLEMS

For all problems in this chapter, assume an ideal operational amplifier model (see Sec. 1.2).

1.1. *(Sec. 1.3)* For both parts of Fig. P1.1, find v_O in terms of v_1 and v_2.

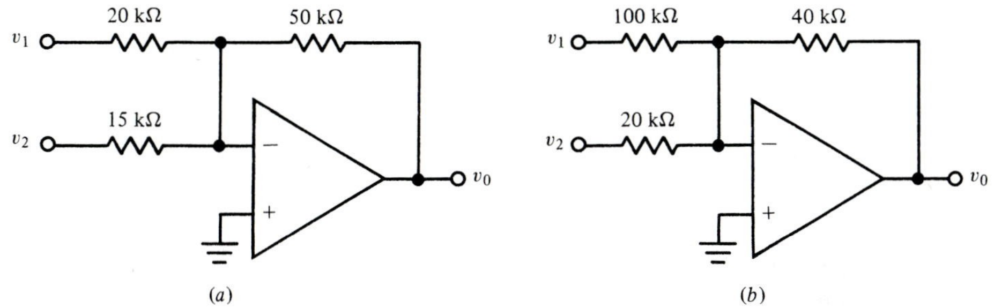

FIGURE P1.1

1.2. *(Sec. 1.3)* Using one operational amplifier and resistors of 100 kΩ or less, but otherwise as large as possible, design an operational amplifier circuit to implement
(a) $v_O = -2v_1 - 5v_2$
(b) $v_O = -4v_1 - 3v_2 - v_3$
(c) $v_O = -2v_1 - 0.5v_2$

1.3. *(Sec. 1.3)* For the circuit of Fig. P1.3 make a table indicating what labeled terminals to connect to the input signal source or the output in order to get all possible (different) amplification factors. Also include the resulting input impedance and what the possible gains are.

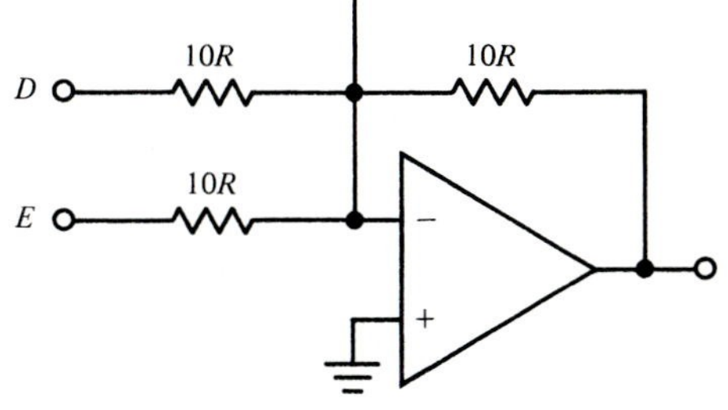

FIGURE P1.3

1.4. *(Sec. 1.3)* Given the circuit of Fig. P1.4, find all possible voltage gains that can be implemented by connecting the lettered resistors to either the input or the output node. Assume we have only one input signal. Make a table listing which nodes are to be connected to the output and which to the input.

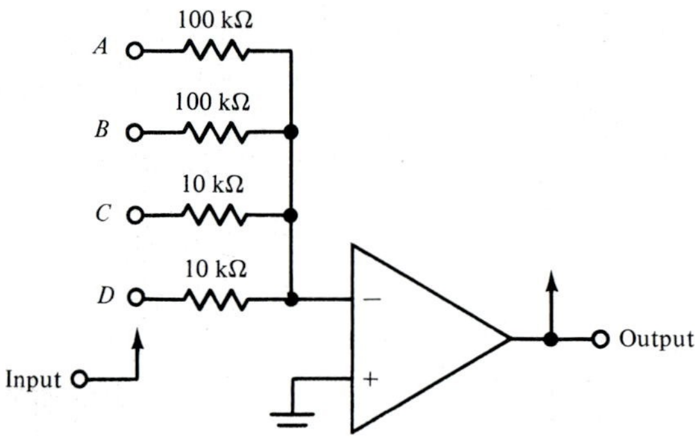

FIGURE P1.4

1.5. *(Sec. 1.4)* Refer to Figs. P1.5 and 1.10*b*. Draw a block-diagram symbol describing the nominal performance of the circuit; also write an expression similar to (1.18).

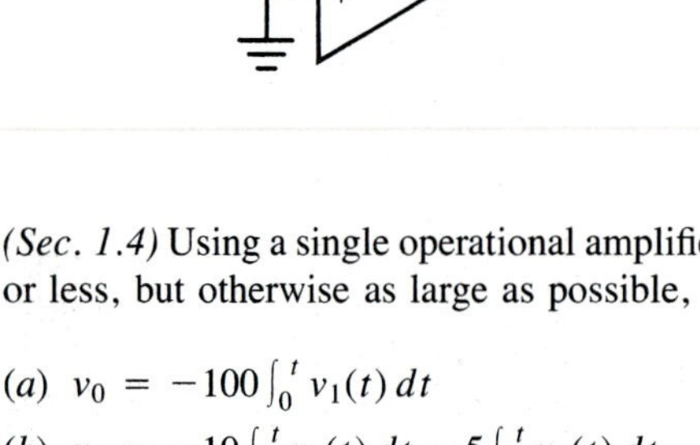

FIGURE P1.5

1.6. *(Sec. 1.4)* Using a single operational amplifier, one capacitor, and resistors of 100 kΩ or less, but otherwise as large as possible, design a circuit to implement

(*a*) $v_O = -100\int_0^t v_1(t)\,dt$

(*b*) $v_O = -10\int_0^t v_1(t)\,dt - 5\int_0^t v_2(t)\,dt$

Assume $v_O = 0$ at $t = 0$.

1.7. *(Sec. 1.4)* Given the situation of Fig. P1.7:

(*a*) Sketch v_O and i_A versus time; assume no other output load on the amplifier.

(*b*) Repeat the above if an output load of 10 kΩ is connected from the amplifier to ground.

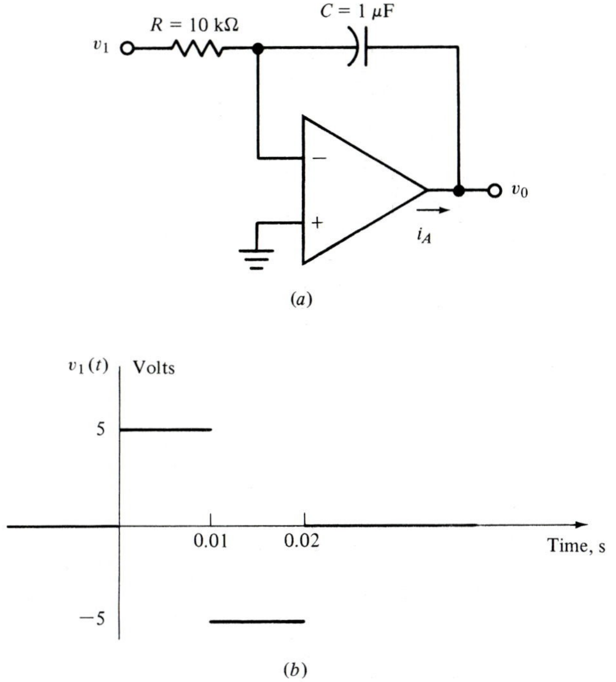

FIGURE P1.7

1.8. *(Sec. 1.4)* Refer to Fig. P1.8; at $t = 0$ switches S_1 and S_2 are thrown from state 1 to state 2. Find $v_O(t)$ for $t \geq 0$, in terms of $v_1(t)$.

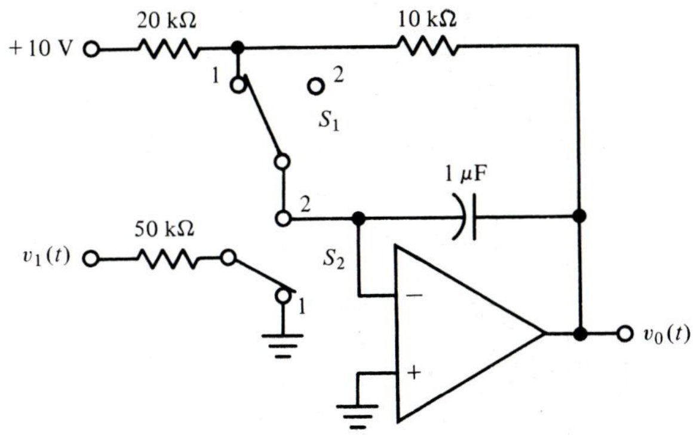

FIGURE P1.8

1.9. *(Sec. 1.4)* For the circuit of Fig. 1.13*e*, make separate sketches for each of the three modes; explain why the initial-condition setting will be faster than for the circuit of Fig. 1.13*a*.

1.10. *(Secs. 1.4 and 1.9)* For each of the circuits of Fig. P1.10:

(*a*) Find the differential equation for v_O in terms of v_1.

(*b*) Find the Laplace transform transfer function

$$H(s) = \frac{V_0(s)}{V_1(s)}$$

Hint: Draw block diagrams of the circuits.

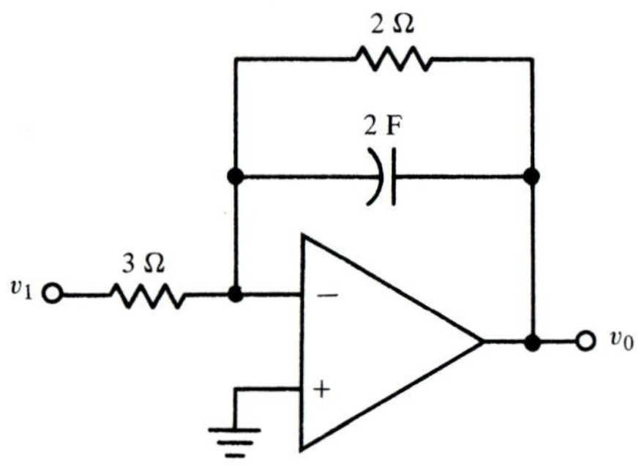

(*a*)

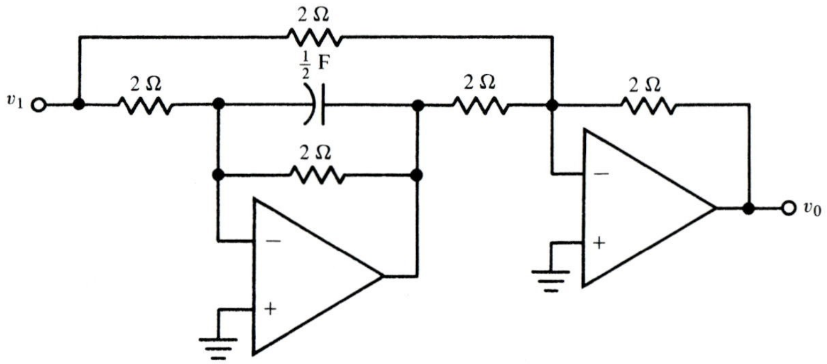

(*b*)

FIGURE P1.10

1.11. *(Sec. 1.4)* Given the circuit of Fig. P1.11 with

$$i_A < 10 \text{ mA} \qquad v_0 < 10 \text{ V} \qquad v_S = A \sin 2\pi ft$$

find the maximum value of A as a function of f.

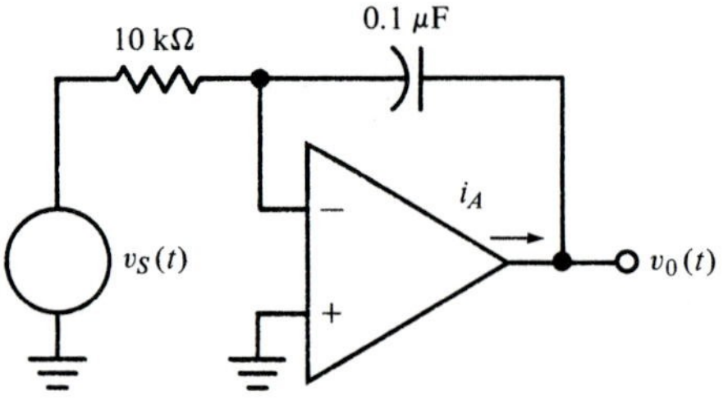

FIGURE P1.11

1.12. *(Sec. 1.5)* For both parts of Fig. P1.12, find v_O in terms of the input variables.

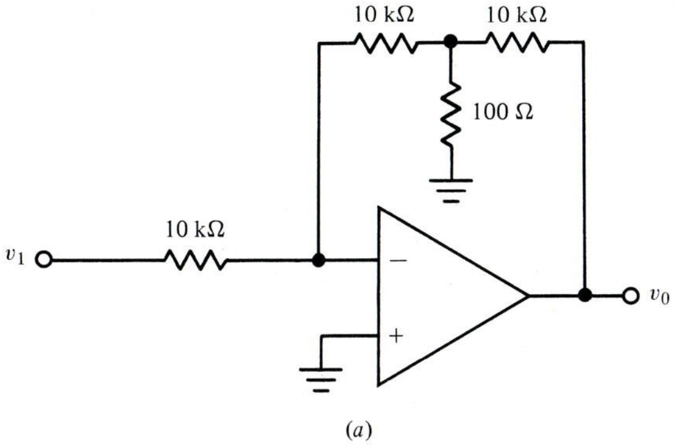

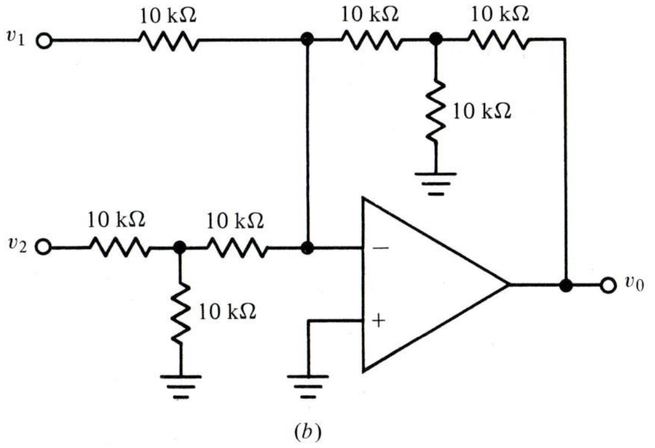

FIGURE P1.12

1.13. *(Sec. 1.5)* In the circuit of Fig. P1.13, find the value of R required to obtain

$$v_O = -80v_1.$$

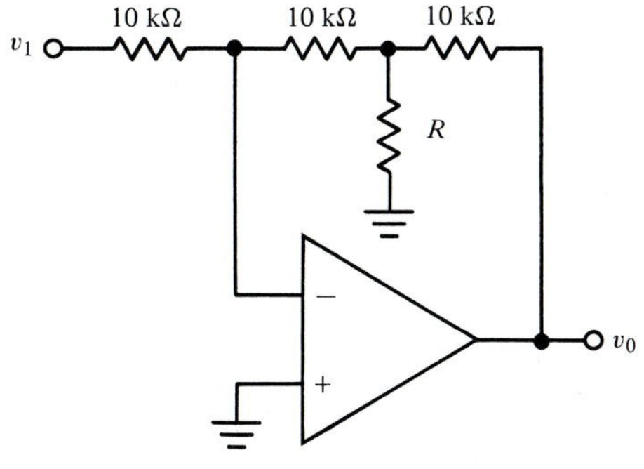

FIGURE P1.13

1.14. *(Secs. 1.5 and 1.9)* Given the circuit of Fig. P1.14:

(*a*) Find the Laplace transform transfer function

$$H(s) = \frac{V_O(s)}{V_I(s)}$$

in terms of the *loaded* potentiometer ratio α, that is,

$$\alpha = \frac{V_C(s)}{V_B(s)} \qquad 0 < \alpha \leq 1$$

(*b*) Repeat part *a* if a 0.002-μF capacitor is connected between points *X* and *Y*.

(*c*) Why is the output amplifier advisable? With the capacitor included, what is the convenient feature of this circuit? Comment on allowable signal level, particularly when α is small.

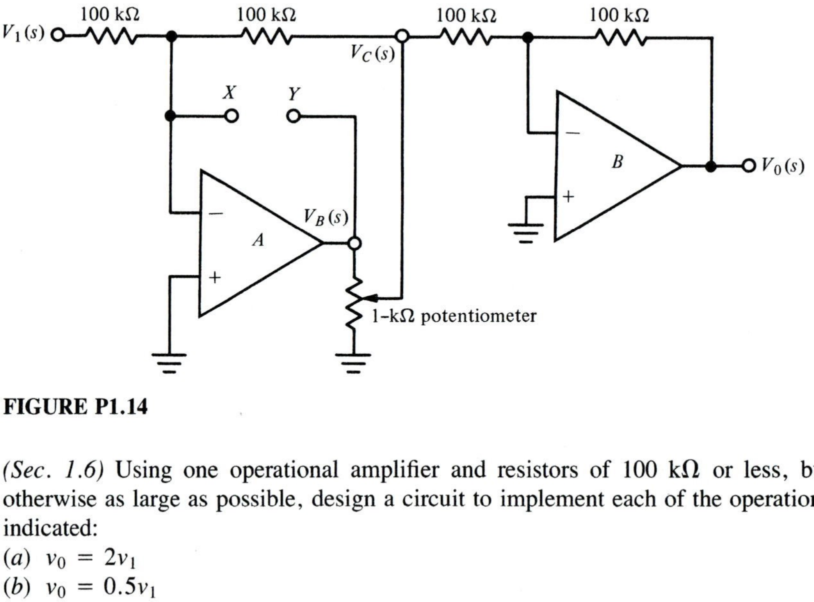

FIGURE P1.14

1.15. *(Sec. 1.6)* Using one operational amplifier and resistors of 100 kΩ or less, but otherwise as large as possible, design a circuit to implement each of the operations indicated:

(*a*) $v_O = 2v_I$

(*b*) $v_O = 0.5v_I$

1.16. *(Sec. 1.6)* For the circuit of Fig. P1.16 (see also Fig. 1.22):

(*a*) Find the Laplace transform transfer function

$$H(s) = \frac{V_O(s)}{V_I(s)}$$

in terms of general *R*'s and *C*.

(*b*) Find the necessary condition on the resistances to obtain a *noninverting integrator*, i.e., to get

$$H(s) = \frac{+K}{s}$$

(*c*) What is the value of *K*?

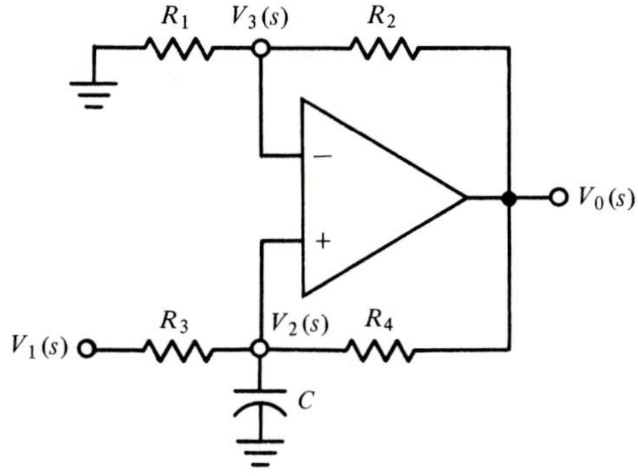

FIGURE P1.16

(*d*) Comment on the importance of accurate resistor values for the accuracy with which the integrator pole is located at $s = 0$.

1.17. *(Secs. 1.6 and 1.8)* For both parts of Fig. P1.17, find v_0 in terms of v_1 and v_2.

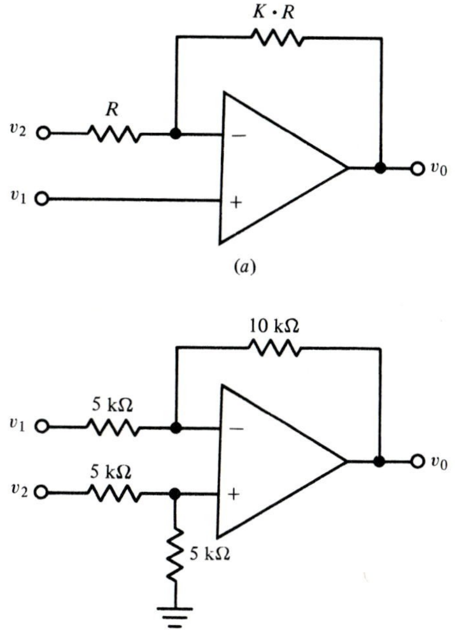

(*b*) **FIGURE P1.17**

1.18. *(Sec. 1.7)* Design a differential amplifier circuit similar to Fig. 1.25*b* which will have a differential voltage gain of 20 and an input impedance (between input terminals) of 30 kΩ.

1.19. *(Sec. 1.7)* The differential amplifier in Fig. P1.19 is assumed to have an ideal operational amplifier and ideal, perfectly accurate resistors. However, the differential signal source is associated with unbalanced source resistances, i.e.,

$$R_{S2} = R_{S1} + \Delta R_S$$

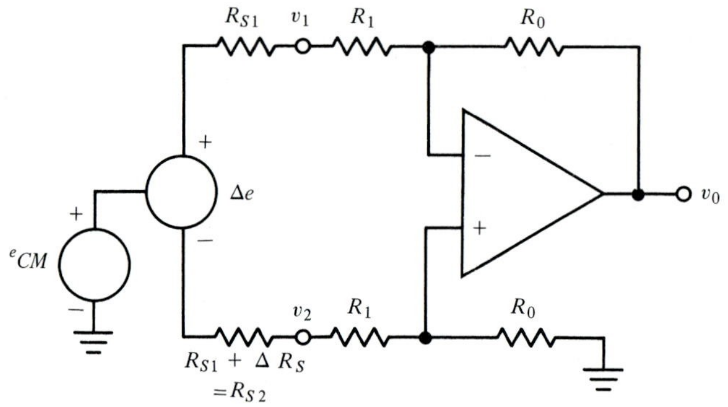

FIGURE P1.19

(*a*) Find v_0 in terms of $e_{cm}, \Delta e, R_0, R_1, R_{s1}$, and ΔR_S.

(*b*) Find the *differential gain*

$$G_D = \frac{v_0}{\Delta e}$$

assuming

$$\Delta R_S \ll R_{S1} \qquad R_{S1} \ll R_1$$

(*c*) Find the *common-mode gain* (due to source resistance unbalance)

$$G_{cm} = \frac{v_0}{e_{cm}}$$

assuming

$$\Delta R_S \ll R_{S1} \qquad R_{S1} \ll R_1$$

(*d*) Let us define a *common-mode rejection ratio* (CMRR) as

$$\text{CMRR} = \frac{G_D}{G_{cm}}$$

Find the CMRR, again assuming

$$\Delta R_S \ll R_{S1} \qquad R_{S1} \ll R_1$$

(*e*) Consider the circuit of Fig. 1.38 with respect to its immunity to the errors caused by R_{S1} and ΔR_S.

Note: In Sec. 2.4, we discuss further errors due to internal unbalance in the operational amplifier itself. Overall differential amplifier performance must be analyzed by considering all anticipated error sources.

1.20. (*Secs. 1.7 and 1.8*) Given the circuit of Fig. P1.20, find the conditions on R_1, R_2, R_3, and R_4 so that

$$v_0 = K(v_2 - v_1)$$

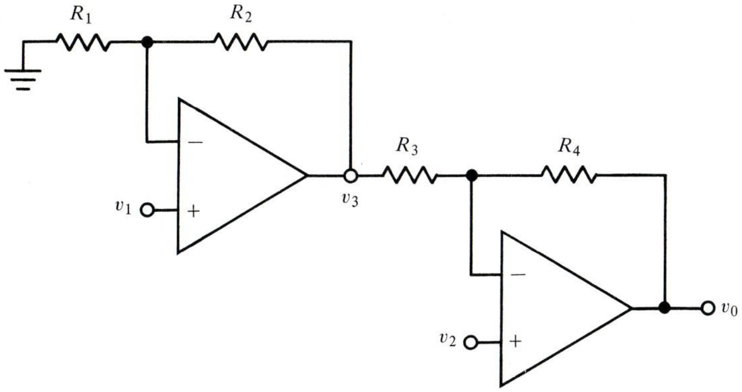

FIGURE P1.20

and find K. Note that the circuit is a differential amplifier under these conditions. Can you think of an advantage of this circuit over that of Fig. 1.25*b*?

1.21. *(Sec. 1.8)* Using one operational amplifier and resistors of 100 kΩ or less, but as large as possible, design circuits to implement each of the operations indicated.
(*a*) $v_O = 3v_1 - 3v_2$
(*b*) $v_O = 2v_1 - v_2$
(*c*) $v_O = 3v_1 + 4v_2$

1.22. *(Sec. 1.8)* Redesign the circuit of Example 1.13 to obtain

$$v_O = -(3v_{I1} + 2v_{I2}) + 10v_{N1}$$

1.23. *(Sec. 1.8)* Given the circuit of Fig. P1.23, find v_O in terms of v_1, v_2, and v_3. Note that this implements combined summing and inverting as per (1.69), but $Z_A(s) \neq Z_B(s)$.

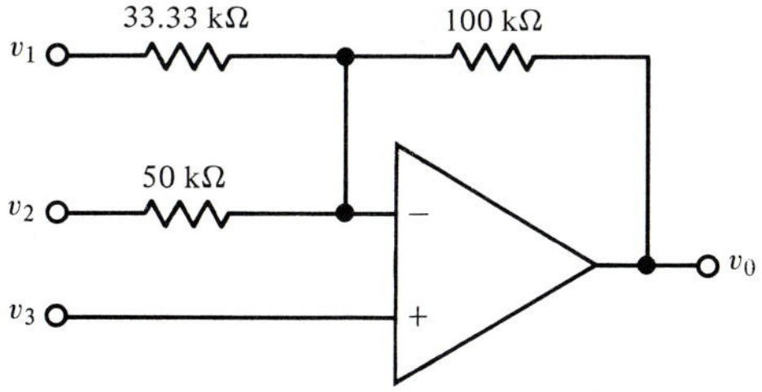

FIGURE P1.23

1.24. *(Sec. 1.9)* Design a circuit using a single operational amplifier to implement

$$V_O(s) = \frac{-[V_1(s) + 4V_2(s)]}{1 + s/10^5}$$

The circuit input impedances should be greater than 5 kΩ and resistive; all resistors should be less than 100 kΩ, but otherwise as large as possible.

1.25. *(Sec. 1.9)* Suppose a signal from a single-ended transducer has the following properties:

Peak value:	Less than 0.1 V (open circuit)
Source impedance:	Less than 5 Ω, but otherwise not well known

Design a low-pass operational amplifier circuit that will amplify the signal 100 times (a polarity inversion is allowable); the circuit should pass all frequencies from direct current on up to around 1 kHz and then exhibit a 6 dB/octave rolloff indefinitely above a −3-dB breakpoint of 2 kHz. All resistors should be no more than 100 kΩ; gain accuracy should be at least 1 percent. Use one operational amplifier.

1.26. *(Sec. 1.9)* For the circuit of Fig. P1.26:

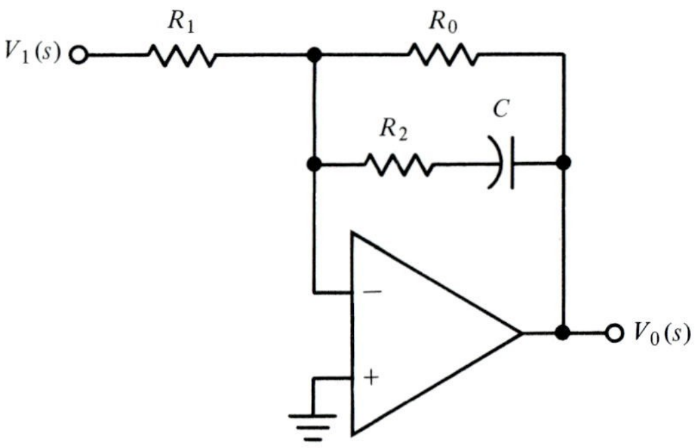

FIGURE P1.26

(*a*) Find the Laplace transform transfer function

$$H(s) = \frac{V_0(s)}{V_1(s)}$$

for general R_0, R_1, R_2, and C.

(*b*) Make a Bode plot (magnitude of H in decibels and phase of H in degrees, both versus log ω) of $H(s = j\omega)$ with

$$R_0 = 100 \text{ k}\Omega \qquad R_1 = 10 \text{ k}\Omega \qquad R_2 = 1 \text{ k}\Omega \qquad C = 0.02\ \mu\text{F}$$

Show all critical frequencies on your plots.

1.27. *(Sec. 1.9)* Given the circuit of Fig. P1.27, find:

(*a*) The Laplace transform transfer function

$$H(s) = \frac{V_0(s)}{V_1(s)}$$

(*b*) The magnitude of the circuit voltage gain at direct current. Can you do this without doing all the work of part *a*?

(*c*) The magnitude of the circuit voltage gain at high frequencies, i.e., at angular frequencies satisfying

$$\omega \gg \frac{1}{R_B C} \qquad \text{rad/s}$$

Again, can you do this without doing all the work of part *a*?

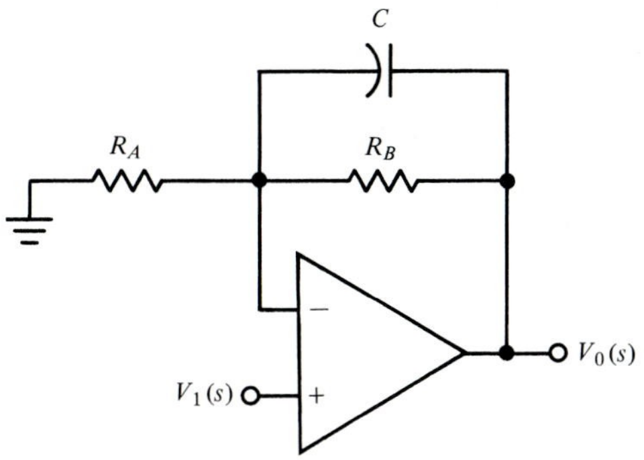

FIGURE P1.27

1.28. *(Sec. 1.9)* Prove that Eq. (1.91) is associated with Fig. 1.32*b*.

1.29. *(Sec. 1.9)*

(*a*) Design a frequency-compensated amplifier to have the following voltage gain characteristics:

Frequency, Hz	Voltage gain, dB
Above 500	40
500	37
50	23
Below 50	20

Assume that the signal source is single-ended with respect to ground and that the signal source should see a load impedance of a constant 50 kΩ. A polarity inversion is permissible; use one operational amplifier.

(*b*) Add a 6 dB/octave high-frequency attenuation with a 2200-Hz break frequency.

1.30. *(Sec. 1.9)* Given the block diagram of Fig. P1.30, show that the resulting transfer function fits Eq. (1.87), and find H_0, T_1, and T_2 in terms of G_1, G_2, and K, where G_1 and G_2 may be positive or negative constants, but K is positive (that is, $-K$ is negative).

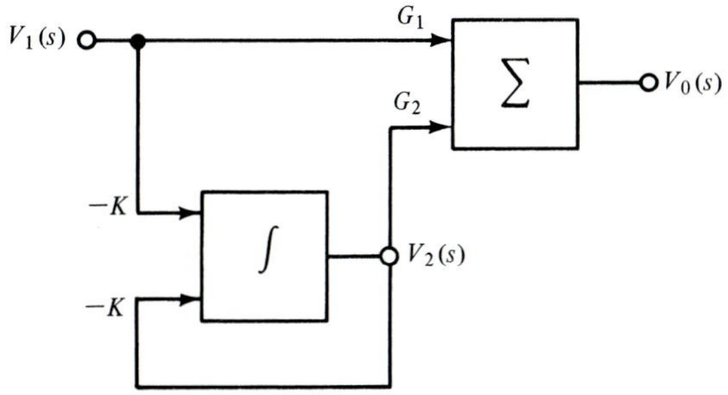

FIGURE P1.30

1.31. *(Sec. 1.9)* Referring to Fig. P1.31:

(*a*) Derive the Laplace transform transfer function

$$H(s) = \frac{V_0(s)}{V_1(s)}$$

(*b*) Find the circuit input impedance.

(*c*) For $R_1 = 100\ \text{k}\Omega$, $R_0 = 100\ \text{k}\Omega$, $R_2 = 10\ \text{k}\Omega$, $C_2 = 2000$ pF, $C_1 = 0.2\ \mu$F, make a Bode plot of $H(s)$, i.e., magnitude in decibels and angle in degrees versus log ω, $s = j\omega$.

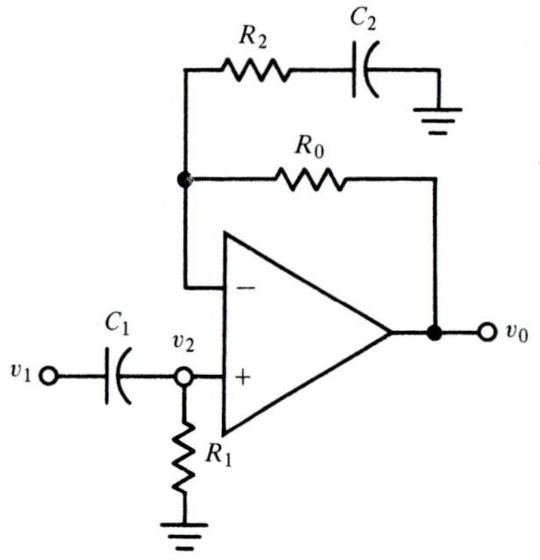

FIGURE P1.31

1.32. *(Sec. 1.10)* Refer to the circuit in Fig. 1.42.

(*a*) Derive Eqs. (1.106*a*) and (1.106*b*); i.e., show that

$$i_L = \frac{-2v_1}{R_1}$$

if

$$R = R_1 + R_2$$

(*b*) If $R = R_1 + R_2$ and the load is a pure resistance of value R_1, find expressions for v_2, v_3, i_A, and v_4.

(*c*) If $R = 20\ \text{k}\Omega$, $R_1 = R_2 = 10\ \text{k}\Omega$, find i_L in terms of v_1.

1.33. *(Secs. 1.7 and 1.10)* Derive Eq. (1.103), associated with Fig. 1.38.

1.34. *(Sec. 1.10)* Referring to Fig. P1.34, find the Laplace transform transfer functions (second-order filter)

$$\underset{bp}{H(s)} = \frac{v_2(s)}{v_1(s)}$$

and

$$\underset{lp}{H(s)} = \frac{v_3(s)}{v_1(s)}$$

Hint: Break up the circuit into basic operational blocks (analog-computer or state-variable approach for implementing transfer functions). (See also Chap. 4.)

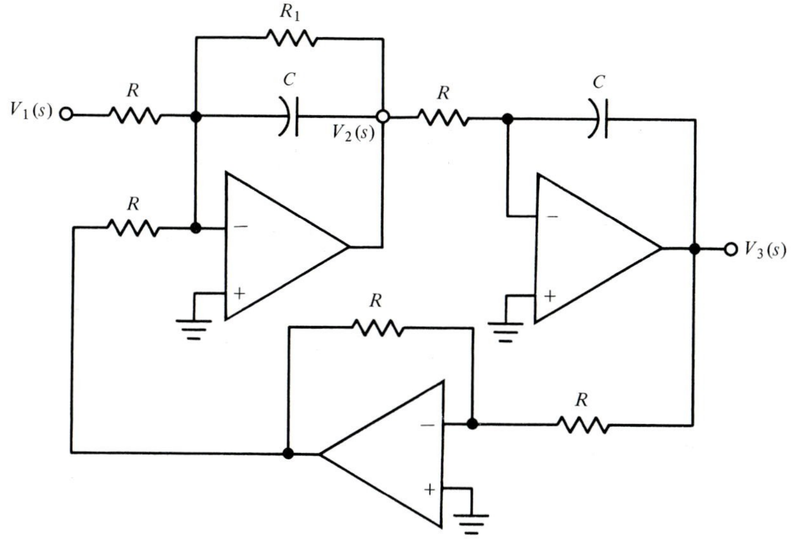

FIGURE P1.34

1.35. *(Sec. 1.10)* This problem deals with *bridge amplifiers*. Referring to Fig. P1.35, suppose we want to measure the *difference* between two resistances (e.g., two elements in a strain gage, or a thermistor bridge).

(*a*) In Fig. P1.35*a*, find v_0 in terms of δ, V, and the circuit resistances; assume $\delta \ll 1, R_0 \gg R$.

(*b*) In Fig. P1.35*b*, find v_0 in terms of δ, V, and the circuit resistances. Do we have to assume δ is small? Note that the power supply has to be floating. Name an advantage of this circuit. (See Fig. P1.35 on next page.)

1.36. *(Sec. 1.10)* Show that the circuit of Fig. P1.36 "gyrates" a capacitance; i.e., show that $Z_{\text{in}}(s)$ looks like an inductance. Find the value of the apparent inductance (Ref. 10). (See Fig. P1.36 on next page.)

1.37. *(Sec. 1.10)* In Sec. 1.10 a number of differentiator and high-pass circuits are described. Suppose we want to implement the transfer function with the following sinusoidal frequency response function:

$$H(j\omega) = \frac{-0.02j\omega}{1 + j\omega/10^4}$$

i.e., it differentiates well below 10^4 rad/s, with a differentiator gain of 0.02.

(*a*) Design a circuit based on Fig. 1.36*c* to implement this transfer function; make all resistors 100 kΩ or less.

(*b*) Add another pole to the transfer function to get a second high-frequency rolloff corner frequency at $\omega_1 = 10^4$ rad/s.

(*c*) Repeat parts *a* and *b*, but use the circuit of Fig. 1.37.

1.38. *(Secs. 1.7 and 1.10)* Refer to the instrumentation amplifier circuit of Fig. 1.38. Use this circuit to design a differential amplifier with a gain adjustable from 10 to 100 by the adjustment of a single resistor; give an equation for the value of the resistor

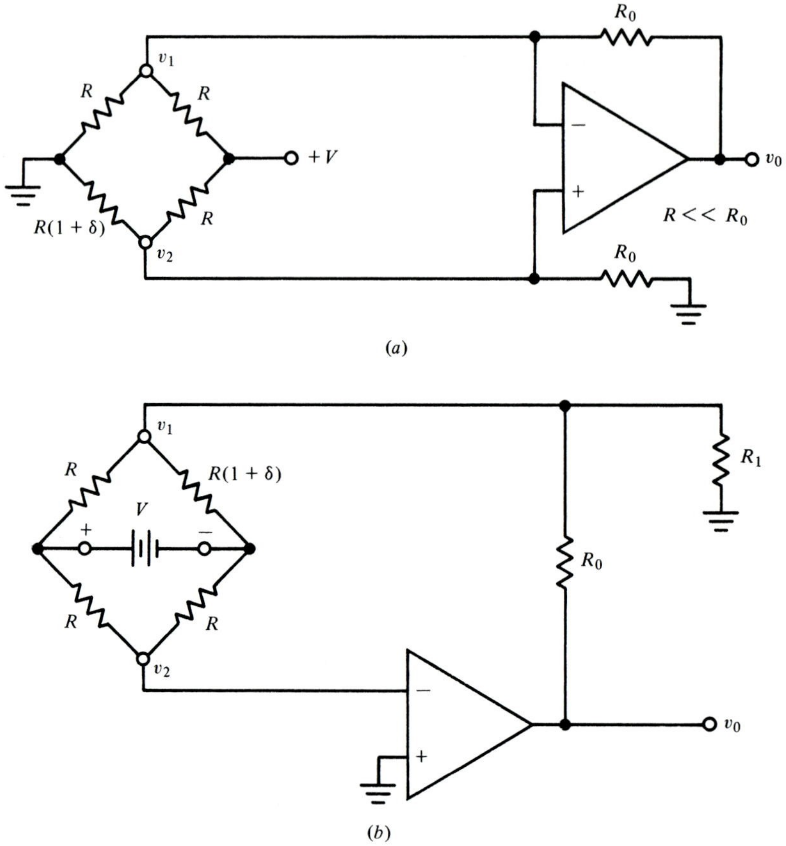

FIGURE P1.35

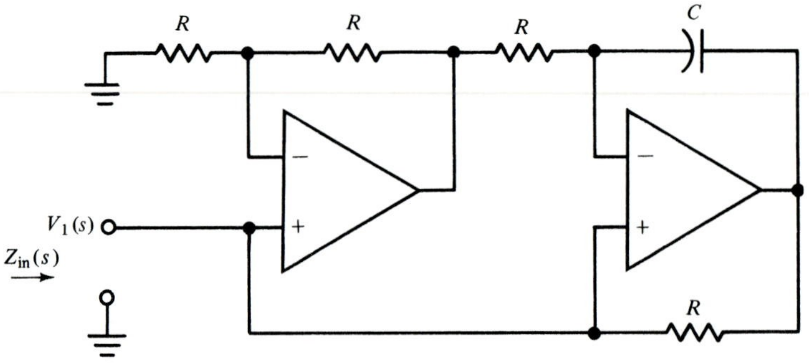

FIGURE P1.36

required to get a gain K, $10 \leq K \leq 100$. List the required values of R for gains of 10, 20, 50, and 100. Make $R_0 = R_2 = R_1 = 20\ \text{k}\Omega$.

1.39. *(Sec. 1.10)* The circuit of Fig. P1.39 was suggested by M. J. Shah (Ref. 11). It uses only one unregulated power supply to drive the operational amplifier. Derive an expression for v_O in terms of v_Z, R_1, R_2, and R_3. Note that the Zener diode current is independent of the power supply voltage.

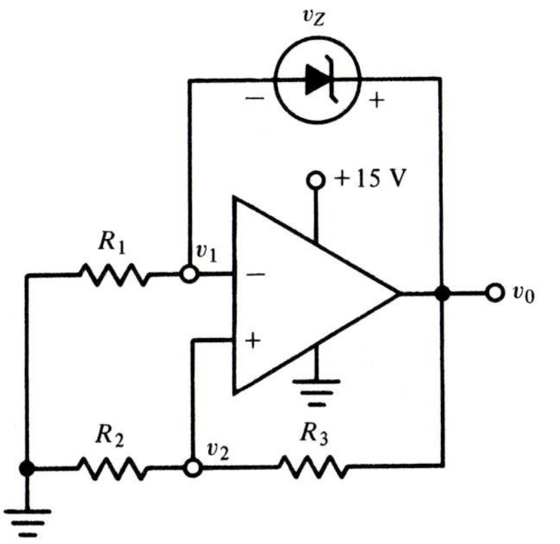

FIGURE P1.39

CHAPTER 2

PERFORMANCE LIMITATIONS: THE NONIDEAL AMPLIFIER

In Chap. 1 we assumed an operational amplifier with the ideal characteristics of Sec. 1.2, i.e., a VCVS with infinite voltage gain, zero output impedance, and infinite input impedance (Fig. 1.4*b*). Of course, such an amplifier cannot be perfectly realized. The implication thus far has been that an actual operational amplifier will not differ from the ideal model enough to significantly affect circuit performance. Table 2.1 lists typical specifications of currently available amplifiers.

This chapter discusses important ways in which modern solid-state operational amplifiers differ from the ideal model and, more importantly, how to estimate the effects of the various types of nonidealness on circuit performance. We first take a look at a general linear model with no dc offsets, internal noise, or nonlinearities (Sec. 2.1). Using this model, we study the effect of finite open-loop gain, finite bandwidth, finite input impedance, and nonzero output impedance on circuit behavior (Sec. 2.2). The effect of these factors on closed-loop stability is discussed in Sec. 2.3. Common-mode effects due to lack of symmetry in the amplifier's response to input at the inverting and noninverting terminals are described, as they relate to differential amplifier performance (Sec. 2.4).

The effect of small, but ever-present, dc unbalances in the amplifier input stage (dc offsets) is analyzed in Sec. 2.5. We use a similar model, involving random noise sources, in Sec. 2.6 to describe the effects of internal amplifier noise.

TABLE 2.1
Typical operational amplifier specifications (integrated-circuit values)

Specification	General-purpose bipolar (uA741)*	Low bias current
Open-loop gain, dc	90–100 dB	115–125 dB
Rated output	±22 V, ±25 mA	±12 V, ±10 mA
Unity-gain frequency f_u	1 MHz	2 MHz
Slew rate S	0.5×10^6 V/s	2×10^6 V/s
Full-power response f_p	7.5 kHz	30 kHz
Input offset voltage e_{os}	±4 mV	±100 μV
Input bias current i_1, i_2	20 nA	1 pA
Input difference current i_{os}	70 nA	0.5 pA
Differential input impedance	0.5 MΩ	10 MΩ
Common-mode input impedance Z_{cm1}, Z_{cm2}	$10^{11}\Omega \parallel 2$ pF	$10^{14}\Omega \parallel 2$ pF
Output impedance, open-loop, Z_0	75 Ω	100 Ω
Common-mode rejection ratio (CMRR)	70 dB	90 dB

* Ref. 16, p. 242.

† Burr-Brown Research Corp., DIFET.

Large-signal output limitations, which are primarily of a nonlinear nature, are discussed in Sec. 2.7, including rated output voltage and current, slewing rate, full-power bandwidth, and settling time.

The use of output power boosting is discussed in Sec. 2.8. The chapter closes with two sections which describe peripheral practical considerations associated with the design of operational amplifier circuits: power supply and grounding systems (Sec. 2.9) and components (Sec. 2.10).

2.1 THE GENERAL NONIDEAL LINEAR MODEL

Figure 2.1 shows the general *linearized (small-signal) model* of an operational amplifier (Fig. 2.1 is essentially Fig. 1.4, but here we imply that a complex frequency domain analysis will be employed). No internal dc offsets, noise sources, or lack of input symmetry are included; these factors are considered later. Also we assume a perfectly linear model, in which all effects can be described in terms of the Laplace transform descriptions of the various model elements. As long as the

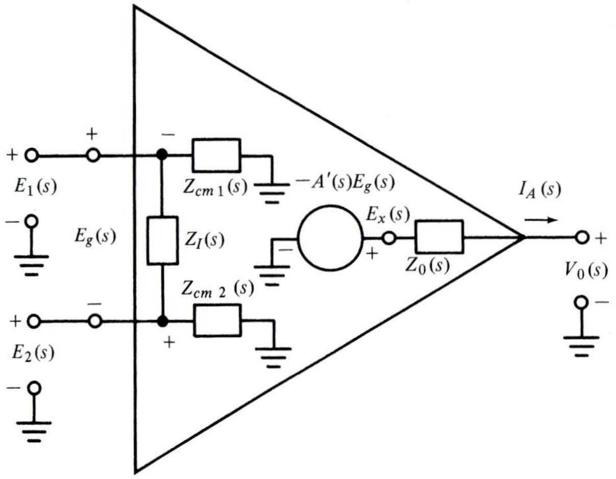

FIGURE 2.1
General linearized (small-signal) model of an operational amplifier.

amplifier *output voltage and current* and *output voltage rate* are relatively small compared to large-signal rated maximum values, our small-signal linear model describes the amplifier's dynamics rather well. *Care must be taken,* however, *in applying* the results of an analysis based on the model of Fig. 2.1 *to large-signal situations*.

In Fig. 2.1, four properties of a nonideal amplifier are included:

1. The *open-loop* (voltage) *gain* $A'(s)$
2. The (open-loop) *output impedance* $Z_0(s)$
3. The *differential input impedance* $Z_I(s)$
4. The two *common-mode input impedances* $Z_{cm1}(s)$ and $Z_{cm2}(s)$

The common-mode input impedances are simply the impedances between each input terminal and ground. Usually these are *very* high, typically greater than $10^8\ \Omega$. The differential impedance between terminals $Z_I(s)$ is also high, but not always as high as the common-mode input impedances. Amplifiers with bipolar-transistor input stages will have a differential input impedance of perhaps $10^6\ \Omega$; amplifiers with FET-input stages may have a very high differential input impedance, approaching the common-mode values (see Table 2.1). Usually these impedances are considered to be pure resistances, although at high frequencies one may need to consider small, but possibly important, common-mode and differential input capacitances.

The amplifier output impedance basically describes the small-signal incremental output resistance of the amplifier output stage. For most analyses, $Z_0(s)$ is treated as a pure resistance. It is perhaps surprising to note (Table 2.1) that typical values of amplifier output impedance are not really very low. We will see, however, that negative feedback, which is usually present around the amplifier, tends to suppress the effect of $Z_0(s)$ on overall performance.

Probably the most significant characteristic of the model of Fig. 2.1 is the open-loop gain function $A'(s)$. As the reader might expect, $A'(s)$ is not infinite, nor does it stay large for all frequencies. Rather, a typical variation of $A'(s)$ with frequency is sketched, somewhat idealized, in Fig. 2.2, where the magnitude and phase of $A'(s = j\omega)$ are plotted separately versus $\log \omega$ (Bode plot). Two important parameters of the plots are usually specified for a given amplifier:

1. The dc or low-frequency open-loop gain A_0, usually given in decibels.
2. The frequency at which the open-loop drops to 1 in magnitude. This is usually called the *unity gain bandwidth* f_u, typically specified in megahertz. In Fig. 2.2, where ω is the independent variable,

$$\omega_u = 2\pi f_u \tag{2.1}$$

ω_u is in radians per second and f_u in hertz.

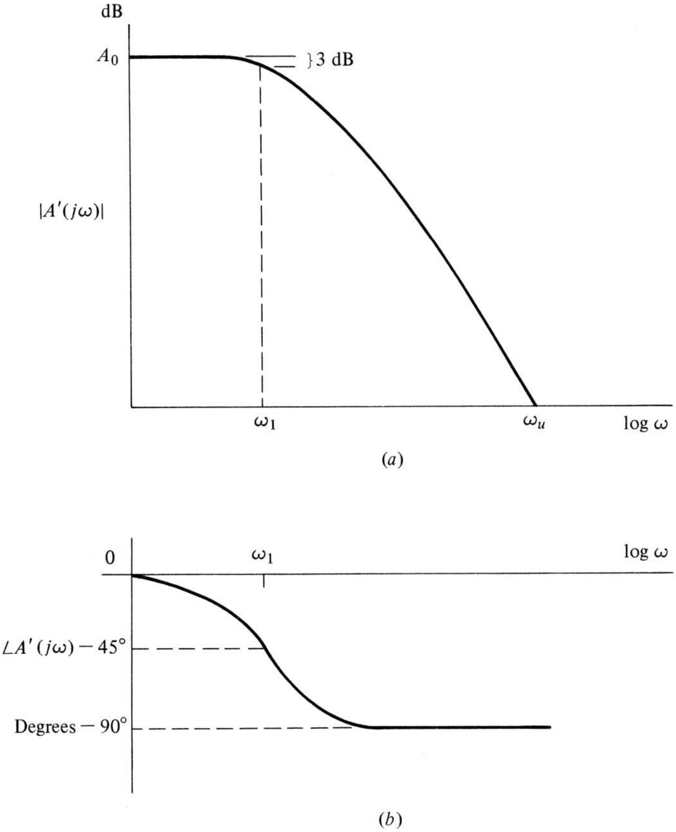

FIGURE 2.2
Typical variation of open-loop gain $A'(j\omega)$ with angular frequency ω (Bode plot). (*a*) Gain in decibels; (*b*) phase angle in degrees.

For the present, we will treat A' as a complex-valued quantity, $A'(s)$ or $A'(s = j\omega)$, as needed.

As implied in Fig. 2.2, the magnitude of A' is usually constant (approximately equal to A_0) from direct current up to some fairly low frequency ω_1; then the magnitude of A' will drop at the rate of 20 dB/decade until beyond the unity gain frequency. Often a *simple one-pole model* is used for $A'(s)$, that is,

$$A'(s) \leq \frac{A_0}{1 + s/\omega_1} \tag{2.2}$$

Of course, A_0 is here expressed as a numeric, not in decibels. The −3-dB frequency ω_1 may thus be estimated from ω_u as follows:

$$|A'(j\omega_u)| = \frac{A_0}{|1 + j\omega_u/\omega_1|} \cong 1$$

assuming $A_0 \gg 1$,

$$\frac{\omega_u}{\omega_1} \cong A_0 \qquad A_0 \gg 1$$

or

$$\omega_1 \cong \frac{\omega_u}{A_0} = \frac{2\pi f_u}{A_0} \qquad \text{rad/s} \tag{2.3a}$$

Of course,

$$f_1 \cong \frac{f_u}{A_0} \qquad \text{Hz} \tag{2.3b}$$

We shall see that f_u is a very important parameter for analyzing small-signal response; A_0 is usually only of direct interest at direct current.

Example 2.1. Given an amplifier with $A_0 = 80$ dB and $f_u = 1$ MHz, find $A'(s)$, ω_1, and f_1.

First find A_0 as a numeric.

$$A_0 = \text{antilog}_{10} \frac{80}{20} = 10^4$$

Also

$$\omega_u = 2\pi \times 10^6 \text{ rad/s}$$

Then, using (2.3),

$$\omega_1 = \frac{2\pi \times 10^6}{10^4} = 628 \text{ rad/s}$$

also

$$f_1 = 100 \text{ Hz}$$

and

$$A'(s) = \frac{10^4}{1 + s/628}$$

Note that in Fig. 2.1

$$V_0(s) = -A'(s)E_g(s) - I_A(s)Z_0(s) \tag{2.4a}$$

$$= A'(s)[E_2(s) - E_1(s)] - I_A(s)Z_0(s) \tag{2.4b}$$

Here we have included the polarity inversion from E_1 to V_0 separately from A'.

2.2 CLOSED-LOOP GAIN, OUTPUT IMPEDANCE, AND INPUT IMPEDANCE

Figure 2.3 shows the general model of Fig. 2.1 as it applies to the inverting amplifier circuit. This circuit is basically that of Fig. 1.15, except we assume that the external networks $Z_F(s)$, $Z_1(s), \ldots, Z_n(s)$ are only *two-terminal networks*. The impedance $Z_g(s)$ is the total extraneous impedance between the inverting amplifier terminal and ground; typically,

$$Z_g(s) = Z_I(s) \parallel Z_{cm1}(s) \tag{2.5}$$

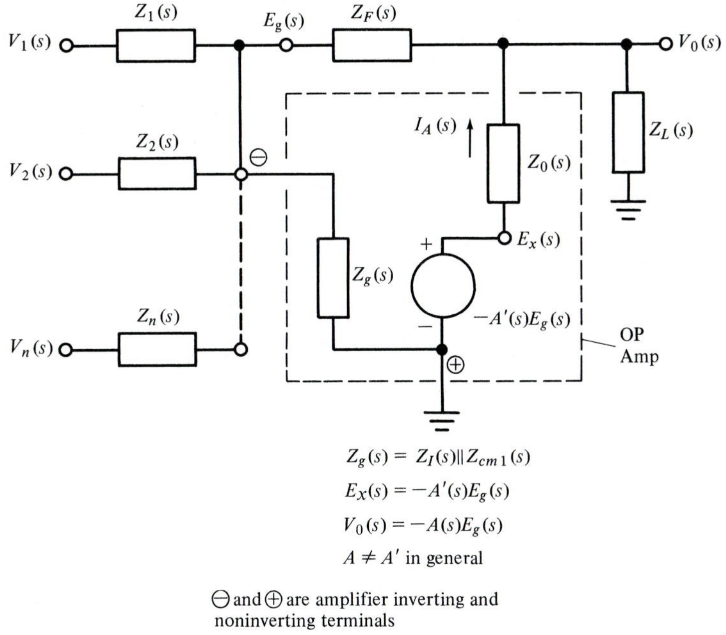

FIGURE 2.3
General linear model of an inverting amplifier circuit.

In some cases, it may also be necessary to include any shunt capacitance associated with this node. Often

$$|Z_I(s)| \ll |Z_{cm1}(s)|$$

at all frequencies of interest, and we can often say

$$Z_g(s) \cong Z_I(s) \tag{2.6}$$

Note that the amplifier open-loop gain $A'(s)$ is the transfer function between $E_g(s)$ and $V_0(s)$ *under no-load conditions;* i.e., the amplifier output terminal has a Thévenin-equivalent voltage source

$$V_0(s) = E_x(s) = -A'(s)E_g(s) \qquad \text{no load} \tag{2.7}$$

associated with it. In general $V_0(s) \neq E_x(s)$, due to the voltage drop across $Z_0(s)$ created by the amplifier output current $I_A(s)$. (Note that we again use the symbol V for the external node voltages of the complete circuit.)

CASE 1: $\mathbf{Z_0(s) = 0}$. This case serves as a starting point for more detailed analyses. It gives approximate results at low frequencies, when the amplifier output current is small. We can write

$$E_g(s) = K_1(s)V_1(s) + K_2(s)V_2(s) + \cdots + K_n(s)V_n(s) + \beta(s)V_0(s) \tag{2.8}$$

Define

$$V_0(s) = -A(s)E_g(s)$$

In later developments, we will reserve the symbol A(s) *for the actual ratio:*

$$A(s) = \frac{-V_0(s)}{E_g(s)} \tag{2.9}$$

In the first case,

$$A(s) = A'(s) \qquad \text{if} \qquad Z_0(s) = 0 \tag{2.10}$$

We also define

$$K_j(s) = \frac{1}{Z_j(s)}\left[\frac{1}{Z_F(s)} + \frac{1}{Z_1(s)} + \cdots + \frac{1}{Z_n(s)} + \frac{1}{Z_g(s)}\right]^{-1} \qquad j = 1, 2, \ldots, n$$

$$= \frac{Z_F(s)}{Z_j(s)}\beta(s) = \frac{Z_A(s)}{Z_j(s)} \tag{2.11}$$

where

$$\begin{aligned}\beta(s) &= \frac{1}{Z_F(s)}\left[\frac{1}{Z_F(s)} + \frac{1}{Z_1(s)} + \cdots + \frac{1}{Z_n(s)} + \frac{1}{Z_g(s)}\right]^{-1} \\ &= \frac{Z_A(s)}{Z_F(s)}\end{aligned} \tag{2.12}$$

and

$$Z_A(s) = \left[\frac{1}{Z_F(s)} + \frac{1}{Z_1(s)} + \cdots + \frac{1}{Z_n(s)} + \frac{1}{Z_g(s)}\right]^{-1}$$

$$= \text{total parallel impedance of all branches terminating at } E_g(s) \quad (2.13)$$

The ratio $\beta(s)$ will be *important* throughout this section; it is the actual *voltage feedback ratio* from the *amplifier output terminal* to the *inverting terminal,*

$$\beta(s) = \frac{E_g(s)}{V_0(s)} \qquad V_1(s) = V_2(s) = \cdots = V_n(s) = 0 \qquad (2.14)$$

Note also that from (2.12) and (2.13), $|\beta| < 1$; it may typically be a small fraction.

From (2.7), (2.8), (2.10), and (2.14) we obtain

$$V_0(s) = \frac{-A(s)}{1 + A(s)\beta(s)}\left[\sum_{j=1}^{n} K_j(s)V_1(s)\right]$$

or

$$V_0(s) = \frac{-A(s)\beta(s)}{1 + A(s)\beta(s)}\left[\frac{1}{\beta(s)}\sum_{j=1}^{n} K_j(s)V_j(s)\right] \qquad (2.15)$$

From (2.11),

$$\frac{K_j(s)}{\beta(s)} = \frac{Z_F(s)}{Z_j(s)} \qquad j = 1, 2, \ldots, n$$

Thus

$$V_0(s) = \frac{-A(s)\beta(s)}{1 + A(s)\beta(s)}\left[Z_F(s)\sum_{j=1}^{n}\frac{V_j(s)}{Z_j(s)}\right] \qquad (2.16)$$

or

$$V_0(s) = \frac{-1}{1 + 1/A(s)\beta(s)}\left[Z_F(s)\sum_{j=1}^{n}\frac{V_j(s)}{Z_j(s)}\right] \qquad (2.17)$$

These last two equations provide a lot of information. First, we note that the transfer function between any input and the output is given by

$$\frac{V_0(s)}{V_j(s)} = \frac{A(s)\beta(s)}{1 + A(s)\beta(s)}\left[\frac{-Z_F(s)}{Z_j(s)}\right] \qquad (2.18a)$$

$$= \frac{-Z_F(s)/Z_j(s)}{1 + 1/A(s)\beta(s)} \qquad (2.18b)$$

Thus each transfer function has unwanted poles at the roots of

$$1 + A(s)\beta(s) = 0$$

Hopefully these roots will have large negative values and will correspond to unimportant high-frequency poles. The possibility of instability is, however, suggested. By using the techniques of feedback system analysis (Bode plots, Nyquist plots, root locus, etc.), we can predict the stability of the closed-loop system. Normally Bode plots are the most useful.

We see from (2.18*b*) that

$$\frac{V_0(s)}{V_j(s)} \cong \frac{-Z_F(s)}{Z_j(s)} \qquad \text{at } |A(s)| \to \infty \tag{2.19}$$

and for $|A(s)\beta(s)| \gg 1$, the circuit behaves in the manner predicted by the ideal-amplifier equations developed in Chap. 1!

Also (2.16) may be rewritten

$$V_0(s) = -\left[1 - \frac{1}{1 + A(s)\beta(s)}\right] Z_F(s) \sum_{j=1}^{n} \frac{V_j(s)}{Z_j(s)} \tag{2.20}$$

From this, we can see that the fractional *error* in $V_0(s)$ [i.e., the amount by which $V_0(s)$ differs from the ideal of (2.19)] may be estimated by

$$\text{Fractional error} = \epsilon(s) = \frac{1}{1 + A(s)\beta(s)} \tag{2.21}$$

That is, if $V_{0D}(s)$ is the desired output,

$$\epsilon(s) = \frac{V_{0D}(s) - V_0(s)}{V_{0D}(s)}$$

Typically $A(s)$ is a large positive quantity and $|A(s)\beta(s)| \gg 1$. Thus

$$\epsilon(s) \cong \frac{1}{A(s)\beta(s)} \tag{2.22}$$

Throughout this section we will see that the quantity $A(s)\beta(s)$ is of primary importance in estimating the effects of finite $A(s)$.

CASE 2: *EFFECT OF NONZERO* $Z_0(s)$. If $Z_0(s)$ is not negligibly small, then $A(s) \neq A'(s)$. If we can find $A(s)$ in terms of $A'(s)$, $Z_0(s)$, and other quantities, the above results still apply. It is possible to show that

$$A(s) = A'(s)\left[a_I(s) - \frac{A_T(s)}{A'(s)}\right] \tag{2.23a}$$

$$\cong A'(s)a_I(s) \tag{2.23b}$$

where $a_I(s)$ is an attenuation ratio caused by the voltage drop through $Z_0(s)$.

$$\begin{aligned} a_I(s) &= \frac{Z_F(s)Z_L(s)}{Z_0(s)Z_L(s) + Z_0(s)Z_F(s) + Z_F(s)Z_L(s)} \\ &= \frac{Z_B(s)}{Z_0(s)} \end{aligned} \tag{2.24}$$

where

$$Z_B(s) = Z_F(s) \parallel Z_0(s) \parallel Z_L(s) \tag{2.25}$$

The ratio $a_T(s)$ is a feed-forward effect, representing the effect of any voltage at $E_g(s)$ on $V_0(s)$ in the absence of the amplifier itself.

$$\begin{aligned} a_T(s) &= \frac{Z_0(s)Z_L(s)}{Z_0(s)Z_F(s) + Z_0(s)Z_L(s) + Z_F(s)Z_L(s)} \\ &= \frac{Z_B(s)}{Z_F(s)} \end{aligned} \tag{2.26}$$

Note that if $Z_0(s) = 0$, as in case 1, then $a_T = 0$, $a_I = 1$, and

$$A(s) = A'(s) \qquad Z_0(s) = 0$$

The factor $a_T(s)$ measures the undesirable lack of isolation between operational amplifier input and output. Normally this factor may be neglected, except possibly at very high frequencies, and (2.23*b*) is used to estimate $A(s)$.

Equation (2.23*b*) may be used in (2.18) to (2.22) to estimate the effect of $Z_0(s)$ and $Z_L(s)$ on overall performance. Note that since $A(s)$ is not now equal to $A'(s)$, the closed-loop poles at the roots of $1 + A(s)\beta(s)$ depend on $a_I(s)$, and *thus* $Z_L(s)$ *can affect stability*. In particular, if $Z_L(s)$ includes a large capacitive component at high frequencies, it is possible for the circuit to become unstable, even though it is stable without $Z_L(s)$.

Closed-Loop Output Impedance

In Fig. 2.4 we apply a test current $I(s)$ to the amplifier output terminal; the resulting value of $V_0(s)$ allows us to find the closed-loop output impedance (all inputs grounded). Using the symbols from the previous development, noting that

$$E_x(s) = -A'(s)E_g(s) \tag{2.27a}$$

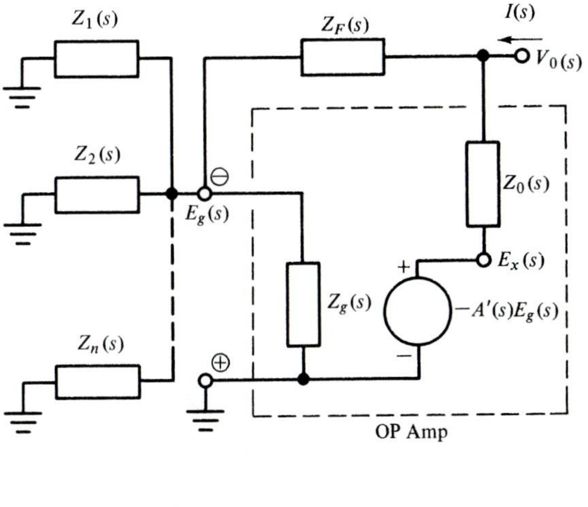

FIGURE 2.4
Circuit for finding the closed-loop output impedance.

and

$$E_g(s) = \beta(s)V_0(s) \tag{2.27b}$$

we write a nodal equation

$$I(s) = \frac{V_0(s) - E_x(s)}{Z_0(s)} + \frac{V_0(s) - E_g(s)}{Z_F(s)} \tag{2.28}$$

Then

$$E_x(s) = -A'(s)E_g(s) = -A'(s)\beta(s)V_0(s) \tag{2.29}$$

and

$$I(s) = \frac{V_0(s)[1 + A'(s)\beta(s)]}{Z_0(s)} + \frac{V_0(s)[1 - \beta(s)]}{Z_F(s)} \tag{2.30}$$

or

$$Y_{\text{out}}(s) = \frac{I(s)}{V_0(s)} = \frac{1 + A'(s)\beta(s)}{Z_0(s)} + \frac{1 - \beta(s)}{Z_F(s)} \tag{2.31}$$

Thus we see that the output impedance is the parallel combination of two impedances, i.e.,

$$Z_{\text{out}}(s) = \frac{Z_0(s)}{1 + A'(s)\beta(s)} \Bigg\| \frac{Z_F(s)}{1 - \beta(s)} \tag{2.32}$$

Since the magnitude of β is always less than 1 and $Z_0(s)$ is usually relatively small, then

$$Z_{\text{out}}(s) \cong \frac{Z_0(s)}{1 + A'(s)\beta(s)} \tag{2.33}$$

Over the useful frequency range of the complete operational amplifier circuit, we expect

$$|A'\beta| \gg 1$$

and we have

$$Z_{\text{out}}(s) \cong \frac{Z_0(s)}{A'(s)\beta(s)} \tag{2.34}$$

Note that from our definition of $A'(s)$, it is a large positive quantity at direct current, and the closed-loop output impedance is usually a very small fraction of $Z_0(s)$. Note also that it is the value of $A'(s)$, *not* $A(s)$, that is of interest in (2.34).

Example 2.2. Given an operational amplifier with the following static (dc) characteristics:

Open-loop, unloaded voltage gain	$A_0 = A_{\text{dc}} = 10^6$ (or 120 dB)	
Output resistance	$R_0 = 10^3\ \Omega$	
Differential input impedance	$R_j = 10^6\ \Omega$	

find the dc gain error and circuit output impedance. Assume common-mode input impedances are very high (over $10^8\ \Omega$).

In the simple resistive inverter circuit of Fig. 2.5,

$$R_F = 10^5\ \Omega \qquad R_1 = 10^4\ \Omega$$

Equation (1.5) predicts (for an ideal amplifier) that

$$\frac{V_0}{V_1} = -\frac{R_F}{R_1} = -10$$

More precisely, using the approach of Sec. 2.2 ($Z_L = \infty$),

$$\beta = \frac{10^4 \parallel 10^5 \parallel 10^6}{10^5} = 0.09009$$

Note that here the effect of $Z_g \cong Z_i = 10^6\ \Omega$ could be ignored without any serious loss of accuracy. Also

$$a_I = \frac{10^3 \parallel 10^5}{10^3} = 0.9901$$

and

$$A = a_I A' = 0.9901 \times 10^6$$

Thus at direct current,

$$A\beta \cong 9 \times 10^4$$

or a more precise estimate of the gain is by (2.18):

$$\frac{V_0}{V_1} = -\frac{A\beta}{1 + A\beta} \times 10$$

$$= -0.99989 \times 10$$

$$= -9.9989$$

The fractional error from (2.22) is

$$\epsilon = \frac{1}{A\beta} = \frac{1}{9 \times 10^4} = 0.0011 = 0.11\%$$

Note that our ideal amplifier model of Chap. 1 is not so bad! The output impedance from (2.34) is

$$Z_{\text{out}}(s) = \frac{10^3}{A'\beta} = 0.011\ \Omega$$

again, very low.

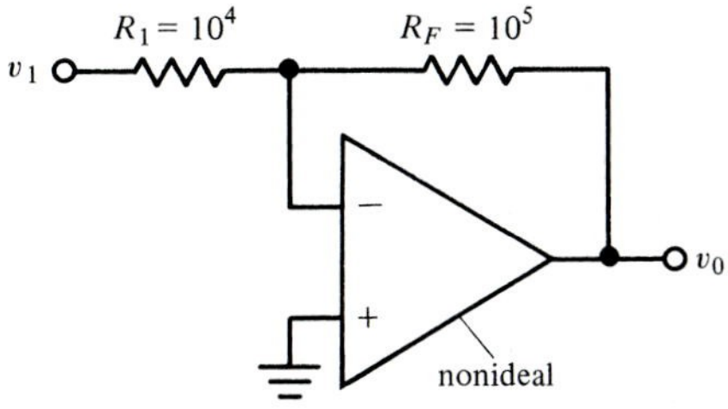

FIGURE 2.5
Simple inverter circuit.

Suppose we load the circuit of Fig. 2.5 with a 1000-Ω load. This changes a_I, that is,

$$a_i = \frac{10^3 \parallel 10^3 \parallel 10^5}{10^3} \cong 0.5$$

Now

$$A = 0.5 \times 10^6$$

$$A\beta \cong 4.5 \times 10^4$$

and the fractional gain error is twice as large (but still very small), $\epsilon \simeq 0.22$ percent.

Before leaving the inverting amplifier circuit for a while, we note that the *input impedance* at each input node V_j may still normally be assumed to equal the associated input summing impedance Z_j. Thus in the last example, the circuit input impedance is assumed to be 10^4 Ω. At very high frequencies, the input impedance will be Z_j in series with some other impedance, due to the fact that the summing junction is no longer a virtual ground.

The Noninverting Amplifier Circuit

In Fig. 2.6 we have the linear nonideal amplifier model embedded in a basic noninverting amplifier circuit. The analysis of this circuit is quite similar to that of the inverting amplifier, and we present only the final results. We first define the following:

$$E_x(s) = -A'(s)[E_g(s) - V_1(s)] \tag{2.35}$$

$$V_0(s) = -A(s)[E_g(s) - V_1(s)] \tag{2.36}$$

$$Z_1'(s) = Z_1(s) \parallel Z_{cm1}(s) \tag{2.37}$$

$$Z_A(s) = Z_1'(s) \parallel Z_F(s) \parallel Z_I(s) \tag{2.38}$$

$$Z_B(s) = Z_0(s) \parallel Z_F(s) \parallel Z_L(s) \tag{2.39}$$

Note that typically $Z_{cm1}(s) \gg Z_1(s)$; thus

$$Z_1'(s) \cong Z_1(s) \tag{2.40}$$

except possibly *at high frequencies,* when the *shunt capacity component* of Z_{cm1} may be significant. In words,

Z_A is the total shunt impedance at node E_g

Z_B is the total shunt impedance at node V_0

An analysis of the circuit of Fig. 2.6 yields

$$\frac{V_0(s)}{V_1(s)} = \frac{Z_1'(s) + Z_F(s)}{Z_1'(s)} \frac{A(s)\beta(s)}{1 + A(s)\beta(s)} \tag{2.41}$$

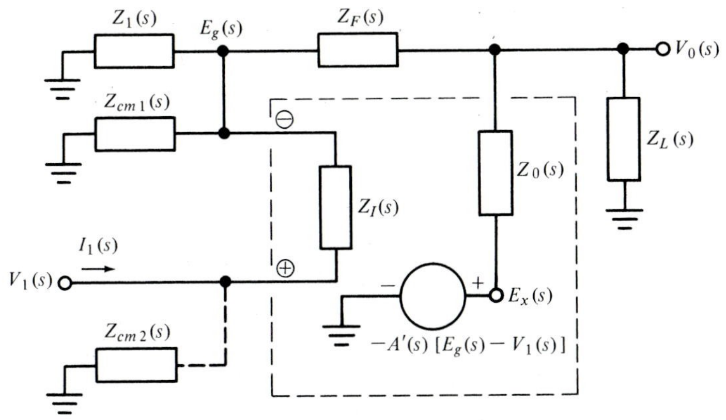

FIGURE 2.6
General linear model for a noninverting amplifier circuit.

where again $\beta(s)$ is a *voltage feedback ratio*

$$\beta(s) = \frac{Z_A(s)}{Z_F(s)} \tag{2.42}$$

Also if we neglect feed-forward effects around the amplifier, i.e., assume that

$$Z_O(s) \ll |Z_F(s)A'(s)| \tag{2.43}$$

then again

$$A(s) = a_1(s)A'(s) \tag{2.44}$$

where

$$a_I(s) = \frac{Z_B(s)}{Z_O(s)} \tag{2.45}$$

These last results are similar to those for the inverting amplifier. The *loop gain* $A(s)\beta(s)$ affects the closed-loop transfer function in the same way as in the inverting circuit, and the actual gain is equal to the desired gain multiplied by

$$\frac{A(s)\beta(s)}{1 + A(s)\beta(s)} = \frac{1}{1 + 1/A(s)\beta(s)} \tag{2.46}$$

Thus the fractional error is again

$$\epsilon(s) \cong \frac{1}{A(s)\beta(s)} \qquad A(s)\beta(s) \gg 1 \tag{2.47}$$

Again, an examination of stability and gain errors can be made from an analysis of the loop gain $A(s)\beta(s)$.

Input Impedance of the Noninverting Amplifier

The input impedance of the noninverting amplifier is (Fig. 2.6)

$$Z_{\text{in}}(s) = \frac{V_1(s)}{I_1(s)} \tag{2.48}$$

but

$$I_1(s) = \frac{V_1(s) - E_g(s)}{Z_I(s)} + \frac{V_1(s)}{Z_{cm2}(s)} \tag{2.49}$$

From (2.36) we have

$$I_1(s) = \frac{V_1(s) + V_0(s)/A(s) - V_1(s)}{Z_I(s)} + \frac{V_1(s)}{Z_{cm2}(s)}$$

$$= \frac{V_0(s)}{Z_I(s)A(s)} + \frac{V_1(s)}{Z_{cm2}(s)}$$

but if $A(s)\beta(s) \gg 1$,

$$V_0(s) \cong \frac{Z_1'(s) + Z_F(s)}{Z_1'(s)} V_1(s)$$

and

$$I_1(s) \cong \left[\frac{Z_1'(s) + Z_F(s)}{Z_1'(s)} \frac{1}{Z_I(s)A(s)} + \frac{1}{Z_{cm2}(s)}\right] V_1(s)$$

Thus

$$Z_{\text{in}}(s) \cong [Z_I(s)A(s)\beta(s) \parallel Z_{cm2}(s)] \tag{2.50a}$$

where

$$G_D(s) = \frac{Z_1'(s) + Z_F(s)}{Z_1'(s)} \cong \text{design closed-loop gain} \cong \frac{1}{\beta(s)} \tag{2.50b}$$

We note that the input impedance is thus normally very large compared to $Z_I(s)$, since typically $A(s) \gg G(s)$ and $Z_{cm2} \gg Z_I$.

Output Impedance of the Noninverting Amplifier

With respect to output impedance, the noninverting circuit of Fig. 2.6 is essentially the same as the inverting amplifier of Fig. 2.4. That is, the output impedance may be found from the results of (2.32), (2.33), and (2.34). In summary,

$$Z_{\text{out}}(s) \cong \frac{Z_0(s)}{A'(s)\beta(s)} \tag{2.51a}$$

$$\cong \frac{Z_0(s)G_D(s)}{A'(s)} \tag{2.51b}$$

Note in (2.51) we use $A'(s)$, not $A(s)$.

The Unity Gain Follower

Note that the inequality of (2.43) does not apply to the unity gain follower circuit of Fig. 2.7. Let us make a special analysis of this case. Here we see that the feedback ratio is also approximately unity, i.e.,

$$\beta(s) \cong 1 \tag{2.52}$$

The nodal equation at $V_0(s)$ is

$$\frac{A'(s)}{Z_0(s)}[V_1(s) - V_0(s)] = \frac{V_0(s)}{Z_x(s)} - \frac{V_1(s)}{Z_I(s)} \tag{2.53}$$

where

$$\begin{aligned} Z_x(s) &= Z_0(s) \parallel Z_L(s) \parallel Z_I(s) \parallel Z_{cm1}(s) \\ &\cong Z_0(s) \parallel Z_L(s) \qquad |Z_I|,\ |Z_{cm1}| \gg |Z_0| \end{aligned} \tag{2.54}$$

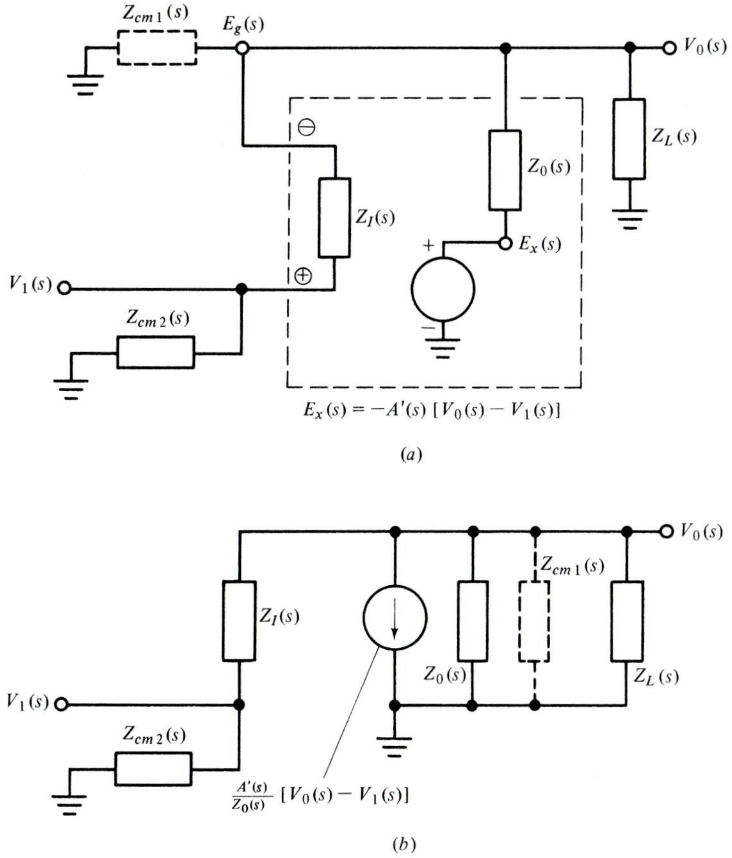

FIGURE 2.7
A general linear model for the unity gain follower circuit. (*a*) Adaptation from Fig. 2.6; (*b*) a Norton equivalent.

or

$$\frac{V_0(s)}{V_1(s)} = \frac{1 + Z_0(s)/A'(s)Z_I(s)}{1 + Z_0(s)/A'(s)Z_x(s)} \tag{2.55}$$

Typically $A'(s)Z_I(s) \gg Z_0(s)$ and $Z_I(s) \gg Z_x(s)$, and we have

$$\begin{aligned}\frac{V_0(s)}{V_1(s)} &\cong \frac{1}{1 + Z_0(s)/A'(s)Z_x(s)} = \frac{1}{1 + 1/A(s)} \\ &= \frac{A(s)}{1 + A(s)}\end{aligned} \tag{2.56}$$

where we define

$$A(s) = A'(s)\frac{Z_x(s)}{Z_0(s)} \tag{2.57}$$

Thus our general expression, (2.41), still holds for this special case where $\beta(s) = 1$ and $A(s)$ is defined by (2.57).

The input impedance is easily found from (2.50*a*) to be

$$Z_{\text{in}} \cong A(s)Z_I(s) \parallel Z_{cm2}(s) \tag{2.58}$$

The output impedance is also found from (2.32), but with $\beta(s) = 1$, so that now

$$Z_{\text{out}}(s) = \frac{Z_0(s)}{1 + A'(s)} \tag{2.59a}$$

or

$$Z_{\text{out}}(s) \cong \frac{Z_0(s)}{A'(s)} \qquad A'(s) \gg 1 \tag{2.59b}$$

Example 2.3. Consider the noninverting amplifier circuit of Fig. 2.8; note that the values of R_F and R_1 are chosen to obtain a noninverting gain of 10. Otherwise, the circuit is very similar to that of Example 2.2. If we assume the same nonideal amplifier model as in Example 2.2, i.e.,

$$\begin{aligned}A_0 &= 10^6 \\ R_0 &= 10^3\ \Omega \\ R_I &= 10^6\ \Omega \qquad \text{(neglect } R_{cm}\text{)} \\ R_F &= 9 \times 10^4\end{aligned}$$

then we can estimate the fractional gain error by (2.47). In this case

$$\beta = \frac{10^4 \parallel 9 \times 10^4 \parallel 10^6}{9 \times 10^4} \cong 0.1 = \frac{1}{G_D(s)}$$

Including the 10^3-Ω output load,

$$\begin{aligned}a_I &= \frac{10^3 \parallel 10^3 \parallel 9 \times 10^4}{10^3} \\ &\cong 0.5\end{aligned}$$

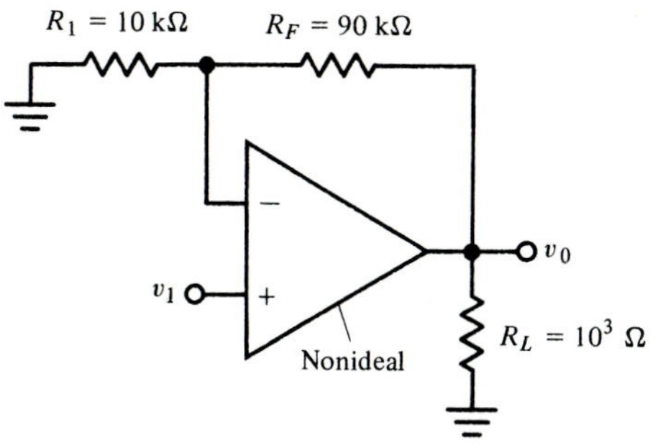

FIGURE 2.8
Circuit of Example 2.3.

Thus

$$A \cong 0.5 \times 10^6$$

at direct current, and the fractional error is

$$\frac{1}{A\beta} = \frac{1}{0.5 \times 10^5} = 2 \times 10^{-5} = 2 \times 10^{-3} \text{ percent}$$

The circuit input impedance may be estimated from (2.50a) as follows:

$$Z_{\text{in}} \cong \frac{AR_1}{G} \bigg\| R_{cm2}$$

where

$$G = 10 = \frac{R_1 + R_F}{R_1}$$

Thus

$$Z_{\text{in}} \cong \frac{0.5 \times 10^6 \times 10^6}{10}$$
$$= 0.5 \times 10^{11}\ \Omega$$

in parallel with R_{cm2}.

In many simple amplifier configurations, the feedback ratio $\beta(s)$ is relatively easy to calculate. Essentially we are trying to find the voltage feedback ratio from the amplifier output terminal to the inverting input terminal. This ratio is, in general, established by some three-terminal network (Fig. 2.9a) connecting these two points, i.e.,

$$\beta(s) = \frac{E_g(s)}{V_O(s)} \tag{2.60}$$

For specific circuits, the resulting value of $\beta(s)$ may be found; for many approximate analyses, it is all right to neglect the effect of $Z_I(s)$ and the common-mode impedances. However, particularly for accurate high-frequency analysis, the effects of the capacitive components of these impedances may need to be included.

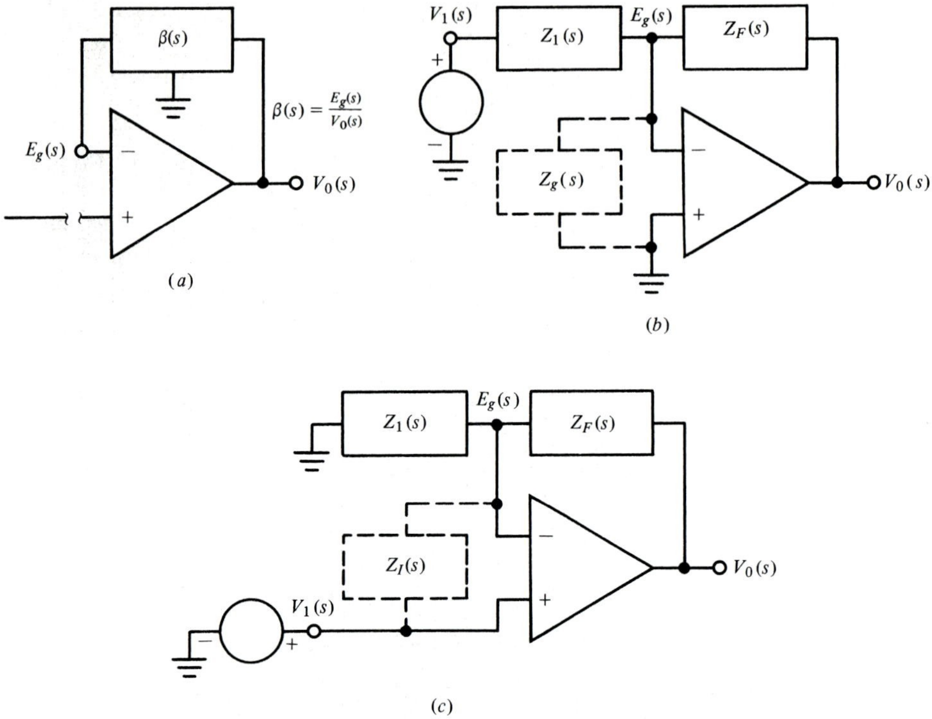

FIGURE 2.9
Finding $\beta(s)$ in typical circuits. (*a*) General two-port feedback network; (*b*) estimating $\beta(s)$ in a simple inverting amplifier circuit; (*c*) estimating $\beta(s)$ for a noninverting amplifier.

For both the single-input inverting amplifier circuit (Fig. 2.9*b*) and the non-inverting amplifier circuit (Fig. 2.9*c*), we find that

$$\beta(s) = \frac{Z_1(s)}{Z_1(s) + Z_F(s)} \tag{2.61}$$

where, if needed, the effects of the amplifier input impedance may be included by including $Z_g = Z_I \parallel Z_{cm1}$ in the calculation of Z_1.

It is often useful to be aware of the fact that for the *single-input* inverting amplifier (Fig. 2.9*b*),

$$\beta(s) = \frac{1}{1 - G_I(s)} = \frac{1}{1 + |G_I(s)|} \tag{2.62}$$

where $G_I(s)$ is the desired or design value of the *closed-loop gain,* i.e.,

$$G_I(s) = \frac{-Z_F(s)}{Z_1(s)} \tag{2.63}$$

[Note the inclusion of the minus sign with $G_I(s)$.] Thus an inverting amplifier with a design gain of magnitude 10 ($G_I = -10$) at some frequency will have a corresponding value of

$$|\beta| = \frac{1}{1-(-10)} = \frac{1}{11}$$

For an amplifier with a high open-loop gain, the value of $|\beta|$ is *approximately* the reciprocal of the magnitude of the design gain, since

$$|\beta| = \frac{1}{|1-G_I|} \cong \frac{1}{|G_I|} \qquad |G_I| \gg 1$$

For the *noninverting* amplifier (Fig. 2.9*c*), we still have

$$\beta(s) = \frac{Z_1(s)}{Z_1(s)+Z_F(s)}$$

where again the effects of $Z_I(s)$ may warrant inclusion. Here, however, if we define $G_N(s)$ as the design value of (noninverting) closed-loop gain, i.e.,

$$G_N(s) = \frac{Z_1(s)+Z_F(s)}{Z_1(s)} \tag{2.64}$$

then for the noninverting amplifier

$$\beta(s) = \frac{1}{G_N(s)} \tag{2.65}$$

Thus a noninverting amplifier with a design gain of magnitude 10 will have $|\beta| = 0.1$.

Many articles on amplifier design will say that the magnitude of β is equal to the reciprocal of the magnitude of the desired closed-loop gain. This is true for the noninverting amplifier and also approximately true for the single-input inverting circuit *if* the closed-loop gain is large compared to 1. Note, however, the following example.

Example 2.4. Consider the simple two-input inverting amplifier of Fig. 2.10. An estimation of β follows directly from (2.12), that is,

$$\beta = \frac{R_1 \parallel R_2 \parallel R_F}{R_F}$$

$$= \frac{R_1R_2}{R_2R_F + R_1R_F + R_1R_2}$$

Defining

$$G_1 = \frac{-R_F}{R_1}$$

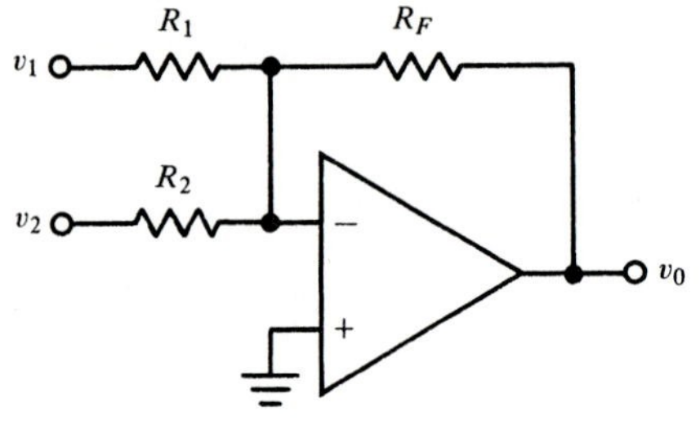

FIGURE 2.10
Circuit for Example 2.4, a two-input inverting amplifier.

and

$$G_2 = \frac{-R_F}{R_2}$$

we can show that

$$\beta = \frac{1}{|G_1| + |G_2| + 1} \cong \frac{1}{|G_1| + |G_2|} \tag{2.66}$$

which is an extension of (2.62) to the two-input case.

Example 2.5. Refer to Fig. 2.5. Assume that the operational amplifier has zero output impedance and a negligibly large differential input impedance; the forward gain of the operational amplifier is to be assumed to have the form

$$A(s) = \frac{A_0}{1 + s/\omega_1}$$

with

$$\omega_1 = \frac{2\pi f_u}{A_0}$$

where A_0 is the dc open-loop gain and f_u the unity gain frequency (in hertz) of the operational amplifier.

Let

$$R_F = GR_1$$

that is, G is the desired magnitude of the closed-loop circuit gain. Find

$$H(s) = \frac{V_0(s)}{V_1(s)}$$

in terms of A_0, f_u, and G. What do you get as $A_0 \to \infty$?

Here

$$A(s) = \frac{A_0}{1 + s/\omega_1}$$

$$\beta = \frac{R}{R + GR} = \frac{1}{1 + G}$$

and

$$A = \frac{K}{1\ as}$$

where

$$K = \frac{A_0}{G+1}$$

$$a = \frac{1}{\omega_1}$$

So the transfer function of the circuit is

$$H(s) = -G\frac{A}{1+A}$$

$$= -G\frac{-G\ [K/(1+K)]}{1+s/\omega_1(K+1)}$$

as

$$A_0 \to \infty,\ K \to \infty$$

and

$$H(s) \to -G$$

the -3-dB frequency is

$$\omega_0 = \frac{\omega_1}{K+1} \tag{2.67}$$

2.3 CLOSED-LOOP STABILITY; BODE PLOT METHODS; COMPENSATION

In the last section, we saw [e.g., in Eq. (2.15)] that the closed-loop transfer function of a linear operational amplifier circuit has unwanted poles at roots of

$$1 + A(s)\beta(s) = 0 \tag{2.68}$$

It is important to determine where the roots of (2.68) lie in the s plane. Specifically, for the circuit to be stable, they must have negative real parts; moreover, these spurious roots should have a large magnitude and relatively small imaginary parts, if unwanted oscillatory transient components in the circuit's response are to be minimized. The well-known Bode plot methods of estimating stability apply very nicely to the analysis of closed-loop operational amplifier circuit behavior; a discussion of this method appears in most control system texts (see, e.g., Ref. 2, Sec. 3.4, or Ref. 3, Sec. 7.9). We summarize the important points here.

Given a linear system transfer function $H(s)$, the Bode plots consist of:

1. Magnitude of $H(s = j\omega)$ in the form $20\log_{10}|H(j\omega)|$ versus $\log\omega$, that is, decibels versus $\log\omega$.
2. Phase of $H(j\omega) = \angle H(j\omega)$ versus $\log\omega$, usually in degrees, but possibly in radians.

Reference 3, Sec. 7.9, summarizes the technique for making approximate plots from a knowledge of critical frequencies of $H(s)$. In the analysis of operational amplifier circuits, we are interested in Bode plots of $H(s) = A(s)\beta(s)$.

To determine closed-loop stability via a Bode plot, we are primarily interested in two frequencies (Fig. 2.11):

1. The *unity-gain crossover frequency* ω_0, that is, the frequency where the magnitude curve first crosses the 0-dB point
2. The *phase crossover frequency* ω_p, where the phase curve first crosses $-180°$ (or $-\pi$ rad)

The *criterion for stability* may be stated in various ways:

1. If

$$\omega_0 < \omega_p \tag{2.69}$$

i.e., if the unity-gain crossover frequency is less than the phase crossover frequency.

2. If the phase is less negative than $-180°$ at the unity-gain crossover frequency: the difference between $-180°$ and the phase at ω_0 is termed the *phase margin* (ϕ_m, Fig. 2.11). Thus the phase margin

$$\phi_m = \angle A(j\omega_0)\beta(j\omega_0) + 180° > 0° \tag{2.70}$$

3. If the gain is less than 0 dB at the phase crossover frequency. This difference is termed the *gain margin* (G_m, Fig. 2.11). Thus the gain margin

$$G_m = -20\log_{10}|A(j\omega_p)\beta(j\omega_p)| > 0 \text{ dB} \tag{2.71}$$

Note that $\angle A\beta$ is usually negative at all frequencies, hence the form of (2.70). Also $|A\beta| > 0$ dB below ω_0; however, G_m is usually defined as positive for stability, hence the form of (2.71).

The above criteria merely ensure stability. Going beyond this, we are usually interested in whether the roots of $A(s)\beta(s) + 1 = 0$ influence the desired closed-loop response; e.g., the frequency response may exhibit peaking, and the transient response will have some form of oscillatory behavior. To ensure against this, the phase margin should be reasonably high; a common rule of thumb that is often applied is that, to ensure good transient behavior,

$$\phi_m > 45°$$

i.e., the phase of $A\beta$ should be about $-135°$ at ω_0. This means that typically the slope of the magnitude curve (dB versus $\log\omega$) will be less than or equal to 20 dB/decade near ω_0.

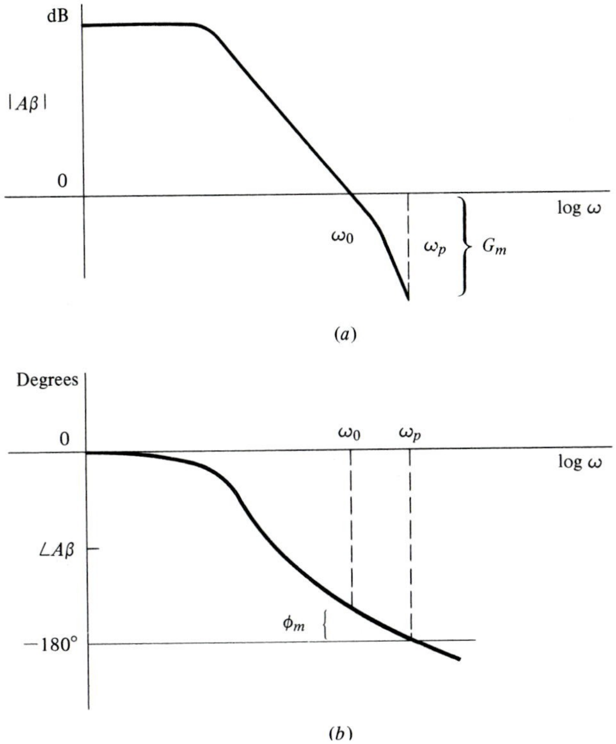

FIGURE 2.11
Typical Bode plots of loop gain $A\beta$. (*a*) Magnitude in decibels; (*b*) phase angle in degrees.

It is common practice to analyze operational amplifier circuits by plotting the magnitude curves of $A(j\omega)$ and $1/\beta(j\omega)$ on the same graph. Thus at the point of intersection of the two curves, $\omega = \omega_0$, $|A\beta| = 1$ (0 dB), and the *rate of closure* of the two curves indicates the approximate phase margin. If the rate of closure is less than 20 dB/decade, we have a good phase margin. If it is greater than or equal to 40 dB/decade, we have instability. In between, we have stability, but probably poor phase margin (Fig. 2.12).

If the above criteria are not met, then normally oscillation will result, since there will be roots of

$$A(s)\beta(s) + 1 = 0 \tag{2.72}$$

with positive real parts.

Example 2.6. In Fig. 2.12, we have a simple gain-of-10 inverting amplifier (Fig. 2.12*c*). For this circuit

$$\beta = \frac{1}{11}$$

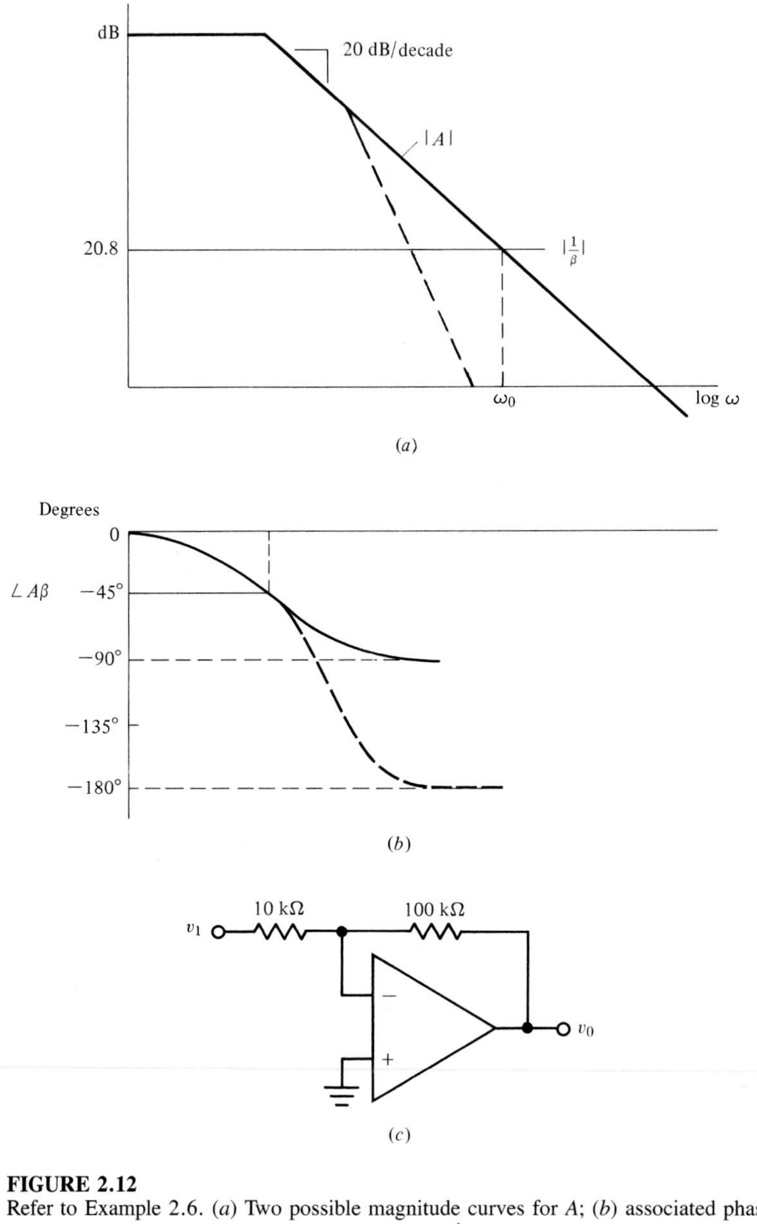

FIGURE 2.12
Refer to Example 2.6. (*a*) Two possible magnitude curves for *A*; (*b*) associated phase curves; (*c*) simple gain-of-10 inverter, with $\beta = \frac{1}{11}$.

and $1/\beta$ in decibels is merely a constant,

$$\left|\frac{1}{\beta}\right| = 20 \log_{10} 11 = 20.8 \text{ dB}$$

The solid curve (Fig. 2.12*a*) representing a typical $|A|$ shows a rate of closure of 20 dB/decade, indicating stability, with probably a good phase margin. The dotted curve, indicating an extra pole in $A(s)$ which is going to add more phase shift to $A\beta$, indicates a higher rate of closure. If the rate of closure is less than 40 dB/decade, then the circuit is presumably closed-loop stable, but the phase margin may not be good. A further analysis of the phase curves for A and β may thus be warranted. Figure 2.12*b* shows corresponding plots of $\angle A\beta$.

Example 2.7. In Fig. 2.13, we have a Bode plot of a typical $A(j\omega)$, indicating two first-order poles. Two different values of β are shown (assumed constant, i.e., for

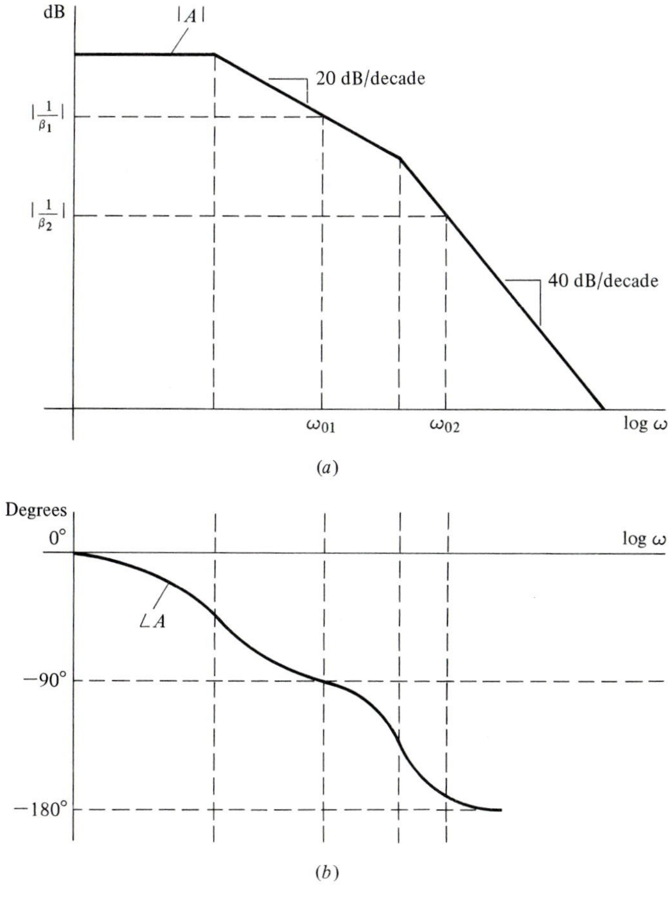

FIGURE 2.13
Refer to Example 2.7. (*a*) Magnitude plots; (*b*) phase plots.

some simple amplifier with resistances for Z_1 and Z_F). Case 1 would presumably be for an amplifier with a relatively high closed-loop gain, case 2 for a lower closed-loop gain. The rate of closure for the two situations indicates that case 1 would lead to good closed-loop stability with good phase margin. Case 2 has a poor phase margin. If β were much larger (lower closed-loop gain), then instability might occur.

At this point, it is useful to note that the simple unity-gain follower circuit, which has a β of 1, provides an acid test for the closed-loop stability of an operational amplifier. Here the amplifier output is connected directly to the inverting terminal (Fig. 2.7). In this case, the phase margin is found by merely examining the phase angle of A at $\omega_0 = \omega_u =$ the unity-gain crossover frequency of the loaded amplifier! Amplifiers which are stable under this condition are termed *short-circuit stable*; they must have $\angle A(j\omega) > -180^\circ$ and preferably in the neighborhood of -135° at ω_u.

It is common practice in the design of integrated-circuit operational amplifiers to provide for the connection of *compensation networks*. These often take the form of a simple series *RC* network that controls the first break frequency in $A'(s)$, for example, ω_1 in Fig. 2.2. By adjusting this break frequency, it is possible to trade bandwidth for closed-loop gain, within the limits of the overall gain-bandwidth product. (See Prob. 2.12 to gain some insight into how this works.)

Figure 2.14 shows some typical compensation schemes that may be used with an amplifier that has no provision for internal compensation; optimum component values are often best determined empirically. (See also Ref. 1, Secs. 3-18 to 3-20.)

Example 2.8. In the circuit of Fig. 2.5, assume that the operational amplifier has zero output impedance and a negligibly large differential input impedance. The forward gain of the operational amplifier is assumed to have the form

$$A(s) = \frac{A_0}{1 + s/\omega_1}$$

with

$$\omega_1 = \frac{2\pi f_u}{A_0}$$

where A_0 is the dc open-loop gain and f_u the unity gain frequency (in hertz) of the operational amplifier. Let

$$R_1 = R \qquad R_F = GR$$

that is, G is the desired magnitude of the closed-loop circuit gain; find

$$H(s) = \frac{V_0(s)}{V_1(s)}$$

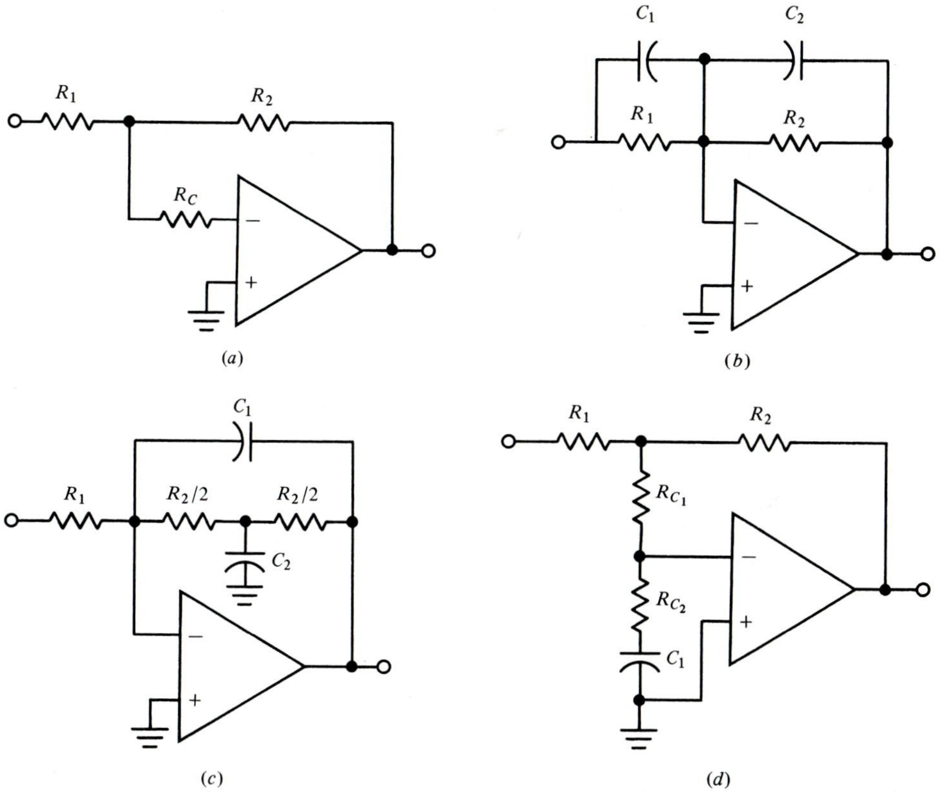

FIGURE 2.14
Typical compensation schemes that may be used to improve the stability and/or frequency response of a simple inverting amplifier. (*a*) Resistor R_c in series with the inverting terminal improves stability, but will generally degrade bandwidth; (*b*) compensating capacitors can be used to correct closed-loop phase margin and also improve bandwidth; C_1 and C_2 are normally adjusted empirically, C_1 often omitted; (*c*) the tee network is easier to adjust and eliminates the capacitative loading of the input source; (*d*) this network has been successfully used in D/A converters to improve the step response by phase-shift correction.

Now

$$\beta = \frac{R}{R + GR} = \frac{1}{1 + G}$$

$$A = \frac{A(s)}{1 + G} = \frac{K}{1 + as}$$

with

$$a = \frac{1}{\omega_1}$$

$$= \frac{A_0}{G + 1}$$

and

$$\omega_1 = \frac{2\pi f_u}{A_0}$$

So

$$\frac{V_0(s)}{V_1(s)} = -G \frac{A\beta}{1 + A\beta}$$

$$= \frac{-K\,[K/(K+1)]}{1 + 1/[1 + s/\omega_1(K+1)]}$$

as

$$A_0 \to \infty \qquad K \to \infty$$

and

$$\frac{V_0(s)}{V_1(s)} \to -G$$

The *−3-dB frequency* is then

$$\omega_0 = \frac{2\pi f_u}{A_0} \left(\frac{A_0 + G + 1}{G + 1} \right)$$

as

$$A_0 \to \infty \qquad \omega_0 \to \frac{2\pi f_u}{G + 1}$$

or

$$f_0 = \frac{f_u}{G + 1} \quad \text{Hz} \qquad \text{as } A_0 \to \infty \tag{2.73a}$$

For a noninverting circuit (Fig. 1.21) a similar development gives a −3-dB frequency

$$f_0 = \frac{f_0}{G} \tag{2.73b}$$

where G is the nominal circuit gain (see Prob. 2.25).

Consider the *unity gain follower* of Fig. 2.7, where the operational amplifier output is strapped directly to the inverting terminal and the signal is applied directly to the noninverting terminal. If the operational amplifier is ideal, then $\beta = 1$. This is the "acid test" of operational amplifier stability: We say the operational amplifier is *short-circuit stable* if the unity gain follower is stable. Here the $|1/\beta|$ line is at 0 dB and is as low as possible, thus bringing it down to possible additional deflections of the $|A|$ curve, due to additional high-frequency poles.

Example 2.9. Referring to Fig. 2.21*c*, show that if the rate of closure of Bode magnitude plots of A and $1/\beta$ is exactly 6 dB/octave, if the curves intersect at $\omega = \omega_0$, and if there are no nearby corner frequencies, then the closed-loop gain of the associated circuit is 3 dB less than the desired gain at $\omega = \omega_0$.

Under the specified conditions

$$|A\beta| = 1 \text{ or } 0 \text{ dB} \qquad \text{and} \qquad \angle A = 90^\circ$$

Therefore

$$A\beta = -j1$$

$$\frac{A\beta}{1 + A\beta} = \frac{-j1}{1 - j1}$$

or

$$\frac{A}{1 + A\beta} = 0.707 \text{ or } -3 \text{ dB}$$

2.4 COMMON-MODE EFFECTS

In the differential amplifier circuit shown in Fig. 2.15, an *ideal* operational amplifier provides an output

$$e_0 = -\frac{R_0}{R_1}\Delta e \qquad \text{ideal} \tag{2.74}$$

where Δe is the differential input signal.

The input *common-mode signal* e_{cm} is ideally rejected. This type of situation is common in many instrumentation systems, where Δe originates from some low-level transducer; e_{cm} may be a dc offset or some form of noise (60-cycle hum or cross-talk from nearby electrical systems), i.e., an *unwanted* signal to be ignored by the amplifier.

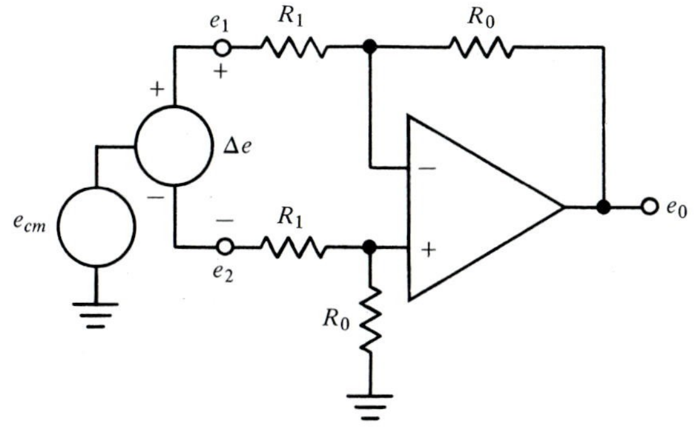

FIGURE 2.15
A differential amplifier circuit, with the input resolved into a differential component Δe and a common-mode component e_{cm}.

Unfortunately, real operational amplifiers exhibit a *common-mode gain* (CMG), which may be thought of as the ratio of output signal to input signal, if both inputs are tied together and excited simultaneously; hence in Fig. 2.16*a*,

$$\text{CMG} = \left|\frac{e_0}{e_i}\right| \tag{2.75}$$

The common-mode rejection ratio (CMRR or CMR) is the ratio

$$\text{CMRR} = \frac{|A_0|}{|\text{CMG}|} \qquad \text{(polarity not specified)} \tag{2.76}$$

where A_0 is the dc or low-frequency value of the normal open-loop differential input gain. *CMRR is often expressed in decibels.*

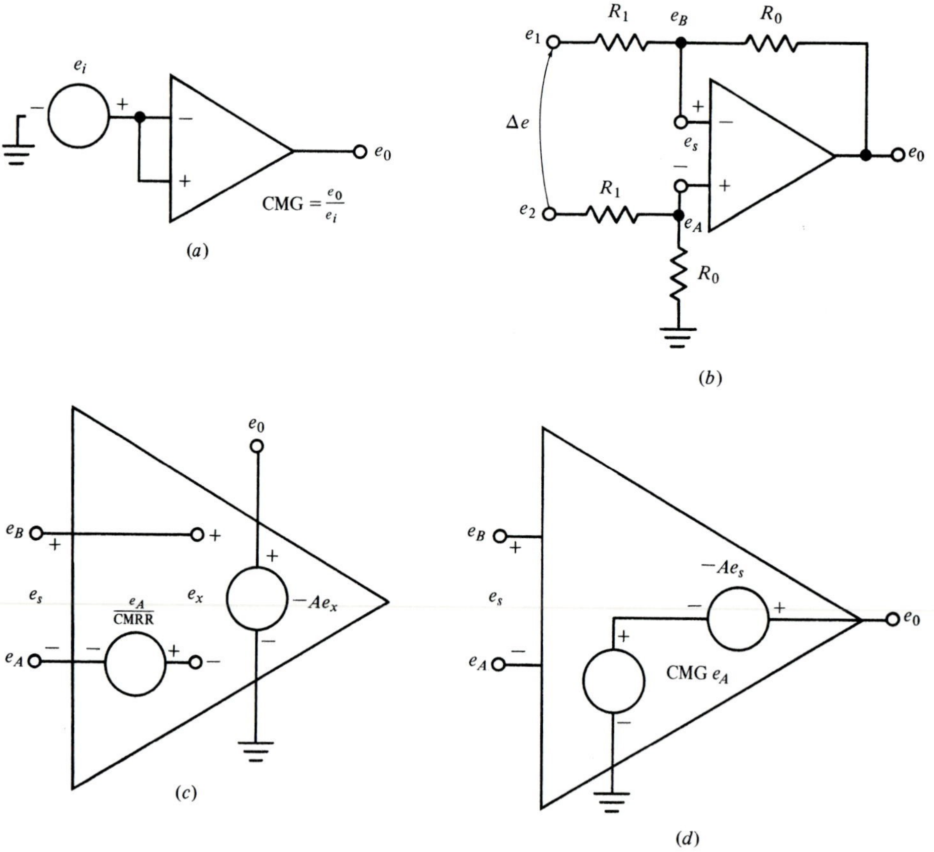

FIGURE 2.16
Analysis of common-mode gain effect. (*a*) Circuit for measuring common-mode gain; (*b*) amplifier circuit; (*c*) operational amplifier mode with common-mode effect in input circuit; (*d*) referral of common-mode gain to the output circuit.

To determine the effect of the common-mode gain on the differential amplifier performance, consider the circuit of Fig. 2.16*b*. The usual model of the amplifier is shown in Fig. 2.16*c* or, equivalently, in Fig. 2.16*d*, which implies

$$e_0 = -Ae_x = -A\left(e_s - \frac{e_A}{\text{CMRR}}\right) \tag{2.77}$$

Also

$$e_{cm} = \frac{e_1 + e_2}{2} = \text{common-mode input signal} \tag{2.78a}$$

$$\Delta e = e_1 - e_2 = \text{differential input signal} \tag{2.78b}$$

$$e_s = e_B - e_A \tag{2.79}$$

From Fig. 2.16*d* we have

$$e_0 = -(Ae_s - \text{CMG}e_A) = -A\left(e_s - \frac{e_A}{\text{CMRR}}\right) \tag{2.80}$$

$$e_B = \frac{e_0 R_1}{R_1 + R_0} + \frac{e_1 R_0}{R_1 + R_0} \tag{2.81}$$

$$e_A = e_2 \frac{R_0}{R_1 + R_0} \tag{2.82}$$

or

$$e_s = \frac{e_0 R_1}{R_1 + R_0} + \frac{\Delta e\, R_0}{R_1 + R_0} \tag{2.83}$$

From (2.80) and (2.83),

$$e_0 = -A\left(\frac{e_0 R_1}{R_1 + R_0} + \frac{\Delta e\, R_0}{R_1 + R_0}\right) + \frac{Ae_A}{\text{CMRR}} \tag{2.84}$$

or

$$e_0\left(1 + \frac{AR_1}{R_1 + R_0}\right) = \frac{-A\,\Delta e\, R_0}{R_1 + R_0} + \frac{Ae_A}{\text{CMRR}}$$

If we let

$$\beta = \frac{R_1}{R_1 + R_0} \tag{2.85}$$

then

$$e_0 = \frac{-A}{1 + A\beta}\left(\frac{R_0}{R_1 + R_0}\,\Delta e - \frac{e_A}{\text{CMRR}}\right) \tag{2.86}$$

For $|A\beta| \gg 1$,

$$e_0 \cong \frac{-1}{\beta}\frac{R_0}{R_1 + R_0}\Delta e + \frac{1}{\beta}\frac{e_A}{\text{CMRR}}$$

$$\cong -\frac{R_0}{R_1}\Delta e + \frac{R_1 + R_0}{R_1}\frac{e_A}{\text{CMRR}} \tag{2.87}$$

From (2.82) and (2.87),

$$e_0 = \frac{-R_0}{R_1}\Delta e + \frac{R_0}{R_1}\frac{e_2}{\text{CMRR}} \tag{2.88}$$

Note that we usually know only the magnitude of CMRR, not its polarity. But also

$$e_2 = e_{cm} - \frac{\Delta e}{2} \tag{2.89}$$

Thus

$$e_0 = \frac{-R_0}{R_1}\Delta e + \frac{R_0}{R_1}\frac{1}{\text{CMRR}}\left(e_{cm} - \frac{\Delta e}{2}\right) \tag{2.90}$$

Typically

$$|\Delta e| \ll |e_{cm}|$$

and

$$e_0 \cong \frac{-R_0}{R_1}\Delta e + \frac{R_0}{R_1}\frac{1}{\text{CMRR}}e_{cm} \tag{2.91a}$$

$$\frac{e_0}{e_{cm}} = \frac{R_0}{R_1}\frac{1}{\text{CMRR}} \tag{2.91b}$$

From Eq. (2.91) we see that the CMRR indicates the relative effect of the differential and common-mode signals on the output.

Noninverting Amplifiers

From Fig. 2.17 it is fairly easy to show that if

$$\beta = \frac{R_1}{R_1 + R_0} \tag{2.92}$$

then

$$\frac{e_0}{e_2} \cong \frac{R_1 + R_0}{R_1}\frac{A\beta}{1 + A\beta}\left(1 + \frac{1}{\text{CMRR}}\right) \tag{2.93a}$$

$$\cong \frac{R_1 + R_0}{R_1}\left(1 + \frac{1}{\text{CMRR}}\right) \qquad |A\beta| \gg 1 \tag{2.93b}$$

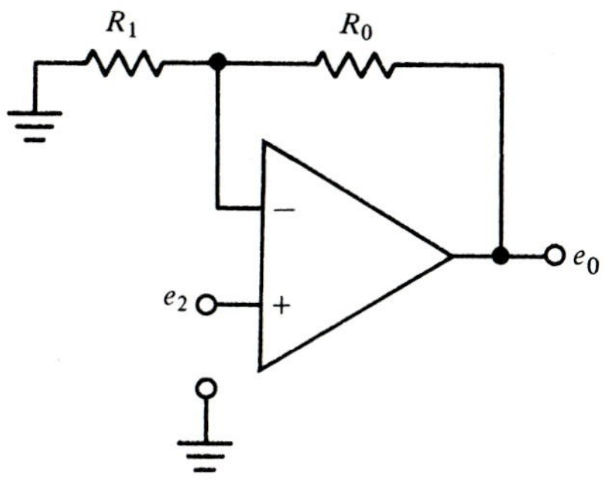

FIGURE 2.17
A simple resistive noninverting amplifier circuit (see Sec. 2.4).

In (2.93) we see an instance where the CMRR can affect the gain expression in a simple noninverting amplifier, leading to another source of gain error in addition to that predicted by the analysis in Sec. 2.2. Indeed, common-mode gain effects may be a principal source of gain inaccuracy, even in single-ended amplifier circuits.

Example 2.10. In Fig. 2.16*b*, let

$$R_0 = 10^5\ \Omega \qquad R_1 = 10^4\ \Omega \qquad \text{CMRR} = 100\ \text{dB}$$

The circuit has a differential gain

$$\frac{e_0}{\Delta e} = \frac{-10^5}{10^4} = -10$$

and a common-mode gain

$$\frac{e_0}{e_{cm}} = \frac{R_0}{R_1}\frac{1}{\text{CMRR}}$$

As a ratio

$$\text{CMRR} = 10^5$$

or

$$\frac{e_0}{e_{cm}} = 10^{-4} \qquad \text{(polarity unknown)}$$

2.5 DC OFFSETS AND TEMPERATURE DRIFT

Thus far we have been assuming that there were no internal dc offsets in the operational amplifier. In fact, there are always offsets, hopefully small, inherent in any operational amplifier, as a result of lack of symmetry in the differential stages, lack of precise bias, etc. Moreover, even though the dc offsets may be small at normal room temperature, at elevated temperatures they may be enhanced appreciably. Power supply voltage drifts may also create offsets. Thus any practical operational amplifier circuit must be designed with a view toward the possible effects of dc offsets.

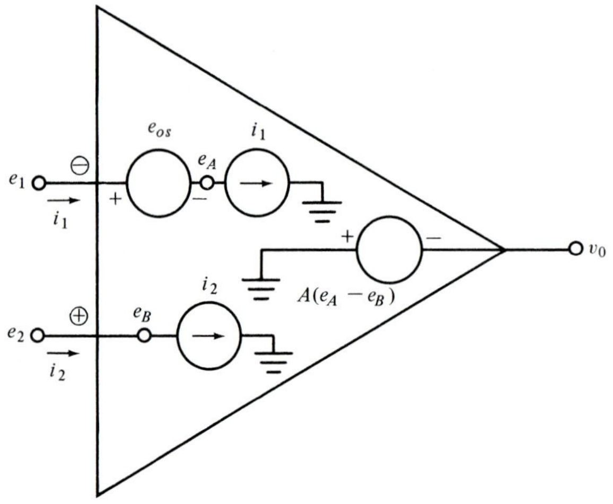

FIGURE 2.18
An equivalent circuit for representing operational amplifier internal dc offsets.

Figure 2.18 illustrates a fairly general amplifier equivalent circuit for use in treating dc offsets. Offset effects are usually described by the following parameters:

e_{os} = differential dc offset voltage (referred to the amplifier input), usually called the *input offset voltage*. It is the differential dc input voltage required to provide zero output voltage, with no other input signal and zero resistance in either input terminal path to ground.

i_1, i_2 = *input bias currents*. These are residual input currents required by the input stage transistors (base current if bipolar, gate current if FET). They are the direct biasing currents required at either input to produce zero output voltage with no input offset voltage.

In most operational amplifiers, i_1 and i_2 will be similar in magnitude and polarity and will tend to track each other with temperature or power supply voltage changes. That is, one may define an *input offset current*

$$i_{os} = i_1 - i_2 \tag{2.94}$$

Typically i_{os} is a small fraction of either i_1 or i_2. Thus, it is possible to minimize the effects of the input bias currents by using compensation techniques to be described below.

Note that we usually have no way of predicting the polarity of the offset components; hence, our estimates of dc offset effects will be worst-case bounds, which is a sensible approach. The designer should be careful to note whether the specified offsets are guaranteed maximum values or merely typical!

Figure 2.19 shows the typical environment where an operational amplifier may be operating. Resistances R_I and R_N represent the *Thévenin-equivalent dc resistances* associated with the input terminals. The feedback network provides a dc path between v_0 and e_1 which has associated with it a *dc feedback ratio* β_{dc}. That is, the Thévenin-equivalent voltage in the inverting amplifier branch has the value $\beta_{dc} v_0$. As an example, Fig. 2.20 shows a simple inverting amplifier circuit with input resistance R_1, feedback resistance R_F, and possibly an offset

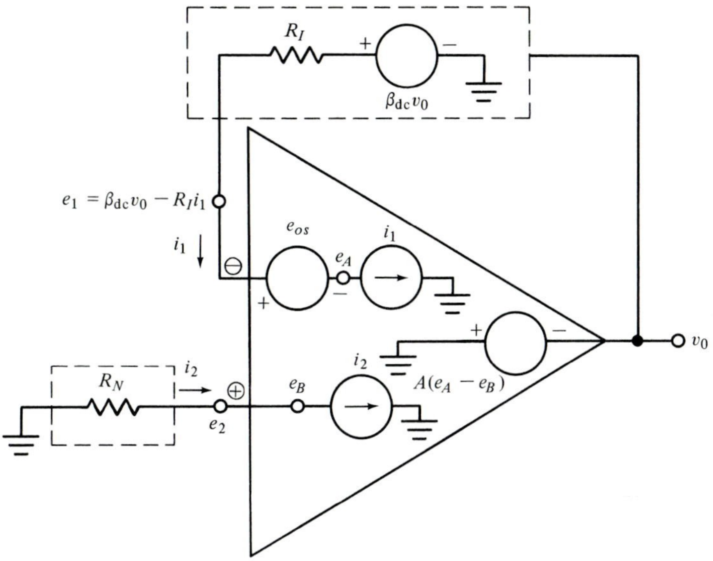

FIGURE 2.19
A general equivalent circuit for analyzing dc offset effects.

compensating resistance R_2 as the noninverting terminal connection to ground. Here R_2 is used solely to permit some compensation for input offset currents, as we shall see later. In Fig. 2.20, it is fairly easy to show that

$$\beta_{dc} = \frac{R_1}{R_1 + R_F} \tag{2.95}$$

$$R_I = R_1 \parallel R_F = \frac{R_1 R_F}{R_1 + R_F} \tag{2.96}$$

and

$$R_N = R_2 \tag{2.97}$$

In a more general circuit, one would have to find the dc values of β, R_I, and R_N for the specific circuit configuration. We emphasize that only the dc values are of interest, so in finding values for β, R_I, and R_N, all capacitors should be replaced by open circuits and inductors by short circuits.

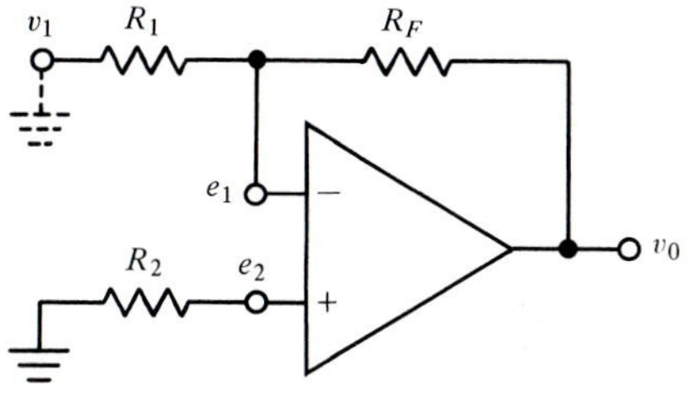

FIGURE 2.20
A simple resistive inverting amplifier circuit; note the compensation resistor in the noninverting terminal.

Basic Error Equations

Referring to Fig. 2.19, Kirchhoff's voltage law gives

$$\beta v_0 - i_1 R_I = e_{os} - R_N i_2 \tag{2.98}$$

The above assumes that the amplifier open-loop gain is very high, and therefore $|e_A - e_B| \cong 0$. We are assuming here that the amplifier is otherwise ideal, and we are not including such effects as finite open-loop gain and finite input impedance. Our results will be normally quite satisfactory, since we do not know the magnitude of the offsets with any great precision anyway.

From (2.98) we find

$$v_0 = \frac{1}{\beta}(e_{os} + i_1 R_I - i_2 R_N) \tag{2.99}$$

From (2.94), we also have

$$v_0 = \frac{1}{\beta}\,[e_{os} + i_1(R_I - R_N) + i_{os}R_N] \tag{2.100}$$

Usually we have no idea about the polarity of the offset components, but we can estimate the worst-case offset as follows:

$$|v_0| \leq \frac{1}{\beta}\,[|e_{os}| + |i_1|(R_I - R_N) + |i_{os}|R_N] \tag{2.101}$$

There are two important special cases based upon (2.101):

CASE 1: $R_N = 0$. If $R_N = 0$, (2.101) becomes

$$|v_0| \leq \frac{1}{\beta}\,(|e_{os}| + |i_1|R_I) \tag{2.102}$$

CASE 2: $R_N = R_I$. We can usually minimize the total dc offset by setting $R_N = R_I$; in this case

$$|v_0| \leq \frac{1}{\beta}\,(|e_{os}| + |i_{os}|R_I) \tag{2.103}$$

For the circuit of Fig. 2.20 we would then find from (2.103) that

$$|v_0| \leq \frac{R_1 + R_F}{R_1}\,(|e_{os}| + R_I|i_{os}|) \tag{2.104a}$$

with

$$R_2 = R_N = R_I = \frac{R_1 R_F}{R_1 + R_F} \tag{2.104b}$$

In (2.101) to (2.104) we note that the offset effects are aggravated by making β smaller. This means, in general, that operational *amplifier circuits* with a *high dc gain* have *more offset*. Also, as indicated by (2.101) and (2.104*a*), the *offset is,* in general, *increased* as we go to circuit designs with a *higher dc impedance* level. The compensation scheme of case 2 above permits us to reduce the effects of current offsets, and it is generally recommended. (Note, however, that we will find in Sec. 2.6 that adding R_2 to the circuit in Fig. 2.20 will often enhance the random noise in the circuit—another design tradeoff.)

DC Offset Balancing

For a given pair of power supply voltages and a given ambient temperature, it is possible to introduce a dc offset balancing signal into the operational amplifier circuit, so that the output voltage is zero in the absence of any normal inputs. Figure 2.21*a* illustrates one simple method. And R_N is chosen so that the conditions of case 2 above are met, i.e.,

$$R_N = R_I$$

An adjustable offset voltage is established at node e_B. Note that the impedance at node e_B is quite low and that adjustments of potentiometer R_B merely introduce a dc offset counterbalance without appreciably disturbing the dc resistance in the noninverting terminal path to ground. Typically R_3 is 10 to 100 Ω; R_4 is chosen to permit e_B to range over the anticipated offsets. Specifically, we must be able to set e_B such that

$$|e_B| > |e_{os}|_{\max} + |i_{os}|_{\max} R_N \tag{2.105}$$

Of course, the circuit of Fig. 2.21*a* will not work if the noninverting terminal must be used for other purposes, e.g., in a differential amplifier or

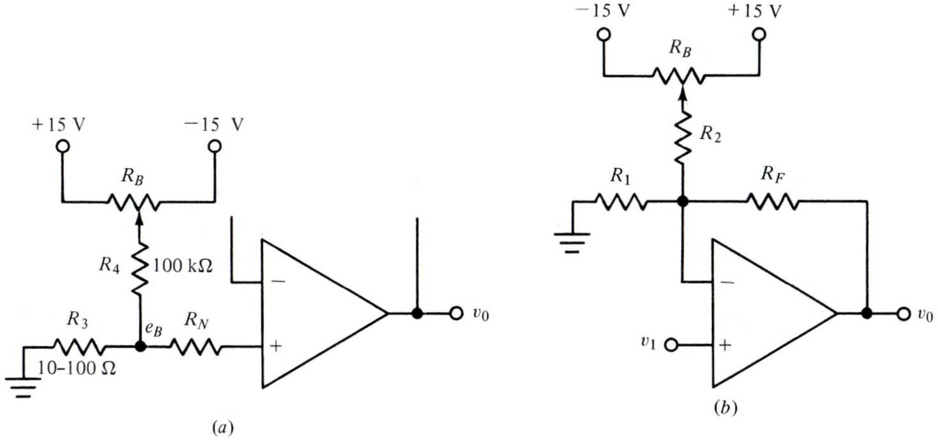

FIGURE 2.21
Typical dc offset-adjusting networks. (*a*) Offset network for the noninverting terminal; (*b*) a similar network for the inverting terminal.

a noninverting circuit. In that case, other alternatives are possible. It is often possible to inject an offset current into the inverting terminal as well, as shown in Fig. 2.21*b*. Here we have a noninverting circuit. By properly choosing R_2 and potentiometer R_B, we can inject sufficient direct current to counteract the amplifier internal offset components. Usually R_2 is chosen to be large compared to the parallel combination of R_1 and R_F. For differential amplifier circuits, it may be preferable to use an operational amplifier that has separate terminals for the addition of a dc balance adjustment network.

We should point out that recent improvements in operational amplifier fabrication have led to the development of amplifiers with very small offsets *at room temperature,* so that the dc offset problem may not be important. However, even these newer units may have significant offsets at elevated temperatures (for example, 150°F).

Offset Drifts

As the ambient temperature in the vicinity of the operational amplifier changes, the dc offset components will change. Even though the dc offsets have been balanced out for a specific set of operating conditions, changes in temperature or power supply voltage will generally cause changes in e_{os}, i_1, i_2, and i_{os}. We may then need to estimate the effect of these changes. Assuming no dc offset at some point, if the offset components undergo a change, the resulting output dc offset in a circuit fitting the general model of Fig. 2.19 will be

$$|\Delta v_0| \leq \frac{1}{\beta_{\text{dc}}} [|\Delta e_{os}| + |\Delta i_1|(R_I - R_N) + |\Delta i_{os}|R_N] \tag{2.106}$$

Again, the compensation scheme of setting $R_N = R_I$ yields

$$|\Delta v_0| \leq \frac{1}{\beta_{\text{dc}}} (|\Delta e_{os}| + |\Delta i_{os}|R_I) \tag{2.107}$$

Table 2.2 shows some typical dc offset parameters for today's operational amplifiers. Special units with low offset current are available; however, they often have reduced bandwidth, so that a design compromise may be involved. It is safe to say that almost daily we will see announcements of new amplifiers with improved offset drift characteristics. Note that, for amplifiers with bipolar transistors in the input stage, an offset-current drift-current coefficient in amperes per degree Celsius is often stated, which may be fairly valid over a reasonable temperature range. For amplifiers with FET input stages, more typically the offset-current components will double for a 10°C change. Both types will typically exhibit voltage offset drifts of 0.5 to 20 μV/°C.

Example 2.11. Refer to the inverting amplifier circuit of Fig. 2.20. Suppose we want to design a gain-of-10 circuit. Use the specifications for the poor-quality IC

TABLE 2.2
Typical amplifier dc offset specifications

	Discrete bipolar	Discrete, FET input	Monolithic IC (low-cost)	Monolithic IC, FET input
Offset voltage e_{os}†	±0.2 mV	±1 mV	±5 mV	±3.5 mV
Input bias current i_1 or i_2†	±20 nA	−20 pA‡	50 nA‡	−10 pA‡
Input offset current i_{os}†	±2 nA	±10 pA	±5 nA‡	±5 pA
Temperature drift of voltage offset $\partial e_{os}/\partial T$	±0.5 μV/°C	±20 μV/°C	±5 μV/°C	±10 μV/°C
Temperature drift of input current $\partial i_1/\partial T$	±0.2 nA/°C	§	± 0.5 nA/°C	§
Temperature drift of input offset current $\partial i_{os}/\partial T$	±0.02 nA/°C	§	±0.05 nA/°C	§

† At 25°C.

‡ Often i_1 and i_2 have known polarity, but the polarity of i_{os} is not known.

§ Doubles over 10°C change.

amplifier given below. Pick $R_F = 100\ \text{k}\Omega$ and estimate the inherent offset with and without R_N. Here $\beta = \frac{1}{11}$, $R_F = 10^5\ \Omega$, $R_I = 10^4\ \Omega$, and thus $R_I = 9.09\times10^3\ \Omega$. Let

$$\frac{\partial i_1}{\partial T} = \frac{\partial i_2}{\partial T} = \pm 1\ \text{nA}/^\circ\text{C} \qquad i_1, i_2 = \pm 500\ \text{nA}$$

$$\frac{\partial i_{os}}{\partial T} = \pm 0.5\ \text{nA}/^\circ\text{C} \qquad i_{os} = \pm 50\ \text{nA}$$

$$\frac{\partial e_{os}}{\partial T} = \pm 5\ \mu\text{V}/^\circ\text{C} \qquad e_{os} = \pm 5\ \text{mV}$$

First, with $R_N = 0$, from (2.102).

$$|v_0| \le 11(5\times10^{-3} + 5\times10^{-7}\times 9.09\times10^3)$$
$$\le 0.105\ \text{V}$$

Picking $R_N = R_I = 9.09\times10^3\ \Omega$,

$$|v_0| \le 11(5\times10^{-3} + 5\times10^{-8}\times 9.09\times10^3)$$
$$\le 0.06\ \text{V}$$

The compensation scheme *is* worthwhile here.

If we make $R_F = 10^4\ \Omega$, then $R_I = 9.09\times10^2\ \Omega$; with $R_N = 0$,

$$|v_0| \le 11(5\times10^{-3} + 5\times10^{-7}\times 9.09\times10^2)$$
$$\le 0.060\ \text{V}$$

and with $R_N = R_I$,

$$|v_0| \leq 0.055 \text{ V}$$

that is, R_N does little good.

If the circuit was otherwise balanced so that $v_0 = 0$ at, say, 25°C, then at 50°C

$$\Delta e_{os} = \pm 1.25\ \mu\text{V}$$

$$\Delta i_1 = \Delta i_2 = \pm 25 \text{ nA}$$

$$\Delta i_{os} = 12.5 \text{ nA}$$

With $R_F = 100$ kΩ and $R_N = 0$, the output voltage drift can be estimated from (2.106).

$$|\Delta v_0| \leq 11(125 \times 10^{-6} + 9.09 \times 10^3 \times 25 \times 10^{-9})$$

$$\leq 3.88 \text{ mV}$$

The addition of $R_N = R_I$ reduces the drift to

$$|\Delta v_0| \leq 11(125 \times 10^{-6} + 9.09 \times 10^3 \times 125 \times 10^{-9})$$

$$\leq 2.62 \text{ mV}$$

a modest improvement.

Chopper-Stabilized and Varactor Diode Amplifiers

Since the early days of vacuum-tube operational amplifiers, the problem of dc offset and offset drift has always been present. At present, one can often use the types of low-offset units described in Table 2.2 with satisfactory results. There are still situations, however, where wide temperature variations or extreme requirements for dc offset reduction make it necessary to employ special amplifiers. These usually use modulated carrier techniques to transmit the dc and very-low-frequency components, while the high-frequency components may be transmitted through the amplifier with ac-coupled techniques (see Ref. 4, Sec. 4.4). Two principal types of units are readily available in discrete form: those using a vibrating reed or solid-state *chopper* channel for dc offset sensing and stabilization, and *varactor* input amplifiers. These units are relatively expensive; however, integrated-circuit amplifiers with chopper stabilization are just now becoming available. This should make it easier to employ these techniques where required.

2.6 AMPLIFIER INTERNAL NOISE EFFECTS

In addition to the dc offsets discussed in the previous section (which could be considered a form of unwanted signal, or noise), an amplifier's output will always contain additional spurious components, which we collectively call *noise*. In many cases, much of this noise comes from *external sources*. In this category are spurious signals coupled into the input circuit or ground return (60-Hz hum or other electromagnetically coupled induced noise), thermally generated Johnson noise in circuit resistances, or noise introduced from the power supply (e.g., due to excessive ripple). We have more to say about techniques for eliminating these forms of external interference in Sec. 2.9. In this section we are concerned with *internal amplifier noise* produced by random phenomena, primarily in the amplifier input stage. We do not describe the noise mechanisms in detail; Refs. 4 and 5 may be consulted for more information on the nature of the noise sources. We merely categorize the noise sources as follows:

1. Broadband white noise with an essentially constant noise spectrum, primarily due to Schottky and Johnson noise in the input transistor stage.
2. Low-frequency ($1/f$) noise with a power spectrum that is inversely proportional to frequency. Typically this noise is not important above about 10 Hz, but it can be important at very low frequencies in special applications.
3. Burst or popcorn noise with a power spectrum of the form

$$\frac{K}{1 + (f/f_b)^2}$$

   Typically this noise is not important above 200 to 300 Hz. It is not always apparent, but has been found in some silicon planar-diffused bipolar integrated-circuit transistors (see Ref. 10).

Figure 2.22 shows a useful model for representing internal amplifier noise effects. The noise phenomena are represented in terms of equivalent input noise voltage and current sources e_n and i_n. Also important is β, the feedback ratio between the output terminal and the inverting input terminal. (Now in general we must consider the frequency-dependent nature of β.) The Thévenin-equivalent impedances in the inverting and noninverting terminal branches Z_I and Z_N are also important. Note that the symbol Z_I has a different meaning here than in Secs. 2.1 to 2.3, where Z_I stood for the differential input impedance of the operational amplifier.

Spectral Description of Noise

At this point we must explore the means by which noise is specified. We assume that the noise signals E_n and I_n (Fig. 2.22) are *gaussian stationary random processes,* which may be *specified in terms* of either their *correlation function* or their *power spectral density* (see, e.g., one of the many good texts on random process theory, such as Ref. 9). We cannot elaborate on these concepts here, but we present principal design equations. Figure 2.23*a* shows a typical sample

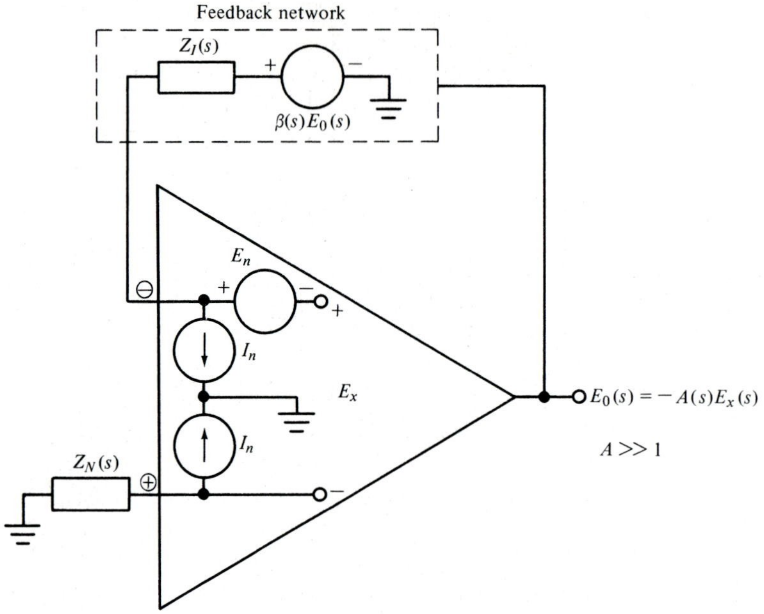

FIGURE 2.22
A general equivalent circuit for analyzing operational amplifier noise effects.

of a random noise signal. Here we assume that the noise is stationary (i.e., the statistical properties do not change with time); thus we can describe the general nature of the noise signal by its *spectral* properties. We also assume that the signals can be described as having a gaussian amplitude distribution with zero average or mean value (Fig. 2.23*b*); hence we are able to describe the overall noise intensity rather well by specifying its rms value (i.e., the standard deviation of the noise amplitude). Actually, the rms value that is observed for a given noise signal depends on the bandwidth (frequency range) observed. For the present discussion, let us describe the noise by a *one-sided power spectral density*.

Let $G(f)$ be the one-sided power spectral density of a stationary random signal $x(t)$ with zero average value. If we observe the process with a unity gain "brickwall" filter which passes frequencies from zero up to some upper frequency F, then we will observe (Fig. 2.24*a*) the rms value

$$X_{\text{rms}} = \sqrt{\int_0^F G(f)\,df} \tag{2.108}$$

or conversely,

$$G(f) = \frac{d}{df}\,x_{\text{rms}}^2 \tag{2.109}$$

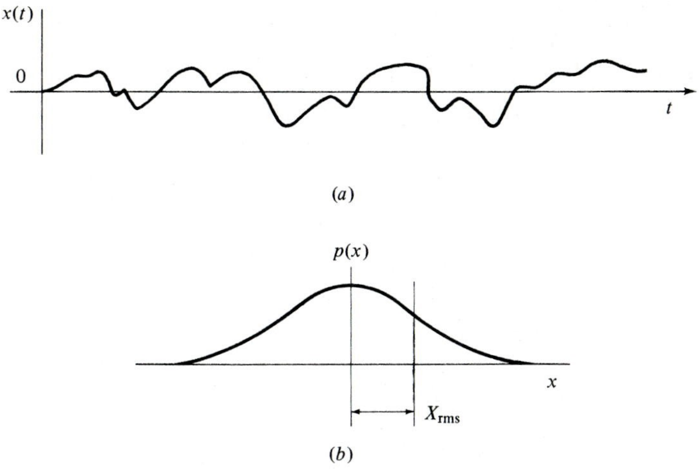

FIGURE 2.23
A typical noise signal. (*a*) A sample waveform; (*b*) probability density of the amplitude.

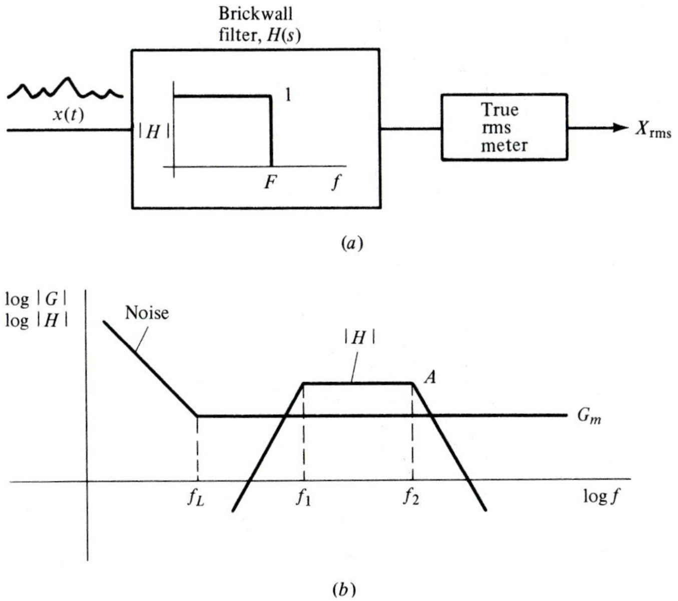

FIGURE 2.24
Effect of filtering on noise. (*a*) Use of a brick wall filter to measure power spectral density, Eqs. (2.108) and (2.109); (*b*) rejection of noise outside the passband f_1 to f_2, Eq. (2.112).

Usually we are interested in the *apparent rms value* of a random signal, as observed over a *finite frequency range,* i.e., from f_1 to f_2, since no physical system can respond to all frequencies. In this case, the apparent rms value of the signal is given by

$$X_{\text{rms}} = \sqrt{\int_{f_1}^{f_2} G(f)\,df} \tag{2.110}$$

Figure 2.24*b* shows a typical power spectral density, with a low-frequency component that increases below some frequency f_L (typically 1 to 10 Hz) and a "white noise" component at higher frequencies. If a filter with transfer function

$$H(j\omega) = H(j2\pi f)$$

is used to observe the noise, then the apparent rms value of the noise will be dependent on the filter bandwidth. In this case, if the filter output is y, then

$$y_{\text{rms}} = \sqrt{\int_{f_1}^{f_2} |H(j2\pi f)|^2 G(f)\,df} \tag{2.111}$$

We can often approximately evaluate this last interval from a knowledge of the frequency range where the major contribution of noise is present. For example, in Fig. 2.24*b* the filter function H passes primarily frequencies in the range from f_1 to f_2, with gain A and $f_1 > f_L$. Thus,

$$y_{\text{rms}} \cong \sqrt{\int_{f_1}^{f_2} |H|^2 G\,df} \tag{2.112a}$$

$$\cong \sqrt{(f_2 - f_1)A^2 G_m} \cong A\sqrt{f_2 G_m} \tag{2.112b}$$

where A and G_m are as indicated in Fig. 2.24*b*. This last estimate will be somewhat in error, but often we are looking for an order of magnitude anyway.

Note: The rms noise is proportional to

$$\sqrt{f_2 - f_1} \cong \sqrt{f_2} \qquad f_1 \ll f_2$$

General Design Equations

Using the equivalent noise model of Fig. 2.22, we can show that the rms noise voltage at the amplifier output is

$$E_{0,\text{rms}} = \sqrt{\int_{f_1}^{f_2} A_n^2 |G_e + G_i(|Z_I|^2 + |Z_N|^2)|\,df} \tag{2.113}$$

where G_e = input noise voltage power spectral density associated with generator E_n, V^2/Hz
G_i = input noise current power spectral density associated with I_n, A^2/Hz
A_n = noise gain of circuit, V/V

The factor

$$G_e + G_i(|Z_I|^2 + |Z_N|^2) \tag{2.114}$$

is an *equivalent input noise voltage power spectral density* which takes into account both the voltage and the current components of the input noise. We see from (2.114) that the amplifier output noise will be affected by the impedance level of the associated circuit and that the noise can be reduced by keeping

$$G_i(|Z_I|^2 + |Z_N|^2) < G_e \tag{2.115}$$

The noise gain A_n is in general

$$A_n = \frac{1}{\beta}\frac{A\beta}{1 + A\beta} \tag{2.116}$$

In many cases, the effective bandwidth of the noise gain is small compared to the operational amplifier bandwidth itself, and we can assume $A \gg 1$ throughout the frequency range of interest. If this is the case,

$$A_n \cong \frac{1}{\beta} \qquad A \gg 1 \tag{2.117}$$

As we have seen before, $1/\beta$ is approximately proportional to the circuit voltage gain [see, e.g., (2.61), (2.65), and Example 2.4]. Hence amplifier circuits with a high voltage gain will have a greater output noise level, as expected. It is obvious from (2.113) that one should shape the amplifier closed-loop transfer function to transmit only those signals that really have important spectral content and to reject noise outside of the frequency range of interest.

One should not normally attempt to evaluate an integral of the form of (2.113) with $f_1 = 0$, since that implies that an infinitely long time of observation is involved. In particular, if the noise has a $1/f$ component, then since

$$\int_0^F \frac{K}{f}\,df = \infty$$

we will get an estimate of infinite rms noise!

A more reasonable approach is to use a fairly small value of f_1, for example, 0.001 or 0.01 Hz, in evaluating integrals of the (2.113) type. Lower values of f_1 imply very long observation times, while other types of drifts due to aging, temperature changes, etc., are probably more important in creating very-low-frequency error sources, and a dc offset analysis is more appropriate.

Also one need not use a value of f_2 greater than the frequency where A_n becomes small. At high frequencies, one should limit f_2 to the frequency where $|A\beta|$ finally drops below 1.

Amplifier Noise Specifications

There are a variety of ways in which manufacturers specify the noise properties of their amplifiers. Most will give separate voltage and current parameters. Some will give details of the spectral density and its variations with frequency. More commonly, however, one finds specifications for the effective rms value of the noise over a range of frequencies. In addition, one will find separate figures for a low-frequency range (for example, 0.01 to 10 Hz), where the noise is presumably of the $1/f$ variety, and a mid-frequency range (for example, 10 Hz to 10 kHz), where the noise spectral density is typically white, i.e., of constant intensity. We attempt here to interpret some typical specifications in terms of the terminology of this section.

Table 2.3 lists a typical set of manufacturer's specifications for amplifier input noise. Implied in these specifications is a noise voltage power spectral density of the form

$$G_e(f) = \frac{K_{e1}^2}{f} + K_{e2}^2 \tag{2.118a}$$

and a current power spectral density

$$G_i(f) = \frac{K_{i1}^2}{f} + K_{i2}^2 \tag{2.118b}$$

i.e., there is a $1/f$ component implied in the low-frequency specification and a constant component implied in the mid-frequency specification. This may be only approximately true, but this is about all we can infer from the given specifications. Note that it is common to give the low-frequency specification in terms of a peak-to-peak value, but the mid-frequency specification is usually given in terms of an rms value. Lacking any other information, we may assume that the peak-to-peak value is 6 to 7 times the rms value; let us use 6.5 times here. (Peak-to-peak value is usually the range wherein the signal lies P percent of the time, where P is typically between 99 and 99.9 percent.) If we assume a $1/f$ variation for the low-frequency spectrum and a white spectrum for the mid-frequency spectrum, then we can find the constants in (2.118) as follows:

Let E_{n1} and I_{n1} be the low-frequency range (f_1 to f_2) peak-to-peak specifications of noise voltage and current, and let E_{n2} and I_{n2} be the mid-frequency

TABLE 2.3
Typical amplifier noise specifications*

Frequency range	Voltage, μV	Current, pA
0.01 to 10 Hz	0.5 p-p	20 p-p
10 Hz to 10 kHz	2 rms	2 rms

* Low-frequency specification is typically given in terms of peak-to-peak value, mid-frequency in terms of rms value.

range (f_3 to f_4) rms specifications. Then

$$E_{n1} = \sqrt{\int_{f_1}^{f_2} \frac{K_{e1}^2}{f} (6.5)\, df} \qquad \text{peak to peak}$$

or

$$K_{e1} = \frac{E_{n1}}{6.5\sqrt{\ln(f_2/f_1)}} \tag{2.119a}$$

Similarly,

$$K_{i1} = \frac{I_{n1}}{6.5\sqrt{\ln(f_2/f_1)}} \tag{2.119b}$$

also

$$E_{n2} = \sqrt{\int_{f_3}^{f_4} K_{e2}^2\, df} \qquad \text{rms}$$

or

$$K_{e2} = \frac{E_{n2}}{\sqrt{f_4 - f_3}} \cong \frac{E_{n2}}{\sqrt{f_4}} \tag{2.120a}$$

and

$$K_{i2} = \frac{I_{n2}}{\sqrt{f_4 - f_3}} \cong \frac{I_{n2}}{\sqrt{f_4}} \tag{2.120b}$$

Example 2.12. Given the specifications of Table 2.3, estimate the power spectral density of the amplifier noise voltage and current. Here we want to take the original specifications and convert them to the constants of (2.118). The low-frequency noise constants K_{e1} and K_{i1} are found by using (2.119); that is,

$$K_{e1} = \frac{0.5 \times 10^{6}}{6.5\sqrt{\ln(10/0.01)}}$$

$$= \frac{0.6 \times 10^{-6}}{6.5 \times 2.63}$$

$$= 2.92 \times 10^{-8}\ \text{V}$$

and

$$K_{ij} = \frac{20 \times 10^{-12}}{6.5 \times 2.63}$$

$$= 1.17 \times 10^{-12}\ \text{A}$$

From (2.120) we estimate the mid-frequency parameters K_{e2} and K_{i2}:

$$K_{e2} = \frac{2 \times 10^{-6}}{\sqrt{10^4 - 10}} = 2 \times 10^{-8}\ \mathrm{V}/\sqrt{\mathrm{Hz}}$$

and

$$K_{i2} = \frac{2 \times 10^{-12}}{\sqrt{10^4 - 10}} = 2 \times 10^{-14}\ \mathrm{A}/\sqrt{\mathrm{Hz}}$$

Note the units of the above parameters. The low-frequency parameters ($1/f$ noise) are correctly specified in volts or amperes; the mid-frequency (white noise) parameters are correctly specified in volts per root-hertz or amperes per root-hertz.

The power spectral density functions are then obtained from (2.118):

$$G_e = \frac{8.4 \times 10^{-16}}{f} + 4 \times 10^{-16}\ \mathrm{V}^2/\mathrm{Hz} \tag{2.121a}$$

$$G_i = \frac{1.4 \times 10^{-24}}{f} + 4 \times 10^{-28}\ \mathrm{A}^2/\mathrm{Hz} \tag{2.121b}$$

These functions may be used in (2.113) to estimate the rms output noise of a particular circuit.

We note here that a model for noise spectral density such as is implied by (2.118) may be considerably in error. In particular, we have not treated burst or popcorn noise in the simple model of Eq. (2.118). Its spectrum is usually of the form

$$\frac{K_{e3}}{1 + (f/f_b)^2}$$

where f_b is typically 200 to 300 Hz (see Ref. 10). In critical low-noise applications, proper selection procedures should be instituted to ensure that units with significant burst noise are rejected.

To make realistic estimates of the noise effects in a specific circuit with a specific amplifier, one should have detailed plots of the input noise voltage and current spectra (usually given in volts per root-hertz and amperes per root-hertz), i.e., plots of $\sqrt{G_e}$ and $\sqrt{G_i}$ versus f. Without such detailed information, one can make only very rough estimates. Reference 5 presents an excellent description of the use of graphical methods for estimating effects, given the details of the input noise spectral density; we do not present these techniques here, since this type of detailed information is often unavailable. We do, however, present one example of a typical analysis, based on equations of the form of (2.113) and (2.118).

Example 2.13. Suppose we have the simple inverting low-pass filter circuit of Fig. 2.25*a*, to be used with the amplifier having the noise specifications of Table 2.3. The circuit impedance level will be determined by the value of R; then C is chosen to give a 10-kHz cutoff frequency. Let us estimate the output rms noise, assuming

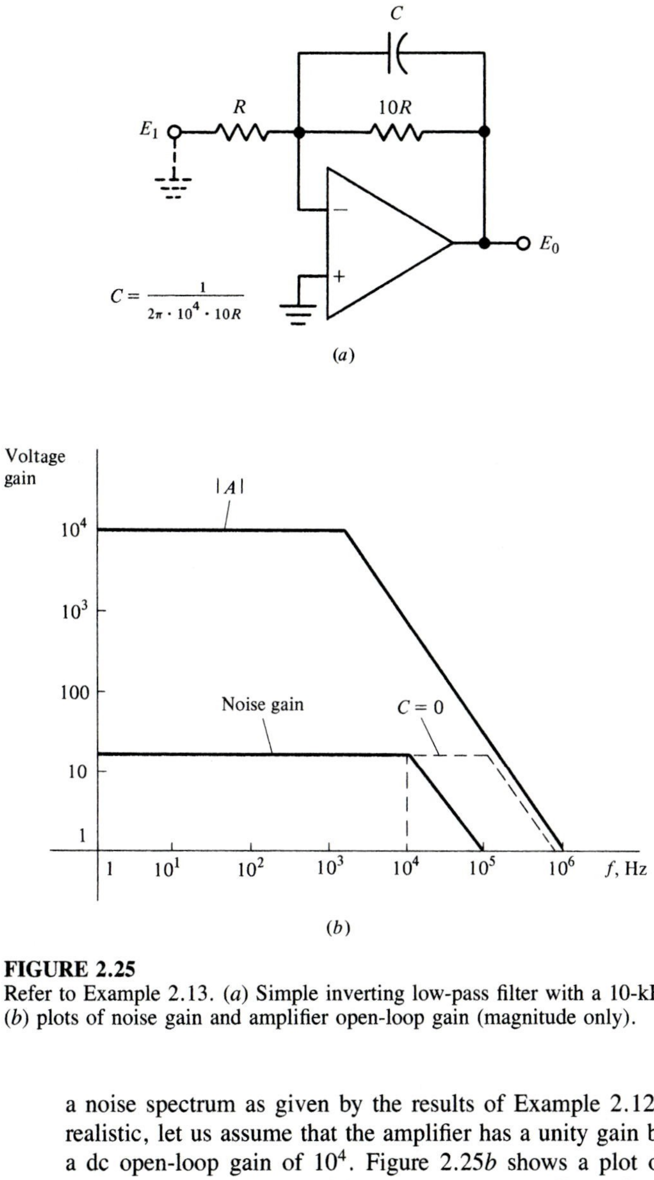

FIGURE 2.25
Refer to Example 2.13. (*a*) Simple inverting low-pass filter with a 10-kHz cutoff frequency; (*b*) plots of noise gain and amplifier open-loop gain (magnitude only).

a noise spectrum as given by the results of Example 2.12, i.e., in (2.112). To be realistic, let us assume that the amplifier has a unity gain bandwidth of 1 MHz and a dc open-loop gain of 10^4. Figure 2.25*b* shows a plot of the noise gain (heavy curve). Note that the amplifier open-loop gain affects the noise gain curve at high frequencies.

To *simplify* the estimation of the noise output, we approximate the noise gain as follows:

$$A_n \cong \begin{cases} 11 & 10^{-3} < f < 10^4 \\ 1 & 10^4 < f < 10^6 \\ 0 & \text{otherwise} \end{cases} \tag{2.122}$$

This seemingly crude approximation will be acceptable here, since we only have a rough idea of the input noise spectrum to begin with.

We will now proceed to evaluate (2.113) for this circuit. To facilitate this, we use the following relationships:

$$\int_{f_a}^{f_b} \frac{K^2\,df}{f} = K^2 \ln \frac{f_b}{f_a} \tag{2.123a}$$

or if

$$f_b = 10N f_a$$

$$\int_{f_a}^{10Nf_a} \frac{K^2\,df}{f} = 2.3K^2 \ln N \tag{2.123b}$$

also

$$\int_{f_a}^{f_b} K^2\,df = (f_b - f_a)K^2 \tag{2.123c}$$

and

$$\int_{f_a}^{f_b} K^2\,df \cong f_b K^2 \qquad f_b > 10 f_a \tag{2.123d}$$

The equivalent series impedance in the inverting terminal branch is approximately

$$Z_I \cong \begin{cases} \frac{10}{11}R & f < 10^4 \\ 0 & 10^4 < f \end{cases} \tag{2.124}$$

Again, we are using a fairly crude representation, since we are seeking a rough estimate.

Using (2.121) through (2.124) in (2.113), we have

$$\begin{aligned} E_0^2 &\cong \left(11^2 \ln \frac{10^4}{10^{-3}} + 1^2 \ln \frac{10^6}{10^4}\right) \times 8.4 \times 10^{-16} \\ &\quad + \left(11^2 \ln \frac{10^4}{10^{-3}}\right) \times 1.4 \times 10^{-24} \left(\frac{10}{11}\right)^2 R^2 \\ &\quad + [11^2(10^4 - 10^{-3}) + 1(10^6 - 10^4)] \times 4 \times 10^{-16} \\ &\quad + 11^2(10^4 - 10^{-3}) \times 4 \times 10^{-28} \left(\frac{10}{11}\right)^2 R^2 \\ &\cong (121 \times 7 \times 2.3 + 1 \times 2 \times 2.3) \times 8.4 \times 10^{-16} \\ &\quad + (121 \times 7 \times 2.3) \times 1.16 \times 10^{-24} R^2 \\ &\quad + (1.21 \times 10^6 + 10^6) \times 4 \times 10^{-16} \\ &\quad + 1.21 \times 10^6 \times 3.31 \times 10^{-28} R^2 \end{aligned}$$

or

$$E_0^2 \cong 1.64 \times 10^{-12} + 2.26 \times 10^{-21} R^2 + 8.84 \times 10^{-10} + 4 \times 10^{-22} R^2$$

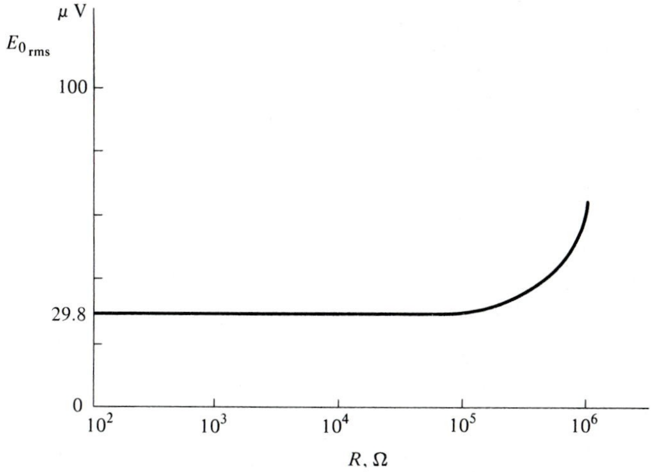

FIGURE 2.26
Effect of impedance level on the output noise produced by the circuit of Fig. 2.25.

Thus

$$E_0^2 \cong 8.86 \times 10^{-10} + 0.266 \times 10^{-20} R^2 \qquad \mathrm{V}^2 \tag{2.125}$$

Figure 2.26 shows a plot of E_0 versus R. We note that for values of R less than about 10^5 Ω, the output noise is 29.8 μV rms. As R is increased above the value, the effect of the input *current noise component becomes important.*

From Example 2.13 we see that the choice of impedance level affects the noise level. Note also that if we introduce a dc offset compensating resistor ($\frac{10}{11}R$) in the noninverting terminal, we increase the noise level, i.e., we replace R^2 by $2R^2$ in (2.125).

If the circuit of Fig. 2.25 must be built with large resistance values, then it is advisable to use an amplifier with a lower noise current. Amplifiers with an FET or varactor input stage are usually preferred for this case. Note also that if the capacitor is omitted in Fig. 2.25, the output noise will be greater (see Prob. 2.19). It is important to restrict the bandwidth of amplifiers to only the frequency range where useful signal components lie in order to reduce the effects of amplifier noise.

2.7 LARGE-SIGNAL LIMITATIONS

In this section we discuss a variety of nonideal amplifier performance measures that relate to the amplifier large signal. These include output limitations such as slewing rate, full-power bandwidth, settling time, overload recovery, and rated common-mode input voltage. Normally, the specifications are derived from large-signal testing of typical units, and they may not apply with high accuracy to all circuit configurations.

Rated Output

We mentioned the basic limitations on amplifier output load-driving capabilities as early as Chap. 1; they seldom can be overlooked. These ratings specify the *maximum* allowable output *voltage swing* (at low frequencies). The rating is usually given under the assumption that an output load is connected which causes the *maximum* or peak *output current* to flow at the rated peak output voltage. Often the output voltage swing is a result of the output current-limiting effects, and the two output signal ratings are interdependent. Complete specifications should include a graph of peak voltage versus peak load current. In many cases, however, the manufacturer will only specify the rated output current (assumed $\pm$) over a maximum nominal output voltage range, for example, ± 20 mA and ± 10 V. The ± 10-V range is typical of most contemporary operational amplifiers, but the current rating will vary greatly, depending on the intended range of applications (and cost) of the unit.

If the rated output is exceeded, the amplifier will go into an *overload* condition. If the amplifier output stage is properly designed (usually called short-circuit overload protection), then no permanent damage will occur; but there will, of course, be a temporary high distortion of the output signal, and there may be an appreciable overload recovery time. Note that if an amplifier is allowed to remain in an overload condition for a long time, internal heat buildup may cause the input stage to develop significant internal dc offsets, which may take a long time to subside.

Slewing Rate and Full-Power Response

These limitations are primarily associated with internal rate limitations in the amplifier. Typically the amplifier input stage can supply only a certain maximum amount of current to charge associated stray capacitances. The general result is that the large-signal dynamic behavior of the amplifier is degraded, as compared with the small-signal response. The circuit of Fig. 2.27 may be used to measure this effect; test results are usually summarized by two specifications: slewing rate and full-power bandwidth.

If the circuit of Fig. 2.27*a* is used with a ± 10-V square wave (with fast rise time), we get a rigorous test for slewing rate. As shown in Fig. 2.27*b*, the amplifier output waveform will exhibit a rate limitation. The maximum rate of change of the output voltage under these conditions is often taken as a measure of slewing rate, i.e.,

$$S = \left|\frac{de_0}{dt}\right|_{\max} \qquad \text{V/s} \tag{2.126}$$

Usually the positive and negative maximum rates will differ, in which case the lower slewing rate should be specified.

Full-power bandwidth may also be measured with the circuit of Fig. 2.27*a*, but using a variable-frequency sine wave input and adjusting the input amplitude

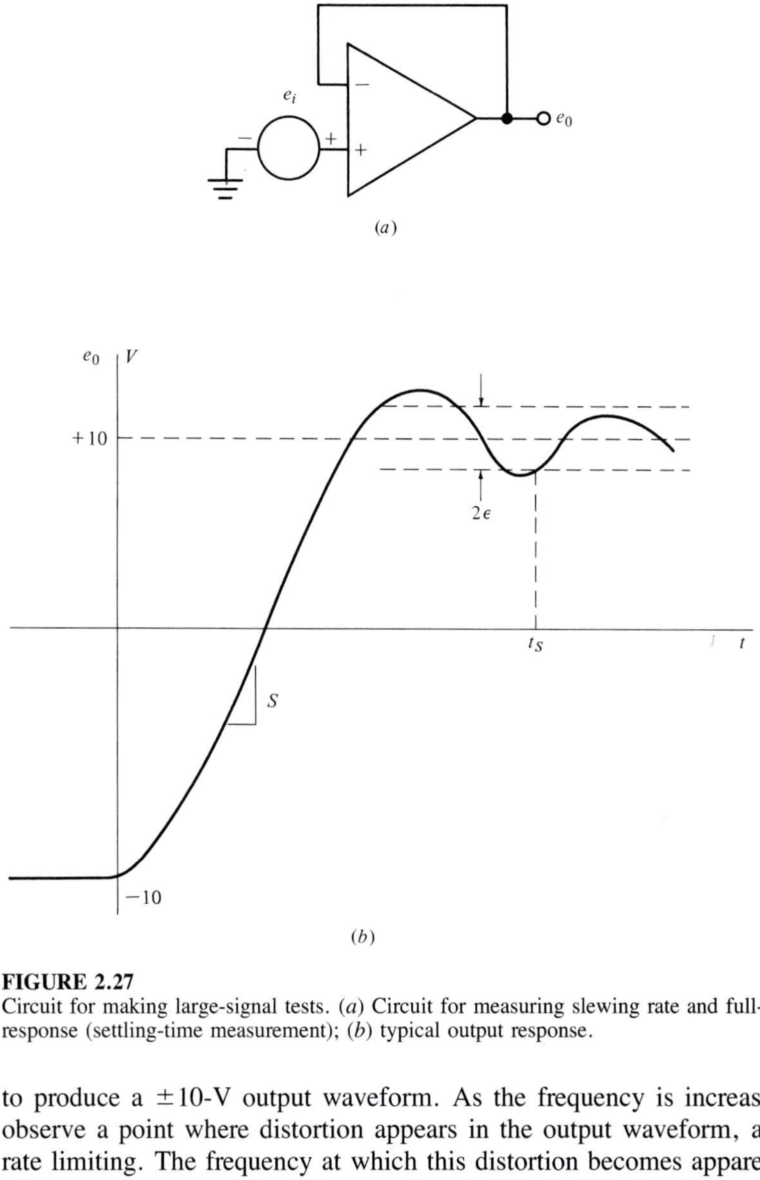

FIGURE 2.27
Circuit for making large-signal tests. (*a*) Circuit for measuring slewing rate and full-power response (settling-time measurement); (*b*) typical output response.

to produce a ±10-V output waveform. As the frequency is increased, one will observe a point where distortion appears in the output waveform, as a result of rate limiting. The frequency at which this distortion becomes apparent is usually taken as a measure of the *full-power bandwidth* f_p. The slewing rate provides an *upper bound* on this figure, i.e.,

$$f_p \leq \frac{S}{2\pi E_{op}} \qquad \text{Hz} \tag{2.127}$$

where E_{op} is the peak output swing (for example, 10 V or less). Hence, (2.127) may be used to estimate the maximum large-signal frequency that may be accommodated at a given output voltage level.

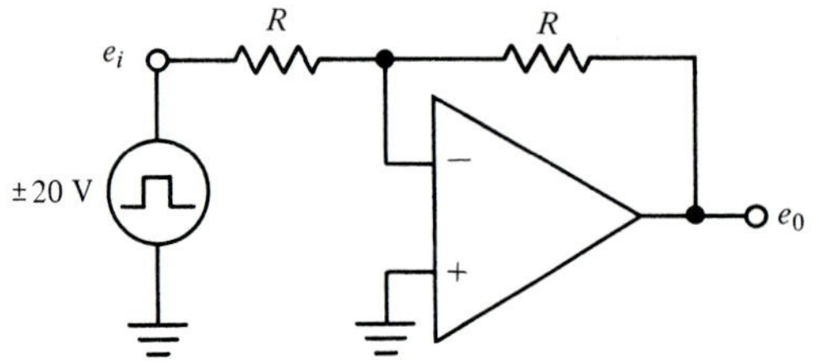

FIGURE 2.28
Measuring overload recovery; input is driven to produce 100 percent overload.

Note that (2.127) is an inequality; in many amplifiers approximate equality holds, while in others which have been specially designed for fast step response, f_p may be somewhat less than the value implied by (2.127).

Example 2.14. Given a ±10-V amplifier with a slewing rate of 10^6 V/s, estimate the full-power bandwidth. Also estimate the maximum frequency at which one can output a 2-V peak sine wave.

From (2.127), we find f_p with $E_{op} = 10$ V:

$$f_p = \frac{10^6}{2\pi \times 10} = 15.9 \text{ kHz}$$

Similarly, with $E_{op} = 2$ V,

$$f_p = 80 \text{ kHz}$$

Settling Time and Overload Recovery

In the slewing rate test of Fig. 2.27*b*, we often observe a small oscillatory transient associated with the large-signal step response of the amplifier. The *settling time* t_s is the total time t_s required for the output to settle to within a specified percentage of its final value. Thus, in Fig. 2.27*b*, if the desired percentage is 0.1 percent (of 10 V), then ϵ is 0.01 V. Settling time is an important specification for amplifiers which are to be used in a switching mode (e.g., in D/A converters, sample-hold circuits, etc.).

Additional useful information is conveyed by a specification of overload recovery time. This is the time the amplifier takes to return to linear operation after being driven into overload. For example, if the pulse response of the circuit of Fig. 2.28 is measured with the input increased so that overdrive is produced, then additional delay due to overload recovery will be observed in the output response. Typically, the overload recovery is measured using 100 percent overdrive.

Maximum Common-Mode Input Voltage

This is a specification of the maximum magnitude of voltage that may be applied to either input terminal of a differential-input operational amplifier. In Fig. 2.16*a*, this would be the maximum value of e_i that still yielded reasonably linear operation. Typical values will be ±10 to 11 V.

2.8 OUTPUT POWER BOOSTING

Occasionally one encounters a situation where an otherwise suitable operational amplifier cannot provide the required output load driving. That is, either the available output current is insufficient or the allowable output swing will not accommodate the signal levels anticipated at the output. Frequently this problem can be resolved by using an output power booster (Figs. 2.29 and 2.30).

Here the booster amplifier is a single-ended amplifier with the following general properties:

1. An output voltage driving capability exceeding that required by the load $v_{0,\max}$.
2. An output current driving capability exceeding that required by the load $i_{0,\max}$.
3. A full-power bandwidth (see Sec. 2.7) consistent with the anticipated load and signal frequency content (or else a slew rate that meets anticipated requirements).
4. An input impedance that can be accommodated by the driving operational amplifier.
5. A forward-voltage gain of at least G_v from direct current up to some high frequency well beyond the anticipated range. Here G_v must satisfy

$$G_v > \frac{v_{0,\max}}{v_{A,\max}}$$

where $v_{A,\max}$ is the maximum output voltage of the driving operational amplifier.

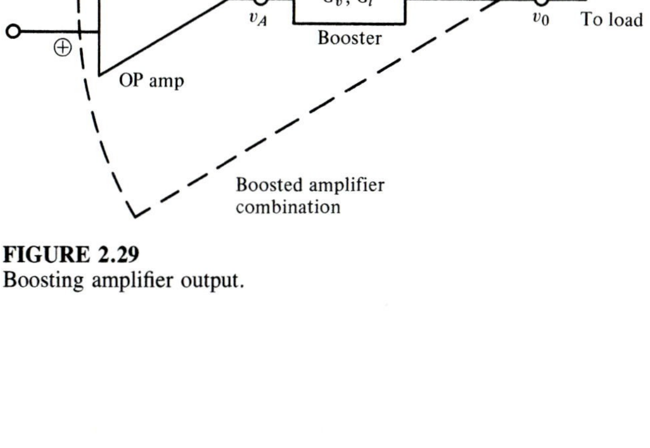

FIGURE 2.29
Boosting amplifier output.

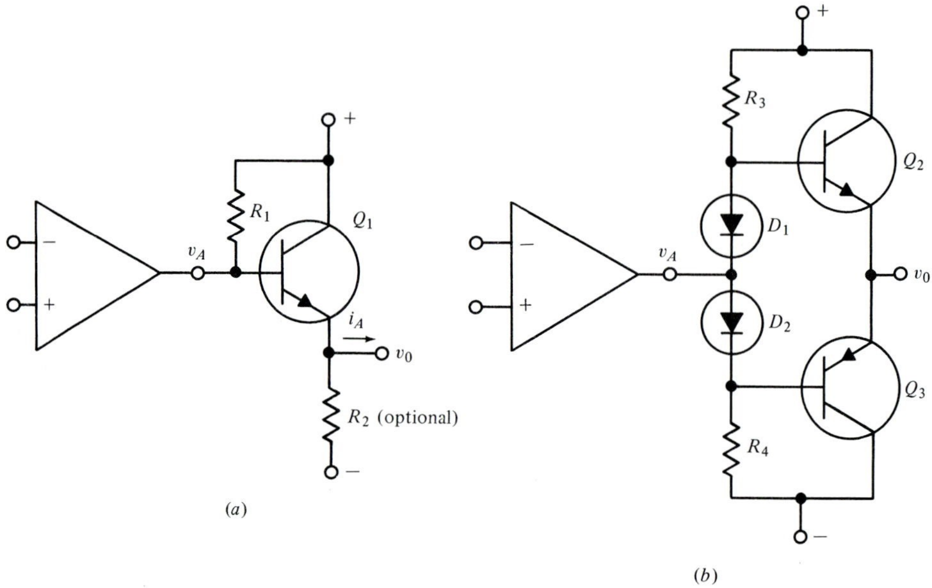

FIGURE 2.30
Some simple booster circuits. (*a*) A transistor emitter follower; (*b*) a complementary-symmetry output booster improves bipolar operation.

6. Implicit in 4, above, is a required booster current gain G_i such that

$$G_i > \frac{i_{O,\max}}{i_{A,\max}}$$

where $i_{A,\max}$ is the maximum rated output current of the driving operational amplifier.

7. The small-signal bandwidth of the booster must be broad enough to prevent significant extra phase shift in the overall loop gain $AG_v\beta$ when the combined operational amplifier–booster combination is placed in an appropriate feedback network, i.e., the overall stability of the final circuit must be preserved.

Note that as long as these general conditions are met, the booster amplifier can be rather sloppy in performance since it is in the feedback loop. In Fig. 2.30*a*, the feedback will suppress the dc offset inherent in the emitter-follower being used as a current booster.

Often only current gain is required of the booster; i.e., the necessary output voltage swing is no more than that of the driving operational amplifier, for example, ±10 V. To meet this situation, many commercial operational amplifier manufacturers provide prepackaged booster amplifiers with only unity voltage gain, but with a fairly high current gain, for example, 10 to 100, and a husky output current, for example, 100 to 500 mA. These units may be used to boost a typical operational amplifier's current driving capability.

Note that if the required voltage swing exceeds that of the operational amplifier, this normally means that the booster amplifier will have to be powered by a new power system with large supply voltages. It is, nevertheless, feasible to use the boosting scheme of Fig. 2.29 if the feedback network properly restricts the terminal potentials of the driving operational amplifier. Thus, for example, a ±10-V operational amplifier could conceivably drive a ±100-V power booster amplifier driving some sort of servomotor.

Figure 2.30 shows two fairly simple boosting circuits that often will do the job. (Design details are not given, since they depend on the specific application.) The emitter-follower circuit of Fig. 2.30*a* is particularly handy. It may be used whenever the output current is essentially of one polarity. The *npn* transistor shown would be used when i_A is positive. Note that v_0 could be bipolar. The emitter load resistor R_2 would be optional; it serves to "keep the circuit alive" under no-load conditions. Base resistor R_1 is chosen to set the base current so as to keep transistor Q_1 in the active region.

In some situations, we may need a booster that can both deliver and absorb current. In this event, we can often use some form of complementary-symmetry amplifier stage. Figure 2.30*b* shows the general form of such a stage with a unity voltage gain. If the required voltage gain is greater than 1, something more complex is required. The two simple boosters of Fig. 2.30 provide a current gain of the order of the transistor current gain, usually called β_F or h_{FE}. Higher current gains can be achieved using Darlington-connected transistors or some form of multistage circuit. High-power output stages using MOS devices have become most popular in recent years. One can also use a high-power unity gain follower circuit as the final output stage of an operational amplifier circuit.

2.9 POWER SUPPLIES, GROUNDING, AND SHIELDING

The topics in this section could well have an entire book devoted to them. We present here only some fundamental points that should be considered in selecting a power supply and laying out grounds and shielding. References 1 and 6 to 8 provide considerable detail.

Power Supplies for Operational Amplifiers

We have already indicated that most operational amplifiers normally must be powered by a pair of symmetric dc voltages, for example, ±15 V dc; a common ground point should be available (Fig. 1.2). We have more to say about grounding below. The power supply voltages should be equal in magnitude (matched to 1 percent, e.g.); more important, the voltages should be well regulated over a *wide range of frequencies*. Long-term drift of the power supply potentials will create dc offset drifts in the amplifiers. Many manufacturers will specify *power supply rejection ratios* indicating how much the input offset components e_{os} and i_{os} will drift with supply voltage. If the two voltages tend to track each other, i.e., if the magnitudes stay equal but change, then the drift effects will be reduced.

Good high-frequency regulation is very important. For broadband operational amplifiers, the power supply internal impedance should be very low (e.g., less than 0.1 Ω) over the entire frequency range where the amplifiers have any appreciable gain. Most operational amplifiers have class B output stages; i.e., they draw a low power supply current when their output current is zero, but they will draw a fairly large current when their output current is high. Further, the current demand will not be symmetric. In Fig. 2.31, we see a sketch of a typical situation. If the amplifier

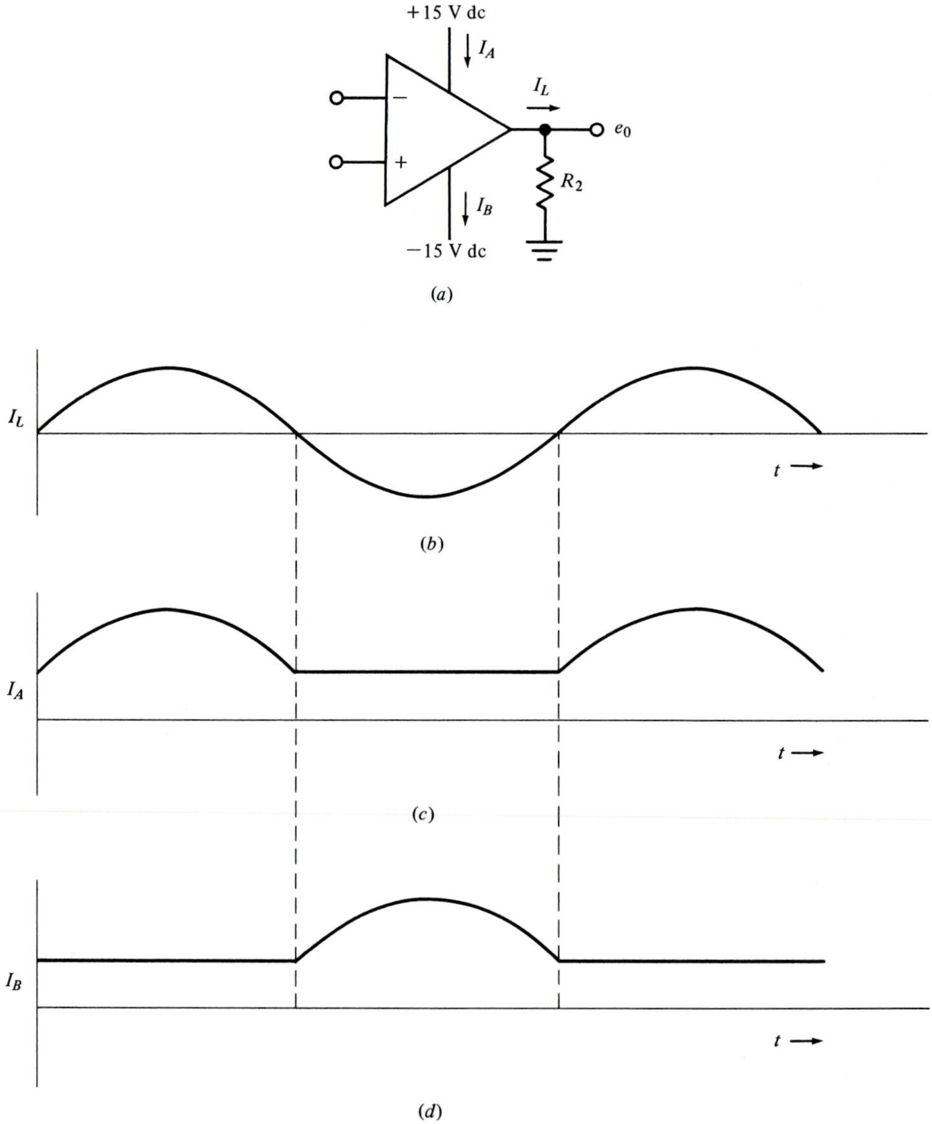

FIGURE 2.31
Power supply demands. (*a*) Output circuit of an operational amplifier; R_L is total output load. (*b*) Typical amplifier output current waveform; (*c*) and (*d*) are associated power bus currents.

load current is a bipolar sinusoid (Fig. 2.31*b*), then alternating pulses of current will be drawn from the power supplies (Fig. 2.31*c* and *d*). The output waveform can conceivably have a frequency in the 10- to 100-kHz range, even at high signal levels; hence the power supply current waveforms will contain relatively strong high-frequency components. Any power supply internal impedance will cause a high-frequency ripple to be induced in the power supply buses.

From the above discussion, we see that it is advisable to minimize the high-frequency impedance of power supply connections. Figure 2.32 shows a recommended technique. A separate heavy (e.g., No. 10) copper bus is run from a common tie point near the power supply to each part of a system where a collection of amplifiers is located. That is, we do not string a long "party line" with a large number of amplifiers all being fed from the same bus. Typically each group of three to six amplifiers would have its own power bus. In this way we reduce the common paths for power supply currents. In addition, it is advisable to

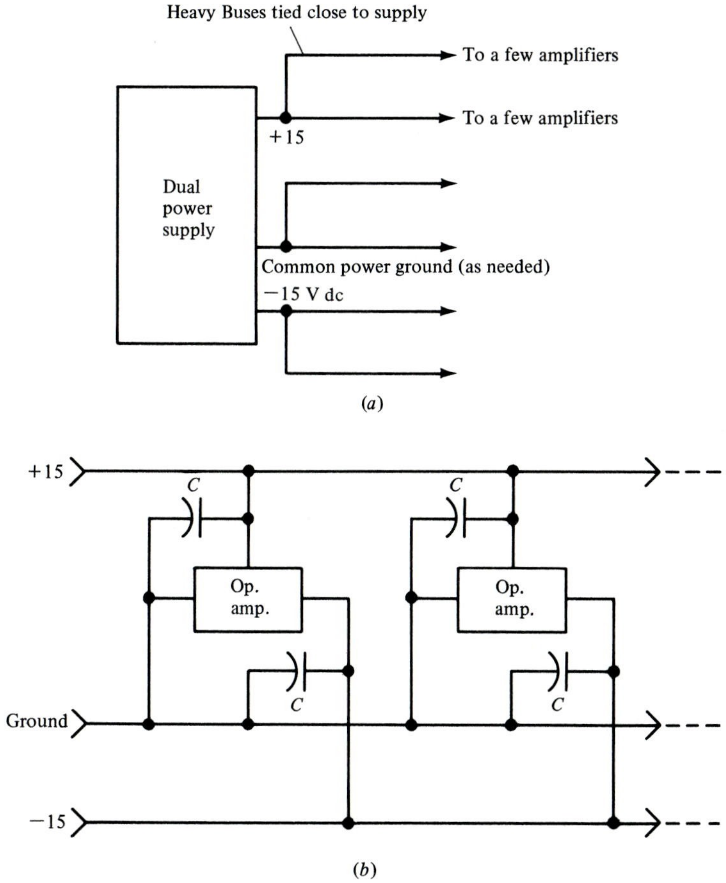

FIGURE 2.32
Technique for minimizing impedance of power supply connections. (*a*) Use of separate buses for each group of amplifiers; (*b*) decoupling at each amplifier location via capacitors.

place bypass capacitors at each amplifier location to shunt high-frequency currents around the busing system. The capacitors should provide a good rf bypass, e.g., a 0.01-μF ceramic type would be typical (Fig. 2.32*b*).

Shielding

The subject of shielding for long low-level signal lines is a complex one. If possible, it is best to use a differential system with both signal lines floating symmetrically with respect to signal ground, e.g., as in Fig. 2.33*a*. Twisted pairs make fairly noise-immune signal paths. Better yet, it is helpful to enclose the signal lines in a coaxial shield, which is then grounded *only at one point,* to avoid ground loops. It is better if the shield itself carries no signal current. Thus, even though we are transmitting single-ended signals, as in Fig. 2.33*b*, it is still better to use a separate signal ground return inside of the shield.

It is important to ground the shield properly. Often the best ground point must be determined experimentally, since the point of grounding can strongly influence the common-mode voltage induced from external fields. Normally the

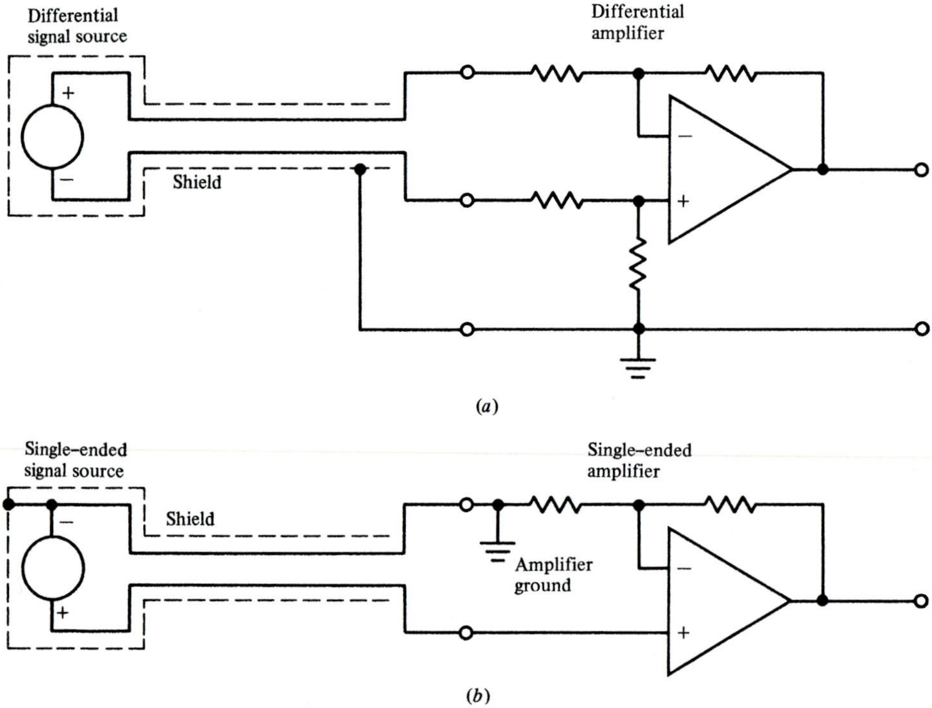

FIGURE 2.33
Shielding against externally induced noise. (*a*) Differential amplifier and floating signal source; note shield is grounded at only one point. (*b*) Single-ended signal source; note use of separate ground return for signal.

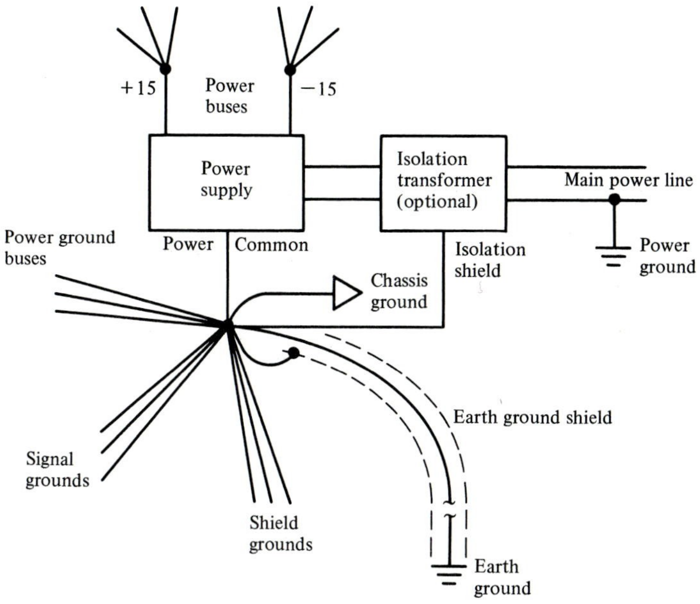

FIGURE 2.34
General layout of a grounding system to eliminate ground loops.

shield should be connected to the signal source ground for single-ended sources (Fig. 2.33*b*).

Ground Systems

Figure 2.34 shows a typical plan for an effective grounding system which has many desirable features. It contains an *earth ground* (use the power-line ground, if nothing else is available), preferably shielded until it enters the earth to keep it from being an antenna! A *common tie point* is used where all ground lines are brought together in a fanlike array, to avoid ground loops. Note that the chassis ground is separate from all other grounds except at this point. Shield grounds are brought in individually, as are signal grounds and power bus grounds. Figure 2.34 represents perhaps an extreme case, but the general technique is useful in avoiding cross-talk problems in complex signal-processing systems. Much misery can be avoided if an adequate grounding system is planned as part of the initial design effort.

2.10 COMPONENTS: RESISTORS AND CAPACITORS

Operational amplifiers may be incorporated into a wide variety of circuit designs. Discrete and hybrid (thin-film/thick-film and IC) packages are more commonly used for precision circuits, although all-monolithic circuits are feasible.

Resistors

The three most common types are carbon, metal-film, and wire-wound. Carbon-composition resistors are used in low-precision discrete circuits. They have a poor temperature coefficient and seldom have the tolerance and aging characteristics required for precision circuits. Metal-film resistors have good temperature coefficients (+10 ppm/°C) and can usually be obtained with stable, accurate values (for example, 0.1 to 1 percent) at modest cost. For very-high-accuracy resistors intended for use in precision instruments and analog computers, wire-wound (non-inductive) resistors often must be employed, even though they are considerably more expensive.

For high-frequency circuits (for example, 10 kHz and above) one must worry about stray reactive components. Here the metal-film units are preferred; typically they will have little parasitic inductance (0.01 μH) and a parallel end-to-end capacitance of perhaps 0.5 pF for a $\frac{1}{2}$-W unit.

In microcircuit applications, resistors may be fabricated by standard monolithic IC methods (diffused and pinch types) or by thin- or thick-film methods. Here one cannot usually achieve high absolute-value accuracy, so that the clever designer will attempt to take advantage of the fact that resistance *ratios* can usually be fairly well matched. Table 2.4 shows some typical parameters of microcircuit resistors. Note that, when required, thin-film resistors may be given precision values by temperature annealing and laser trimming techniques.

Capacitors

Accurate, low-loss capacitors are fairly expensive, particularly in the larger values. One must look carefully into specifications of dissipation coefficient (or power factor) and temperature coefficient. Large-value (over 0.1-μF) low-loss precision units preferably have polystyrene dielectric, although Mylar and Teflon are often acceptable. In the smaller sizes, one may choose from Mylar, mica, NPO

TABLE 2.4
Typical microcircuit resistor parameters*

Type	Range (20% tolerance)	Matching, %
P^+ or N^+ –	500 Ω to 10 kΩ	< 2
Diffused thick-film	1–10 MΩ	0.5
n^+ polysilicon	< 1kΩ	0.5
Thin-film	500 Ω to 100 kΩ	< 0.012†
Thick-film	1–10 MΩ	0.5‡

* Provided in part by Dr. Mojtaba Nowrozi, Burr-Brown Corp., Tucson, Ariz.

† May be trimmed to 0.05%.

‡ May be trimmed to 0.5%.

ceramic, glass, and polycarbonate units. Cheap ceramics are probably not suitable for any precision work. Electrolytic capacitors are seldom useful, since they are unipolar.

The dissipation coefficient, or power factor, is a measure of the effective resistance component in the capacitor. If the coefficient is D, then the effective series resistance at f Hz is

$$R_s = \frac{D}{2\pi f C}$$

The corresponding effective parallel resistance is

$$R_p = \frac{1}{2\pi f C D}$$

Dissipation factors may typically range from 1×10^{-4} for Teflon or polystyrene to 50×10^{-4} for ceramic and polycarbonate dielectrics (see Ref. 1, Table 3.1, and Ref. 4, Table 8.1).

2.11 COMPUTER-AIDED DESIGN

Computer-aided design (CAD) of circuits has evolved to a high level. Software packages like MICRO-CAP and SPICE are available at many industrial and educational computer facilities (Refs. 11 and 12).

MICRO-CAP II (Ref. 11) is a PC-compatible software package with 10 prepackaged operational amplifier models. These models include dc input voltage and current, differential-input impedance, output impedance, dc open-loop gain, and a three-pole dynamic model. The model parameters may be easily modified, or entirely new models may be developed.

In recent years, accurate models of operational amplifiers have also been presented using the SPICE integrated-circuit analysis program (Refs. 12 to 15). These models represent the amplifier with a high level of modeling detail and provide the amplifier designer with the ability to predict amplifier behavior with precision. They are very sophisticated; they may provide up to 10 poles, four zeros, and a large number of nonlinearities. The CAD approach permits full-scale simulation or CAD of highly detailed amplifier models. Of course, a reasonable amount of computer effort is required, but the savings in prototype development of IC operational amplifiers surely warrants the CAD approach.

In Fig. 2.35, we see the results of a SPICE simulation of a simple gain-of--10 inverter, using a simple model with the parameters of the UA741C IC operational amplifier (Ref. 16). Note that the operational amplifier model may be embedded in a SPICE subcircuit and used repeatedly in circuit modeling. More information on the use of this type of operational amplifier is available in PSPICE documentation (Ref. 17).

```
*******14-JUN-90 *******  SPICE 2G.5 (10AUG81)  *******13:07:45*****

* JUNE 14, 1990,    UA741 OP.AMP. MODEL          *

****      INPUT LISTING                      TEMPERATURE =    27.000 DEG C

*************************************************************************


* CIRCUIT GAIN OF -10 *
VS 1 0 AC 1.0 0.0
RIN 1 2 1.0E03
RF 2 3 1.0E04
X1 2 0 3 OPAMP2
.SUBCKT OPAMP2 1 2 4
*  UA 741C  MODEL OPEN-LOOP GAIN = 2E5 OR 100 DB  *
* UNITY-GAIN FREQUENCY =  6.0 MHZ  *
* BREAK FREQ. = 3.33333E04 HZ      *
RG1 1 0 50MEG
RG2 2 0 50MEG
RI 1 2 2MEG
G1 3 0 2 1 2.0
R1 3 0 1K
C1 3 0 26.6U
G2 4 0 3 0 1.333333
ROUT 4 0 75
.ENDS OPAMP2
.WIDTH OUT=80
.AC DEC 10 100. 1.0E07
.PRINT AC VDB(3) VP(3)
.PLOT AC VDB(3) VP(3)
.END
```

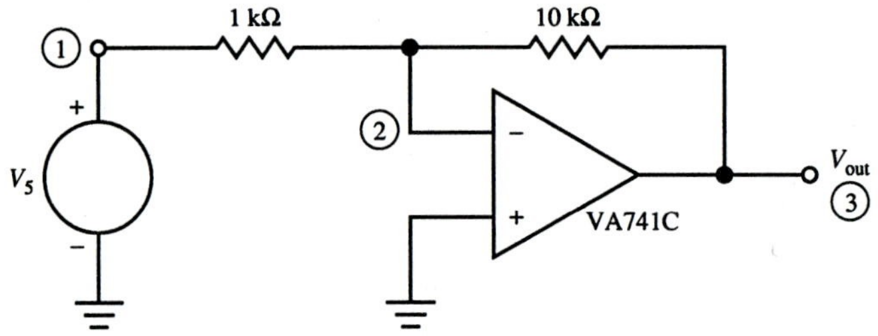

FIGURE 2.35
SPICE analysis of a UA741C operational amplifier circuit. (*a*) Problem listing of a gain-of-−10 inverter. (*continued on next page*)

```
* JUNE 14, 1990,    UA741 OP.AMP. MODEL         *

****     AC ANALYSIS                       TEMPERATURE =   27.000 DEG C

******************************************************************************


     FREQ          VDB(3)         VP(3)


   1.000E+02     2.000E+01     1.799E+02
   1.259E+02     2.000E+01     1.799E+02
   1.585E+02     2.000E+01     1.799E+02
   1.995E+02     2.000E+01     1.799E+02
   2.512E+02     2.000E+01     1.799E+02
   3.162E+02     2.000E+01     1.798E+02
   3.981E+02     2.000E+01     1.798E+02
   5.012E+02     2.000E+01     1.797E+02
   6.310E+02     2.000E+01     1.797E+02
   7.943E+02     2.000E+01     1.796E+02
   1.000E+03     2.000E+01     1.795E+02
   1.259E+03     2.000E+01     1.793E+02
   1.585E+03     2.000E+01     1.792E+02
   1.995E+03     2.000E+01     1.789E+02
   2.512E+03     2.000E+01     1.787E+02
   3.162E+03     2.000E+01     1.783E+02
   3.981E+03     1.999E+01     1.779E+02
   5.012E+03     1.999E+01     1.773E+02
   6.310E+03     1.998E+01     1.767E+02
   7.943E+03     1.998E+01     1.758E+02
   1.000E+04     1.996E+01     1.747E+02
   1.259E+04     1.994E+01     1.733E+02
   1.585E+04     1.991E+01     1.716E+02
   1.995E+04     1.985E+01     1.695E+02
   2.512E+04     1.977E+01     1.669E+02
   3.162E+04     1.964E+01     1.637E+02
   3.981E+04     1.945E+01     1.598E+02
   5.012E+04     1.915E+01     1.551E+02
   6.310E+04     1.872E+01     1.497E+02
   7.943E+04     1.812E+01     1.436E+02
   1.000E+05     1.731E+01     1.372E+02
   1.259E+05     1.627E+01     1.306E+02
   1.585E+05     1.501E+01     1.242E+02
   1.995E+05     1.355E+01     1.184E+02
   2.512E+05     1.193E+01     1.132E+02
   3.162E+05     1.019E+01     1.087E+02
   3.981E+05     8.360E+00     1.050E+02
   5.012E+05     6.471E+00     1.020E+02
   6.310E+05     4.543E+00     9.949E+01
   7.943E+05     2.589E+00     9.746E+01
   1.000E+06     6.182E-01     9.581E+01
   1.259E+06    -1.363E+00     9.445E+01
   1.585E+06    -3.351E+00     9.333E+01
   1.995E+06    -5.344E+00     9.238E+01
   2.512E+06    -7.339E+00     9.156E+01
   3.162E+06    -9.335E+00     9.082E+01
   3.981E+06    -1.133E+01     9.012E+01
   5.012E+06    -1.333E+01     8.944E+01
   6.310E+06    -1.533E+01     8.872E+01
   7.943E+06    -1.732E+01     8.793E+01
   1.000E+07    -1.932E+01     8.703E+01
```

FIGURE 2.35 (continued)
(*b*) Listing of gain and phase. (*continued on next page*)

```
*: VDB(3)
+: VP(3)

    FREQ      VDB(3)

(*)------------- -2.000D+01    -1.000D+01     0.000D+00     1.000D+01  2.000D+
                      - - - - - - - - - - - - - - - - - - - - - - - - - - - -

(+)-------------  5.000D+01     1.000D+02     1.500D+02     2.000D+02  2.500D+
                      - - - - - - - - - - - - - - - - - - - - - - - - - - - -
 1.000D+02  2.000D+01 .             .             .       +     .
 1.259D+02  2.000D+01 .             .             .       +     .
 1.585D+02  2.000D+01 .             .             .       +     .
 1.995D+02  2.000D+01 .             .             .       +     .
 2.512D+02  2.000D+01 .             .             .       +     .
 3.162D+02  2.000D+01 .             .             .       +     .
 3.981D+02  2.000D+01 .             .             .       +     .
 5.012D+02  2.000D+01 .             .             .       +     .
 6.310D+02  2.000D+01 .             .             .       +     .
 7.943D+02  2.000D+01 .             .             .       +     .
 1.000D+03  2.000D+01 .             .             .       +     .
 1.259D+03  2.000D+01 .             .             .       +     .
 1.585D+03  2.000D+01 .             .             .       +     .
 1.995D+03  2.000D+01 .             .             .       +     .
 2.512D+03  2.000D+01 .             .             .       +     .
 3.162D+03  2.000D+01 .             .             .       +     .
 3.981D+03  1.999D+01 .             .             .       +     .
 5.012D+03  1.999D+01 .             .             .       +     .
 6.310D+03  1.998D+01 .             .             .      +      .
 7.943D+03  1.998D+01 .             .             .      +      .
 1.000D+04  1.996D+01 .             .             .      +      .
 1.259D+04  1.994D+01 .             .             .      +      .
 1.585D+04  1.991D+01 .             .             .     +       .
 1.995D+04  1.985D+01 .             .             .    +        .
 2.512D+04  1.977D+01 .             .             .    +        .
 3.162D+04  1.964D+01 .             .             .   +         .            *
 3.981D+04  1.945D+01 .             .             .  +          .            *
 5.012D+04  1.915D+01 .             .             .+            .            *
 6.310D+04  1.872D+01 .             .             +             .           *
 7.943D+04  1.812D+01 .             .           + .             .          *
 1.000D+05  1.731D+01 .             .         +   .             .         *
 1.259D+05  1.627D+01 .             .        +    .             .        *
 1.585D+05  1.501D+01 .             .      +      .             .      *
 1.995D+05  1.355D+01 .             .    +        .             .    *
 2.512D+05  1.193D+01 .             .   +         .             .  *
 3.162D+05  1.019D+01 .             . +           .             *
 3.981D+05  8.360D+00 .             .+            .           * .
 5.012D+05  6.471D+00 .             .+            .        *    .
 6.310D+05  4.543D+00 .             +             .     *       .
 7.943D+05  2.589D+00 .            +.             .   *         .
 1.000D+06  6.182D-01 .            +.             .*            .
 1.259D+06 -1.363D+00 .           + .           * .             .
 1.585D+06 -3.351D+00 .           + .        *    .             .
 1.995D+06 -5.344D+00 .           + .      *      .             .
 2.512D+06 -7.339D+00 .           + .   *         .             .
 3.162D+06 -9.335D+00 .          +  .*            .             .
 3.981D+06 -1.133D+01 .          +* .             .             .
 5.012D+06 -1.333D+01 .        * +  .             .             .
 6.310D+06 -1.533D+01 .      *   +  .             .             .
 7.943D+06 -1.732D+01 .   *      +  .             .             .
 1.000D+07 -1.932D+01 .*        +   .             .             .
                      - - - - - - - - - - - - - - - - - - - - - - - - - - - -
```

FIGURE 2.35 (continued)
(*c*) Bode plot of gain and phase.

REFERENCES AND BIBLIOGRAPHY

1. Korn, G. A., and T. M. Korn: *Electronic Analog and Hybrid Computers,* 2d ed., McGraw-Hill, New York, 1972.
2. Gupta, S. L., and L. Hasdorff: *Fundamentals of Automatic Control,* Wiley, New York, 1966.
3. Huelsman, L. P.: *Basic Circuit Theory with Digital Computations,* Prentice-Hall, Englewood Cliffs, N.J., 1972.
4. Tobey, G. E., T. G. Graeme, and L. P. Huelsman: *Operational Amplifiers,* McGraw-Hill, New York, 1971.
5. "Noise and Operational Amplifier Circuits," *Analog Dialog,* vol. 3, no. 1, March 1969.
6. Morrison, R.: *Grounding and Shielding Techniques in Instrumentation,* Wiley, New York, 1972.
7. Stewart, E. L.: "Grounds, Grounds, and More Grounds," *Simulation,* August 1965, pp. 121–128.
8. ———: "Noise Reduction on Interconnect Lines," *Simulation,* September 1965, pp. 149–155.
9. Papoulis, A.: *Probability, Random Variables and Stochastic Processes,* McGraw-Hill, New York, 1965.
10. Broderson, A. J., E. R. Chenette, and R. C. Jaeger: "Noise in Integrated-Circuit Transistors," *IEEE Journal of Solid-State Circuits,* vol. SC-5, no. 2, pp. 63–66, April 1970.
11. Roden, M. S.: *A Student Edition of MICRO-CAP II,* Benjamin/Cummings, Menlo Park, Calif., 1988.
12. Vladimirescu, A., A. R. Newton, and D. O. Pederson: *SPICE User's Guide* (Version 2.1G or later), University of California, Berkeley, 1980.
13. Till, J.: "Flexible Op-Amp Model Improves SPICE," *Electronic Design,* June 22, 1989.
14. Buxton, J.: "OP-42 Advanced SPICE Model," Precision Monolithics, Inc., Applications Notes AN-110 and AN-117, 1989.
15. Tuch, B.: "An Analog Macromodel Craze, Op Amps Lead the Way," *Electronic Products,* May 1990.
16. Franco, S.: *Design with Operational Amplifiers and Integrated Circuits,* McGraw-Hill, New York, 1988.
17. *PSPICE User's Guide,* Version 4.03, MicroSim Corp., 1987.

PROBLEMS

2.1. *(Sec. 2.1)* Given the following operational amplifier specifications:

Open-loop gain at dc	90 dB
Unity gain frequency	3 MHz

find $A'(s)$, using Eq. (2.2).

2.2. *(Sec. 2.2)* Suppose a simple unity gain follower is constructed from an operational amplifier with these specifications:

Open-loop gain at dc	90 dB
Unity gain frequency	3 MHz

Assume zero output impedance and a negligibly large differential input impedance; use a simple one-pole model of the amplifier. Make a Bode plot of the magnitude of the circuit closed-loop gain; what is the -3-dB frequency of the circuit?

2.3. *(Sec. 2.2)* For this problem, and the two to follow, assume that the following operational amplifier characteristics are important:

Unity gain frequency f_u	3 MHz
Open-loop gain at dc A_0	100 dB

Differential input impedance Z_I	5 MΩ
Output impedance Z_0	500 Ω

Refer to the circuit of Fig. P2.3; let

$$R_F = 200 \text{ k}\Omega$$
$$R_1 = 2 \text{ k}\Omega$$

(*a*) Find the percent closed-loop gain error at direct current.
(*b*) Find the closed-loop circuit output impedance at direct current.
(*c*) Repeat parts *a* and *b* if $R_F = 100$ kΩ.
(*d*) If $v_1 = -0.5$ V dc and $R_F = 10$ kΩ, find Δv_0 = change between v_0 with no load and v_0 with 1-kΩ output load connected to ground. (*Hint:* Use the results of parts *a* and *b*.)
(*e*) Find the circuit input impedance.

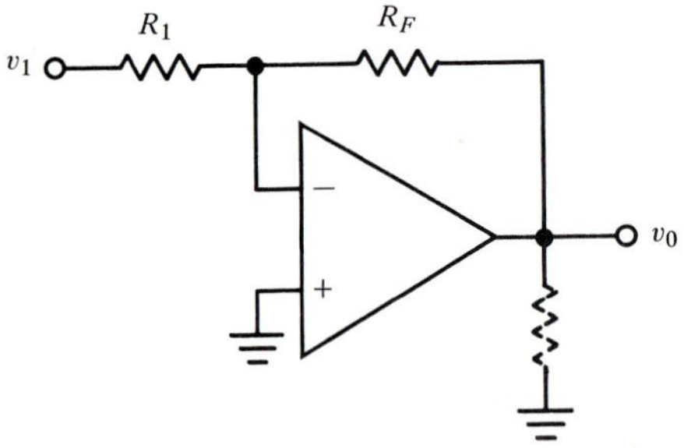

FIGURE P2.3

2.4. *(Sec. 2.2)* Use the amplifier specifications of Prob. 2.3 and the circuit of Fig. P2.4, with

$$R_F = 5 \text{ k}\Omega \qquad R_1 = 1 \text{ k}\Omega$$

(*a*) Find the percent closed-loop gain error at direct current.
(*b*) Find the closed-loop circuit output impedance.
(*c*) Find the circuit input impedance.

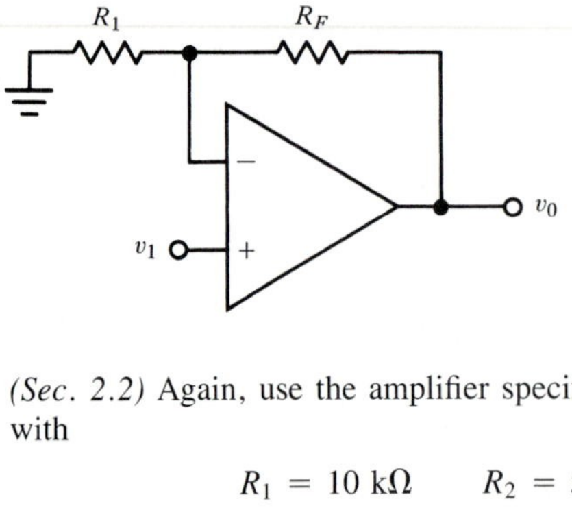

FIGURE P2.4

2.5. *(Sec. 2.2)* Again, use the amplifier specifications of Prob. 2.3; refer to Fig. P2.5, with

$$R_1 = 10 \text{ k}\Omega \qquad R_2 = 5 \text{ k}\Omega \qquad R_F = 100 \text{ k}\Omega$$

(*a*) Find the percent closed-loop gain error at direct current (note that it is the same for both inputs).
(*b*) Find the circuit output impedance.

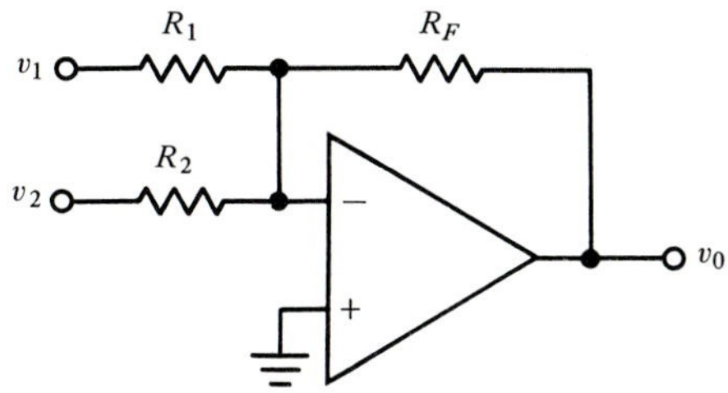

FIGURE P2.5

2.6. *(Sec. 2.2)* Refer to the circuit of Fig. P2.6; here we assume that the input impedance of the operational amplifier is negligibly high below 100 kHz. This circuit can then be used to estimate the open-loop gain of the operational amplifier. We also assume that the output loading is negligible. The following test results are obtained:

1-V dc input:

v_0, V dc	v_x, V dc
−0.997	0.000997

1-V ac input:

Frequency, Hz	v_0, V rms	v_x, V rms
10	0.997	0.00111
20	0.997	0.00141
50	0.997	0.00268
100	0.997	0.00508
200	0.997	0.0100
500	0.994	0.0249
1,000	0.986	0.0493
2,000	0.955	0.0955
5,000	0.798	0.200
10,000	0.554	0.277
20,000	0.316	0.316
50,000	0.132	0.330
100,000	0.0665	0.333

(*a*) Estimate the open-loop gain of the amplifier; i.e., estimate the dc value of the open-loop gain A_0 and the unity-gain frequency f_u.
(*b*) Note that it is difficult to measure the voltages in some frequency ranges. At low frequencies, v_x is small in magnitude; at high frequencies; v_0 is small. Careful shielding is required to make meaningful measurements. Suggest alternative

circuits that could be used at very low and very high frequencies to improve the signal levels.

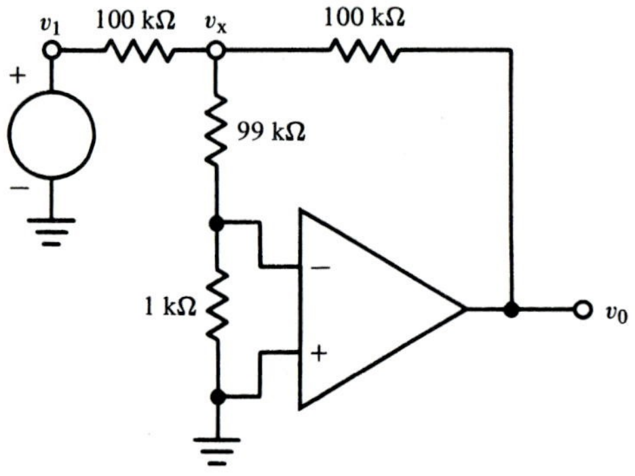

FIGURE P2.6

2.7. *(Sec. 2.3)* Design an operational amplifier circuit to implement

$$v_O = -(v_1 + 10v_2)$$

i.e., a two-input summer-inverter. Assume perfect resistors, but all less than 100 kΩ in value. Assume the operational amplifier has zero output impedance and negligibly large differential input impedance.

(*a*) Specify the minimum value of dc open-loop gain required to achieve an overall gain accuracy of 0.1 percent at direct current.

(*b*) Determine the minimum unity gain frequency in hertz for the operational amplifier required to ensure that the −3-dB frequency of the actual circuit is at least 100 kHz (assume a simple one-pole model for A).

(*c*) Using the values of A_0 and f_u obtained above, at what frequency (in hertz) is the circuit accurate to 1 percent?

2.8. *(Sec. 2.3)* Refer to Fig. P2.8. Assume the operational amplifier has a negligibly large differential input impedance, but with

Open-loop gain at dc	100 dB
Unity gain frequency	1 MHz

(*a*) Find the fractional error in the voltage gains at direct current, compared to the ideal amplifier case; express the answer as a percentage.

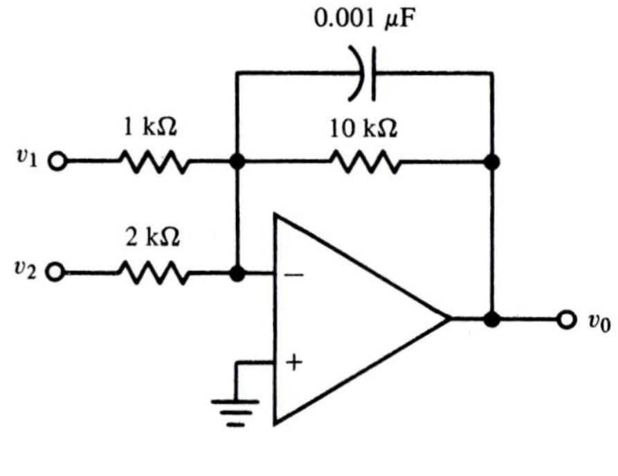

FIGURE P2.8

(*b*) Make a Bode plot (magnitude and phase) of $1/\beta$.
(*c*) Make a Bode plot of A on the same graphs as part *b*.
(*d*) What are the gain and phase margins?
(*e*) At what frequencies in hertz are the magnitudes of the circuit gains in error by 1 percent? At what frequency are they in error by 10 percent? (Give approximate answers.)
(*f*) Does the fact that the operational amplifier is nonideal greatly affect the performance of this circuit as a summer, amplifier, and low-pass filter?

2.9. *(Sec. 2.3)* Refer to the integrator circuit of Fig. P2.9 with

$$R = 2\ \text{k}\Omega \qquad C = 1\ \mu\text{F}$$

Assume that we have an operational amplifier with the following specifications:

Unity gain frequency f_u (that is, $\omega_u = 10^7$ rad/s)	1.59 MHz
Open-loop gain at direct current	90 dB
Differential input impedance	Negligibly large
Output impedance	0 Ω

(*a*) Make a Bode plot of A and $1/\beta$; put both magnitude curves on one plot, both phase curves on another.
(*b*) Make a separate Bode plot for $A\beta$.
(*c*) Estimate the frequency range where the integrator gain accuracy is 1 percent or better (be sure to look at both low and high frequencies).
(*d*) Find the overall circuit transfer function

$$H(s) = \frac{V_0(s)}{V_1(s)}$$

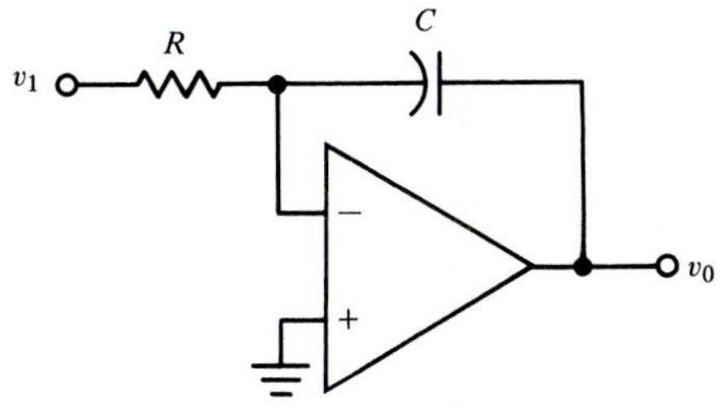

FIGURE P2.9

2.10. *(Sec. 2.3)* Refer to Fig. P2.10; assume an ideal operational amplifier, except that

$$A(s) = A'(s) = \frac{10^4}{(1 + s/\omega_1)(1 + s/\omega_2)(1 + s/\omega_3)}$$

where $\omega_1 = 10$ rad/s
$\omega_2 = \omega_3 = 10^5$ rad/s

(*a*) Use Bode plot methods to find the phase margin of this circuit if it is designed to have a gain of 1; repeat for gains of 10 and 100. Are any of these cases unstable? Be sure to plot carefully; in particular, plot $|A|$ fairly precisely at the corner frequencies, rather than using a simple asymptotic corner plot. In each case, what is the approximate bandwidth of the circuit (for the stable cases)?

(*b*) Some operational amplifiers (notably low-cost IC varieties) permit the user to adjust the frequency response. Typically, ω_1 can be made smaller by adding an extra *RC* network. Repeat part *a* if ω_1 is moved down to 1 rad/s. Comment on advantages and disadvantages of the *compensation*.

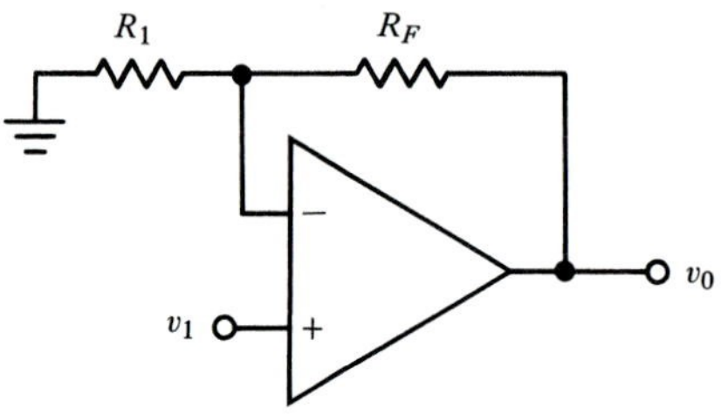

FIGURE P2.10

2.11. *(Sec. 2.3)* Refer to the differentiator circuit of Fig. 1.36*b*. Discuss the potential problems of using this circuit without modification. Use Bode plots to describe potential conditions for instability. Discuss the reasons why the modified circuit of Fig. 1.36*c* improves the stability situation.

2.12. *(Sec. 2.4)* Given an operational amplifier with the following specifications:

Common-mode rejection ratio	80 dB
Common-mode input impedance	10 MΩ

(*a*) Design a differential amplifier circuit (one operational amplifier) with a differential gain of 20. Pick the resistors so that the largest value is 100 kΩ.
(*b*) Neglecting the common-mode input impedances, estimate the common-mode gain of the circuit.
(*c*) If the common-mode input impedances are both equal to 10 MΩ, will this affect the differential and common-mode gains of the circuit, and if so, how much? (*Hint:* Refer to Sec. 1.8.)
(*d*) If one of the common-mode input impedances is 10 MΩ and the other is 15 MΩ, repeat part *c*.

2.13. *(Sec. 2.5)* Refer to Table 2.2. Estimate the maximum values of e_{os} and i_{os} at 100°C for all four types of amplifiers. For the FET-input amplifiers, be sure to observe footnote †; that is, the magnitudes of the currents double every 10°C.

2.14. *(Sec. 2.5)* Refer to Fig. P2.14; modify the design to improve the sensitivity of the circuit to dc offsets. Also add circuitry to permit balancing out up to 10 mV

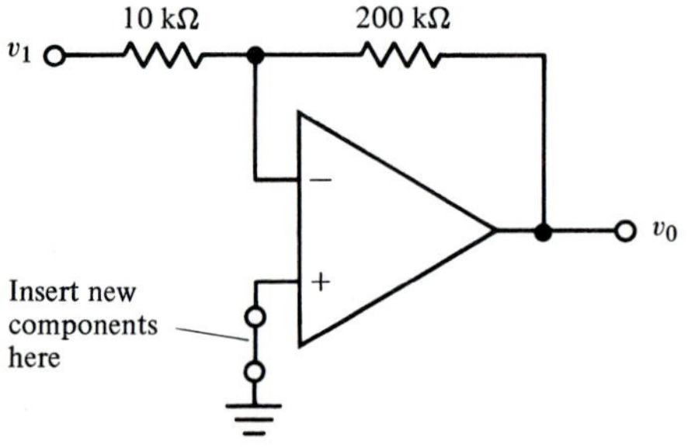

FIGURE P2.14

of input offset, in accordance with Eq. (2.105). Use standard RTMA 10 percent resistor values in your design, and assume ±15 V dc is available.

2.15. *(Sec. 2.5)* In the circuit of Fig. 2.20,

$$R_F = 50\ \text{k}\Omega \qquad R_1 = 5\ \text{k}\Omega$$

(*a*) Pick R_2 for minimum dc offset.
(*b*) Make a table of estimated output dc offsets for this circuit, using the typical parameters of Table 2.2. Consider all four amplifier types. Consider the case of $R_2 = 0$ and R_2 equal to the optimum value from part *a*.
(*c*) Assume that some form of offset adjustment is available, so that at 25°C the offset is zero. For each amplifier of Table 2.2, find the drift that will occur if the ambient temperature changes 10°C. Consider cases where $R_2 = 0$ and R_2 equals the value from part *a*.
(*d*) Repeat all the above, but with

$$R_F = 1\ \text{M}\Omega \qquad R_1 = 100\ \text{k}\Omega$$

2.16. *(Sec. 2.5)*
(*a*) Using the monolithic (low-cost) IC amplifier of Table 2.2, design a gain-of-50 inverting amplifier. Use the largest resistance values that yield an output dc offset drift of 0.5 mV/°C.
(*b*) Repeat the above design, using the FET-input IC amplifier of Table 2.2.

2.17. *(Sec. 2.5)* Refer to the integrator mode control circuit in Fig. 1.13*a*; we want to have one input with a gain of 10 s^{-1} for integrating. The circuit uses ideal switches, resistors, and capacitors; however, the operational amplifier has the following specifications:

Dc offset voltage	±2 mV
Dc input bias current	±5 μA
Dc input offset current	±0.2 μA

(*a*) Using all resistors of 100 kΩ, design the circuit and estimate the effects of the dc offsets on the three modes of operation; i.e., how much dc offset and/or drift do they create?
(*b*) Can you improve the situation with regard to dc offset effects by changing the values of the components to achieve a different impedance level?

2.18. *(Sec. 2.5)* Refer to Fig. P2.18; here we are interested in the effects of operational amplifier dc offset on the performance of an integrator. Basically, we find that in the absence of an external input (v_1 grounded), the output will drift with a certain maximum rate of change. Assume that the operational amplifier has dc offsets e_{os}, i_1, i_2, and i_{os}, as defined in Sec. 2.5.

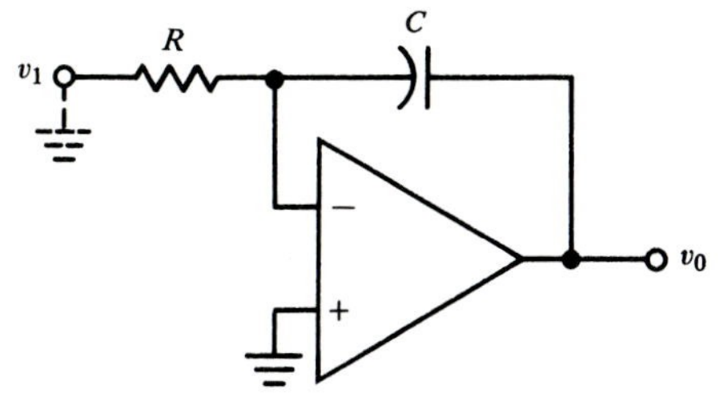

FIGURE P2.18

(*a*) Derive an expression for the maximum rate of change of v_O with time, if $v_I = 0$.
(*b*) Would the addition of a resistance in the noninverting input branch of the amplifier improve this drift rate? If so, what should its value be, and how much is the drift with the resistor added?
(*c*) How would you add an offset adjustment network to the circuit?

2.19. *(Sec. 2.6)* Using the circuit of Example 2.13, find the rms output noise for the following cases; compare your answers to the result of the example.
(*a*) $R = 10^4$, but C is adjusted to give a closed-loop bandwidth of 1 kHz.
(*b*) $R = 10^4$, $C = 0$.

2.20. *(Sec. 2.6)* Given an operational amplifier with the following noise specifications, estimate the power spectral density functions G_e and G_f.

Frequency range	Voltage, μV	Current, pA
0.001 Hz to 5 Hz	1 p-p	10 p-p
5 Hz to 5 kHz	2 rms	2 rms

2.21. *(Sec. 2.7)* Given that a particular operational amplifier has a minimum slew rate of 20×10^6 V/s and is rated at an output of ±10 V, estimate the full-power bandwidth of the amplifier, in hertz.

2.22. *(Sec. 2.7)* Refer to Fig. P2.22. Given that a particular operational amplifier has a rated full-power bandwidth of 20 kHz at ±10 V, sketch the approximate response of the circuit to an input ideal step of +2 V.

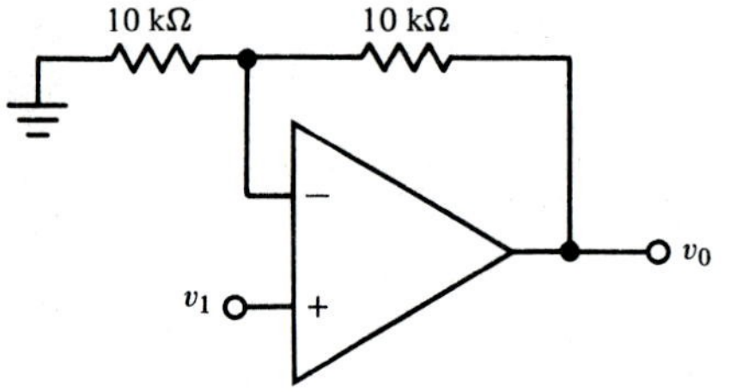

FIGURE P2.22

2.23. *(Sec. 2.7)* Refer to Fig. P2.23. Given that a particular operational amplifier has the ratings

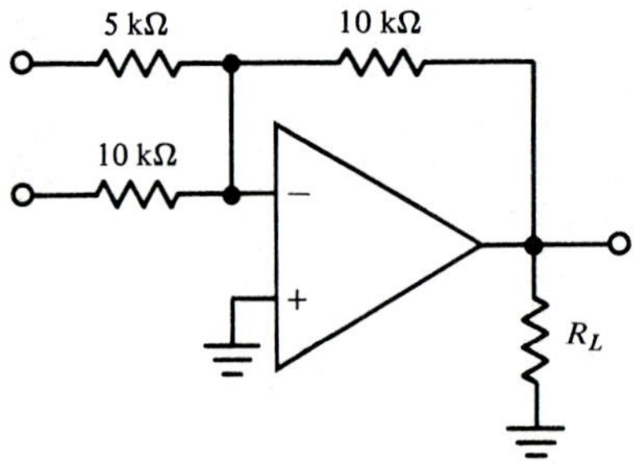

FIGURE P2.23

Peak output voltage	± 10 V
Peak output current	± 20 mA

(*a*) Find the minimum value of R_L that can be driven over the full output $\pm$ voltage range; be sure to include all current drains on the amplifier.

(*b*) Repeat part *a* if R_L is connected to -15 V dc instead of to ground.

2.24. *(Sec. 2.9)* Refer to Fig. P2.24. Suppose we have an operational amplifier rated at ± 20 mA and ± 10-V output. The required currents drawn from the power buses depend on the output load current. In this case we know that

$$I_+ = \begin{cases} 0.005 + I_L & I_L > 0 \\ 0.005 & I_L < 0 \end{cases}$$

and similarly we know I_- (currents in amperes). If the parasitic inductance in the power supply bus is 0.1 μH and the output of the amplifier is a 10-V peak sinusoid of frequency 100 kHz, find the peak ripple voltage induced in the parasitic inductances if the load R_L is drawing maximum rated output current.

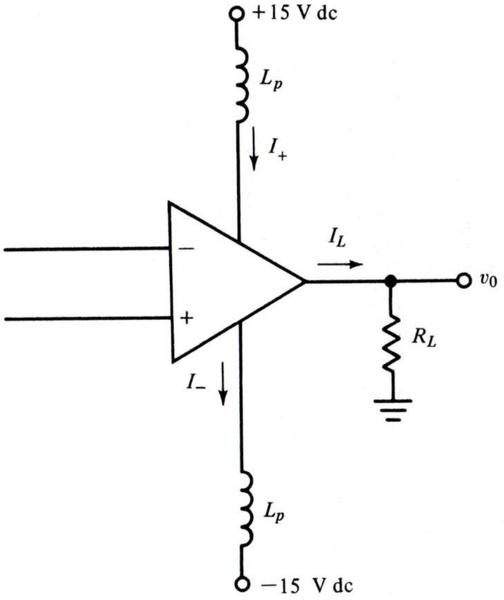

FIGURE P2.24

2.25. Use a development similar to that of Eq. (2.73*a*) to prove Eq. (2.73*b*).

CHAPTER 3

NONLINEAR CIRCUIT APPLICATIONS

In the previous two chapters, we have attempted to present a detailed picture of how an operational amplifier may be used as a general circuit element and how nonideal effects in the amplifier may affect circuit performance. To this point, our examples have all been associated with so-called linear applications, in which simple *RC* networks were used to provide the desired performance of a complete circuit. If this were all we ever did with operational amplifiers, they would still be eminently useful components. In this chapter we go a step further and describe some of the ways in which operational amplifiers may be combined with not only linear but also nonlinear circuit elements (primarily diodes and transistors) to obtain an extremely broad range of circuit functions.

In this realm of application, the clever circuit designer can use principles of feedback to suppress undesirable properties of a nonlinear element and at the same time take advantage of desirable properties. The operational amplifier is used to do the things it does best, e.g., sensing current magnitude and direction without loading a node and providing a low-impedance output circuit.

The number of possibilities available to the designer is boundless. We do not attempt to provide a complete catalog of all special or trick circuits that have been devised. We do, however, present a number of well-known nonlinear circuits and describe their function and design. Our hope is to thereby whet the readers' appetites and convince them that almost any analog signal-processing operation they can think of can be implemented.

Section 3.1 describes the general nonlinear design approach, based primarily on ideal models for the operational amplifier and the associated nonlinear devices;

nonideal properties are discussed throughout the chapter, where appropriate. In Sec. 3.2, we discuss some important first examples, the feedback limiter and the comparator. Section 3.3 discusses the general design of piecewise-linear function generators and includes examples of absolute-value circuits and peak-detection and holding operators. In Sec. 3.4, the design of logarithmic operators introduces the use of the exponential law of *pn* junction behavior. The last two sections deal with well-known applications of nonlinear operational amplifier circuits. Section 3.5 describes how nonlinear function generators may be used to implement analog squaring, multiplying, and dividing operations. Section 3.6 discusses techniques that may be used for precision waveform generation.

Again, our aim is not to provide a comprehensive presentation of every circuit ever designed; manufacturer's applications notes are often very helpful, as are the many popular industrial journals on electronic design (e.g., Refs. 1 and 2). References 3 and 4 also provide many examples and design variations.

3.1 GENERAL NONLINEAR DESIGN APPROACH

For most of our initial discussions, we still assume an ideal operational amplifier (Chap. 1, Sec. 1). We embed such an amplifier in general (nonlinear) networks (Fig. 3.1), and we assume the following:

1. The current flowing into the amplifier terminals is very small (essentially zero); $i_g \cong 0$.
2. The voltage between the amplifier input terminals is very small; typically (but not always) the noninverting terminal is grounded, and $e_g \cong 0$.

The above conditions are assumed to be true at every instant of time; the amplifier output will attempt to act upon the accompanying network to *force* the above conditions to be true. If it *cannot,* due to output current, voltage, or slewing rate limitations, then the amplifier will *overload*. This condition is normally not

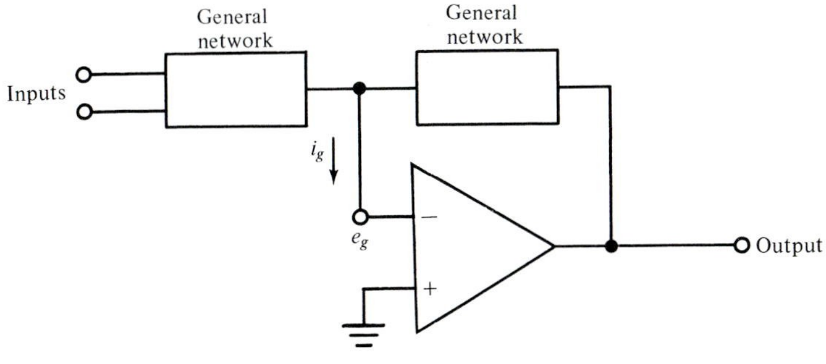

FIGURE 3.1
General nonlinear operational amplifier circuit (grounded noninverting terminal).

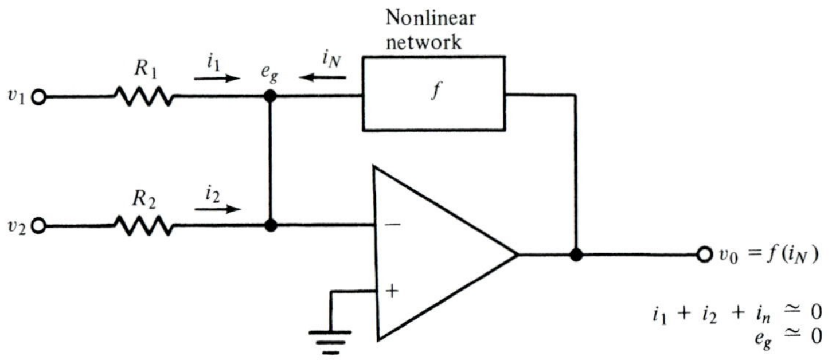

FIGURE 3.2
Refer to Example 3.1; input-output characteristic is controlled by the short-circuit conductance f.

permitted even momentarily, since most operational amplifiers have a relatively slow overload recovery time.

Example 3.1. The circuit of Fig. 3.2 is a general nonlinear operator which also performs a summing of two or more inputs if desired. The circuit behavior is dependent on the voltage-current characteristic

$$v_0 = f(i_N) \tag{3.1}$$

where v_0 is the amplifier output terminal voltage and i_N is the short-circuit current that flows as indicated in Fig. 3.2. The amplifier forces $e_g = 0$, and thus

$$i_1 + i_2 + i_N = 0 \tag{3.2}$$

Also since $e_g = 0$,

$$i_1 = \frac{v_1}{R_1} \qquad i_2 = \frac{v_2}{R_2}$$

and thus

$$v_0 = f\left(-\frac{v_1}{R_1} - \frac{v_2}{R_2}\right) \tag{3.3}$$

Nonlinear Device Models

In many parts of this chapter, we use some form of circuit model for nonlinear devices, primarily diodes and transistors. To make our analyses tractable, for the most part we use the so-called ideal diode model. Figure 3.3 illustrates what we mean by this for a simple *pn* junction diode. This model is often known as the "demon-with-a-switch," for fairly obvious reasons. This model describes the diode as follows:

1. Zero forward resistance if $i > 0$; diode ON.
2. Infinite reverse resistance if $v < 0$; diode OFF.

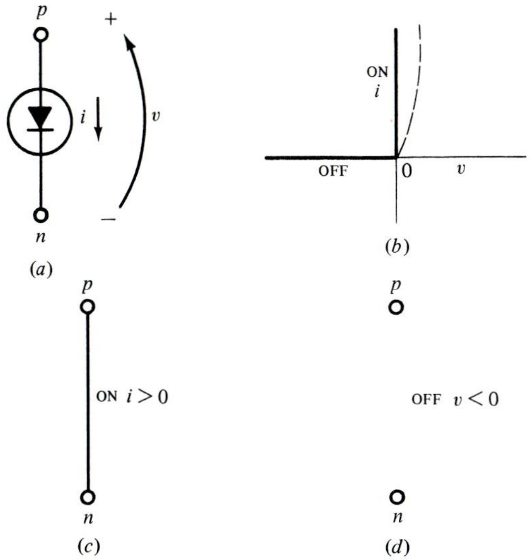

FIGURE 3.3
Ideal diode model. (*a*) Diode symbol and terminal conventions; (*b*) ideal "demon-with-a-switch" characteristic (solid line) and more realistic characteristic including forward drop (dotted line); (*c*) ON state equivalent circuit; (*d*) OFF state equivalent circuit.

(Current and voltage polarities are as shown in Fig. 3.3*a*.) Hence the ideal model assumes that the diode is either a perfect short circuit or a perfect open circuit, depending on the direction of current flow that the rest of the circuit attempts to establish. Note that *v is never positive,* and *i is never negative.*

This ideal model is, of course, only approximately true. An actual diode will have a forward voltage of 0.3 to 0.7 V if any appreciable current is flowing (dotted line, Fig. 3.3*b*). There will be a small reverse current, although at room temperature often this can be neglected. Nevertheless, the ideal diode model is extremely useful for making a first-cut analysis of how a circuit will operate. Once the general regions where diodes are ON or OFF are established, then more careful models may be employed, e.g., to estimate the effects of forward drop and leakage current. In this chapter we also neglect the dynamics of the switching of a diode from one state to another. In practice one may have to examine the effects of diode storage time and junction capacitance on circuit speed. Reference 5 provides a good discussion of diode and transistor circuit properties; or see any modern electronics text. It is also encouraging to note that, by means of the high amount of negative feedback normally provided by the operational amplifier, many of the nonideal properties of diodes and transistors are suppressed!

Figure 3.4 shows similar ideal diode models for an *npn* bipolar transistor; for a *pnp* type, the diode polarities are reversed. Again, we assume ideal diodes and neglect the diode forward drop and reverse leakage. Even with these omissions, we still retain a model that is generally very useful for analyzing the behavior of the transistor when it is embedded in a complex circuit. Four basic modes of

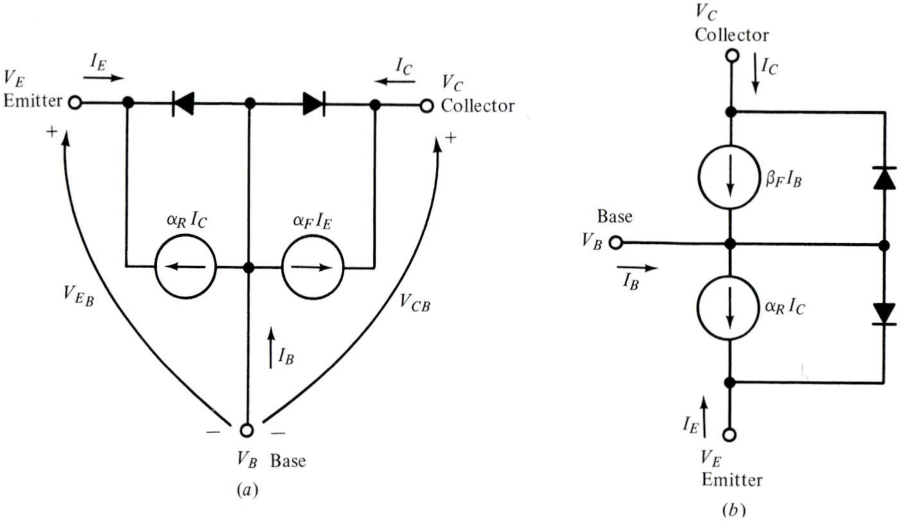

FIGURE 3.4
Ideal bipolar transistor equivalent circuit (diode polarity shown for *npn*). (*a*) Common-base model with emitter and collector diodes and ICIS current sources; (*b*) common-emitter equivalent.

operation (corresponding to the four possible ON and OFF states of diodes) are apparent (Fig. 3.5):

1. *Normal or active:* This state is the one normally used for amplification. It occurs when the emitter-base junction is forward-biased and the collector-base junction is reverse-biased, best stated as:

npn	***pnp***
$V_{CB} > 0$ $I_E < 0$	$V_{CB} < 0$ $I_E > 0$

In this mode, there is an effective current generator in the collector circuit, that is, $I_C \cong -\alpha_F I_E = \beta_F I_B$. Here $\alpha_F < 1$ and $\beta_F = \alpha_F/(1 - \alpha_F)$.

2. *Cutoff:* In this state, which is one of the *two switching modes* of a bipolar transistor, both diodes are reverse-biased, i.e.,

npn	***pnp***
$V_{EB}, V_{CB} > 0$	$V_{EB}, V_{CB} < 0$

The actual boundary of this region is where I_E tends to reverse polarity from the active region, which requires at least some reverse biasing of the emitter-

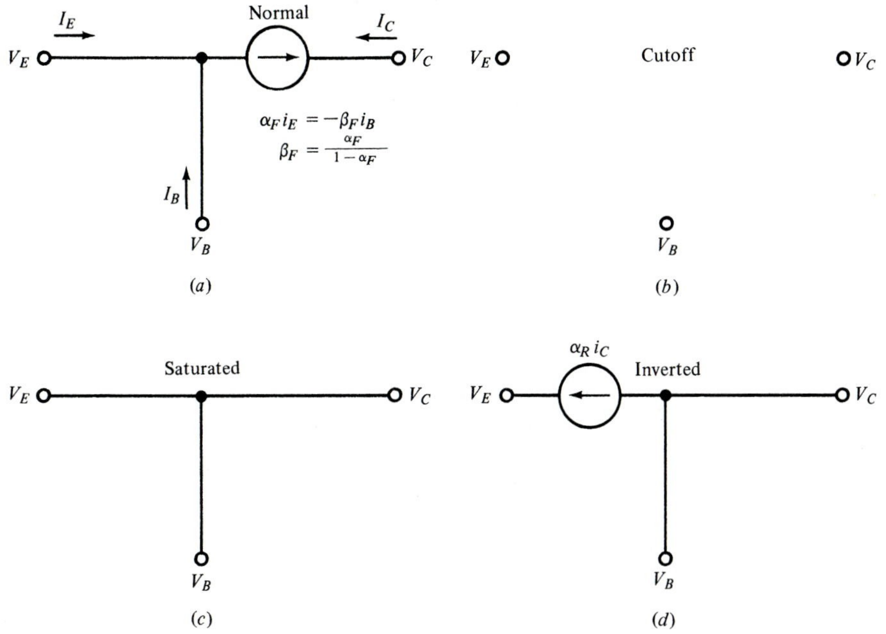

FIGURE 3.5
The four major bipolar transistor states. (*a*) Normal or active mode; (*b*) cutoff mode; (*c*) saturated or ON state; (*d*) inverted mode.

base junction (for example, 0.1 to 0.2 V). In the cutoff mode the three transistor terminals have very little current flowing into them.

3. *Saturated or ON:* This is the second switching mode of a bipolar transistor. In this state, both diodes are forward-biased, and V_{CE} is smaller than either V_{EB} or V_{BC} in magnitude. The state is usually described by:

npn	***pnp***
$V_{CB} < 0$	$V_{CB} > 0$
$I_E < 0$	$I_E > 0$

In this mode V_{CE} may be reduced to less than 0.1 V if the base current satisfies

$$|I_B| \gg \frac{|I_C|}{\beta_F}$$

4. *Inverted:* This mode is not frequently encountered, but it does occur in many nonlinear and switching applications. It occurs when the collector-base junction is forward-biased and the emitter-base junction is reverse-biased. The reverse-current gain is usually very poor, compared with the normal mode.

One commonly overlooked problem in both the cutoff and inverted modes is that typically the emitter-base reverse breakdown voltage of most high-speed transistors is relatively low, for example, 2 to 3 V.

For most of our discussions in this chapter, we use the ideal models of the diode and the bipolar transistor described above. Where the effects of junction forward drops and reverse leakage currents are worth noting, we do so in at least a qualitative way. More will be said about both bipolar and field-effect transistors (FETs) in Chap. 5, when we discuss precision switching of analog signals. In later sections of this chapter, we discuss a representative collection of nonlinear operational amplifier circuits that may be analyzed with considerable confidence by using the ideal diode models of the junction diode or the bipolar transistor. The key to success in analyzing these circuits is to test all possible diode ON or OFF states and to deduce the points where a transition is made from one state to another.

Example 3.2. Consider the simple diode dead-space circuit of Fig. 3.6. We want to know the transfer characteristic, that is, v_O versus v_I. We can analyze the circuit as follows:

1. Diodes D_1 and D_2 OFF

$$v_O = 0 \tag{3.4a}$$

2. Diode D_1 ON, D_2 OFF

$$v_O = \frac{v_I + E}{3} \tag{3.4b}$$

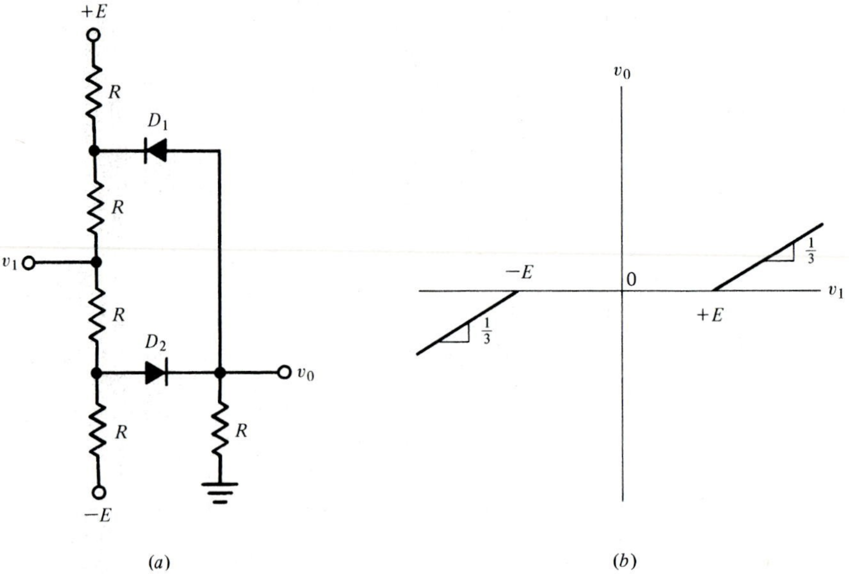

FIGURE 3.6
Refer to Example 3.2. (*a*) Simple passive dead-space circuit; (*b*) transfer characteristic.

3. Diode D_2 ON, D_1 OFF

$$v_O = \frac{v_I - E}{3} \tag{3.4c}$$

4. The diodes are never both ON at the same time (by inspection).

Figure 3.6*b* shows a sketch of the transfer characteristic obtained from Eqs. (3.4), where we observe that the segments of the three equations are chosen to obtain a continuous, single-valued curve. Note that if the circuit were not symmetric, the analysis would be more involved but qualitatively the same.

Example 3.3. Figure 3.7*a* shows a simple transistor inverter circuit; suppose we want to estimate the transfer characteristic, v_O versus v_I. Again, we can analyze the circuit with a simple model (Fig. 3.7*b*).

1. Diodes D_1, D_2 OFF

$$v_O = E \tag{3.5a}$$

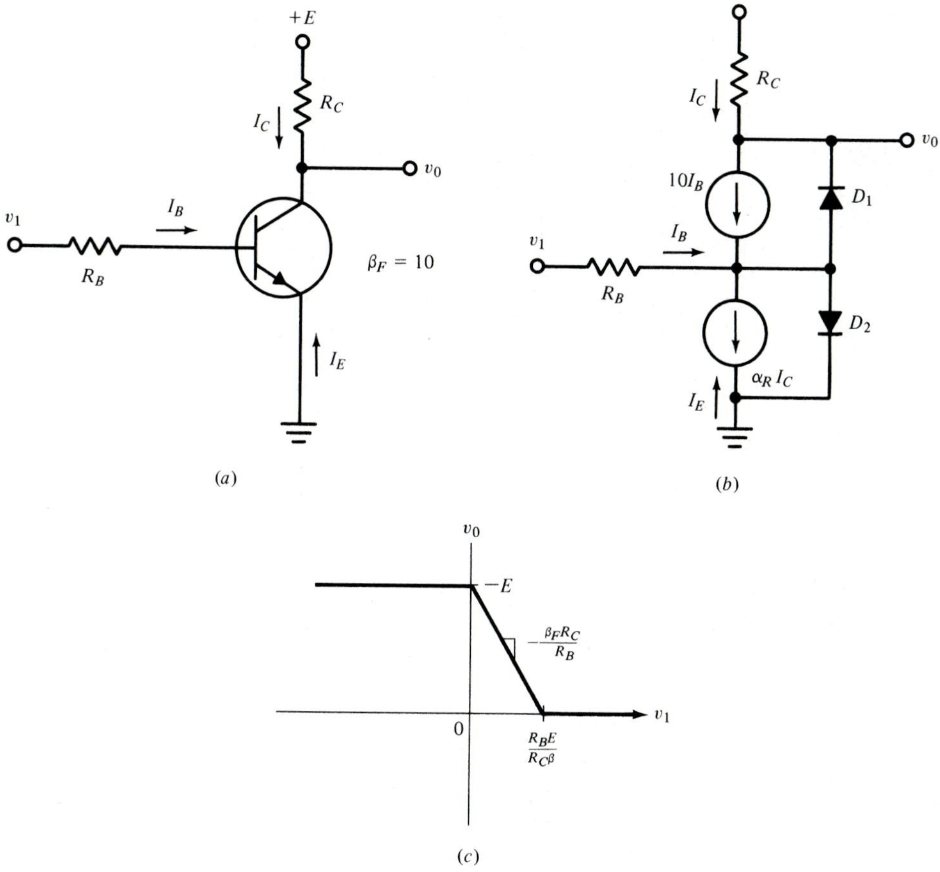

FIGURE 3.7
Refer to Example 3.3. (*a*) Simple transistor inverter; (*b*) ideal model, based on modes of Fig. 3.5; (*c*) transfer characteristic predicted by model of part *b*.

2. Diode D_2 ON, D_1 OFF

$$v_O = E - \frac{R_C \beta_F}{R_B} v_I \tag{3.5b}$$

3. Diodes D_1 and D_2 both ON

$$v_O = 0 \tag{3.5c}$$

Figure 3.7*c* shows a sketch of the desired input-output transfer characteristic, which is that of a grounded-emitter inverter.

Example 3.4. Figure 3.8 shows a very useful four-diode bridge circuit (note that the output load R_O could be the input resistor of an inverting amplifier circuit). All

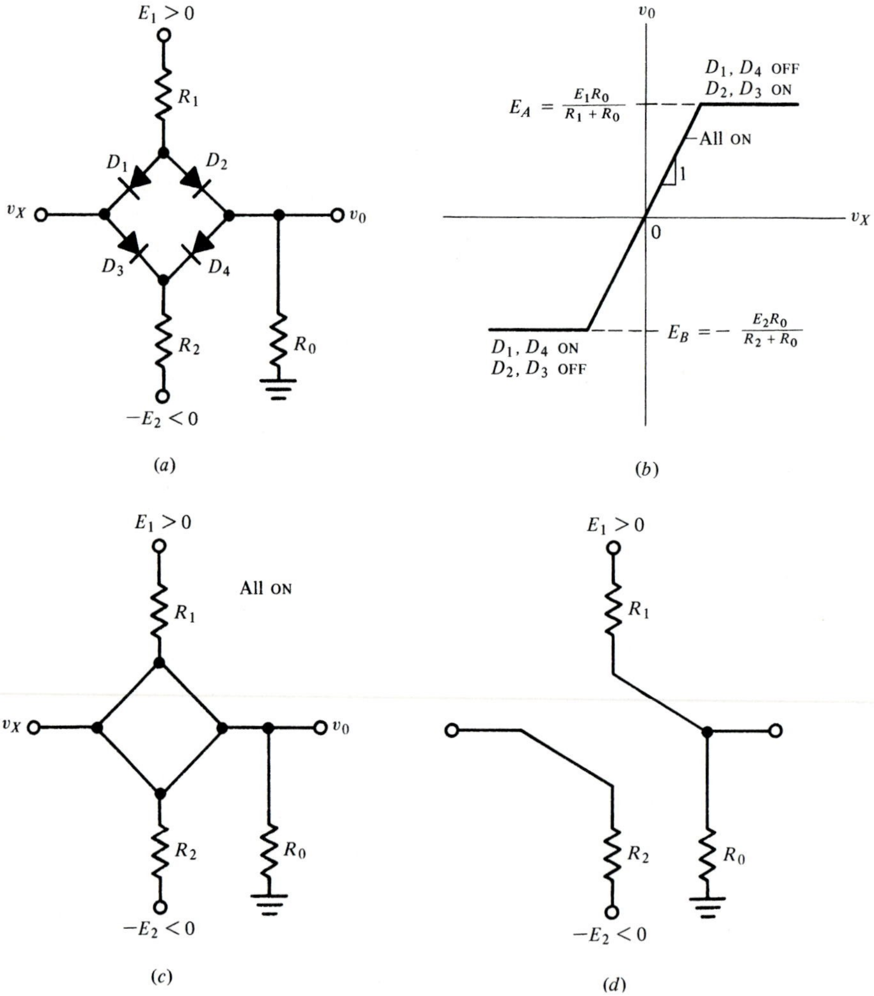

FIGURE 3.8
Refer to Example 3.4. (*a*) A four-diode bridge circuit; (*b*) transfer characteristic predicted by ideal diode model; (*c*) equivalent circuit with all four diodes ON; (*d*) equivalent circuit with D_2 and D_3 ON, D_1 and D_4 OFF.

output load current must be provided from the outside legs of the bridge; the input signal controls the way in which current branches. Using the ideal diode model, one could look at all 16 possible diode states; such detail is not usually required. Some thought will reveal that only three possibilities need be considered in detail:

1. All four diodes ON (Fig. 3.8*c*)

$$v_0 = v_X \tag{3.6a}$$

2. Diodes D_2, D_3 ON; D_1, D_4 OFF (Fig. 3.8*d*)

$$v_0 = \frac{E_1 R_0}{R_1 + R_0} = E_A \tag{3.6b}$$

3. Diodes D_1, D_4 ON; D_2, D_3 OFF

$$v_0 = \frac{E_2 R_0}{R_2 + R_0} = E_B \tag{3.6c}$$

The resulting sketch which combines these three cases is shown in Fig. 3.8*b*. Note that we have a limiter action. The limiting levels are given by (3.6*b*) and (3.6*c*). The four-diode bridge circuit is one well worth remembering. Besides being a fairly precise limiter, it can be used as an analog switch; i.e., levels E_1 and E_2 can be reversed in polarity by a switching waveform so as to completely block transmission. Note also that the limiting levels are dependent on the load R_0.

3.2 FEEDBACK LIMITERS AND COMPARATORS

Limiting is a very important nonlinear operation. Figure 3.9 describes the *limiter operator* with a possible block symbol (Fig. 3.9*a*) and a sketch of the transfer characteristic (Fig. 3.9*b*). In general we want the ideal characteristic:

$$v_0 = K v_1 \qquad L_- \le v_0 \le L_+, \ \frac{L_-}{K} \le v_1 \le \frac{L_+}{K} \tag{3.7a}$$

$$v_0 = L_+ \qquad v_1 > \frac{L_+}{K} \tag{3.7b}$$

$$v_0 = L_- \qquad v_1 < \frac{L_-}{K} \tag{3.7c}$$

Of course,

$$L_- \le L_+ \tag{3.7d}$$

Such an operation is ideally a zero-memory operation; i.e., (3.7) is satisfied regardless of the frequency content of the input. In actuality such an operation can be implemented with reasonable accuracy for signals with spectral content ranging from direct current up to some upper frequency determined by the internal dynamics of the circuit elements used to implement it, typically 10 to 100 kHz.

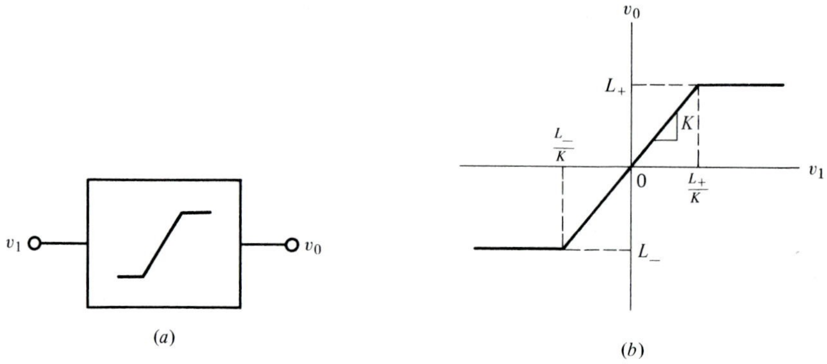

FIGURE 3.9
The limiter operator. (*a*) A typical block-diagram symbol; (*b*) important parameters; K is linear gain, and L_+ and L_- are the positive and negative limiting levels, respectively.

A related operator is a *comparator*, shown in Fig. 3.10. This is essentially a limiter with a very high (ideally infinite) gain K. In Fig. 3.10 we have shown a comparison-level E_R as well, to indicate that often in this application, we want to compare two signals to each other and decide which is larger. Comparators form an essential part of all analog-digital interfaces, and we discuss this application further in Chap. 6. For the present, we want to describe how to design one. The ideal comparator can be described by

$$v_O = \begin{cases} L_+ & v_I > E_R \\ 0 & v_I = E_R \\ L_- & v_I < E_R \end{cases} \tag{3.8}$$

Now consider Fig. 3.11. This is an operational amplifier circuit which uses a breakdown or Zener diode (we use the latter term throughout this text). The i-v

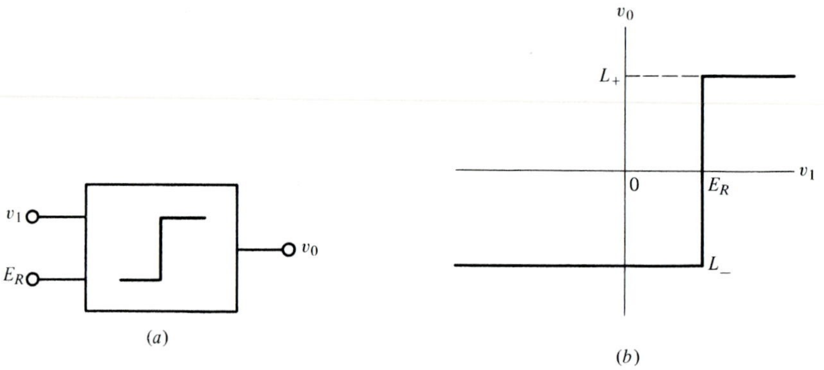

FIGURE 3.10
The comparator characteristic. (*a*) A typical block-diagram symbol; (*b*) important parameters; E_R is the comparison threshold, and L_+ and L_- are the two nominal output levels. Often one level will be near zero, e.g., when the circuit is to drive some standard digital logic gate.

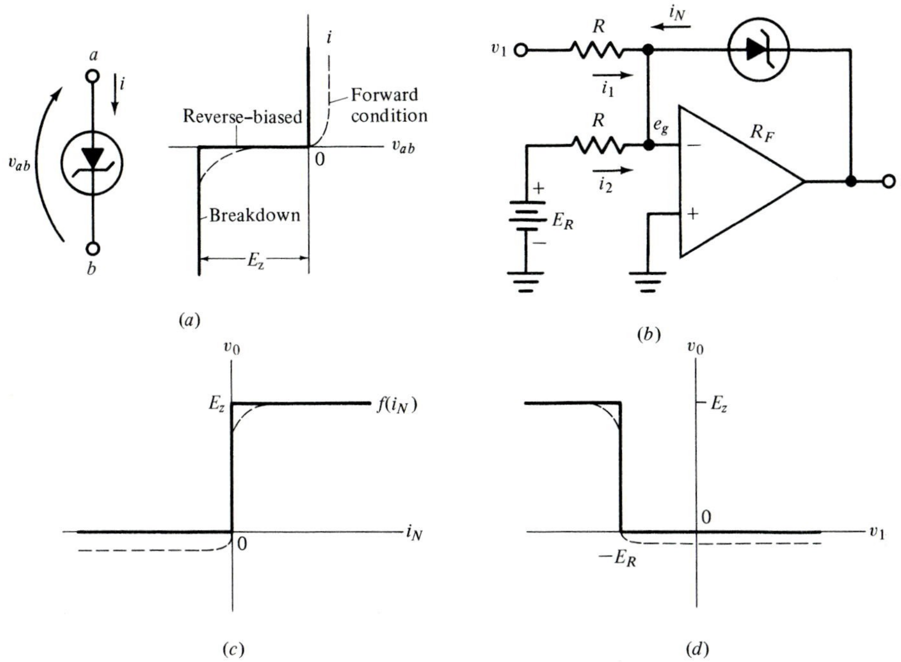

FIGURE 3.11
A simple comparator made with an operational amplifier and a Zener (breakdown) diode. (*a*) An ideal model of the Zener diode: (*b*) the basic circuit; (*c*) the conductance characteristic of the diode, $f(i_N)$; (*d*) the circuit transfer characteristic, based on $f(i_N)$.

characteristic of such a diode is sketched in Fig. 3.11*a*, where the solid line is the ideal characteristic and the dotted line indicates the more realistic, nonideal behavior. The principal deviations from ideal will be a forward drop of perhaps 0.6 V when the diode is forward-biased, a small reverse leakage current, and often a considerable amount of rounding of the characteristics near the reverse breakdown point. If the circuit of Fig. 3.11*b* is constructed with this type of diode as a feedback element, we can analyze the circuit performance along the lines of Example 3.1; i.e., see Fig. 3.2 and Eq. (3.2). Figure 3.11*c* shows the relationship between v_0 and i_N; again

$$i_1 + i_2 + i_N = 0$$

and the summing junction voltage e_g is forced to be zero by the operational amplifier (or else the amplifier is in overload). Also (since we assume an ideal operational amplifier)

$$i_1 = \frac{v_1}{R} \qquad i_2 = \frac{E_R}{R}$$

and

$$v_0 = f(i_N) = f\left(\frac{-v_1 - E_R}{R}\right)$$

With the Zener diode as a feedback element, then, using the ideal model of the diode,

$$f(i_N) = \begin{cases} E_z & i_N \geq 0 \\ 0 & i_N < 0 \end{cases}$$

Thus we have the overall transfer characteristic of Fig. 3.11*d*

$$v_0 = \begin{cases} E_z & v_1 \leq -E_R \\ 0 & v_1 > -E_R \end{cases} \tag{3.9}$$

This is evidently a comparator circuit; except for the polarity inversion, it matches the sketch of Fig. 3.10*b*. In actuality, the output will be a few tenths of a volt negative when v_1 exceeds the threshold level $-E_R$ because of the diode forward drop. The transition between the two output levels is not discontinuous,

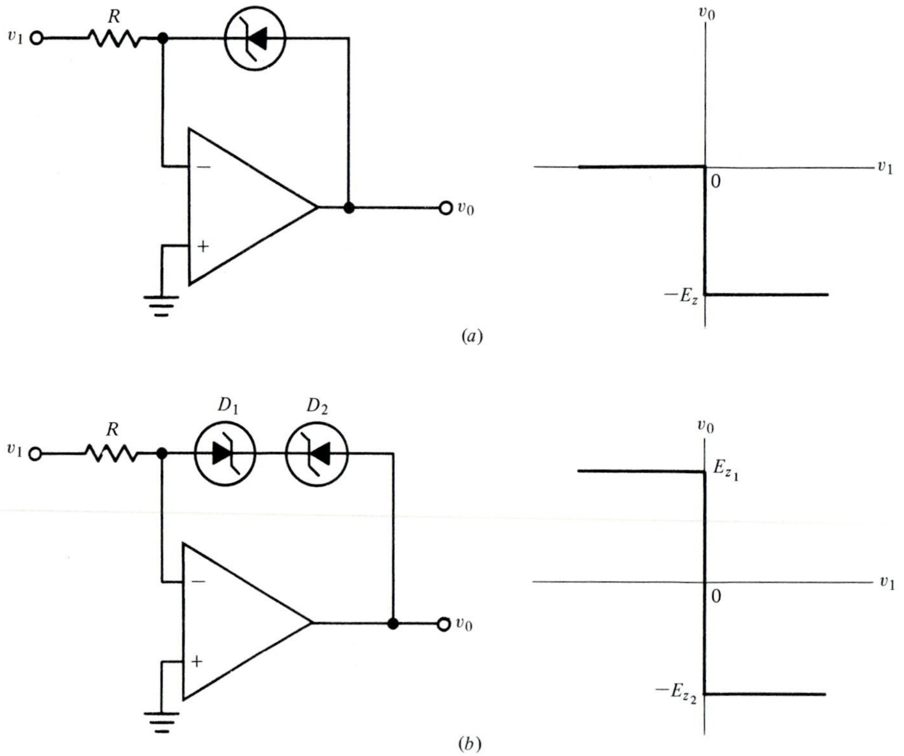

FIGURE 3.12
Variations of the Zener diode comparator circuit. (*a*) Characteristic obtained from Fig. 3.11, but with diode polarity reversed; (*b*) characteristic obtained by two Zener diodes in series opposition, or by a single double-anode Zener diode.

but is made with a very steep slope, i.e., the open-loop gain of the operational amplifier.

Figure 3.12 shows some variations of the simple Zener diode comparator circuit. By reversing the diode (Fig. 3.12*a*) we get the characteristic shown. (Here we have also removed the reference-comparing voltage E_R and its associated resistance, so that the comparison is with respect to 0 V.) In Fig. 3.12*b* we employ two Zener diodes in series opposition to get the indicated bipolar characteristic. Indeed, one can even purchase so-called *double-anode Zener diodes* with both junctions incorporated in a single device.

Figure 3.13 shows a logical extension of the simple Zener diode comparator circuit to a limiter amplifier. With only the one input resistor from v_1, we have

$$v_0 = -\frac{R_F v_1}{R_1} \qquad -E_{z2} < v_0 < E_{z1} \tag{3.10}$$

However, the two Zener diodes limit the output level as indicated. This transfer characteristic agrees with that of (3.7); of course, the gain has a negative polarity in the amplifying region. The addition of a second input v_2, as shown in Fig. 3.13*a*, permits forming a weighted sum of two inputs with output limiting, i.e., we have a *summer-inverter limiter* circuit. In general, then, the two Zener diodes act as a *level clamp* or *feedback limiter* to limit output excursions, but otherwise do not appreciably affect circuit operation.

In this circuit, of course, the limiting level is controlled by the Zener diode breakdown voltage, and thus it is difficult to adjust it precisely. Limiting is fairly *hard;* i.e., the slope of the transfer characteristic in the limiting regions is low. Some Zener diodes, however, will exhibit considerable rounding near the beginning of breakdown, and thus the limiting does not begin sharply.

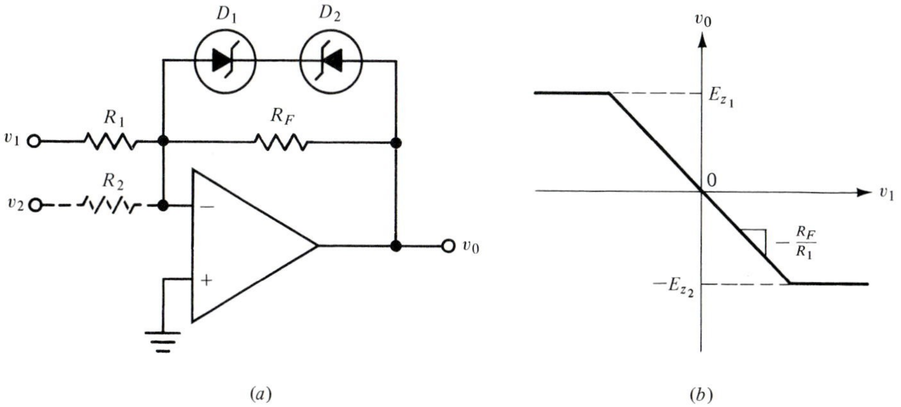

FIGURE 3.13
A simple limiter amplifier circuit made with an operational amplifier and two Zener diodes. (*a*) The basic circuit (note that summing can also be performed by adding an auxiliary input v_2); (*b*) the nominal transfer characteristic.

Figure 3.14*a* shows a different type of limiter amplifier circuit which uses ordinary *pn* junction diodes. Its analysis is facilitated by noting that three combinations of diode states are possible:

1. Diodes D_1 and D_2 both OFF (Fig. 3.14*c*). Here obviously the diode networks have no effect on the summing junction current, and

$$v_O = -\frac{R_F}{R_1} v_1 \tag{3.11}$$

2. Diode D_1 ON, D_2 OFF (Fig. 3.14*d*). If we make v_1 sufficiently positive, eventually node e_1 reaches 0 V and can get no more negative since D_1 then turns ON. Thus we have a *breakpoint* when

$$v_O = -\frac{E_1 R_B}{R_A} \tag{3.12a}$$

which establishes a nominal limiting level. Readers should study this until they fully understand where this last relationship comes from. As v_1 becomes more positive, the effective feedback resistance around the amplifier is

$$R_B' = R_B \parallel R_F \simeq R_B \tag{3.12b}$$

From Fig. 3.14*d*, we can deduce that

$$v_O = -R_B'\left(\frac{v_1}{R_1} + \frac{E_1}{R_A}\right) \tag{3.12c}$$

This last equation is not itself too important; note that (3.12*a*) gives the breakpoint at the corner (i.e., the nominal limiting level). Also we see that the slope of (3.12*c*) is

$$\frac{dv_O}{dv_1} = -\frac{R_B'}{R_1} \simeq -\frac{R_B}{R_1}$$

Thus if R_B is made small compared to R_F, then the slope in the limiting region can be made fairly small, and

$$R_B' \cong R_B \qquad R_F \gg R_B$$

3. Diode D_2 ON, D_1 OFF. The analysis of this case is similar to the last; we find that there is another corner or breakpoint at

$$v_O = \frac{E_2 R_C}{R_D} \tag{3.13a}$$

which establishes another nominal limiting level. The slope in the limiting region (upper left of Fig. 3.14*b*) is again found by defining

$$R_C' = R_C \parallel R_F \simeq R_C \tag{3.13b}$$

and then

$$\frac{dv_O}{dv_1} = -\frac{R_C'}{R_1} \simeq -\frac{R_C}{R_1} \tag{3.13c}$$

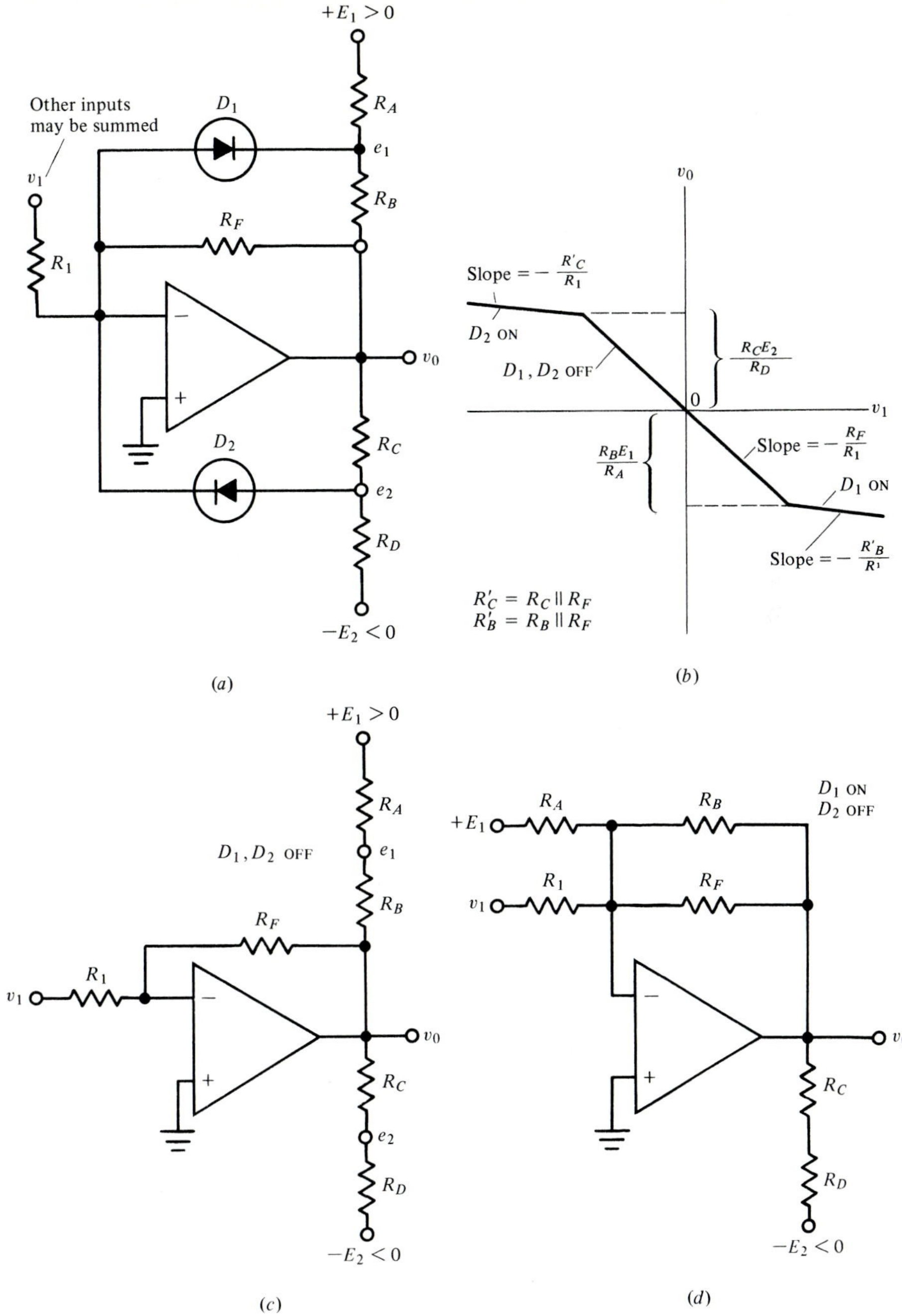

FIGURE 3.14
A general-purpose limiter made with ordinary diodes and resistor networks. (*a*) The basic circuit (note that dc power voltages are required); (*b*) the major features of the transfer characteristic; (*c*) equivalent circuit in the normal amplifying region, both diodes OFF; (*d*) equivalent circuit in the negative-limit region, D_1 ON, D_2 OFF. Note that the slopes in the limiting regions are highly dependent on the choice of resistance values; limiting levels are easily adjusted by adjusting resistance ratios.

Now we have a circuit that limits reasonably well if R_B and R_C are small compared to R_F. The circuit is, however, what is normally called a *soft limiter;* i.e., the limiting action is not sharp, there is some rounding at the corners, and the slope in the limiting regions is not as low as might be desired. The circuit does have the advantage that it is relatively easy to adjust the limiting levels by adjusting the resistance ratios and/or the supply voltages, as indicated in (3.12*a*) and (3.13*a*). Of course, the *removal of the feedback resistor* R_F makes the circuit into a *comparator;* note that the limiting levels are still *found by (3.12a) and (3.13a),* and the slopes in the limiting region are given by (3.12*c*) and (3.13*c*) but with R'_B and R'_C replaced by R_B and R_C, respectively.

Example 3.5. Suppose we want a limiter amplifier to have an inverting gain of 20, to limit at ±5 V, and to have slopes in the limiting region of less than 0.1 V/V. Assume ±15 V dc is available. The largest resistor is to be 100 kΩ. Figure 3.15*a* shows a possible design. Hopefully it is clear that R_F is the largest resistor; thus $R_F = 100\ \text{k}\Omega$, and

$$R_1 = \frac{R_F}{20} = \frac{10^5}{20} = 5\ \text{k}\Omega$$

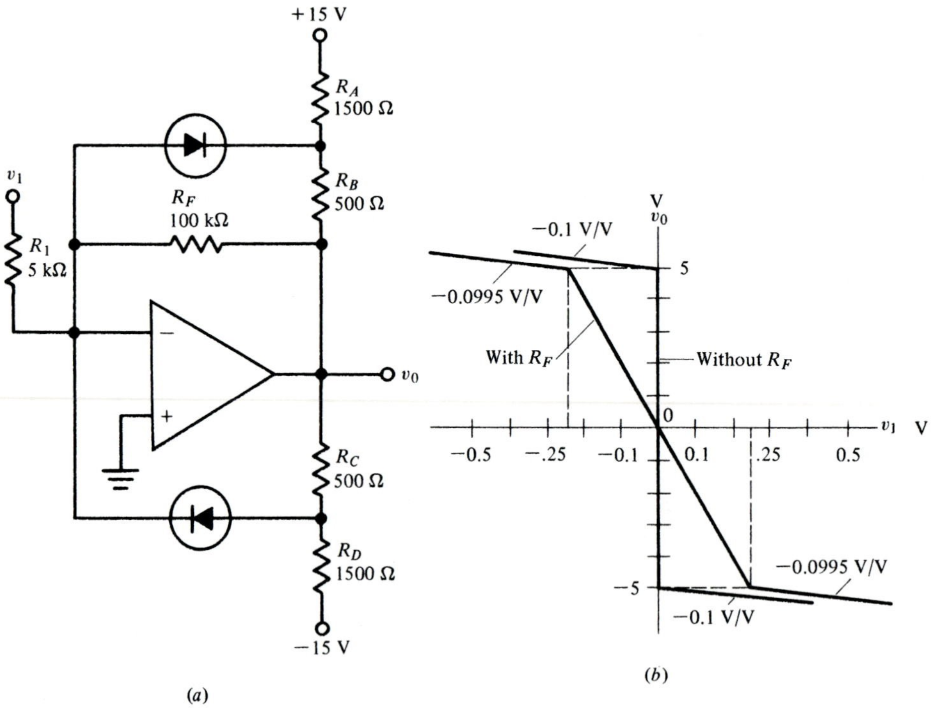

FIGURE 3.15
Refer to Example 3.5. (*a*) A proposed design to meet the specifications; (*b*) the resulting nominal characteristic, with and without R_F.

We establish the ratio of 3 to 1 for the limit-setting resistor from (3.12*a*) and (3.13*a*), that is,

$$-5 = \frac{-E_1 R_B}{R_A} = \frac{-15 R_B}{R_A}$$

or

$$\frac{R_B}{R_A} = \frac{1}{3}$$

and similarly

$$\frac{R_C}{R_D} = \frac{1}{3}$$

Note that the +15-V dc supply is used in establishing the negative limiting level, and vice versa. We now must find suitable values for the limiting resistors to obtain the desired slope. If we note that

$$R'_B < R_B \qquad \text{and} \qquad R'_C < R_C$$

we can simplify our job by setting

$$\frac{R_B}{R_1} = \frac{R_C}{R_1} = 0.1$$

or

$$R_B = R_C = 500\ \Omega$$

and thus

$$R_A = R_D = 1500\ \Omega$$

Actually, the slope will be a bit less than 0.1 V/V in the limiting region.

Figure 3.15*b* shows the transfer characteristic of the final design (based on an ideal diode model). Included are more precise estimates of the slopes in the limiting region:

$$R'_B = R'_C = 10^5\ \Omega \parallel 500\ \Omega = 497.5\ \Omega$$

and thus the slopes are

$$-\frac{497.5}{5000} = -0.0995\ \text{V/V} \simeq -0.1\ \text{V/V}$$

Also, the limiting levels will be slightly in error, due to the forward characteristic of the diodes, and there will be some rounding at the corners.

Figure 3.15*b* also shows the transfer characteristic which results if R_F is removed. Note that the nominal limiting levels are unchanged, and the slopes in the limiting regions are exactly −0.1 V/V, but now the transition between the two levels is abrupt (limited only by the open-loop gain of the operational amplifier), and we have a zero-threshold comparator.

Zero Limiter or Half-Wave-Rectifier Circuits

In Fig. 3.16, we see that an ordinary diode may be used to make a crude zero limiter or half-wave-rectifier operator. Ideally, we want $v_O = -v_I$ in only one

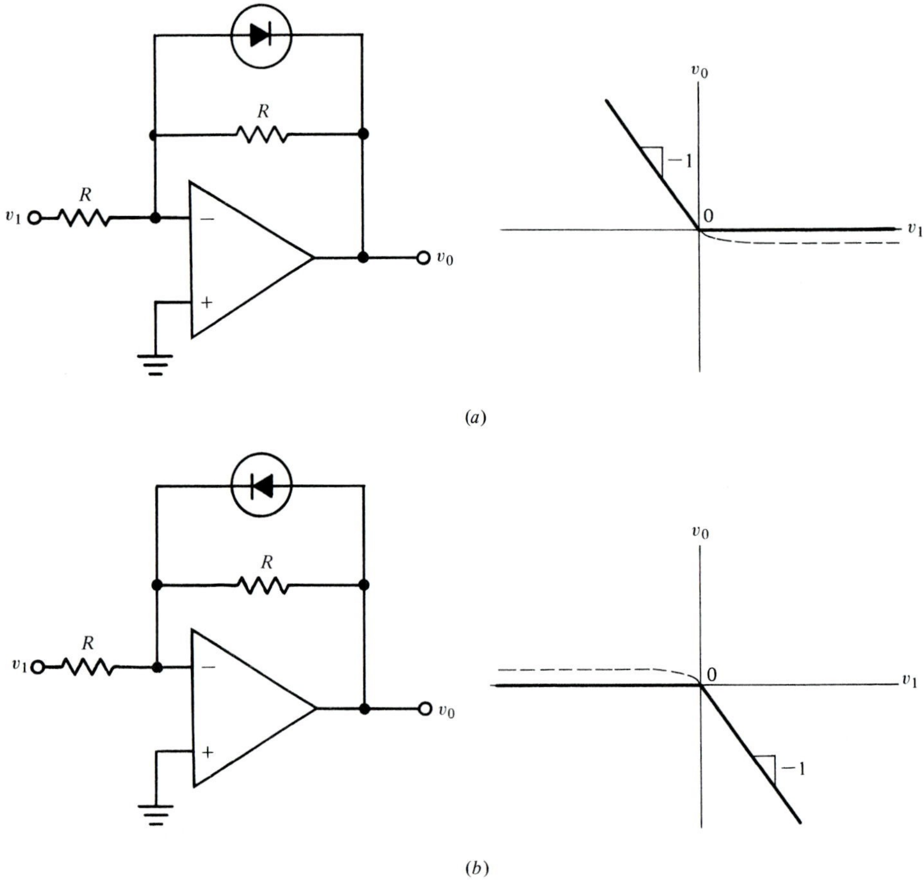

FIGURE 3.16
Crude zero limiters made with only one diode; diode forward drop will not permit precise zero limiting. Circuits (*a*) and (*b*) show the two polarities available.

quadrant and zero in the other. As indicated by the dotted line, the forward drop of the diode will allow the output to be a few tenths of a volt in the limiting region, instead of zero. The *precision limiter circuit* of Fig. 3.17*a embeds* the *limiting diode* D_1 in a *feedback loop* to suppress its nonideal characteristics. Now D_1 merely serves to sense the polarity of the operational amplifier output voltage. Note also that the circuit output v_O is not taken directly from the operational amplifier. This circuit is worthy of some study. When the input is negative, the diode D_1 is ON, since v_A tends to be positive. We can look upon the forward drop of D_1 as an offset in the operational amplifier output stage; the operational amplifier (assuming it is not in overload) will do whatever is required to keep $e_g = 0$. Since v_A is positive and e_g is 0, diode D_2 is OFF. Thus for negative-valued input, the feedback resistor and the input resistor establish summing junction currents at

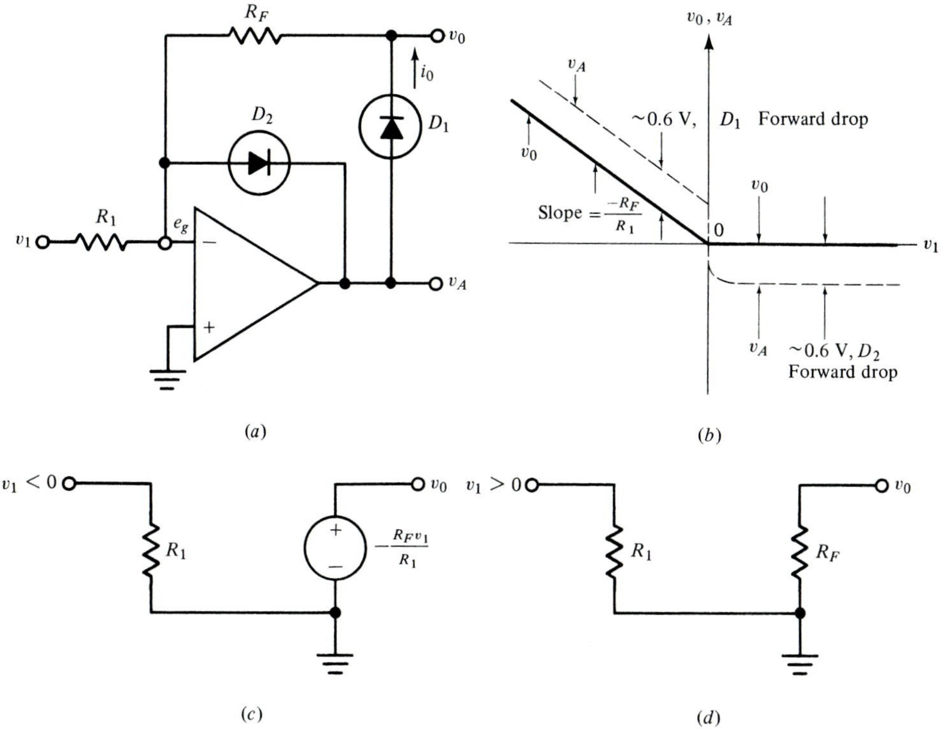

FIGURE 3.17
A precision zero limiter, made by embedding the diode D_1 in the feedback loop; D_2 merely catches the operational amplifier to prevent overload. (*a*) The basic circuit; (*b*) the resulting precise limiting characteristic; note the difference between v_O and v_A (the dotted line is v_A); (*c*) the overall equivalent active circuit in the nonlimiting mode; note the low output impedance; (*d*) the equivalent circuit in the limiting mode; note the high output impedance.

e_g such that

$$v_0 = \frac{-R_F}{R_1} v_1 \qquad v_1 < 0 \tag{3.14a}$$

If, however, v_1 becomes positive, then v_A becomes negative; v_0 cannot follow along, since D_1 can only conduct in a direction which makes i_0 positive. Thus the feedback resistor is left essentially unconnected to the amplifier output but still connected to the virtual ground at e_g, and

$$v_0 = 0 \qquad v_1 > 0 \tag{3.14b}$$

Some form of catching network is needed to keep the amplifier from going into overload when $v_1 > 0$; in Fig. 3.17*a*, silicon diode D_2 serves to catch the output and clamp v_A to a small negative value, i.e., the forward drop across D_2, which is typically 0.6 V (see dotted line in Fig. 3.17*b*).

This circuit is quite accurate! The corner of the transfer characteristic and, at low frequencies, the ideal limiter characteristic will be implemented with an accuracy of better than 0.1 V. At higher frequencies, diode capacitance will cause some errors in the neighborhood of the corner; these errors can be minimized by using fast diodes for both D_1 and D_2 and small resistance values. Reversing both diodes will cause the circuit to limit for negative inputs and amplify for positive ones.

Note that the output impedance of the circuit of Fig. 3.17*a* varies with circuit conditions. When v_O is positive (D_1 ON), the output impedance will be very low because of the negative feedback provided by the amplifier. When $v_I > 0$, v_O will have an open-circuit voltage very close to zero, but will have the feedback resistance R_F as a Thévenin-equivalent output resistance. Thus, if any external load attached to the circuit has an internal offset voltage that tends to pull v_O away from ground, v_O may not stay equal to zero. Figure 3.17*c* and *d* show the approximate equivalent circuits for this limiter in both modes of operation. It is thus often desirable to buffer the output of this circuit with an additional amplifier stage, so that external loads will not generate unwanted behavior. In the next section of this chapter, we will see how the circuit of Fig. 3.17*a* can be augmented to make an absolute-value or full-wave-rectifier operator.

Example 3.6. Suppose we want to build an ac voltmeter which will provide a voltage equal to the rms value of sinusoidal input signals. We assume that the input is a pure sinusoid with zero average value, i.e.,

$$v_I = A \sin \omega t$$

Assume that v_I can be as large as 10 V rms; thus the peak value is

$$A \leq \sqrt{2} \times 10 = 14.14 \text{ V} \quad \text{peak}$$

Let us use the circuit of Fig. 3.18 for this job. The output amplifier B provides a low-impedance output, but more importantly it is used to filter the output so that only a steady-state or dc voltage appears at v_O. The filtering time constant is $R_F C_F$, which must be chosen so that

$$R_F C_F \gg \frac{1}{\omega}$$

in order to get good filtering. Suppose we have ± 10-V amplifiers to work with. We have to be sure that neither overloads.

Figure 3.18*c* shows a sketch of v_2, which is the output of the half-wave rectifier. To keep amplifier A from overloading, we must ensure that

$$\frac{AR_2}{R_1} \leq 10 - \text{diode drop} \qquad \text{V}$$

In this case, then, we pick

$$R_2 = \frac{R_1}{2}$$

e.g., we let $R_1 = 10\ \text{k}\Omega$ and $R_2 = 5\ \text{k}\Omega$. The average value of v_2 is

$$v_{2,\text{av}} = \frac{AR_2}{\pi R_1} = \frac{A}{2\pi}$$

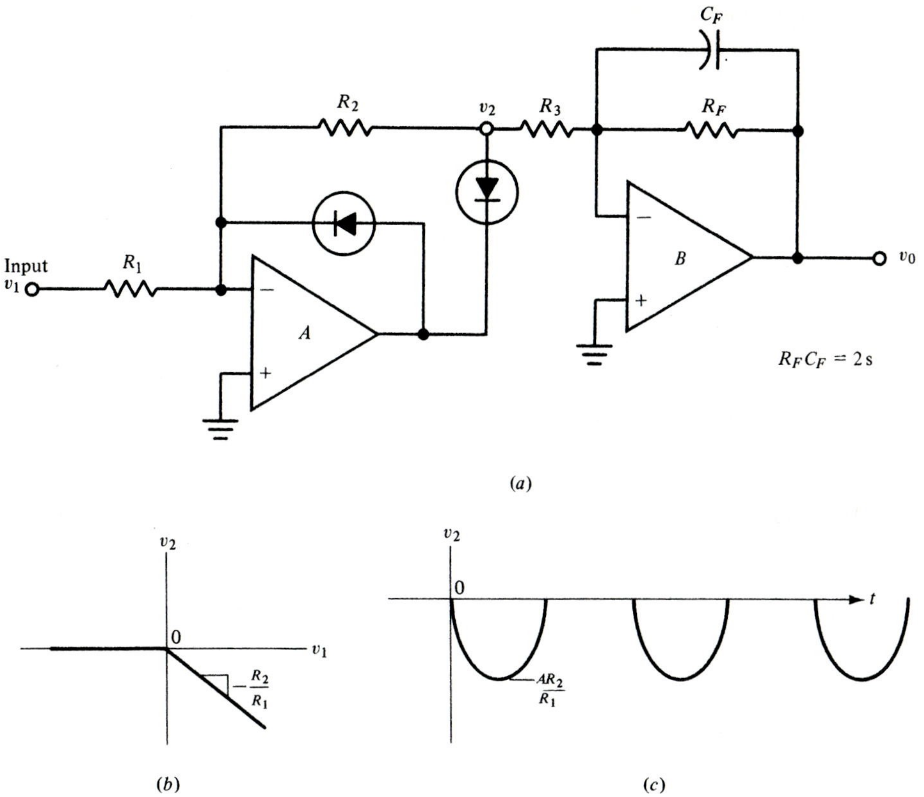

FIGURE 3.18
A basic ac voltmeter circuit. (*a*) Two-amplifier half-wave rectifier with filter capacitor C_F; (*b*) v_2 versus v_1; (*c*) v_2 versus time. Output amplifier performs averaging.

Now the rms value of v_1 is

$$v_{1,\text{rms}} = \frac{A}{\sqrt{2}} = 0.707A$$

Thus we pick the dc gain of the second stage (i.e., that dc gain from v_2 to v_0) so that

$$\frac{v_{0,\text{av}}}{v_{2,\text{av}}} = \frac{A/\sqrt{2}}{A/2\pi} = \pi\sqrt{2} = 4.44$$

or

$$R_3 = \frac{R_F}{4.44} = 0.225R_F$$

Suppose we know that the input sinusoids have a minimum frequency of 10 Hz; then

$$\omega \geq 20\pi = 62.8 \text{ rad/s}$$

and we want

$$R_F C_F \gg \frac{1}{62.8} = 0.0159 \text{ s}$$

Let us say

$$R_F C_F = 2 \text{ s}$$

To keep C_F of reasonable size, we will pick

$$R_F = 1 \text{ M}\Omega \qquad C_F = 2\ \mu\text{F} \qquad R_3 = 225 \text{ k}\Omega$$

Note that we have not said anything about possible dc offset problems in the amplifiers, which might dictate a different resistance level, with smaller resistors and a larger capacitor. The circuit will hopefully do the specified job for all input signals above 10 Hz, up to say 10 kHz with a fairly fast amplifier *A* (note that amplifier *B* can be slow).

Improved Limiters

Although the zero limiter of Fig. 3.17 is quite precise, the other general limiter circuits we have discussed (e.g., Figs. 3.12 to 3.14) often do not perform as well as desired. Figure 3.19 shows an improved Zener diode limiter comparator circuit which makes a sharper and faster transition between states. The use of low-leakage diodes for D_3 and D_4 reduces the effect of leakage currents on the circuit in the nonlimiting mode. The slope in the limiting region will be

$$\frac{dv_O}{dv_I} = \frac{-(R_d + R_z)}{R_1}$$

where R_d = diode forward resistance and R_z = Zener diode breakdown resistance.

The limiting levels cannot be adjusted without changing the Zener diode breakdown voltages; also note that now the limiting levels are $+(E_{z1} + E_{f3})$ and $-(E_{z2} + E_{f4})$, where E_{z1} and E_{z2} are the Zener breakdown voltages and E_{f3} and E_{f4} are the forward drops of D_3 and D_4.

The biasing resistors R_A and R_B are chosen to establish a few milliamperes of current through the Zener diodes, keeping them always in the breakdown mode.

Figure 3.20 shows a transistor hard limiter. The circuit operates very much like the limiter of Fig. 3.14, but the slopes in the limiting region are reduced by the forward current gain of the transistors. As in Fig. 3.14, the limiting levels are given by

$$\text{Positive limit} = \frac{+E_2 R_C}{R_D} \tag{3.15a}$$

$$\text{Negative limit} = \frac{-E_1 R_B}{R_A} \tag{3.15b}$$

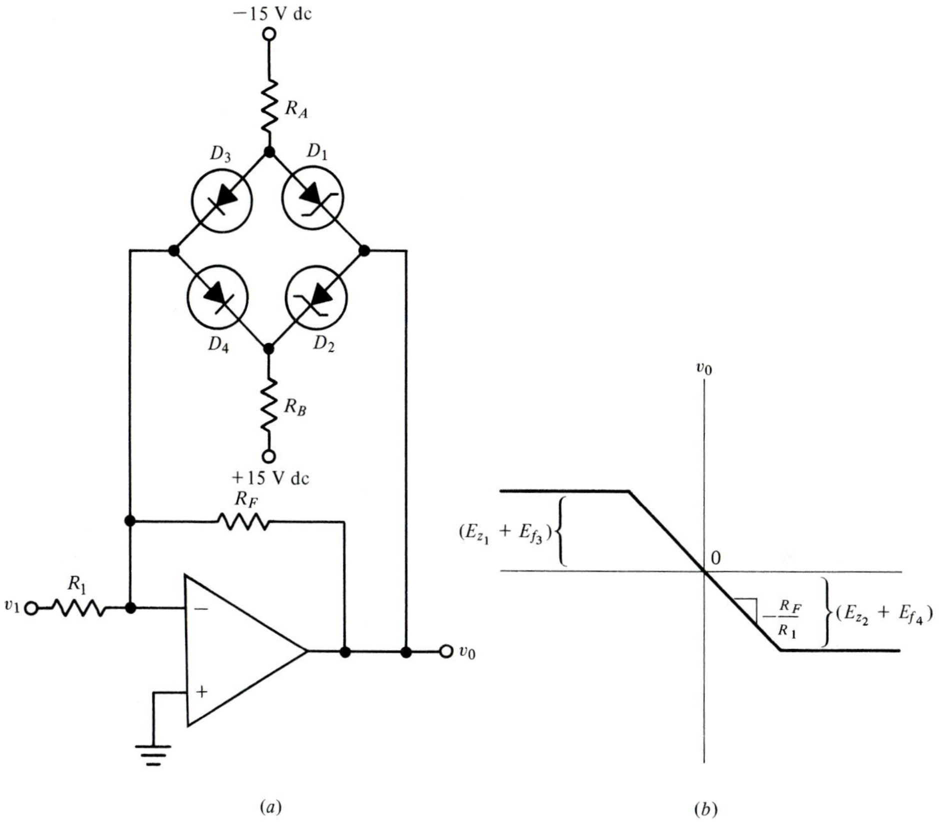

FIGURE 3.19
An improved Zener diode limiter (or comparator) circuit. (*a*) The current provided by the outer bridge legs via R_A and R_B provides for sharper corners and faster switching speeds; (*b*) approximate transfer characteristic.

i.e., they are the same as those of the limiter of Fig. 3.14. Also the slope in the nonlimiting region is

$$\text{Nonlimiting slope} = \frac{-R_F}{R_1} \tag{3.15c}$$

Of course, a comparator may be made by removing R_F.

In the circuit of Fig. 3.20, however, the limiting slopes are greatly reduced. In this circuit, the limiting is initiated by the emitter-base diodes of Q_1 and Q_2, as one or the other goes into the active mode. In the active mode, the emitter current is $\beta + 1$ times the base current provided by the feedback network. The net effect of the forward current gain of the transistors is to reduce the limiting slopes by the factor β. Thus the limiting slopes become

$$\text{Slope in positive limit} \cong \frac{-R_C}{R_1\beta_2} \tag{3.15d}$$

$$\text{Slope in negative limit} \cong \frac{-R_B}{R_1\beta_1} \tag{3.15e}$$

Since the transistor betas can be 50 to 200, the slopes are relatively small, and we have so-called *hard* limiting. Note diodes D_1 and D_2. They are usually needed to prevent emitter-base reverse breakdown when the transistors are cut off, since in that state the bases may move several volts away from ground while the emitters are held at ground potential. Also note that Q_1 and Q_2 should be a reasonably well-matched complementary-symmetry pair.

Example 3.7. If the circuit of Fig. 3.20 is used instead of that of Fig. 3.15, i.e., we use transistors instead of diodes but leave everything else the same, then the

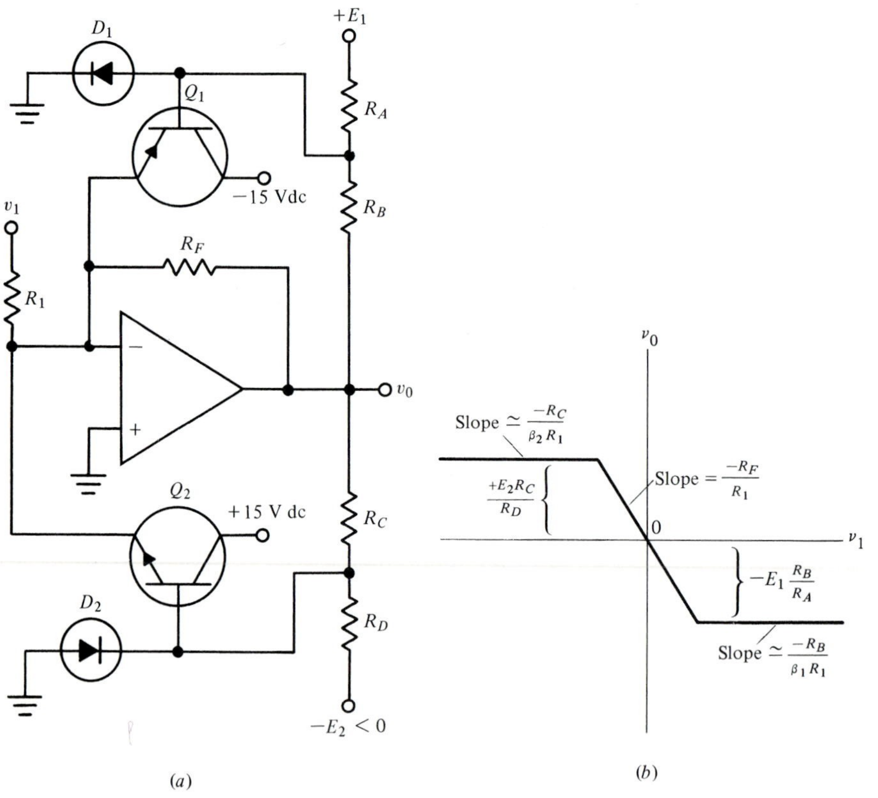

FIGURE 3.20
A transistor hard limiter, based upon the circuit of Fig. 3.14, but using the emitter-base junctions of two bipolar transistors as the limit-sensing elements. (*a*) Compare the basic circuit to Fig. 3.14; (*b*) the resulting transfer characteristic has much lower limiting slopes, which have been reduced by the factor β of the transistors. Diodes D_1 and D_2 are often needed to prevent reverse emitter-base breakdown of the transistors.

slopes in the limiting regions will be less. For example, if the transistor betas were 50 or more, the slopes would be about 2 mV/V.

The Precision Bridge Limiter

The transistor limiter just discussed (Fig. 3.20) will serve for most accurate limiting applications. However, for high-accuracy limiting (e.g., better than 50-mV accuracy) the precision bridge limiter circuit of Fig. 3.21 is probably the best,

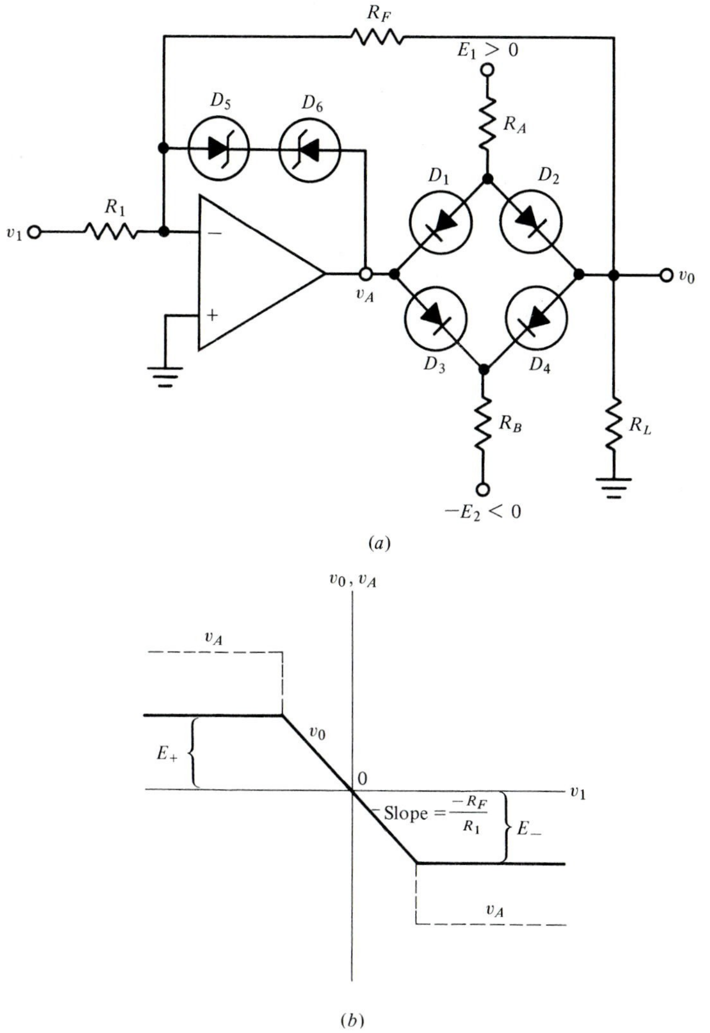

FIGURE 3.21
A precision limiter designed around a four-diode bridge embedded in a feedback loop. (*a*) In the basic circuit Zener diodes D_5 and D_6 merely serve to prevent amplifier overload; (*b*) the nominal transfer characteristic is dependent on the values of both R_L and R_F; see Eq. (3.16). Static accuracy of 50 mV is possible.

particularly at low frequencies. This is another interesting example of how a nonlinear network (R_A, R_B, and D_1 through D_4) can be embedded in a feedback loop, to make its properties more ideal. The limiting levels are accurately established by the four-diode bridge, which is similar to the one described in Example 3.3 and Fig. 3.8. The Zener diode catching network is merely used to keep the amplifier from going into overload when the bridge limits, since during limiting the output of the amplifier is effectively disconnected from the feedback loop. The only tricky aspect of analyzing this circuit is the calculation of limiting levels (which, incidentally, will be fairly exact). Note that R_L and R_F together act as the output load of the bridge. Thus the limiting levels are found to be (refer to Fig. 3.21)

$$E_+ = \frac{R_0 E_1}{R_0 + R_A} \tag{3.16a}$$

and

$$E_- = \frac{-R_0 E_2}{R_0 + R_B} \tag{3.16b}$$

where

$$R_0 = R_F \parallel R_L \tag{3.16c}$$

Thus the load being supplied by the circuit affects the limiting levels, and a buffer amplifier may be advisable. One might, for example, follow the circuit by an inverting amplifier, with R_L as the input resistor.

Suppression of Leakage Currents and Stray Capacitances

In Fig. 3.22, we show a comparator made by adding leakage-suppression diodes to a clamp or limiting circuit. The network D_1, D_2, and R_A through R_D functions as a basic limiter, similar to Fig. 3.14. Diodes D_3 and D_4, along with R_5, form a tee network in the feedback loop which raises the short-circuit feedback resistance in the nonlimiting mode. This approach might be advisable in building a precision comparator, where the threshold level is to be as accurate as possible. It would suppress leakage currents from diodes D_1 and D_2, which tend to create an offset error in the circuit in a manner similar to operational amplifier offset currents. Other limiters, such as a Zener diode network, may also be used with this approach. Equations (3.11) through (3.13) still apply to the design of such a circuit. The use of relatively low resistance values would also tend to improve circuit speed and reduce the effects of offset currents and parasitic capacitances.

General Use of Feedback Limiters or Clamps

The reader has probably noticed by now that the limiter and comparator circuits thus far discussed have a number of common properties. Figure 3.23 shows the general structure of these circuits; with R_F absent we have a comparator, with it present we have a limiter. The limiter circuits in Figs. 3.12, 3.13, 3.14,

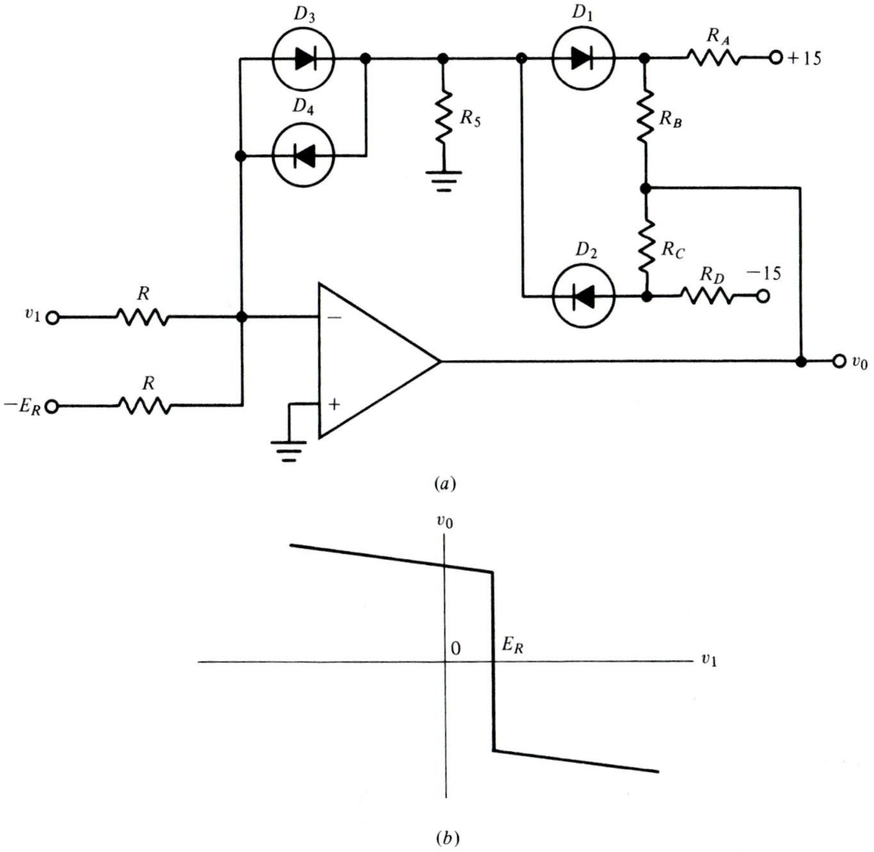

FIGURE 3.22
The use of extra diodes (D_3 and D_4) to suppress leakage currents and establish a precise threshold level. (*a*) The basic circuit, which operates in a manner similar to Fig. 3.14; (*b*) a sketch of the transfer characteristic; the comparison threshold will not be particularly sensitive to leakage current in the diodes.

and 3.20 all follow this same form, and one can usually apply one of these so-called *feedback limiters* between the output and inverting input terminals of an operational amplifier to prevent overload. If the limiting levels are set at the nominal output voltage extremes of the amplifier (±10 V), then the feedback limiter serves as an overload protector but has little, if any, effect on the amplifier circuit as long as the output level is not exceeded. Feedback limiters are also often called *clamps* or *bound circuits;* most feature the short-circuit conductance characteristic of Fig. 3.23*b*. Hard limiters, such as Fig. 3.20, feature higher slopes in the conductance curve.

Note that the precision limiters of Figs. 3.17 and 3.20 use crude limiters to prevent amplifier overload, but actually do the precise limiting by disconnecting the amplifier from the output node in the limiting regions.

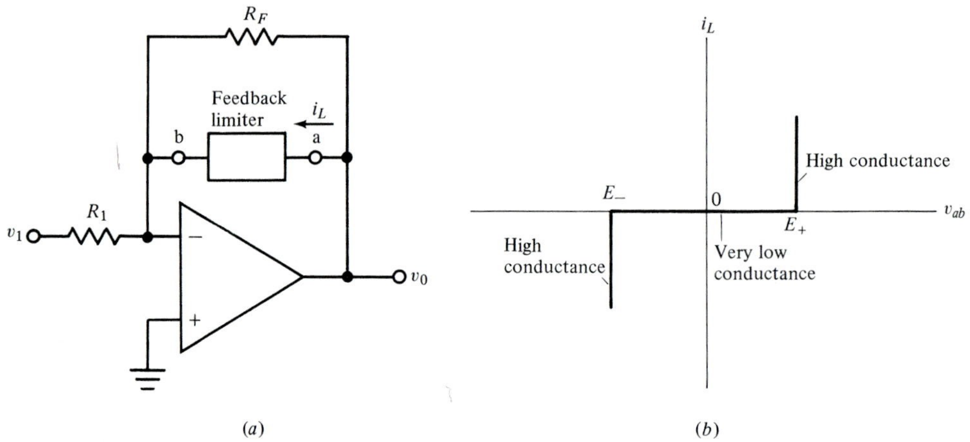

FIGURE 3.23
The general feedback limiter concept. (*a*) Basic circuit structure; (*b*) required conductance characteristic of the feedback limiter.

Improving Comparator Noise Immunity with Hysteresis

Figure 3.24 demonstrates the use of positive feedback to add hysteresis. This provides a latching characteristic which gives the comparator some noise immunity. Without this, a fast high-gain amplifier will often have a tendency to oscillate a bit during the transition between limiting states. We might call this a sort of "squegging" (an archaic term used by electronic designers to describe parasitic oscillations in radio transmitters). The exact behavior of the circuit depends on the type of clamp used (see Prob. 3.11 for common examples). If the amount of positive feedback is small, then the amount of hysteresis E_h is given approximately by

$$E_h = \frac{R_A E_{\text{swing}}}{R_A + R_B} \tag{3.17}$$

Typically we make E_h 10 to 20 mV. The elimination of squegging prevents spurious triggering of logic circuits driven by the comparator. Note that the addition of the hysteresis makes the circuit behave as a bistable or flip-flop circuit; i.e., once the input is returned to zero or removed, it will latch in one state or another, depending on its history. We have then a form of binary memory circuit, although for that purpose we would want to widen the hysteresis (see Sec. 3.3 for more on the topic of bistables).

Many linear IC manufacturers provide *general-purpose comparator function modules* such as the Analog Devices, Inc., Model AD9685, which has a propagation delay time of under 2.7 ns and a *latch setup input* with a setup time of 0.5 ns.

Some function modules have provisions for externally adjusting the hysteresis width by adding an external resistor; see, e.g., the Burr-Brown Model 4082

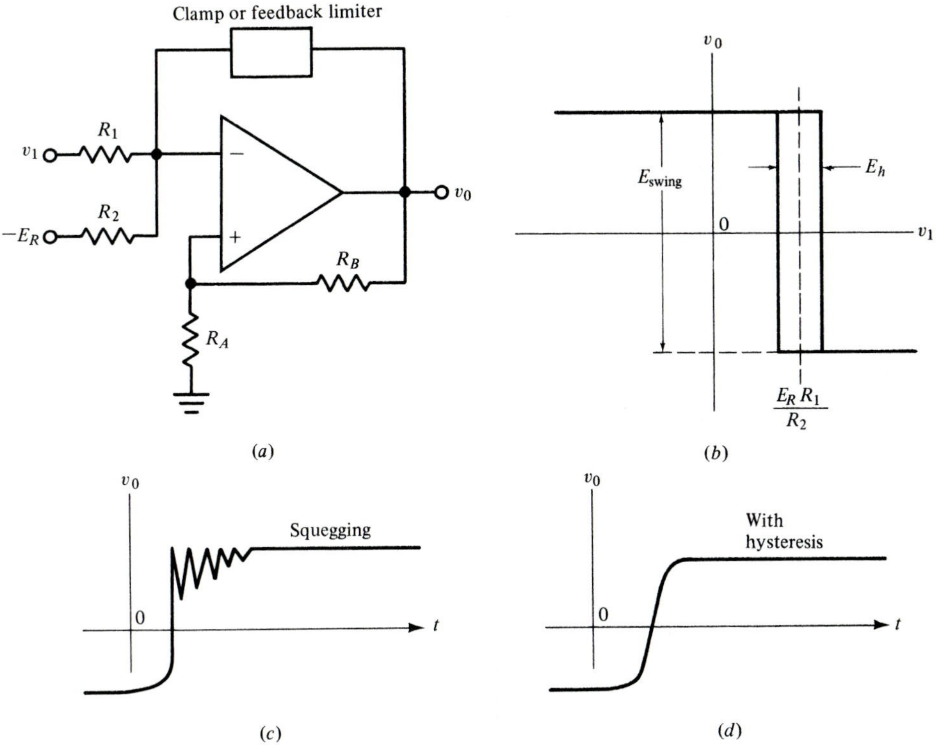

FIGURE 3.24
Improving a comparator noise immunity with hysteresis. (*a*) Resistors R_A and R_B are added to a conventional comparator circuit to provide positive feedback and thus hysteresis; (*b*) sketch of resultant transfer characteristic; E_h is the hysteresis window; (*c*) noisy transition or "squegging" without hysteresis; (*d*) monotone transition waveform with hysteresis.

General Purpose Comparator. This unit will supply up to 100 mA to an external load (GO light, warning device, etc.).

3.3 ADDITIONAL PIECEWISE-LINEAR FUNCTIONS

Figure 3.25 shows two generally useful piecewise-linear functions: the absolute-value function (*a*) and the dead-space operator (*b*). Hopefully the examples we have selected here will demonstrate some techniques that the reader can apply to a broad variety of situations and will provide a basis for developing new circuits as the need arises.

The Absolute-Value Operator

Figure 3.26 shows how the zero limiter or half-wave-rectifier circuit of the previous section (Fig. 3.17) can be augmented to implement the absolute-value, or full-

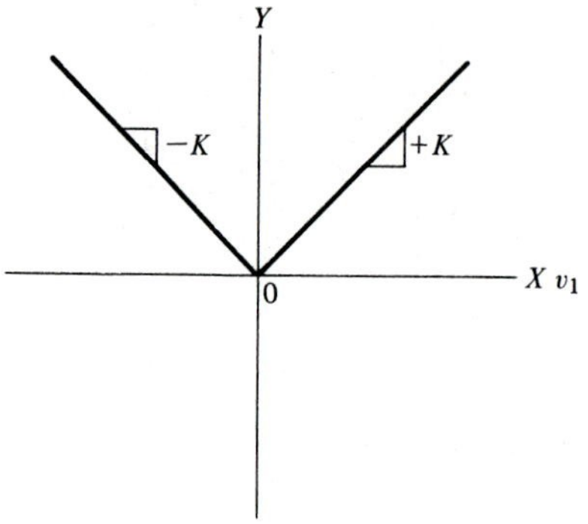

(a)

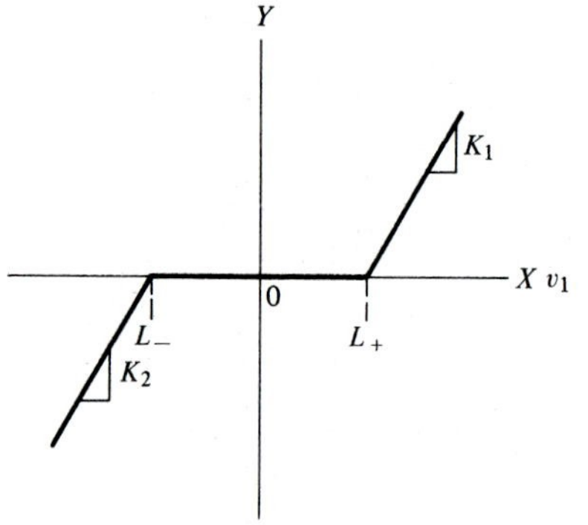

(b)

FIGURE 3.25
Piecewise-linear functions. (*a*) The absolute-value operator; (*b*) the dead-space amplifier.

wave-rectifier, operation. In Fig. 3.26*a*,

$$v_O = K|v_1| \qquad K > 0 \tag{3.18}$$

The circuit operation can be understood by looking at the block diagram of Fig. 3.26*b*; here we see that by summing the output of a zero limiter circuit with the input itself, along with appropriate polarity inversions, we get the desired result. In Fig. 3.26*c*,

$$v_2 = \begin{cases} \dfrac{-R_4 v_1}{R_3} & v_1 \geq 0 \\ 0 & v_1 < 0 \end{cases} \tag{3.19}$$

also

$$v_O = -R_F\left(\frac{v_1}{R_1} + \frac{v_2}{R_2}\right) \tag{3.20}$$

Now if, e.g., we set

$$R_3 = R_4 = R \tag{3.21}$$

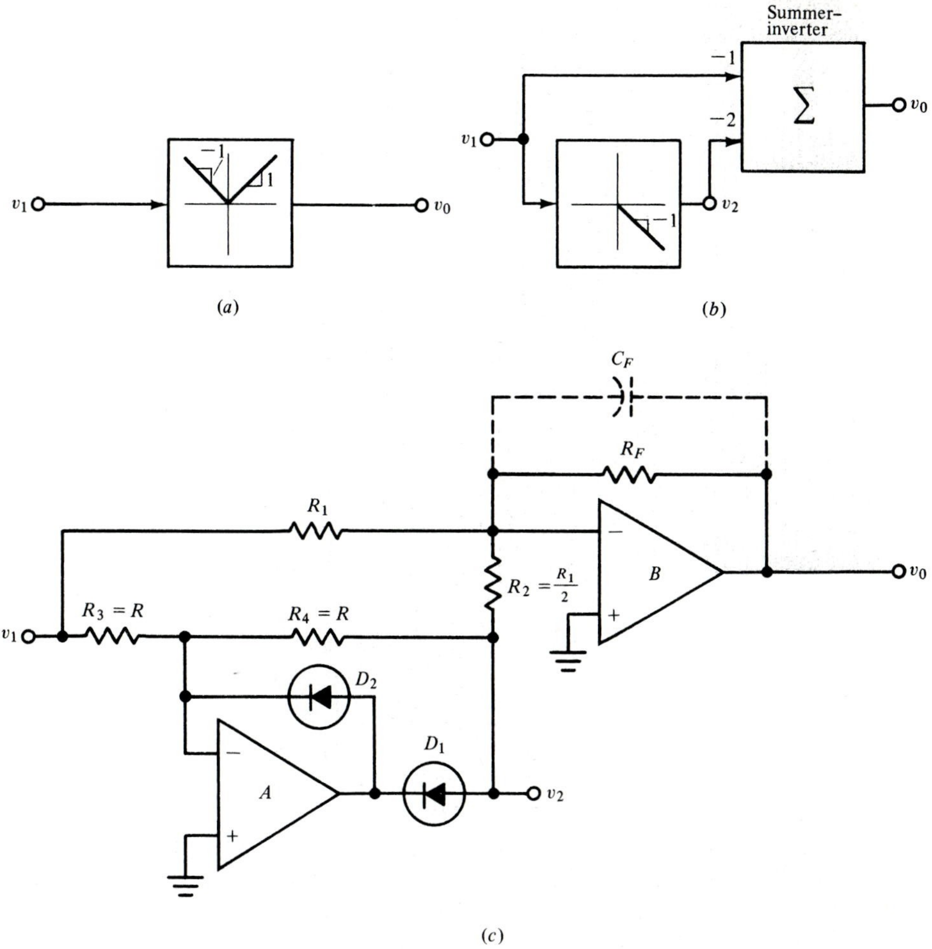

FIGURE 3.26
The absolute-value operator. (*a*) Overall operator symbol; (*b*) equivalent block diagram based on zero limiter of Fig. 3.17; (*c*) circuit details for a two-amplifier implementation.

then

$$v_2 = \begin{cases} -v_1 & v_1 \geq 0 \\ 0 & v_1 < 0 \end{cases} \tag{3.22}$$

With

$$R_2 = \frac{R_1}{2}$$

we have

$$v_0 = \frac{R_F}{R_1}|v_1| \tag{3.23a}$$

which matches (3.18) with

$$K = \frac{R_F}{R_1} \tag{3.23b}$$

Note that if we reverse diodes D_1 and D_2, we get a negative constant K, and v_0 is proportional to minus the absolute value of v_1. Frequently the circuit is constructed with

$$R_1 = R_3 = R_4 = R_F = R \tag{3.24a}$$

and

$$R_2 = \frac{R}{2} \tag{3.24b}$$

in which case

$$v_0 = |v_1| \tag{3.24c}$$

or, with the diodes reversed,

$$v_0 = -|v_1| \tag{3.24d}$$

An absolute-value circuit using two matched resistors is described in Franco, Ref. 16; see also Prob. 3.25.

Measurement of Average Absolute Value

In Fig. 3.26*c*, if we add a capacitor C_F around the output amplifier feedback resistance R_F, then we may use the circuit to estimate the average absolute value of v_1; that is,

$$v_0 \cong |v_1|_{\text{av}} \cong \frac{1}{T}\int_0^T |v_1|\, dt \tag{3.25a}$$

Note that this will be an approximate average, with an averaging time of the order of magnitude of

$$T \cong R_F C_F \tag{3.25b}$$

Thus the circuit is useful for measuring such things as average absolute error (note that it does not measure true rms value for general waveforms; we will show how to do that later). The circuit is thus a precise full-wave rectifier, and the addition of C_F adds a first-order *RC* filter to the rectifier output. The amount of "ripple" in v_0 will, as we might expect, depend on the relative values of $R_F C_F$ and the period of the waveform being rectified. (See Prob. 3.13 for an application of this circuit to sinusoidal ac measurements.)

The Dead-Space Operator

Figure 3.27*a* shows the transfer characteristic of a dead-space operator; such a transfer characteristic may be implemented in a variety of ways. The circuit of Fig. 3.27*b* (designed by G. A. Korn) is one of the more clever approaches. It uses a four-diode bridge to control the feedback conductance. This is a fairly difficult circuit to understand. Note that the left side of the bridge is connected to node e_g, which is always near ground potential; thus (using ideal diode models), currents I_1 and I_2 are essentially constant:

$$I_1 \cong \frac{E_1}{R_1} \tag{3.26a}$$

$$I_2 \cong \frac{E_2}{R_2} \tag{3.26b}$$

All four diodes will be ON whenever

$$-I_1 < I_3 < I_2 \tag{3.27}$$

but

$$I_3 = \frac{V_1}{R_3} \tag{3.28}$$

thus we can identify three valid modes of operation.

Region I (diodes D_2 and D_3 OFF) This region occurs when v_1 becomes sufficiently negative that the upper bridge leg cannot supply I_3 through D_1 alone; the extra current must come via R_F, and hence v_0 becomes nonzero. The boundary of region I determines one breakpoint,

$$\frac{-E_-}{R_3} = \frac{E_1}{R_1}$$

or

$$E_- = \frac{-E_1 R_3}{R_1} \tag{3.29a}$$

Region II (all diodes ON) This region occurs when v_1 is small enough that (3.27) is satisfied. The feedback impedance is now very low, and $v_0 \cong 0$.

Region III (diodes D_1 and D_4 OFF) This is similar in analysis to region I; the positive breakpoint is found to be

$$E_+ = \frac{E_2 R_3}{R_2} \tag{3.29b}$$

In Fig. 3.27*c*, the feedback resistor is replaced by a level clamp to obtain a *dead-space comparator* with the characteristic shown in Fig. 3.27*d*. The dead-space limits E_+ and E_- are still found from (3.29), and the limits L_+ and L_- are

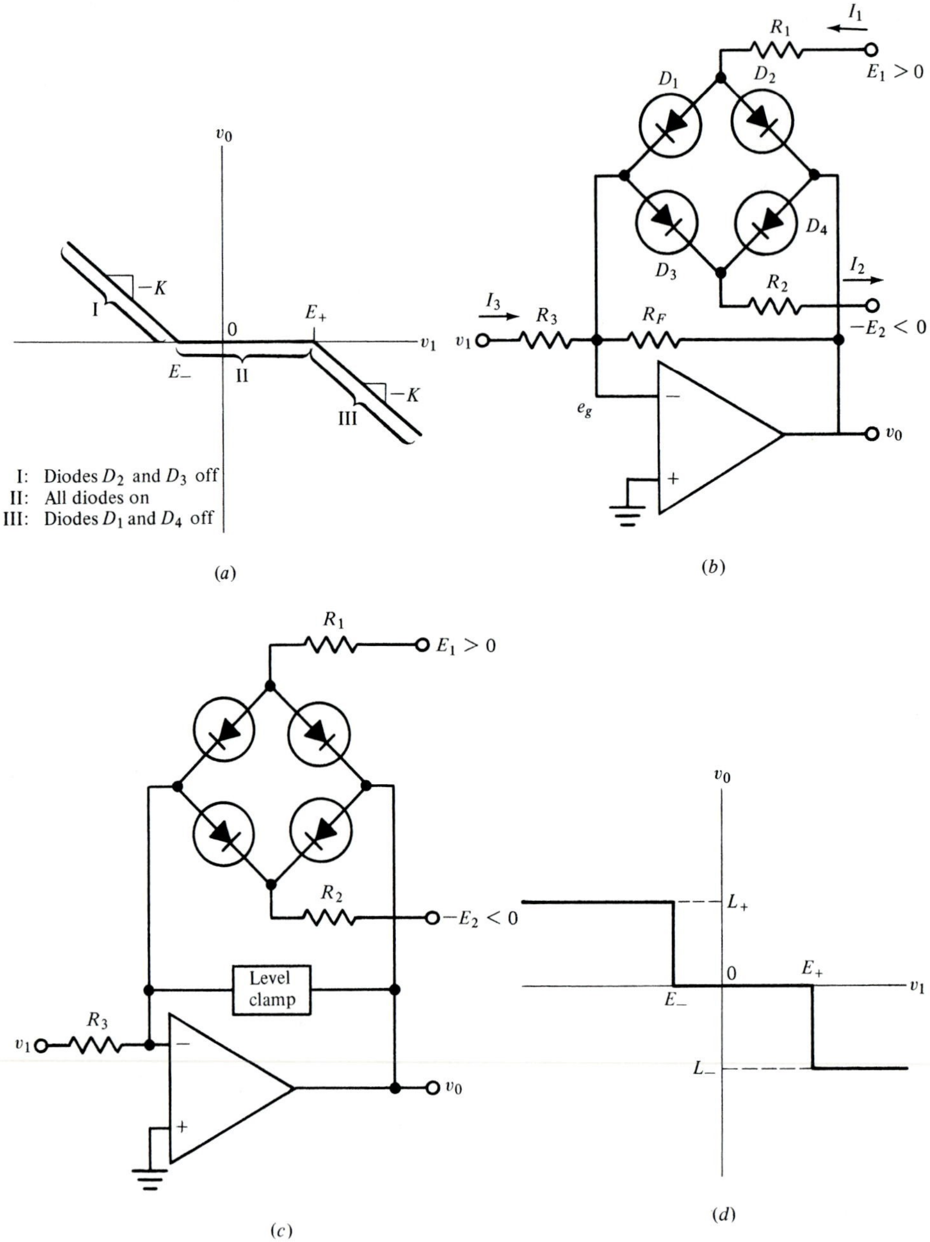

FIGURE 3.27
Dead-space operator. (*a*) General transfer characteristic; (*b*) four-diode bridge implementation; (*c*) modification to provide a dead-space comparator; (*d*) transfer characteristic of a dead-space comparator.

established by the level clamp, which may be made up of Zener diodes or any of the other feedback limiters described in Sec. 3.2.

Example 3.8. Suppose we want the dead-space characteristic of Fig. 3.28*a*, with ±15 V dc available for supplying the diode bridge. Let us arbitrarily pick $R_F = 10\ \text{k}\Omega$. Then to get the desired nonzero slope of −2, we must pick

$$R_3 = \frac{R_F}{2} = 5\ \text{k}\Omega$$

Now (3.29) predicts

$$R_1 = \frac{-15 \times R_3}{-2} = 37.5\ \text{k}\Omega$$

and

$$R_2 = 25\ \text{k}\Omega$$

Implicit Function Generation

The previous discussion described ways to build a network with a desired short-circuit conductance, approximating some nonlinearity by piecewise-linear functions. Figure 3.29 illustrates the general approach, where the network F has a short-circuit conductance described by

$$i = F(v_1) \tag{3.30}$$

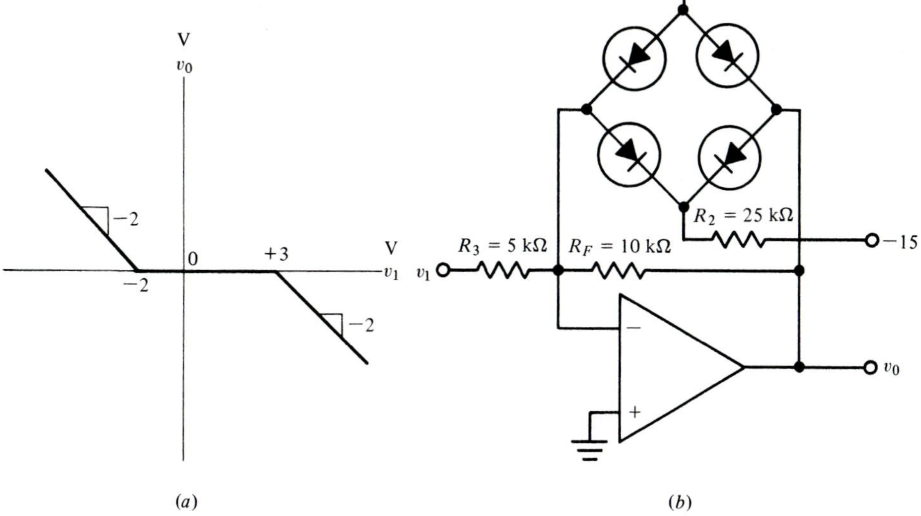

FIGURE 3.28
Refer to Example 3.8. (*a*) Desired transfer characteristic; (*b*) circuit design using a four-diode bridge.

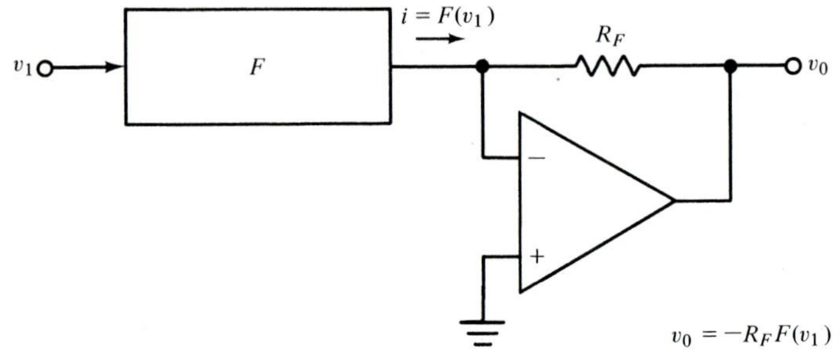

FIGURE 3.29
A general transfer conductance F in an amplifier input path, $v_0 = -R_F F(v_1)$.

In Fig. 3.29 we have then

$$\frac{v_0}{R_F} + F(v_1) = 0$$

or

$$v_0 = -R_F F(v_1) \tag{3.31}$$

and the network F could be implemented by, e.g., the diode limiter circuits just described. Now suppose we embed the network F in an operational amplifier feedback loop, as in Fig. 3.30. This situation is basically an extension of the feedback limiter concepts discussed in Sec. 3.2. In the configuration of Fig. 3.30,

$$F(v_0) + \frac{v_1}{R_1} = 0$$

or

$$v_0 = F^{-1}\left(-\frac{v_1}{R_1}\right) \tag{3.32}$$

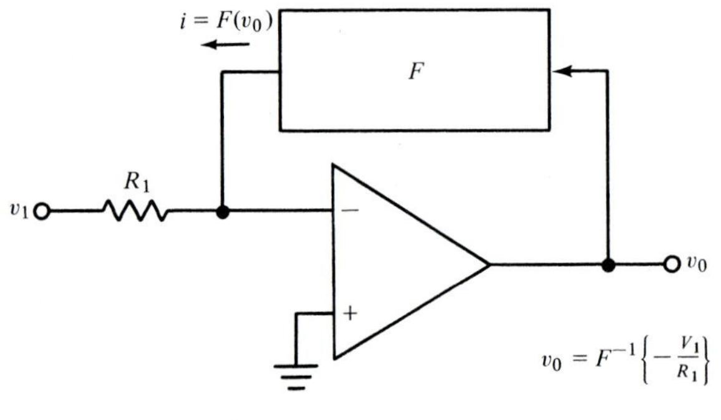

FIGURE 3.30
A general transfer conductance F in an amplifier feedback path, $v_0 = F^{-1}(-v_1/R_1)$.

where F^{-1} denotes the *inverse* function to F, that is,

$$F^{-1}[F(x)] = x$$

We will use this technique of *implicit function generation* to advantage in Sec. 3.5, where we get $\sqrt{X}$ from X^2.

Amplitude Selectors and Peak Detectors

Figure 3.31 shows two ways to implement the maximum-selection operator

$$v_3 = \max(v_1, v_2)$$

Note that the circuit of Fig. 3.31*a* is a simple diode AND gate. For analog signals v_1 and v_2 the output of such a circuit will be approximately equal to the maximum of the two, but there will be an error in the operation resulting from the diode forward drop. In Fig. 3.31*b*, however, we use two precision limiters (see Fig. 3.17) in a *precision selector* circuit that yields the desired maximum-selection operation with high accuracy. Note, unfortunately, that the inputs must now be $-v_1$ and $-v_2$ (and both inputs must be negative in polarity). The 1-MΩ resistor is used as a "keep-alive" resistor to ensure that one diode is always ON. Diodes D_3 and D_4 act as clamps to prevent overload. Note that the circuit always has a low output impedance. Additional diodes (up to perhaps five) can be added to the circuits of Fig. 3.31 to obtain

$$v_3 = \max(v_1, v_2, \ldots)$$

Note also that if all diode and power supply polarities are reversed, then we obtain

$$v_3 = \min(v_1, v_2)$$

Figure 3.32*a* shows an (inverting) *peak detector* circuit (Ref. 4, p. 138). A typical waveform is illustrated in Fig. 3.32*b*. Switch S is used to reset the circuit output to zero before the history of v_1 is observed. Note that $-v_1$ must be provided and must be one polarity. Capacitor C is chosen as a compromise between slewing rate and holding ability; 0.01 μF is typical for signals in the audio-frequency range. Figure 3.33 shows two other useful circuits. Figure 3.33*a* (see Ref. 3, p. 356) shows a noninverting circuit; capacitor C is charged through D_1 by A_1. Amplifier A_2 serves as a buffer between C and the output when D_1 is OFF. Diode D_2 prevents A_1 from overloading. Capacitors C_1 and C_2 may be needed to improve stability and transient response. Amplifier A_1 must have good common-mode rejection and be stable with C as a load. Amplifier A_2 should have a high input impedance (e.g., an FET or varactor input stage). The circuit of Fig. 3.33*b* (Ref. 6) is an inverting circuit which has been used in high-speed applications; it is more stable than the circuit of Fig. 3.33*a*. The circuit slewing rate may be improved by adding a current-boosting emitter follower at point A.

Bistables

Figures 3.34 and 3.35 show two implementations of a bistable or flip-flop (see also Ref. 3, p. 363, and Ref. 4, p. 139). The two-amplifier circuit of Fig. 3.34*a*

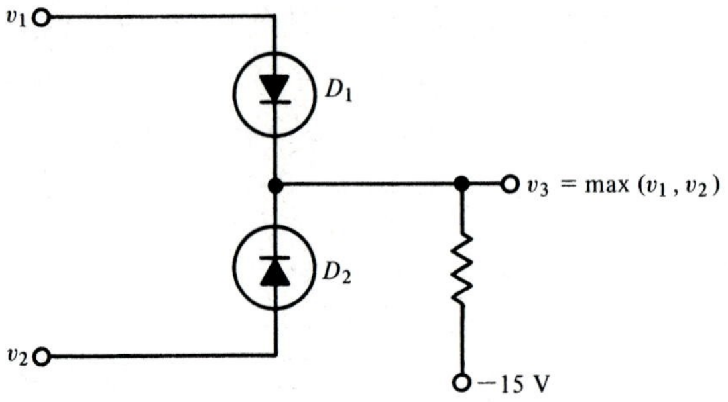

(a)

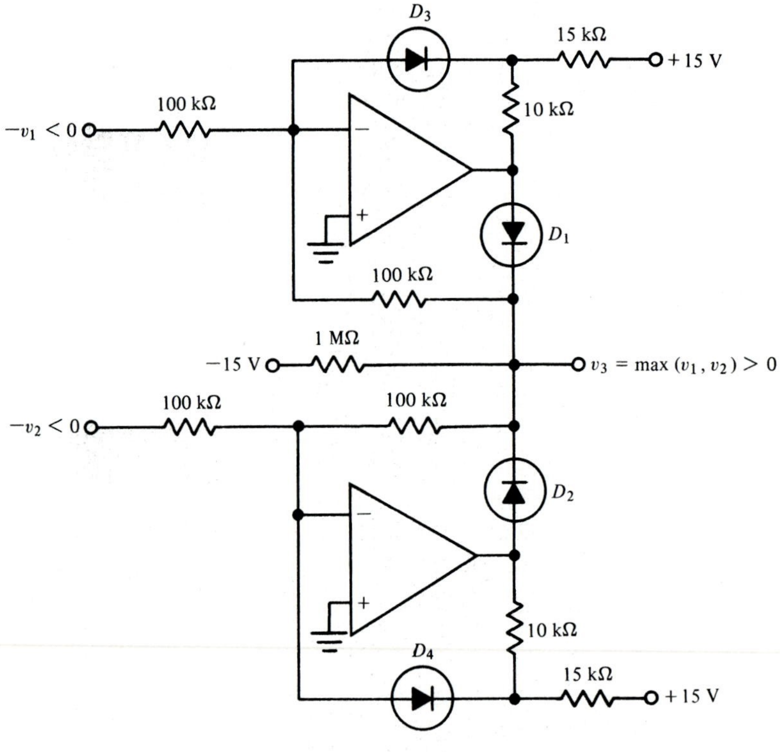

(b)

FIGURE 3.31
Selection circuits. (*a*) A simple diode maximum selector; (*b*) precision level selection using operational amplifiers. Reversal of diodes and power polarities provides min (v_1, v_2).

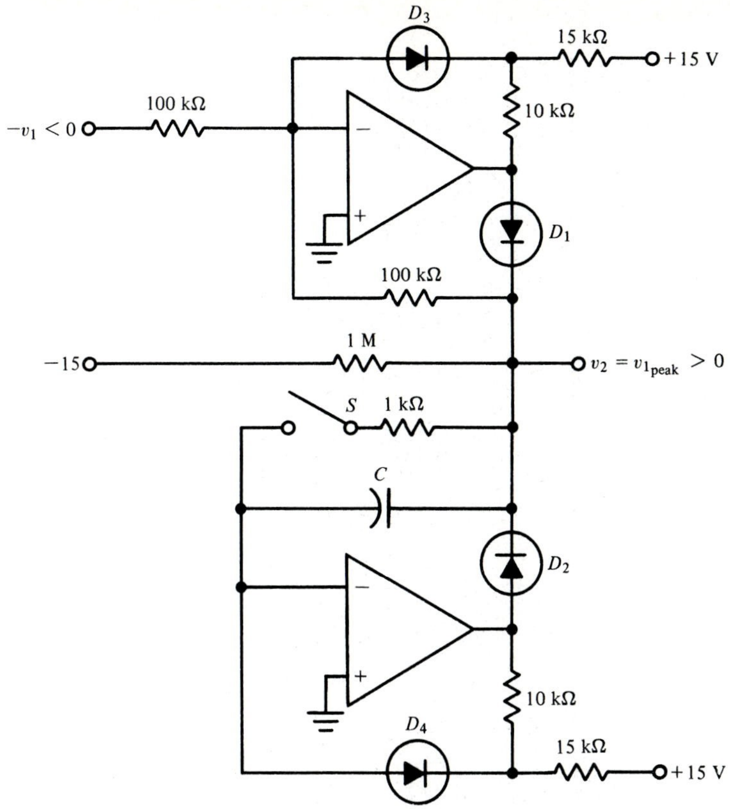

(a)

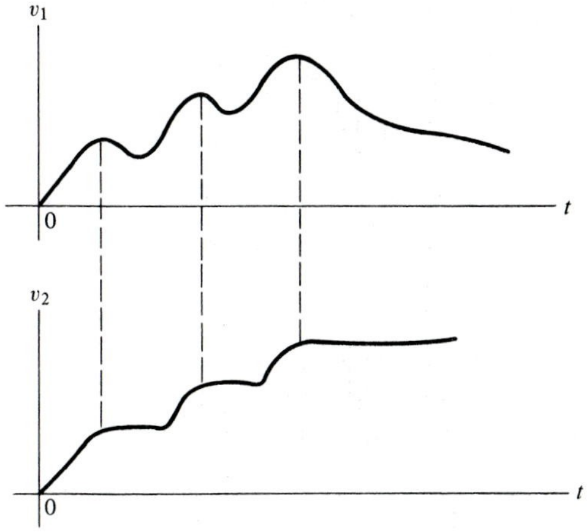

(b)

FIGURE 3.32
Two-amplifier inverting peak detector circuit. (a) The circuit combines two precision limiters with a capacitor to retain the peak value; (b) typical waveforms.

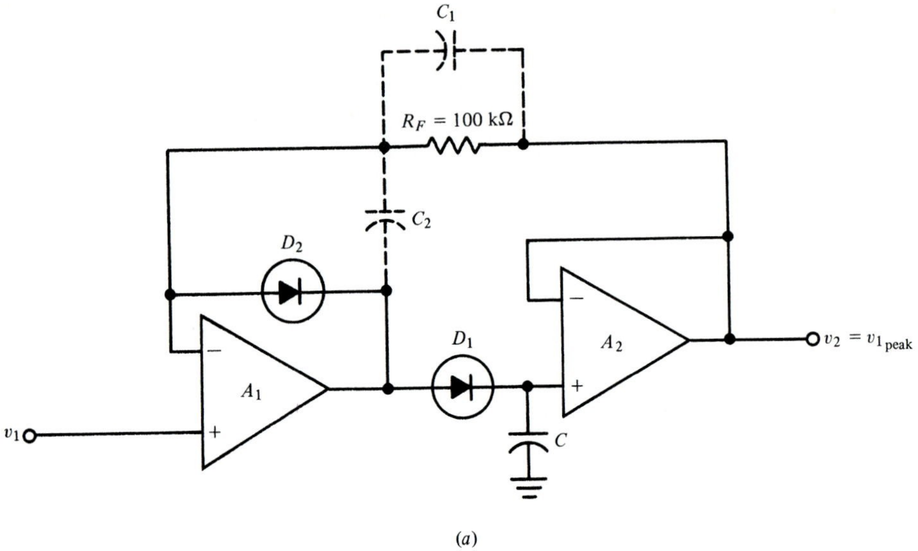

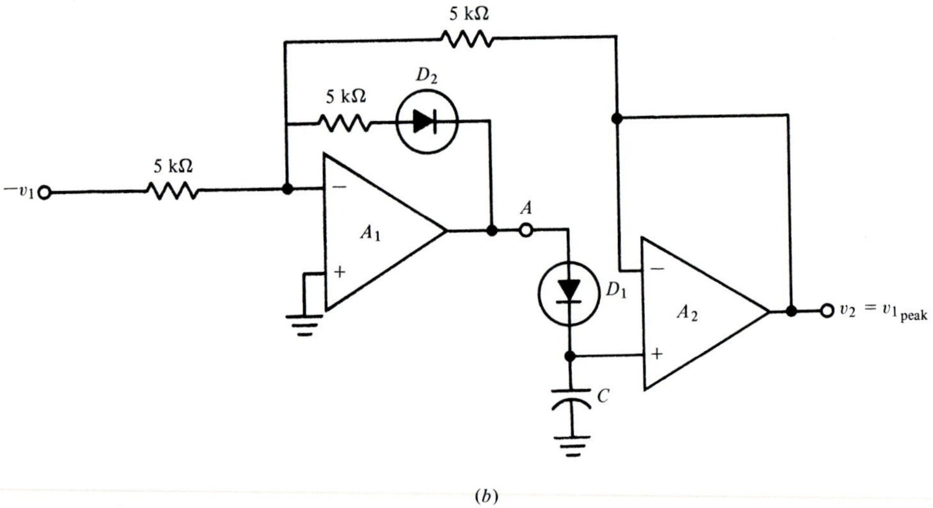

FIGURE 3.33
Two more peak detector circuits. (*a*) A noninverting circuit; C_1 and C_2 may be required for stability; (*b*) a high-speed inverting circuit; slewing rate may be improved by adding current boosting at point *A*.

provides the noninverting transfer characteristic shown in Fig. 3.34*b*. The tops of the characteristic will be fairly flat, even with a simple pair of diode clamps on the output, because of the low-amplitude drive from A_1. If D_1 and D_2 are replaced by transistors (as in Fig. 3.20), then the tops will be very flat. Normally R_5 is made large compared to R_2 and R_3. The dimensions of the hysteresis loop

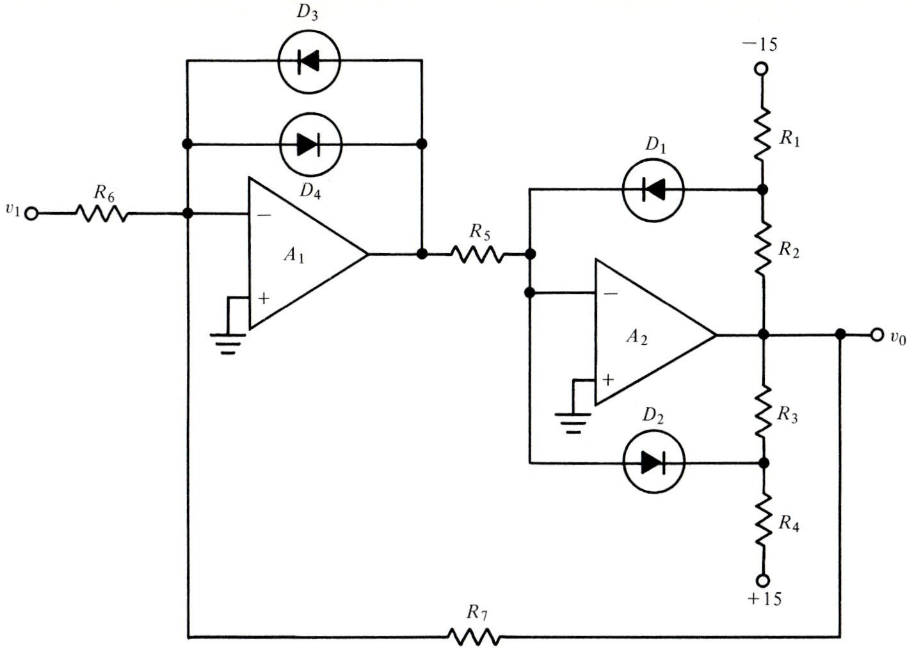

(*a*)

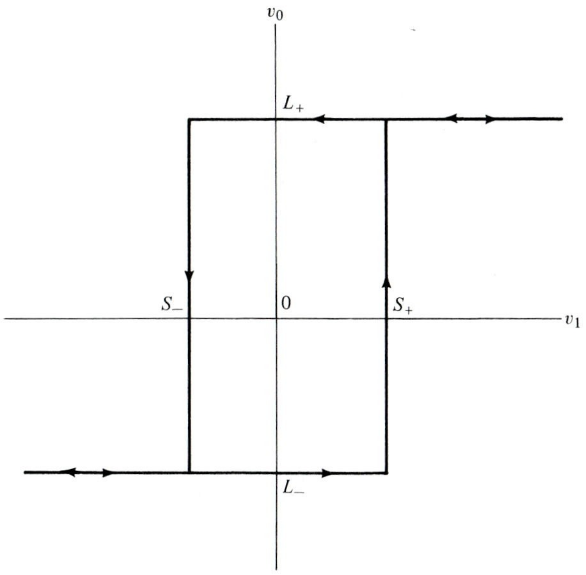

(*b*)

FIGURE 3.34
A two-amplifier bistable circuit. (*a*) Output levels are set by soft limiters on A_2, and diodes D_3 and D_4 limit the output of A_1 and improve output level regulation and circuit speed; (*b*) typical output characteristic; note that it is noninverting.

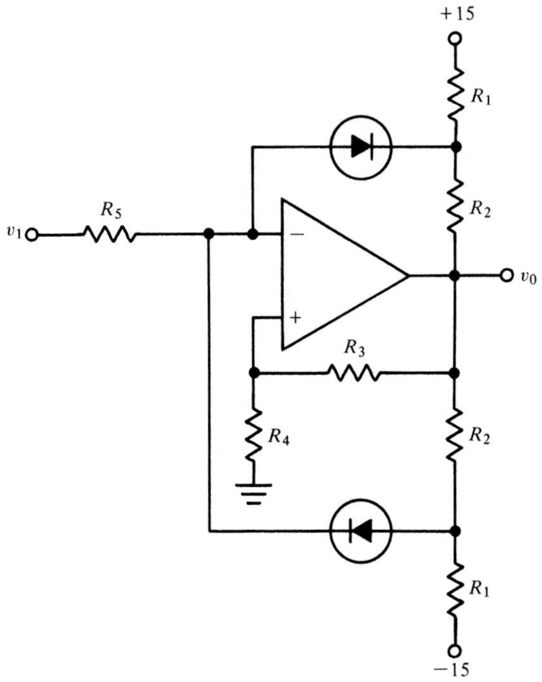

(a)

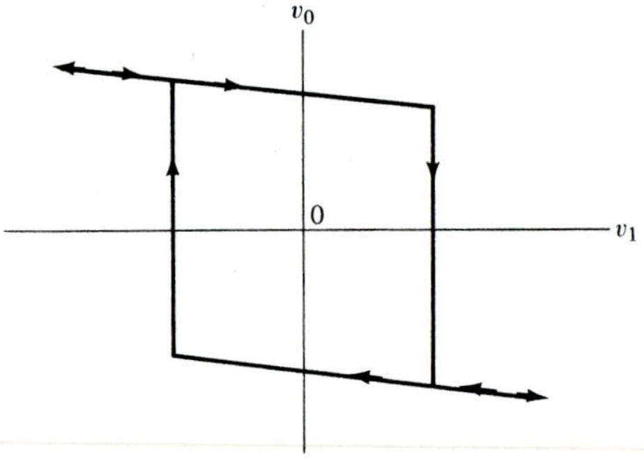

(b)

FIGURE 3.35
A simple one-amplifier bistable. (*a*) Positive feedback via R_3 and R_4 adds hysteresis; (*b*) output characteristic is not well limited and is inverting.

are determined by the following equations:

$$L_+ = \frac{15R_2}{R_1} \tag{3.33a}$$

$$L_- = \frac{-15R_3}{R_4} \tag{3.33b}$$

$$S_+ = \frac{R_6|L_-|}{R_7} = \frac{15R_6R_3}{R_7R_4} \tag{3.33c}$$

$$S_- = \frac{R_6L_+}{R_7} = \frac{-15R_6R_2}{R_7R_1} \tag{3.33d}$$

Example 3.9. The circuit of Fig. 3.34 can be made to implement a 10- by 10-V square hysteresis loop by making

$R_2 = R_3 = 10\ \text{k}\Omega$

$R_1 = R_4 = 15\ \text{k}\Omega$ (adjust to get exact limiting)

$R_6 = R_7$ (typically 10 kΩ)

$R_5 = 100\ \text{k}\Omega$ (not critical)

With these values and ±15-V supplies,

$$L_+ = +10 \qquad L_- = -10$$

$$S_+ = +10 \qquad S_- = -10$$

The one-amplifier circuit of Fig. 3.35 is actually an extension of the technique of using positive feedback to add hysteresis to a comparator (Fig. 3.24). With simple diode clamps, the output levels will vary some with the input. By adding better limiting, e.g., by using a transistor limiter (see Fig. 3.20), one can obtain an almost square characteristic (with a polarity inversion).

We will see in Sec. 3.6 that these bistable circuits are used in the design of precision waveform generators (square and triangle waves).

3.4 LOGARITHMIC OPERATORS

In the past two decades, a number of specialized log and antilog function generation circuits have been developed based on the exponential behavior of a *pn* junction or, more specifically, the exponential behavior of the emitter-base junction of a transistor (Fig. 3.36). It is not our intention here to provide all the information required to design these circuits so that they perform well. The interested reader can learn more about this from other sources (see Ref. 3, sec. 7.4; Ref. 7, sec. 5.3; and Ref. 8). We describe the basic manner in which these circuits work; further, we show a number of ways in which logarithmic operators can be used to implement a variety of nonlinear functions.

In Fig. 3.36, we see a basic log operator circuit. The transistor is connected in the *transdiode* configuration. In this connection, the Ebers-Moll model of a bipolar transistor predicts (Ref. 7, pp. 87ff.)

$$I_C = \alpha I_{ES}(e^{-qv_{EB}/kT} - 1) \tag{3.34}$$

where v_{EB} = emitter-base voltage (v_O in Fig. 3.42)

T = temperature, K

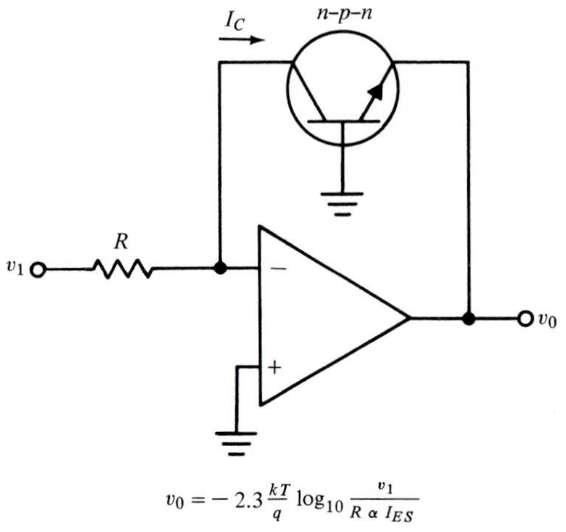

FIGURE 3.36
A basic log operator circuit; the transistor is used in the transdiode connection for improved dynamic range.

k = Boltzmann's constant, 1.38×10^{-23} J/K

q = electron charge, 1.6×10^{-19} C

I_{ES} = reverse saturation current of emitter-base junction, with collector connected to base

α = forward grounded-base current gain of transistor

Note: $kT/q \cong 26$ mV at 27°C.

Whenever $v_{EB} < 100$ mV or so, (3.34) becomes

$$I_C \cong \alpha_F I_{ES} e^{-qv_{EB}/kT} \tag{3.35}$$

Thus in Fig. 3.36, we can show that since $v_{EB} = v_0$,

$$v_0 \cong \frac{-kT}{q} \ln \frac{v_1}{R\alpha I_{ES}}$$

or

$$v_0 \cong \frac{-2.3kT}{q} \log_{10} \frac{v_1}{R\alpha I_{ES}} \tag{3.36}$$

Thus we obtain an output voltage proportional to the logarithm of the input voltage. This equation is obviously very temperature-sensitive. Moreover, as the input signal varies over a wide range, the small-signal or incremental feedback resistance provided by the transistor will change, possibly by several orders of magnitude, i.e.,

$$r_{ac} \cong \frac{kT}{qI_C} = \frac{0.026}{I_C} \quad \Omega \tag{3.37}$$

This means that for small currents (low input signals), the feedback impedance is possibly quite high, and thus any stray capacitance associated with the circuit may interact to yield poor frequency response for small signals. On the other hand, for large signals, the feedback impedance is quite low, and this may lead to stability problems. Of course, the ideal exponential relationship will not hold exactly; in properly selected transistors, however, reasonably accurate "logging" can be performed over 6 to 8 decades of input signal level.

The circuit of Fig. 3.37 shows a more practical version of the logarithmic operator. It may be shown that

$$v_0 \cong -\frac{R_1 + R_2}{R_2}\frac{kT}{q}\ln\frac{i_1}{i_2} \tag{3.38a}$$

or

$$v_0 = -\frac{R_1 + R_2}{R_2}\frac{kT}{q}\ln\frac{v_1 R_3}{v_2 R_5} \qquad v_1, v_2 > 0 \tag{3.38b}$$

Thus we have a *log ratio* operator. Such a circuit might be packaged as a single module, in which case the resistance values would be selected for some nominal

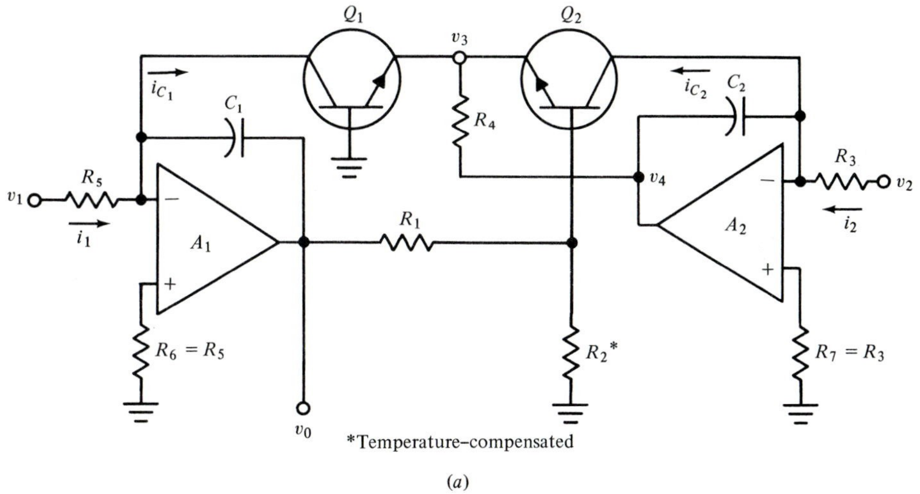

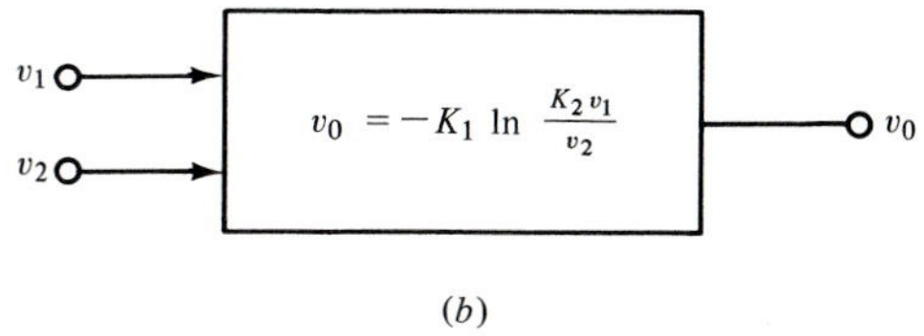

FIGURE 3.37
A practical log operator circuit. (*a*) This two-amplifier arrangement permits finding the log ratio; (*b*) a block-diagram symbol.

range of operation. Capacitances C_1 and C_2 are usually required for stability, and they limit the speed of operation. R_4 permits amplifier A_2 to operate over its normal output range.

This circuit has a transfer characteristic

$$v_0 = -K_1 \ln \frac{K_2 v_1}{v_2} = \frac{-K_1}{0.433} \log_{10} \frac{K_2 v_1}{v_2} \qquad v_1, v_2 > 0 \tag{3.39}$$

with

$$K_1 = \frac{R_1 + R_2}{R_2} \frac{kT}{q} \tag{3.40a}$$

and

$$K_2 = \frac{R_3}{R_5} \tag{3.40b}$$

Often one input, for example, v_2, is held at a reference potential, and the circuit merely generates an output proportional to the log of the input, i.e., it is a *log operator* or *log amplifier*. Appropriate scale factors are chosen. Such a circuit, with careful design and selection of critical components, can be made accurate to perhaps 1 percent over an input dynamic range of perhaps 6 decades. Note that R_2 is usually made from some form of temperature-sensitive element, e.g., a thermistor, and is designed to provide temperature compensation for changes in temperature, in order to hold K_1 in (3.40*a*) reasonably constant. Due to the capacitative compensation, such circuits have rather slow settling times, particularly when the input signals are small.

Example 3.10. Evaluate K_1 and K_2 if, in Fig. 3.37,

$$R_1 = 15.7\text{ k}\Omega \qquad R_3 = 1.5\text{ M}\Omega$$

$$R_2 = 1\text{ k}\Omega \qquad R_5 = 10\text{ k}\Omega$$

$$\frac{kT}{q} = 0.026\text{ V} \qquad V_2 = +15\text{ V dc}$$

From (3.40) we have

$$K_1 = 0.433 \qquad K_2 = 150$$

Thus (3.39) becomes

$$v_0 = -0.433 \ln 10 v_1 = -\log_{10} 10 v_1 \tag{3.41}$$

Figure 3.38 shows the plot of (3.41) over 4 decades of input values (see also Prob. 3.25).

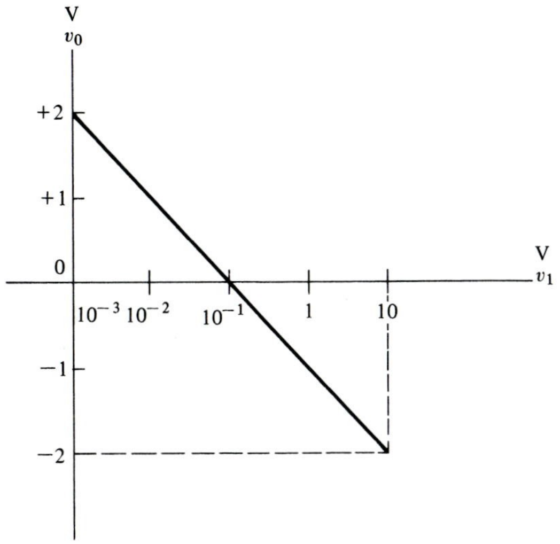

FIGURE 3.38
The transfer characteristic of Example 3.10, showing 4-decade range.

Antilog or Exponentiation Operators

The circuit of Fig. 3.39 does the inverse operation to that of Fig. 3.37. Here we can show that

$$v_O = \frac{v_2 R_6}{R_3} \exp\left(\frac{-q}{kT} \frac{R_2}{R_1 + R_2} v_1\right) \qquad v_1 > 0 \tag{3.42}$$

In this circuit v_2 is usually a fixed reference voltage, but it could be a time-varying signal. The general form of the operation is thus

$$v_O = K_1 v_2 \, e^{-K_2 v_1} \qquad v_1 > 0 \tag{3.43a}$$

with

$$K_1 = \frac{R_6}{R_3} \tag{3.43b}$$

and

$$K_2 = \frac{q}{kT} \frac{R_2}{R_1 + R_2} \tag{3.43c}$$

Again at room temperature (27°C) this last equation becomes

$$K_2 = \frac{R_2}{0.026(R_1 + R_2)} \tag{3.43d}$$

Alternatively, we can express (3.42) in the form

$$v_O = v_2 K_1 \times 10^{-K_3 v_1} \qquad v_1 > 0 \tag{3.44a}$$

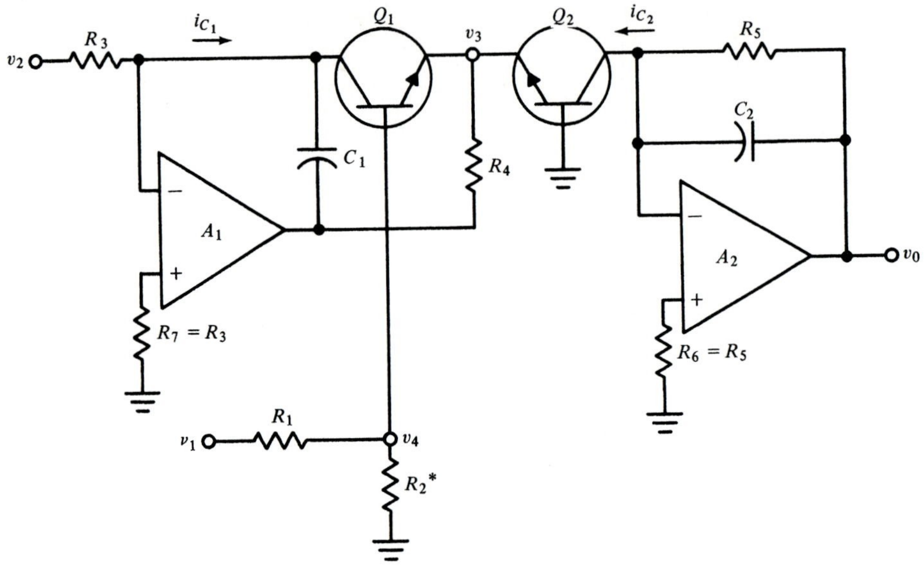

(a)

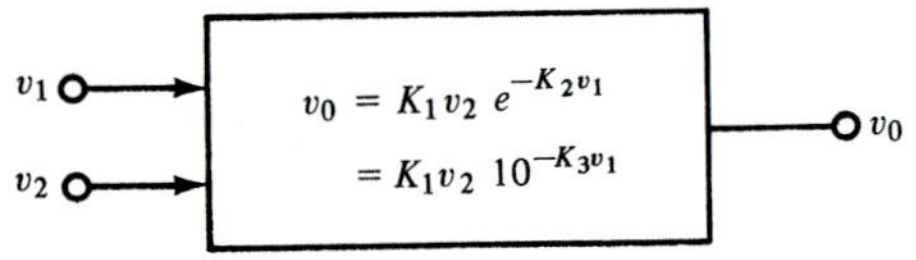

(b)

FIGURE 3.39
A practical exponentiation or antilog operator. (*a*) In this two-amplifier circuit v_2 is often a fixed reference, but it may be time-varying; (*b*) block-diagram symbol.

where K_1 is given by (3.43*b*) and

$$K_3 = \frac{q}{2.3kT}\,\frac{R_2}{R_1 + R_2} \tag{3.44b}$$

$$= \frac{R_2}{0.0598(R_1 + R_2)} \qquad \text{at } 27^\circ\text{C} \tag{3.44c}$$

Example 3.11. In Fig. 3.39, if

$$v_2 = +15 \text{ V dc}$$
$$R_3 = R_7 = 30 \text{ k}\Omega$$
$$R_4 = 2 \text{ k}\Omega \qquad \text{(not critical)}$$
$$R_1 = 15.7 \text{ k}\Omega \qquad R_2 = 1 \text{ k}\Omega$$
$$R_5 = R_6 = 10 \text{ k}\Omega$$

then we have, from (3.44),

$$v_0 = 5 \times 10^{-v_1} \qquad v_1 > 0 \tag{3.45}$$

Figure 3.40 shows a sketch of (3.45).

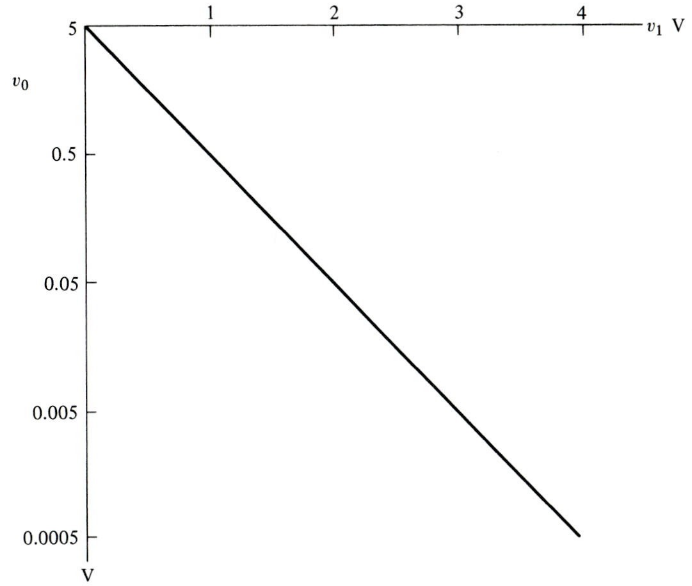

FIGURE 3.40
The transfer characteristic of Example 3.11; note the log scale for the output v_O.

Special Nonlinear Operations

Of course, the log and antilog circuits described above can be combined to implement a wide variety of special nonlinear operations. Consider the block diagrams of Fig. 3.41. Here the blocks labeled "Log" represent circuits similar to Fig. 3.37, and the blocks labeled "Antilog" represent circuits similar to Fig. 3.39. The other elements are simple summing and coefficient operations.

Thus we see in Fig. 3.41*a* how a log and an antilog circuit can be combined to yield a general *exponentiation operator*

$$v_O = K_2 v_1^m \tag{3.46a}$$

In principle, m can be either positive or negative; typically $0.2 < m < 5$. Note also that v_1 is of one polarity, $v_1 > 0$.

In Fig. 3.41*b*, we combine a log ratio circuit with an antilog circuit to obtain a *division operator* or *divider*

$$v_O = K_2 \frac{v_1}{v_2} \tag{3.46b}$$

Again the polarity of the inputs is restricted, typically $v_1 > 0$ and $v_2 > 0$.

Similarly, Fig. 3.41*c* shows how the output of two log operators may be summed, and then an antilog circuit used to form a *multiplication operator* or *multiplier*

$$v_O = K_2 v_1 v_2 \tag{3.47}$$

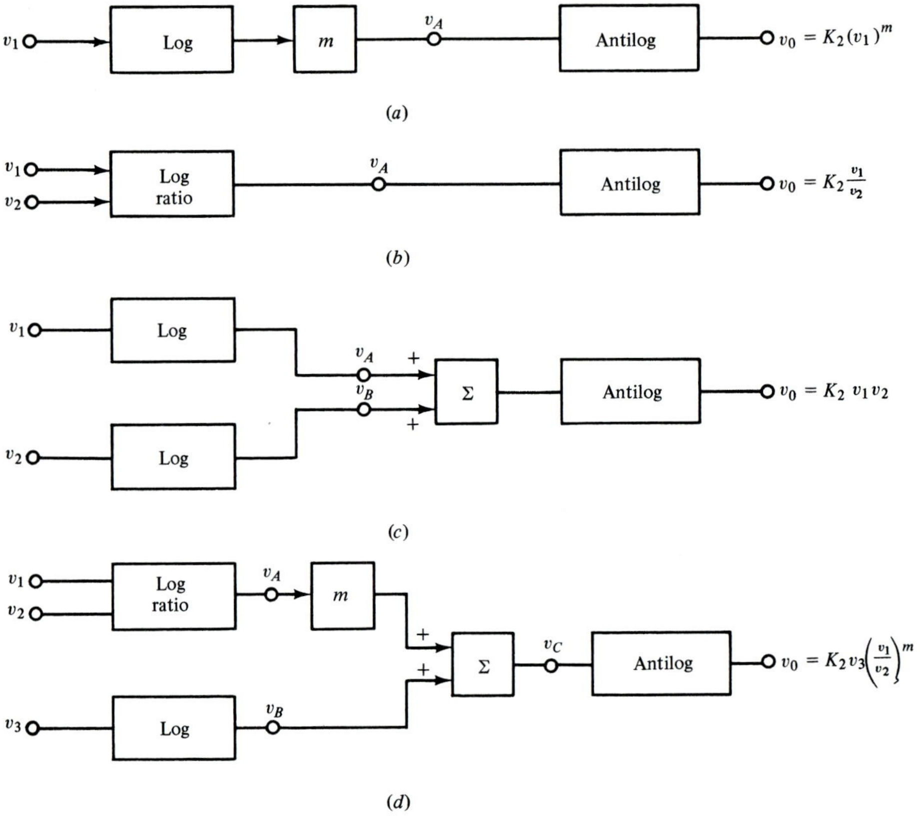

FIGURE 3.41
Combining log and antilog operators. (*a*) One log and one antilog operator do general exponentiation (raising to a power); (*b*) one log ratio and one antilog do one-quadrant division; (*c*) similarly, two log operators and an antilog do one-quadrant multiplication; (*d*) a general multipurpose nonlinear operator; see Table 3.1 for typical capabilities.

Obviously more inputs can be added; again $v_1 > 0$ and $v_2 > 0$ (a one-quadrant multiplier).

As a final example, consider Fig. 3.41*d*. This shows the block diagram of a general *multipurpose nonlinear module* that implements

$$v_0 = K_2 v_3 \left(\frac{v_1}{v_2}\right)^m \tag{3.48}$$

Such a unit can be used in a variety of applications (Refs. 9 and 10). Of course, with $m = 1$ and $v_3 =$ constant, we have a divider; with $m = 1$ and $v_2 =$ constant, we have a multiplier. Moreover, since m is adjustable, this provides a great deal of flexibility.

Table 3.1 shows some of the salient specifications of the Analog Devices Model AD538 Programmable Multifunction Module, which is similar in structure

TABLE 3.1
Characteristics of a typical programmable multifunction module*

General functional expression†	$v_0 = v_y\left(\frac{v_z}{v_x}\right)$
Typical signal range	$0.1\text{ V} < v_x, v_y, v_z < 10\text{ V}$
Range of m (adjust externally with resistor)	$0.2 < m < 5$
Accuracy	Approx. 0.5%, $< 500\ \mu\text{V}$
DC offset	$< 100\ \mu\text{V}$
Small-signal bandwidth, $100\text{ mV} < v_y, v_z$	400 kHz
Slew rate	$1.4\text{ V}/\mu\text{s}$

* Similar to the Real-Time Computational Unit Model AD 538, Analog Devices, Inc., Norwood, Mass.

† Unit will also operate on input currents I_x, I_y, I_z.

to Fig. 3.41*d*. It implements the general function

$$v_0 = \frac{10}{9} v_y \left(\frac{v_z}{v_x}\right)^m \qquad 0.2 \geq m \geq 5.0$$

using log and antilog techniques.

The exponent m is set by adding two external resistors or, in the case of $m = 1$, by making a simple connection. Note the specifications, which indicate an accuracy of about 0.5 percent of full scale or better at 25°C. The principal limitations on this unit are that the inputs must be of one polarity (hence a one-quadrant multiplier or divider may be implemented) and that the speed is somewhat restricted when the inputs are small in magnitude. This is, however, characteristic of such units based on logarithmic operations.

Reference 10 illustrates how this type of operator can be used to implement polynomial approximations to transcendental functions. For example, noting that

$$\sin\theta \cong \theta - \frac{\theta^{2.827}}{2\pi}$$

(which is accurate to 0.25 percent, $0 \geq \theta \geq \pi/2$ rad), we can implement the sine function with one such module and an external summing amplifier (see also Prob. 3.29).

3.5 MULTIPLIERS, DIVIDERS, AND APPLICATIONS

We often need to perform the following operations (Gilbert, 1968):

Multiplication: $$v_m = K_m v_1 v_2 \tag{3.49}$$

Division:

$$v_d = \frac{K_d v_1}{v_2} \tag{3.50}$$

We have already seen how the logarithmic operators of the last section may be used to implement these operations. In this section, we describe additional important methods of doing these operations and how functional modules that can do multiplication and division may be used in a variety of nonlinear circuit applications.

Scale Factors

We normally assume that general-purpose analog signal operators should be able to accommodate analog signals with a ± 10-V range; similarly, their output range is typically ± 10 V. Thus *amplitude scaling* considerations must be used in selecting the gain factors K_m and K_d above and in using the operators for designing signal-processing systems. The proper choice for the multiplier gain factor is fairly obvious: If the analog signal range is $\pm E$, then the multiplier should implement

$$v_m = \frac{v_1 v_2}{E} \tag{3.51}$$

that is, $K_m = 1/E$. This ensures that

$$|v_0| \leq E \tag{3.52}$$

Thus for a ± 10-V analog system, we would normally design the multiplier so that

$$v_m = 0.1 v_1 v_2 \tag{3.53}$$

A truly general-purpose multiplier should feature *four-quadrant* operation; that is, no limitations should be placed on the polarity of either input. Thus for a ± 10-V analog system, (3.52) would be satisfied, and $|v_0| \leq 10$ with the inputs permitted to range into four quadrants: $-10\ \text{V} \leq v_1, v_2 \leq 10\ \text{V}$. Note that the *logarithmic multiplier* described in Sec. 3.4 *does not* meet this requirement.

In the case of a divider, the choice of the gain factor K_d is less obvious; it is dependent on the anticipated range of values of the inputs, in particular on the *minimum magnitude* of the *divisor* signal v_2. Typically, for a 10-V analog system, we would pick $K_d = 10$ and implement

$$v_d = 10\frac{v_1}{v_2} \tag{3.54}$$

Obviously, this choice is based on the assumption that the ratio v_1/v_2 never exceeds 1. Moreover, we usually have to restrict the polarity and range of the divisor v_2. Typically, for a 10-V system, we would see specifications such as

$$v_d = 10\frac{v_1}{v_2} \qquad \left|\frac{v_1}{v_2}\right| \leq 10 \qquad \text{and} \qquad 1\ \text{V} \leq v_2 \leq 10\ \text{V} \tag{3.55}$$

These specifications ensure that overload will never occur, and they are typical of the allowable ranges within which divider accuracy will be guaranteed. The above equation describes a *two-quadrant* divider, in which the numerator input may take on either polarity but the denominator input is restricted to one polarity.

Multiplier and divider performance *specifications* usually fall into two categories: *static accuracy* and *frequency response*. Manufacturers' specifications are usually given in terms that make the multiplier look as good as possible.

Static accuracy is usually given in terms of the percentage of the nominal output range. Thus, for a 10-V multiplier, a 1 percent static accuracy means that the output will not differ from the ideal by more than 0.1 V, as long as the inputs are within their normal limits. *Note that for small outputs,* this could be a relatively *large true percentage error*. Suppose, e.g., that $v_1 = v_2 = 1$ V in (3.53). Then a nominal 1 percent static accuracy specification could mean that the output error might be as much as 0.1 V, which is as large as the desired value; i.e., we might have a 100 percent error relative to the proper value!

Dynamic accuracy or frequency response is also usually specified in an optimistic way. Often one will see a specification of −3-dB bandwidth. This is an almost useless measure of multiplier speed, since errors due to phase shift might be quite high long before the −3-dB frequency was reached.

Consider Fig. 3.42; here the desired waveform is

$$v_d = A \sin \omega t \tag{3.56a}$$

but the actual waveform might be different, due to phase shift, even though the amplitude was correct; e.g., if

$$v_a = A \sin (\omega t - \theta) \tag{3.56b}$$

then the error would be

$$\delta = A[\sin \omega t - \sin(\omega t - \theta)] \tag{3.57}$$

For small θ, $\sin \theta \cong \theta$ and $\cos \theta \cong 1$; thus, using (3.57) and some well-known

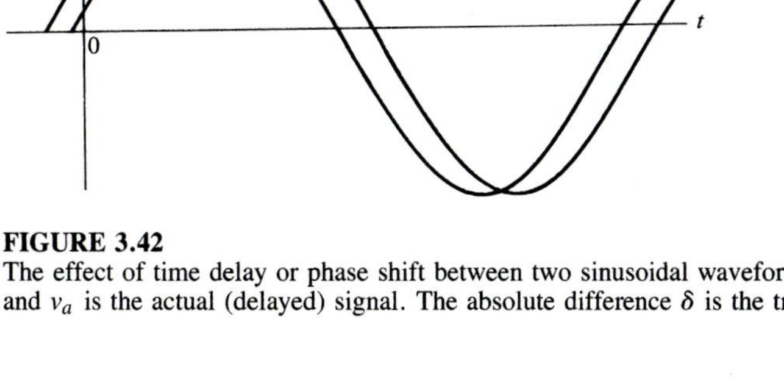

FIGURE 3.42
The effect of time delay or phase shift between two sinusoidal waveforms; v_d is a desired signal, and v_a is the actual (delayed) signal. The absolute difference δ is the true total error.

identities,

$$\delta = A[\sin\omega T - \sin\omega t\cos(-\theta) - \cos\omega t\sin(-\theta)]$$
$$\cong A\theta\cos\omega t \tag{3.58}$$

That is, the peak error is

$$\delta_{\text{peak}} \cong A\theta \tag{3.59}$$

Thus the maximum error is 100θ percent, where θ is in radians, or about 1.8 percent per *degree* of phase error. Suppose now we take a multiplier (or divider) and hold one input constant, while the other is varied sinusoidally. A typical first-order frequency response will have at least 45° phase lag at the -3-dB frequency f_0 Hz. At frequencies well below f_0, we can often estimate the phase shift as

$$\theta \cong \frac{-f}{f_0} \qquad \text{rad} \qquad f \ll f_0 \tag{3.60}$$

Thus we will have typically 1 percent error at $0.01 f_0$ Hz! This well-known result (see Ref. 4, Sec. 3.2) applies to linear operators in general and shows how important knowledge of phase shift in amplifiers and filters may be! Thus one should not expect high accuracy from a multiplier *unless input frequencies are quite small compared to the −3-dB frequency*.

Instead of merely giving the -3-dB frequency, a cooperative manufacturer will give additional information such as slew rate, settling time, and frequencies below which the total error due to phase and/or amplitude response can be assumed to be less than some guaranteed amount.

Analog multiplication has been implemented in a variety of ways (see, e.g., page 205 of the first edition of this text). Currently the current ratioing or variable transconductance method is most common (see Gilbert, Ref. 12). This is available in integrated-circuit form, with high speed and reasonably good accuracy (see Table 3.2).

Multiplier Applications

In the following discussion we assume that we are dealing with ±10-V analog systems and that we have available four-quadrant multipliers that implement

$$v_0 = \frac{v_1 v_2}{10} \tag{3.61}$$

Figure 3.43*a* illustrates our block-diagram symbol for a multiplier. Note that we

TABLE 3.2
Current ratioing or transconductance multipliers*

Static error, %	−3-dB bandwidth	1% bandwidth	Slew rate, V/μs	Accuracy, stability, %/°C
0.5–1	50 kHz	70 kHz	20	0.008
2.0	10 MHz	100 kHz	20	0.02

* Typical of current units (1988), e.g., Burr-Brown MPY 100, 634.

are assuming *no polarity inversion* of the product in (3.61); in many cases one will find that commercial multipliers may actually have a polarity inversion in their transfer characteristic, or they may permit the user to choose the polarity of the generated product. In any event, it should be possible to use either type of multiplier in any of the applications described here simply by using appropriate inverters.

Any of the three types of multipliers mentioned in Table 3.2 may be used in the applications described, although, of course, some will have superior accuracy and/or speed advantages. Nevertheless, they will all tend to behave in about the same way. Typical commercial multipliers will have an output driving capability and impedance similar to an operational amplifier; i.e., they will deliver a ± 10-V signal and an output current of the order of 5 to 10 mA. Input impedances are typically greater than 10 kΩ.

IMPLICIT DIVISION. Often analog division is performed by using a circuit similar to Fig. 3.43*b*; indeed, many manufacturers internally connect the multiplier in the feedback loop of an operational amplifier so that the unit may be set up by the user as either a multiplier or a divider. The circuit of Fig. 3.43*b* may be analyzed as follows:

The feedback path around the amplifier hopefully prevents amplifier overload; if that is true, then, assuming an ideal operational amplifier,

$$\frac{v_3}{R_2} + \frac{v_1}{R_1} = 0 \tag{3.62}$$

However, the multiplier forces

$$v_3 = \frac{v_0 v_2}{10} \tag{3.63}$$

Combining these last two equations, we can obtain

$$v_0 = \frac{-10 R_2 v_1}{R_1 v_2} \qquad v_2 > 0 \tag{3.64}$$

Thus we have implemented division. Note that the overall scale factor may be adjusted by using the resistance ratio. If $R_1 = R_2$, then (3.64) becomes similar to (3.54) except for the polarity inversion.

Note also that it is necessary to restrict the polarity of the divisor; here we must ensure that v_2 is positive in order to get the proper feedback polarity. Thus we have a two-quadrant inverting divider. If a polarity inversion is introduced into the feedback path, then the polarity of the division operation changes, and the input v_2 must be strictly negative (see Prob. 3.31).

Again, it is necessary to restrict the *minimum* value of v_2 so that the output does not overload; e.g., in (3.64) we must ensure that

$$\frac{R_2 v_1}{R_1 v_2} \leq 1 \tag{3.65}$$

for a 10-V analog system.

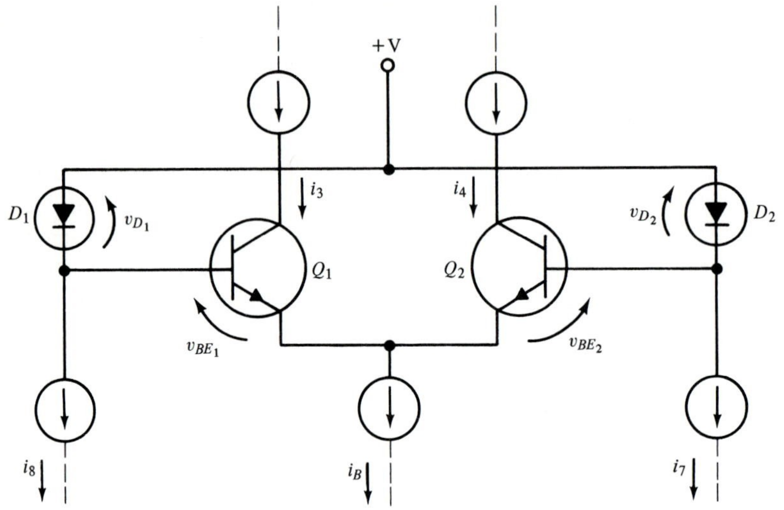

(*a*)

FIGURE 3.43
The current ratioing or transconductance type of multiplier. (*a*) Basic gain cell; (*b*) a complete four-quadrant multiplier circuit. The entire circuit may be implemented on one IC chip. *(continued on next page)*

SQUARE-ROOTING. In Fig. 3.44 we see how a multiplier may be used to perform the square root operation. Here again we have to be careful about the polarity of the input (as we should expect, we cannot take the square root of a negative number). If the circuit input v_1 is positive, we will find that we need to invert the multiplier output in Fig. 3.44, that is,

$$v_2 = \frac{-v_0^2}{10} \tag{3.66}$$

Again, since feedback forces

$$v_2 = \frac{-R_2}{R_1} v_1 \tag{3.67}$$

we can show that

$$v_0 = -K\sqrt{v_1} \qquad v_1 > 0 \tag{3.68a}$$

where

$$K = \sqrt{\frac{10R_2}{R_1}} \tag{3.68b}$$

The minus sign in (3.68*a*) is required for stability (see Prob. 3.30). Note that if

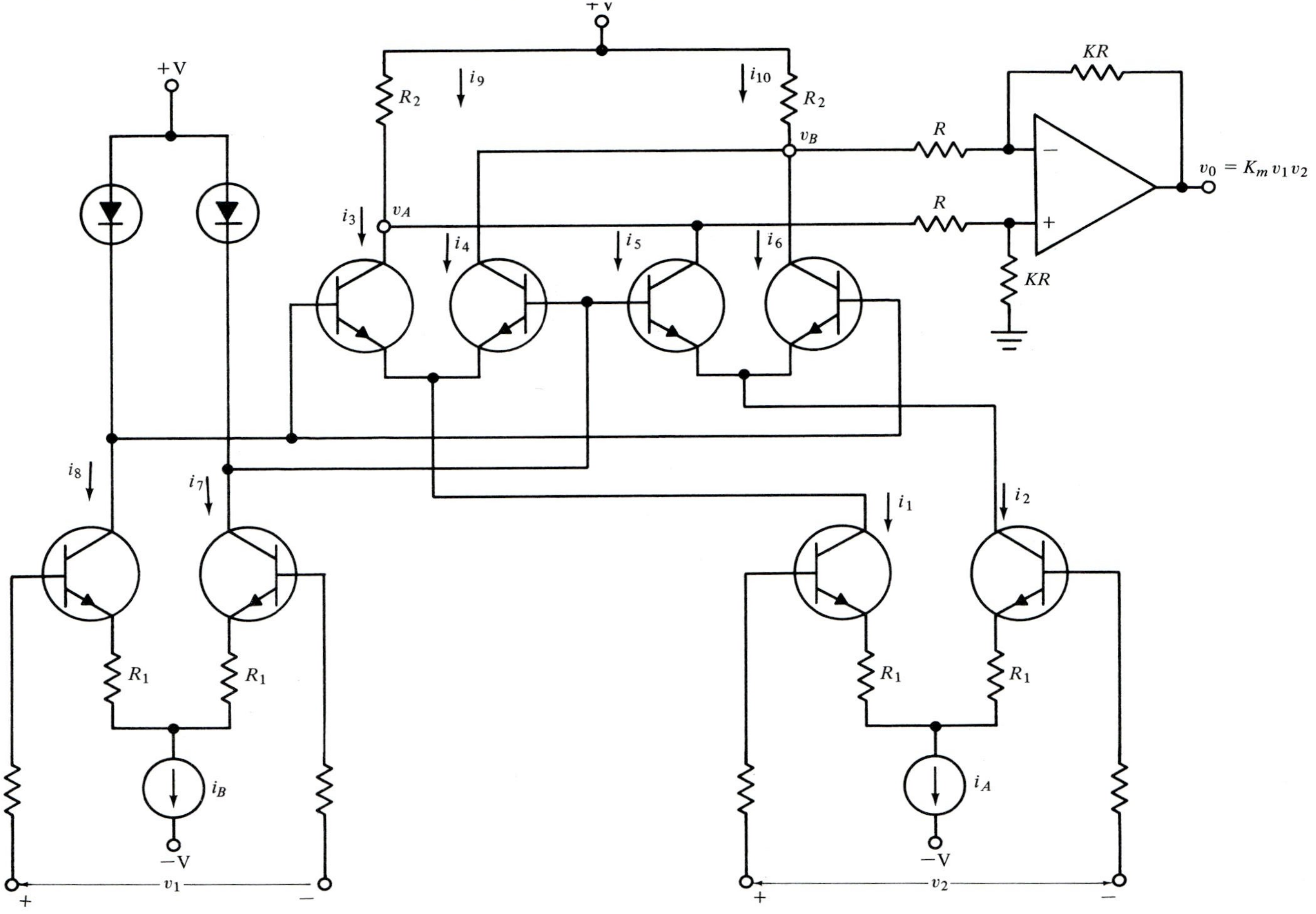

(*b*)

FIGURE 3.43 (continued)

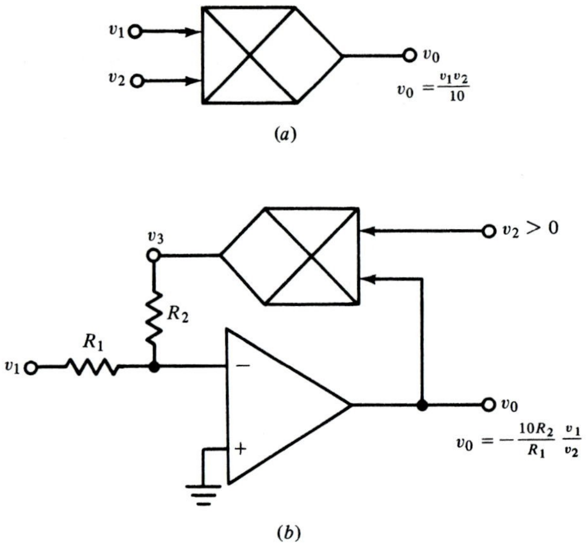

FIGURE 3.44
Multiplication and implicit division. (*a*) Typical multiplier block-diagram symbol; (*b*) the use of a multiplier in a feedback loop permits implicit division.

v_1 is strictly negative, then we can implement

$$v_0 = K\sqrt{-v_1} \qquad v_1 < 0 \tag{3.69a}$$

by removing the inverter following the multiplier in Fig. 3.44*a*; K is still given by (3.68*b*). Figure 3.44*b* shows a typical block-diagram symbol for a square-rooter. Usually we set $R_1 = R_2$ and $K = \sqrt{10}$, and so (3.68*a*) becomes

$$v_0 = -3.16\sqrt{v_1} \qquad v_1 > 0 \tag{3.69b}$$

that is, 10 V in yields 10 V out.

RMS OPERATIONS. In Fig. 3.45 we have a circuit that will form the rms combination of two inputs. The basic operating equations are

$$v_3 = \frac{v_1^2}{10} \tag{3.70a}$$

$$v_4 = \frac{v_2^2}{10} \tag{3.70b}$$

$$v_6 = -\frac{v_0^2}{10} \tag{3.70c}$$

$$v_6 = -\frac{R_2}{R_1}(v_3 + v_4) \tag{3.70d}$$

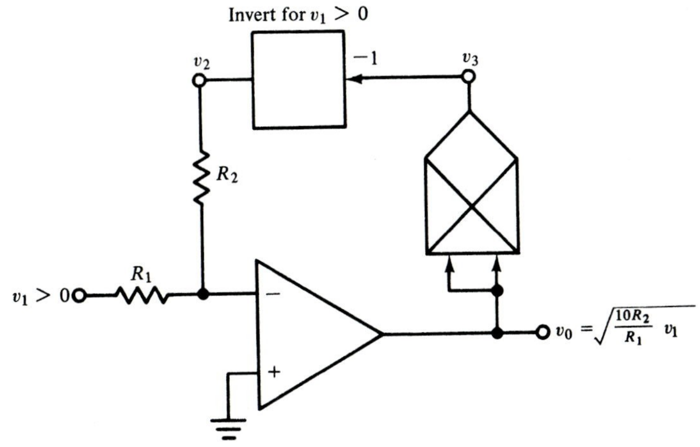

(*a*)

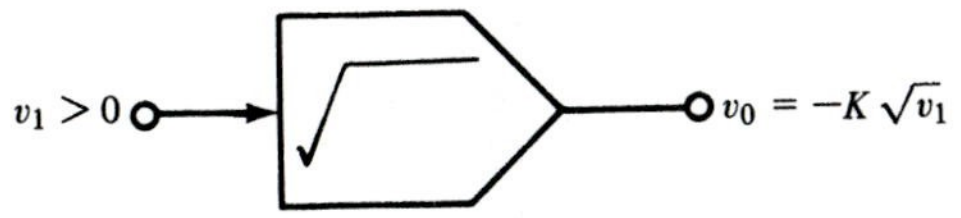

(*b*)

FIGURE 3.45
Implementing the square-root operation. (*a*) A multiplier (or squaring operator) in a feedback loop implements the operation; (*b*) a typical block-diagram symbol. Note that only one polarity of the input may be accommodated; removal of the inverter between v_3 and v_2 permits the use of negative inputs.

From these equations, we obtain

$$v_0^2 = \frac{R_2}{R_1}(v_1^2 + v_2^2)$$

or

$$v_0 = -\sqrt{\frac{R_2}{R_1}}\sqrt{v_1^2 + v_2^2} \tag{3.71}$$

Note that the two inputs may have either polarity. Also, (3.71) must have the minus sign to ensure stability (see Prob. 3.32).

Example 3.12. In Fig. 3.45 if we know that both inputs may range over ± 10 V, then we must pick R_2/R_1 so that the overall transfer characteristic is properly amplitude-scaled, that $v_0 < 10$ V. In this instance

$$v_1^2 + v_2^2 \leq 200$$

and in (3.71) we require

$$\frac{R_2}{R_1} \leq 0.5$$

Thus, picking $R_2 = 50\ \text{k}\Omega$ and $R_1 = 100\ \text{k}\Omega$,

$$v_O = -0.707\sqrt{v_1^2 + v_2^2}$$

which is properly scaled.

Note that proper amplitude scaling is important, in order to obtain full static accuracy. If the multipliers are used in such a way that their inputs are never more than a small fraction of the allowable 10-V range, then the percentage of static accuracy (which is only guaranteed in terms of full-scale output magnitude) may be very poor! The problem is the same as the one encountered in scaling analog computers.

Example 3.13. In Fig. 3.46 we see a design of a system to measure true rms values. Here we want to average out short-term fluctuations of the input. The circuit shown has a 10-s averaging time constant. The input voltage should be limited to an rms value that is perhaps 0.2 to 0.33 times the allowable peak input value, so that the input squarer will normally not overload due to temporary excursions of the input. The circuit of Fig. 3.46 is designed to accommodate an input with an rms value of less than 1 V and to provide a true rms reading regardless of signal waveform. Of course, the bandwidth of the input must be consistent with the speed of the input multiplier. Also we want a gain of 10, that is, $v_O = 10$ V when $v_{1,\text{rms}} = 1.0$ V; this assures that the output square root operation is done fairly accurately.

Note that we have selected the various subsystem gains in an attempt to properly scale. For example, $v_2 = 2v_1$, and v_2 will have a value of about 2 V rms

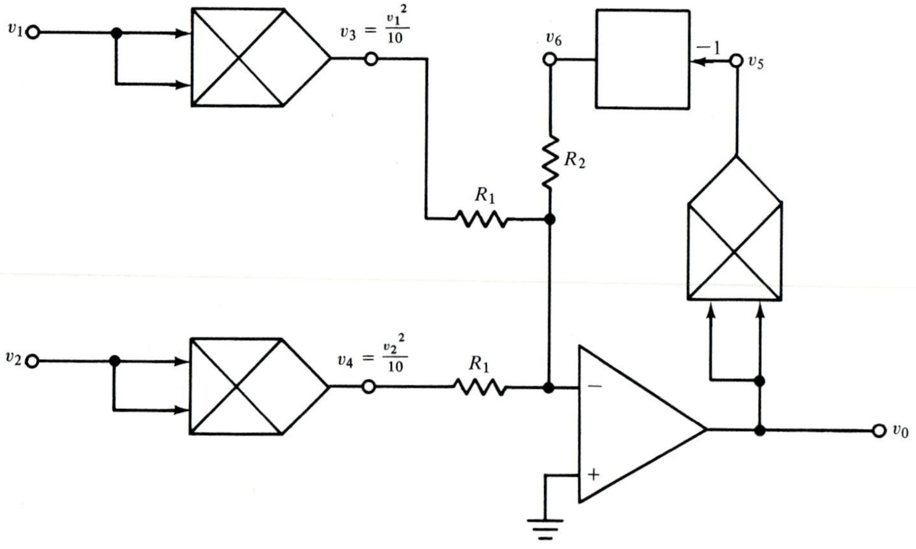

FIGURE 3.46
A two-input rms operator; additional inputs may be added, provided proper amplitude scaling is maintained.

and a peak value of no more than 6 to 10 V in most cases. The input amplifier A_1 provides a high input impedance to the signal being observed and amplifies it slightly. Also

$$v_3 = \frac{v_2^2}{10} = 0.4v_1^2$$

The low-pass filter associated with A_2 averages the square of v_1 with a 10-s time constant, and thus v_4 is a slowly varying quantity with

$$v_4 = \frac{-10^5}{R_2} v_{3,\text{av}} = \frac{-0.4 \times 10^5}{R_2} v_{1,\text{av}}^2$$

We choose R_2 so that

$$v_4 = -10 \text{ V} \qquad \text{when} \qquad v_1 = +1 \text{ V rms}$$

Thus

$$-10 = \frac{-10^5}{R_2} 0.4$$

or

$$R_2 = 4 \text{ k}\Omega$$

The output amplifier should have the proper gain so that $v_O = 10$ V when $v_{1,\text{rms}} =$ 1.0 V. Note that since v_4 is negative, we *do not* want an inverter following multiplier M_2.

In this case

$$v_O = \sqrt{\frac{10R_4}{R_3}} \sqrt{-v_4}$$

and we want

$$v_O = +10 \text{ V} \qquad \text{when} \qquad v_4 = -10 \text{ V}$$

Thus we set

$$R_4 = R_3 = 10 \text{ k}\Omega$$

Since, on average, v_3 is about 0.4 V, it seems that M_1 is considerably underscaled. Note, however, that if v_1 momentarily became 3 V, then v_3 would accordingly be 3.6 V, etc. We have left considerable allowance for temporary excursions of v_1 well above the rms value, which we should do if we have no prior knowledge of the input waveform. If, e.g., we knew that the input to an rms measuring circuit were a random signal with a gaussian or normal amplitude distribution, then typically the input would seldom be more than 3 to 4 times the rms value. Such information could be used to better scale the input gain.

3.6 WAVEFORM GENERATION

This topic could well occupy a full chapter (see, e.g., Chap. 10 of Ref. 3). Our main goal here is to present a few representative techniques for generating commonly needed waveforms (sine square and triangle waves, and controlled pulses). Many circuits have been developed to do these jobs. The ones we have chosen to describe hopefully represent basic design approaches and are relatively easy to build with a minimum of special debugging.

Sine Wave Oscillators

The generation of sinusoidal waveforms with precisely controlled amplitude and low harmonic distortion is an important requirement in many communication systems. For radio frequencies (say above 1 MHz), one usually uses high-Q resonant circuits (LC or quartz-crystal resonators) in some form of simple transistor oscillator circuit. We are more interested in the generation of sinusoids in the frequency range of, say, 0.001 Hz to 1 MHz. Here operational amplifier circuits are generally employed. Below about 10 Hz, we recommend using the method, described below, of generating a triangle waveform and then using a sine function generator or bandpass filter to get the final waveform. Above 10 Hz, circuits similar to the block diagram of Fig. 3.47 are generally satisfactory. Figure 3.48 shows details of typical circuit implementations.

This is the so-called analog-computer approach, also called the *quadrature oscillator*. Both configurations in Fig. 3.47 implement the differential equation

$$\frac{d^2 v_0}{dt^2} + \omega_0^2 v_0 = 0 \tag{3.72}$$

The solution to this equation is

$$v_0 = A \sin(\omega_0 t + \theta) \tag{3.73}$$

where θ depends on the initial conditions.

In actuality, some form of regeneration must be added to ensure that the oscillation will build up. In some cases, the generation will occur due to the inherent phase lag in analog integrators (Ref. 4, Sec. 3.24). It is a good idea, however, to provide for some regeneration. Thus in Figs. 3.47*a* and 3.48*a*, we are actually setting up the equation

$$\frac{d^2 v_0}{dt^2} - \epsilon\omega_0 \frac{dv_0}{dt} + \omega_0^2 v_0 = 0 \tag{3.74}$$

which will build up slowly for small positive ϵ, with a frequency of oscillation that is still close to ω_0. Usually ϵ is adjusted empirically for proper performance. In Figs. 3.47*b* and 3.48*b*, we show an alternative way to implement (3.72), using one inverting and one noninverting integrator. To obtain regeneration, we usually can adjust the location of the pole of the noninverting integrator (i.e., move into the left half of the s plane). In Fig. 3.48 this is done by adjusting R_3. Note that the noninverting integrator is the one previously discussed in Chap. 1 (Fig. 1.22, Prob. 1.16). In both parts of Fig. 3.48 the individual integrator gains establish

$$\omega_0 = \frac{1}{R_1 C_1} \quad \text{rad/s} \tag{3.75}$$

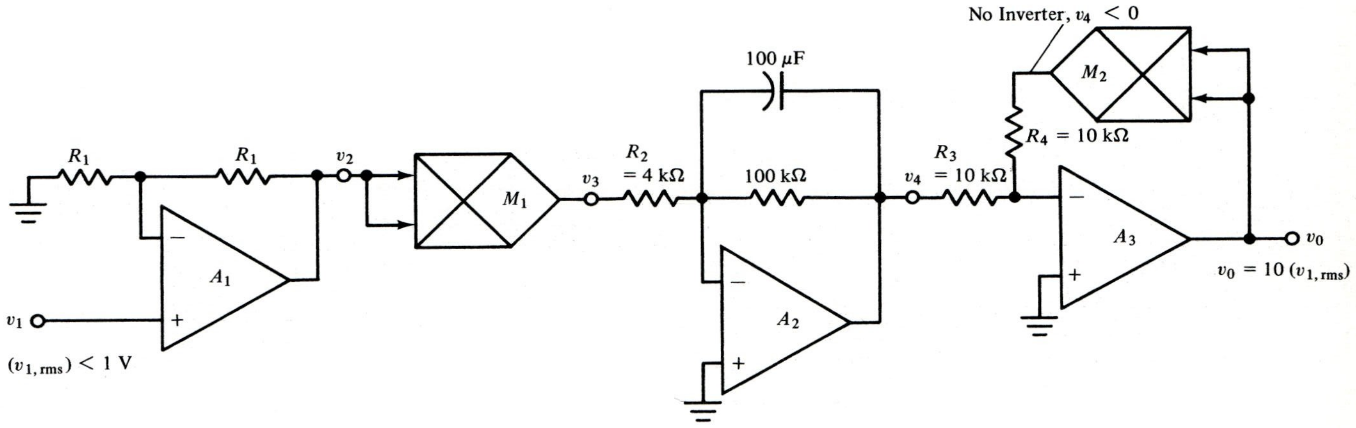

FIGURE 3.47
A true rms voltmeter. The input buffer amplifier is optional. Care must be taken to avoid internal overloads due to momentary excursions of the signals.

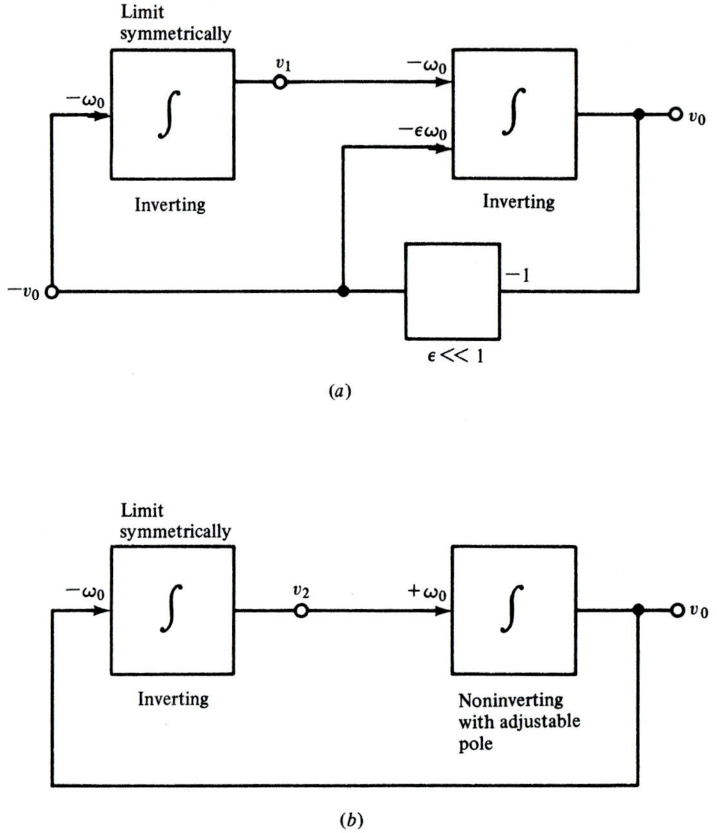

FIGURE 3.48
Analog-computer or sine-loop type of sinusoidal oscillator. (*a*) Two inverting integrators and a unity gain inverter solve the harmonic oscillator differential equation; (*b*) one inverting and one noninverting integrator also may be used.

Both integrator outputs will have approximately the same amplitude, but will be 90° out of phase.

Another requirement is to limit the amplitude of the oscillation, hopefully without adding distortion. One fairly simple limiting method is to provide feedback limiting around the integrator *not* being used to provide the main output. Any distortion due to clipping in this integrator output will then be suppressed by the output integrator. The limiter should be fairly soft and adjustable. The diode feedback limiter illustrated in Fig. 3.48*b* is often used.

Example 3.14. Suppose we want to use the circuit of Fig. 3.49*b* to design a 1-kHz oscillator with a ±10-V amplitude. Using $C_1 = 0.001\ \mu F$, we have from (3.75)

$$R_1 = \frac{1}{\omega_0 C_1} = \frac{1}{2\pi \times 10^3 \times 10^{-9}}$$
$$= 159\ \text{k}\Omega$$

Thus

$$R_2 = 318 \text{ k}\Omega$$

and R_3 is made adjustable in the neighborhood of 318 kΩ to establish the desired amount of regeneration. If the limiter on A_1 is set to clip at ±10 V, v_0 will also have a ±10-V amplitude. With the ±15-V supplies, typical values are $R_5 = 1.7$ kΩ and $R_4 = 1$ kΩ. The value of the two resistors labeled R_4 should be trimmed to get the desired amplitude, with symmetric limiting.

Note that the circuit of Fig. 3.49*a* requires an extra operational amplifier, but it can be set up on an analog computer, which normally does not have noninverting amplifier terminals available. Improved automatic gain control (AGC) circuits for regulating the amplitude of the waveform may be employed (see, e.g., Ref. 3, p. 390).

FREQUENCY-MODULATED SINE-WAVE OSCILLATORS. Figure 3.50 shows a modification to the system of Fig. 3.47 which permits generation of FM waveforms. Normally two multipliers should be used to modulate the loop gain and keep the magnitude of the integrator output properly scaled. The instantaneous angular frequency of oscillation will be

$$\omega_0(t) = \frac{K\,[v_m(t) + A]}{10} \qquad \text{rad/s} \tag{3.76}$$

for a 10-V system. Limiting and regeneration will be added in a manner similar to Example 3.14.

OTHER SINE-WAVE OSCILLATORS. The operational amplifier may be used as the active element in a variety of oscillator circuits. Figure 3.51 shows two common approaches:

1. The Wien bridge oscillator (Fig. 3.51*a*)
2. The phase-shift oscillator (Fig. 3.51*b*)

These types are more fully discussed in Ref. 3, Chap. 10. We present them here mainly for completeness. Although they employ only one amplifier, their design takes some care if well-controlled amplitude with low distortion is to be achieved. Moreover, they do not provide the two-quadrature outputs (sine and cosine waves) available from the three-amplifier circuits discussed above.

The Wien bridge circuit oscillates at approximately

$$\omega_0 \cong \frac{1}{RC} \qquad \text{rad/s} \tag{3.77a}$$

with

$$R_2 \cong 2R_1 \tag{3.77b}$$

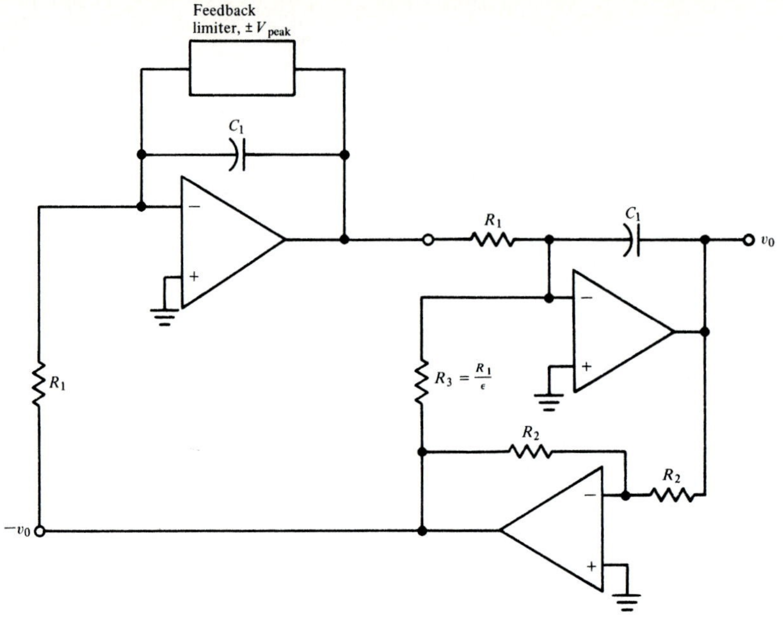

(*a*)

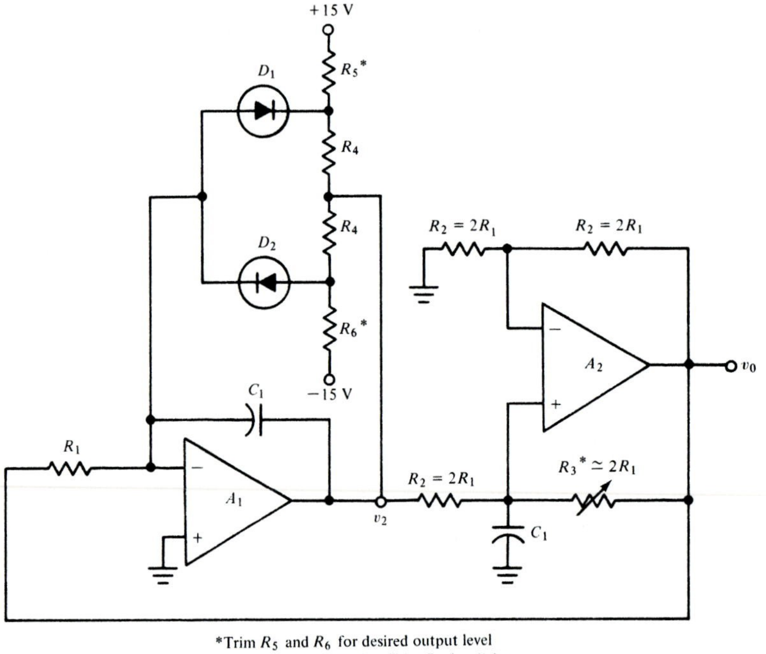

*Trim R_5 and R_6 for desired output level and symmetrical limiting. Trim R_3 for slight regeneration.

(*b*)

FIGURE 3.49
Sinusoidal quadrature oscillator, based upon Fig. 3.47; the main integrator gains should be equal, and regeneration is adjusted by R_3. (*a*) Three-amplifier circuit based on Fig. 3.47*a*; R_3 is usually very large; (*b*) two-amplifier circuit based upon Fig. 3.47*b*. Details of soft limiter are shown; R_3 is adjusted to control regeneration, and resistors R_5 and R_6 are adjusted to control amplitude and to limit symmetrically.

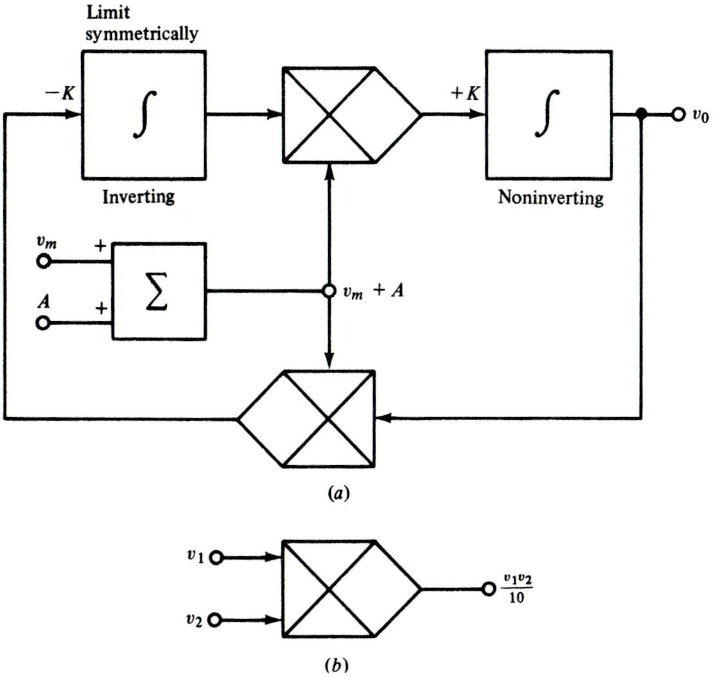

FIGURE 3.50
Generation of frequency-modulated (FM) waveforms. (*a*) General block diagram; (*b*) explanation of multiplier symbol.

The amplitude of oscillation may be regulated by using some form of nonlinear resistor for R_1. For example, if R_1 is a thermistor or simple incandescent lamp, R_1 will increase with the power dissipated in R_1, and this will tend to regulate the amplitude.

The phase-shift oscillator of Fig. 3.50*b* oscillates at approximately

$$\omega_0 \cong \frac{1}{\sqrt{3}\,RC} \tag{3.77c}$$

and the feedback resistance R_1 should be slightly greater than $12R$.

Precision Square and Triangle Wave Generation

Figure 3.52 shows the block diagram of a system for generating square and triangle waves. It uses a precision bistable circuit (see, e.g., Fig. 3.34) with output level $L_+ > 0$ and $L_- < 0$ and switching points $S_+ > 0$ and $S_- < 0$. The inverting integrator output is a triangle wave, as shown in Fig. 3.52*b*, with peak values determined by S_+ and S_-. Similarly, the square wave has peak values determined by L_+ and L_-. We have shown a symmetric case. The period of the waveform is

$$T = \frac{4S}{LK} \quad \text{s} \tag{3.78}$$

where $L = L_+ = |L_-|$ and $S = S_+ = |S_-|$.

Figure 3.52*c* shows a typical situation, where the bistable has $L = S = 10$ V, and $K = 1/RC$; in this case

$$T = 4RC \quad \text{s}$$

and we get ±10-V square and triangle waves.

Figure 3.53 shows a more complete circuit. The frequency of oscillation is

$$f = \frac{1}{4RC} \quad \text{Hz} \tag{3.79}$$

The bistable is based on Fig. 3.34.

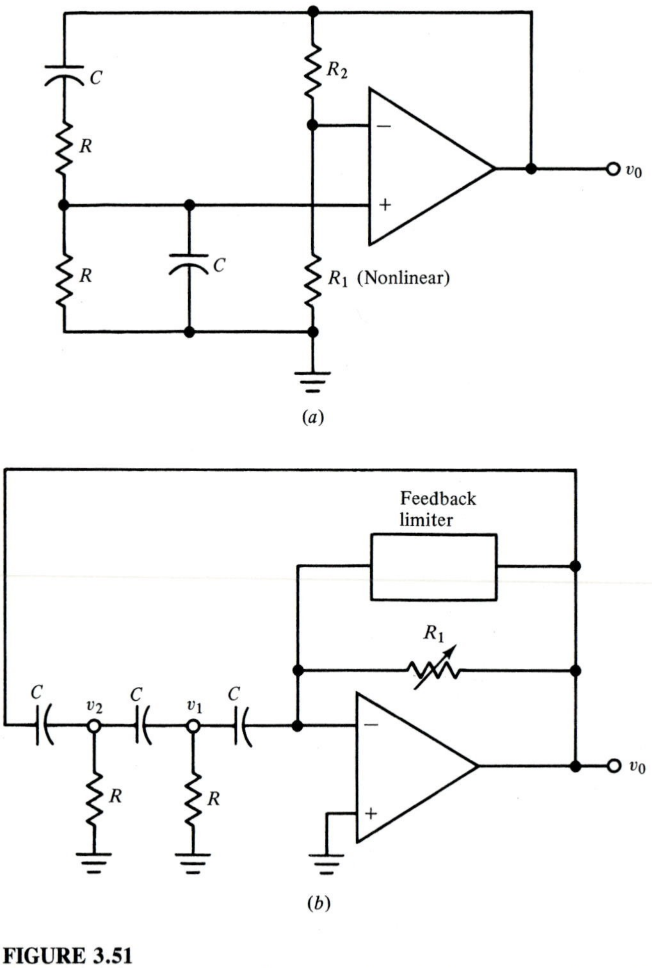

FIGURE 3.51
One-amplifier sinusoidal oscillators. (*a*) Wien bridge type; (*b*) phase-shift type.

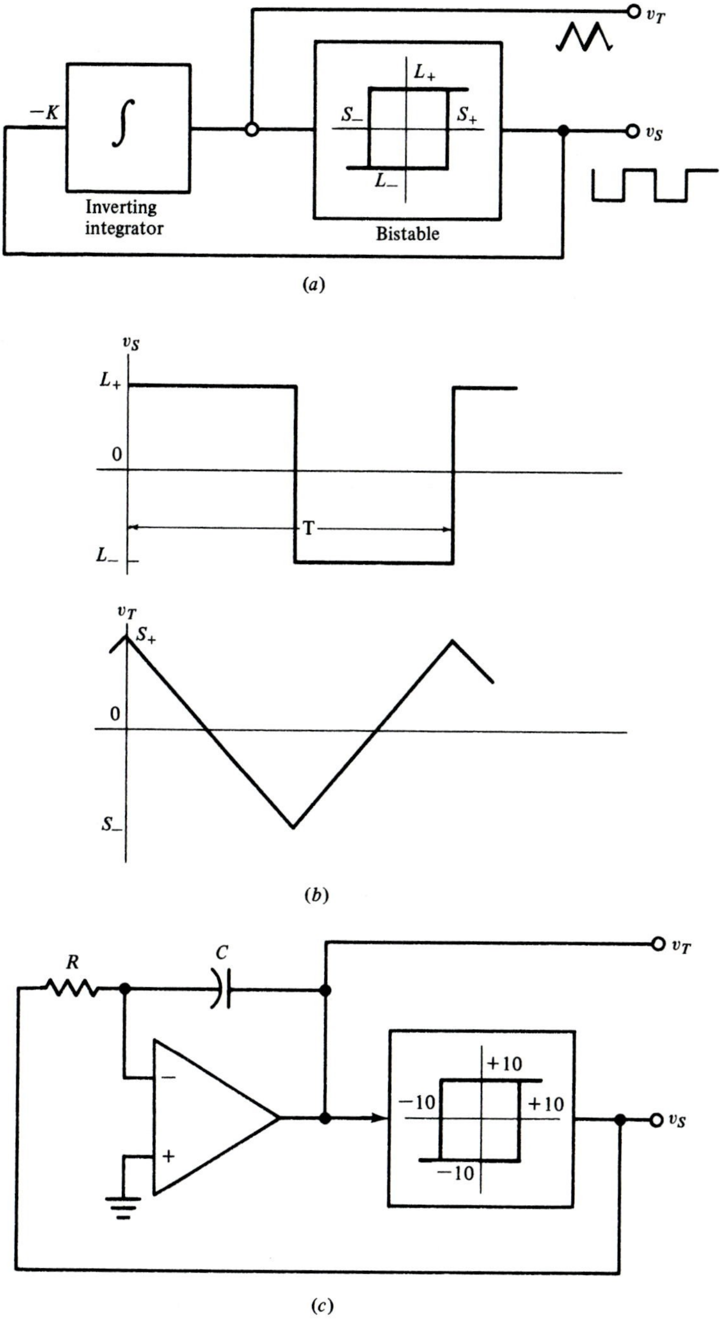

FIGURE 3.52
Generation of square and triangle waves. (*a*) General block diagram; (*b*) typical waveforms; (*c*) typical circuit parameters.

In Fig. 3.54, we see some interesting modifications that may be made to the basic system of Fig. 3.52. First, we have shown the addition of a sine function generator. This could be implemented, e.g., by using triangle to sine wave converters (see Refs. 4 and 16).

Also we have added a four-diode bridge limiter in the oscillator feedback path of Fig. 3.54. This limiter is driven by a source of modulating signal v_m, so that signal v_L is a square wave of amplitude $\pm v_m$. Essentially we are modulating the loop gain of the circuit, and the result is a *VCO* (voltage-controlled oscillator) or *FM oscillator*. An analog multiplier can be used instead of the bridge and inverter, but it is not needed here. The amplitudes of the triangle wave will still be controlled by the bistable switch points S_+ and S_-, and the amplitude of the square-wave output will still be controlled by the bistable output levels L_+ and L_-.

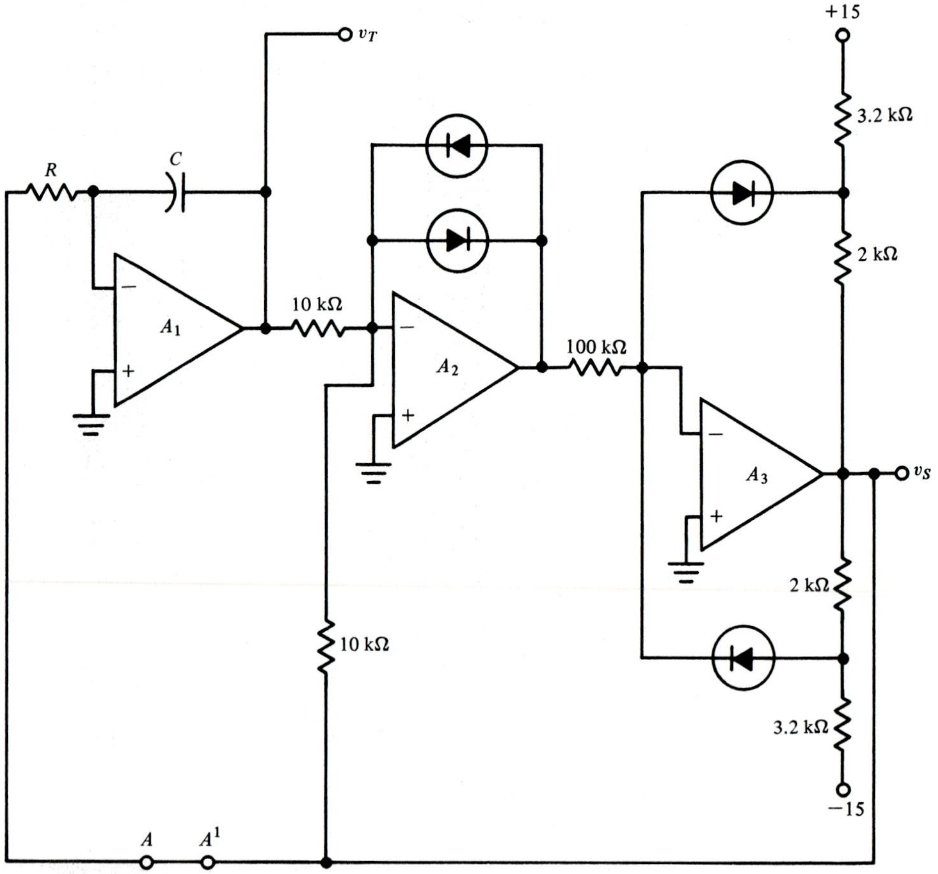

FIGURE 3.53
A complete circuit for generating square and triangle waveforms.

Example 3.15. Suppose we use the basic circuit of Fig. 3.53, but add a four-diode bridge at point A-A'; from (3.78) with $S = 10$, $K = 1/RC$, and

$$L = v_m \frac{R}{R + R_1}$$

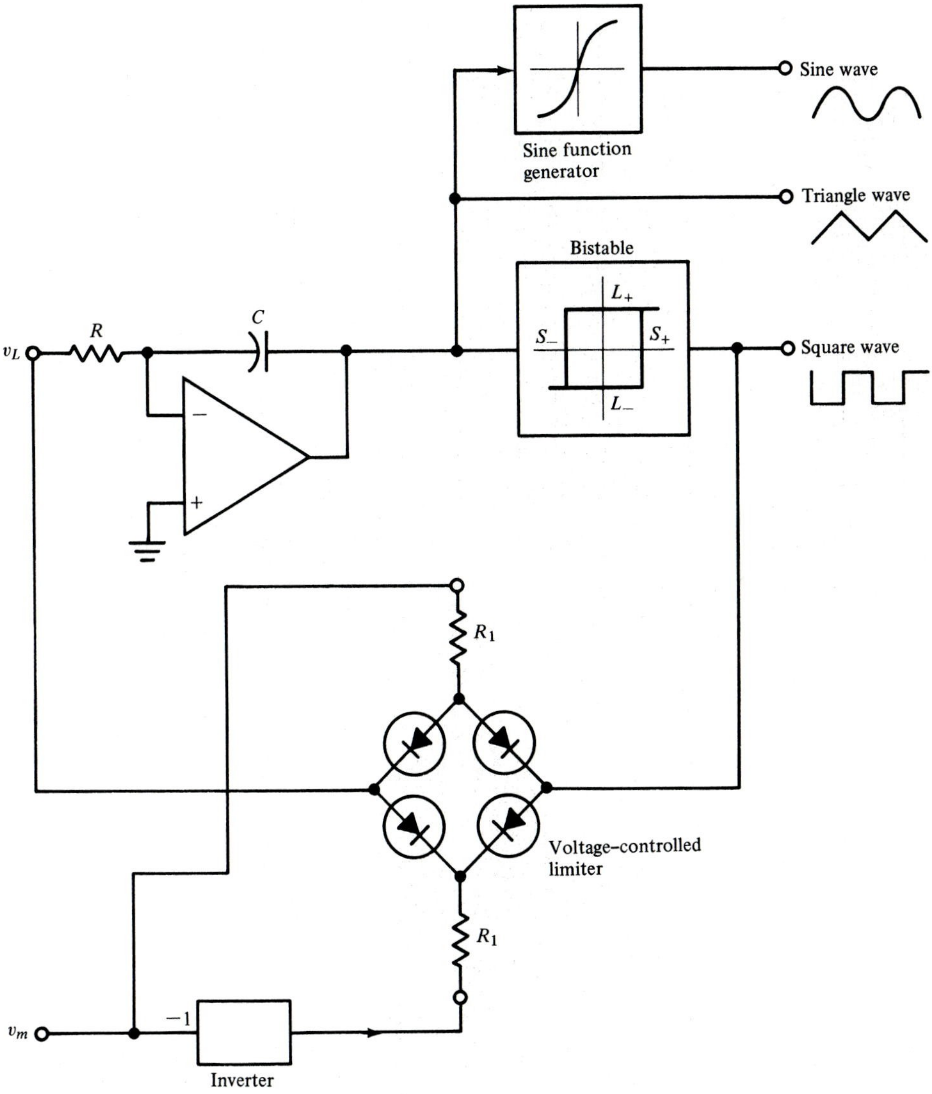

FIGURE 3.54
Modifications to the basic square and triangle wave generator to obtain sine waves as well, and to frequency-modulate via v_m and the four-diode bridge limiter.

then

$$T = \frac{4RC \times 10(R + R_1)}{v_m R}$$

or

$$T = \frac{4C(R + R_1) \times 10}{v_m}$$

Thus the frequency of oscillation is

$$f = \frac{v_m}{40(R + R_1)C}$$

We see that the circuit has very linear frequency modulation and thus makes a very nice VCO.

This general technique for generating square, triangle, and sine waves has become so well established that one can now buy commercial integrated circuits that provide a source of square, triangle, or sine waves which may be amplitude- or frequency-modulated. For example, the EXAR XR-205 monolithic waveform generator can generate 3-V peak-to-peak waveforms with frequencies up to 4 MHz.

REFERENCES AND BIBLIOGRAPHY

1. *Electronics:* A biweekly journal for electronics engineers; in particular, see the Circuit Design features. McGraw-Hill, Inc.
2. *Electronic Design:* A biweekly journal for electronics engineers; in particular, see the articles in the Technology section. Hayden Publ. Co.
3. Tobey, G. E., J. G. Graeme, and L. P. Huelsman: *Operational Amplifiers,* McGraw-Hill, New York, 1971.
4. Korn, G. A., and T. M. Korn: *Electronic Analog and Hybrid Computers,* 2d ed., McGraw-Hill, New York, 1972.
5. Searle, C. L., A. R. Boothroyd, et al.: *Elementary Circuit Properties of Transistors,* Wiley, New York, 1964.
6. Callahan, M. J., R. L. Peck, and J. A. Reagan: "A Lidar Output Energy Digital Detector," *Abstracts, 4th Conference on Laser Radar Studies of the Atmosphere, University of Arizona,* June 27, 1972, pp. 14–16.
7. Clayton, G. B.: *Operational Amplifiers,* Butterworth, London, 1971.
8. Dobkin, R. G.: "Logarithmic Converters," *IEEE Spectrum,* November 1969, pp. 69–72.
9. *Analog Dialogue,* vol. 6, nos. 2 and 3, Analog Devices, Inc.
10. Sheingold, D. H.: "Approximate Analog Functions with a Low-cost Multiplier/Divider," *Electronic Design News,* Feb. 5, 1973, pp. 50–52.
11. Cate, T.: "Modern Techniques of Analog Multiplication," *The Electronic Engineer,* April 1970, pp. 75–79.
12. Gilbert, B.: "A Precise Four-Quadrant Multiplier with Subnanosecond Response," *IEEE Journal of Solid-State Circuits,* December 1968, pp. 365–373.
13. Schwartz, M.: *Information, Transmission, Modulation, and Noise,* 4th ed., McGraw-Hill, New York, 1990.
14. Burr-Brown Corp.: *The Handbook of Linear IC Applications,* Burr-Brown, Tucson, Ariz.

15. Graeme, J. G.: "Single Potentiometer Controls Gain of Precision Rectifier," *Electronic Design News,* Nov. 8, 1979.
16. Franco, S.: *Design with Operational Amplifiers and Analog Integrated Circuits,* McGraw-Hill, New York, 1988.

PROBLEMS

3.1. *(Sec. 3.1)* Refer to Fig. P3.1; the nonlinear network that is placed in the amplifier feedback path is sketched in Fig. P3.1*b*. Assume an ideal operational amplifier.
(*a*) If $R_1 = R_2 = 20\ \text{k}\Omega$ and $v_2 = 0$ V dc, sketch v_O versus v_1, $|v_1| \leq 10$ V.
(*b*) Repeat part *a* if $v_2 = +3$ V dc.
(*c*) Repeat part *a* if $v_2 = +3$ V dc and $R_2 = 10\ \text{k}\Omega$.

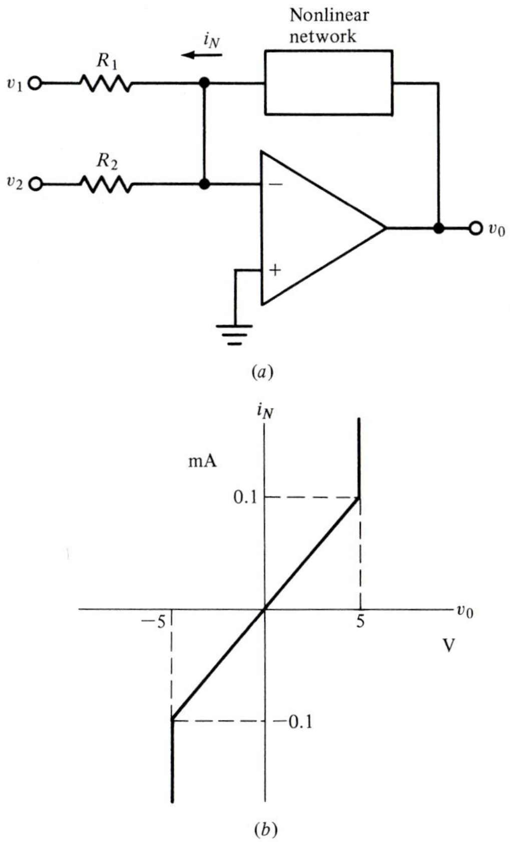

FIGURE P3.1

3.2. *(Sec. 3.1)* Refer to Fig. P3.2, (p. 224). Find v_O versus v_1 for this more general case of Example 3.2.

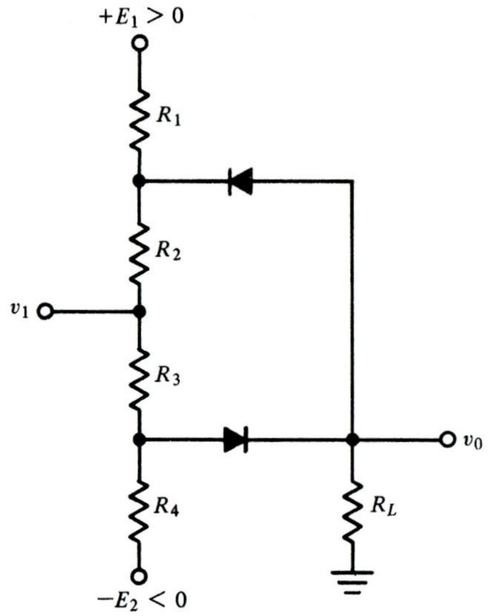

FIGURE P3.2

3.3. *(Sec. 3.1)* Refer to Fig. P3.3; assume an ideal operational amplifier and diodes.

(*a*) Sketch v_0 versus v_1.

(*b*) Repeat part *a*, but with diodes D_3 and D_4 and their associated resistor and power source removed, i.e., with a half-bridge.

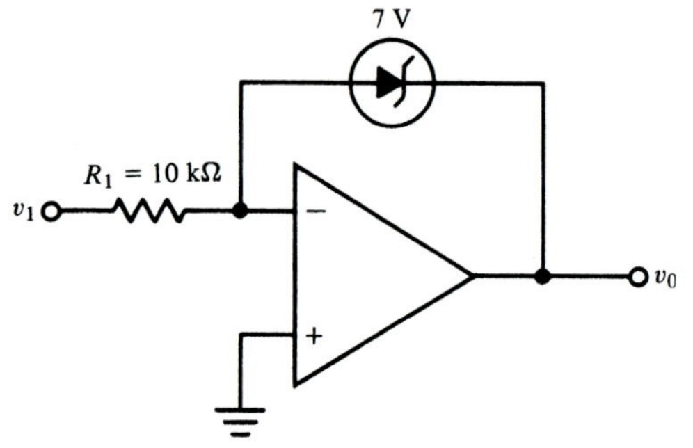

(*a*)

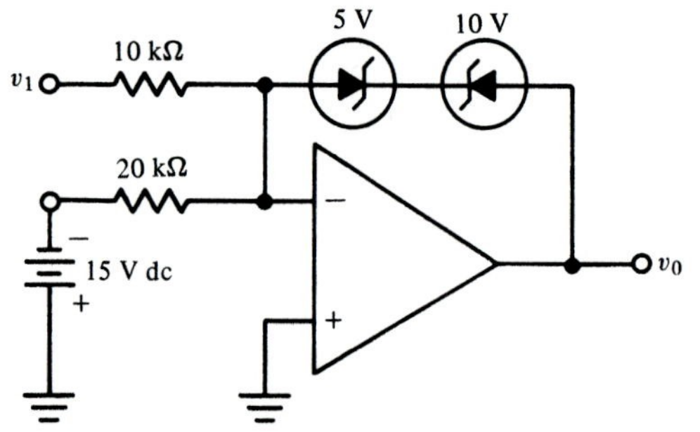

(*b*)

FIGURE P3.3

3.4. *(Sec. 3.2)* Sketch the transfer characteristic, v_O versus v_I, for the circuits of Fig. P3.3. The Zener diode voltages are labeled. Cover the range $-10\text{ V} \leq v_I \leq +10\text{ V}$.

3.5. *(Sec. 3.2)*

(*a*) Sketch the transfer characteristic, v_O versus v_I, for the circuit of Fig. P3.5. Assume ideal diodes and operational amplifier. Cover the range $-10\text{ V} \leq v_I \leq +10\text{ V}$.

(*b*) Repeat the above, but add a 10-kΩ resistor between the operational amplifier output and the inverting terminal.

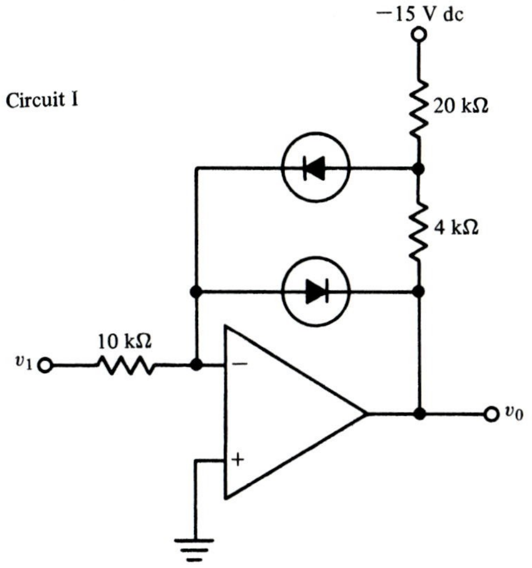

(*a*)

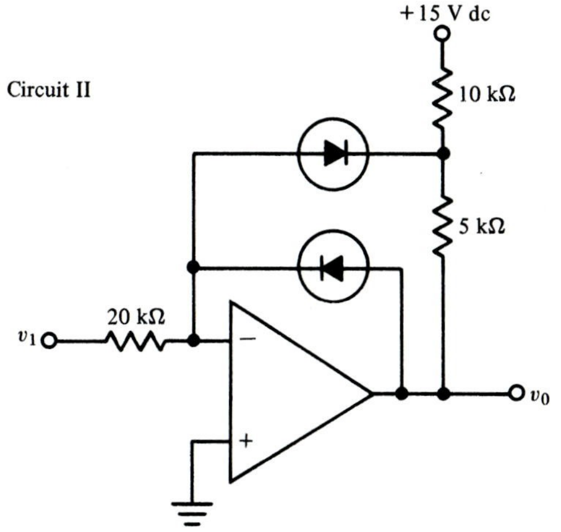

(*b*)

FIGURE P3.5

3.6. *(Sec. 3.2)* Using the circuit of Fig. 3.14, design a limiter amplifier to match the transfer characteristic of Fig. P3.6; use an ideal diode model.

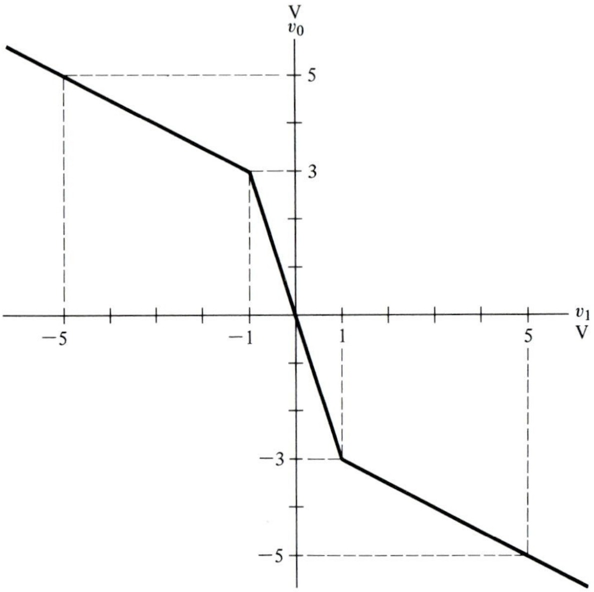

FIGURE P3.6

3.7. *(Sec. 3.2)* Use the basic circuit of Fig. 3.17 to design circuits for the following transfer functions:

$$(a)\ v_O = \begin{cases} -2v_I & v_I < 0 \\ 0 & v_I > 0 \end{cases}$$

$$(b)\ v_O = \begin{cases} -4v_I & v_I > 0 \\ 0 & v_I < 0 \end{cases}$$

3.8. *(Sec. 3.2)* Sketch v_O versus v_I for the circuit of Fig. P3.8, (p. 227). Try to make accurate estimates of slopes, etc. Cover the range $-10\text{ V} \leq v_I \leq +10\text{ V}$.

3.9. *(Sec. 3.2)* Refer to Fig. 3.21; assume ideal diodes and the following parameters:

$$E_1 = +15\text{ V dc} \qquad -E_2 = -15\text{ V dc}$$

$$R_A = R_B = 10\text{ k}\Omega \qquad R_L = 10\text{ k}\Omega$$

$$R_F = 20\text{ k}\Omega \qquad R_1 = 10\text{ k}\Omega$$

Both Zener diodes break down at a reverse voltage of 8 V.

(*a*) Sketch v_O versus v_I, $-10\text{ V} \leq v_I \leq +10\text{ V}$.

(*b*) Repeat part *a* if D_3, D_4, and R_B are removed. Does this answer suggest another way to build a precision half-wave rectifier?

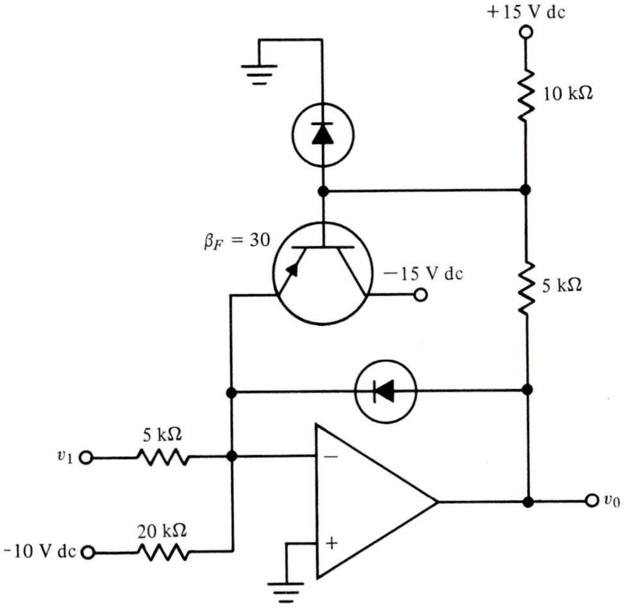

FIGURE P3.8

3.10. *(Sec. 3.2)* Figure P3.10, (see p. 228), shows two different comparator circuits; sketch v_O versus v_I for each circuit, and compare. Assume the Zener diodes have an incremental resistance of 60 Ω in the breakdown region and that the other diodes have an ac resistance of 30 Ω and a forward drop of 0.1 V.

3.11. *(Sec. 3.2)* Figure P3.11, (see p. 229), shows two different level clamps used to form a comparator, with positive feedback for hysteresis. Use the following symbols:

$$K_1 = \frac{R_2}{R_1 + R_2} \qquad K_2 = \frac{R_4}{R_3 + R_4}$$

(*a*) Compare the two circuits; i.e., derive expressions for both circuits for estimating

$$E_+ = \text{positive output level}$$

$$E_- = \text{negative output level}$$

$$E_h = \text{hysteresis width}$$

in terms of the rest of the symbols; verify (3.17) for both circuits. Assume both Zener diodes in circuit II break down at E_z V. *Note:* The sketch of v_O versus v_I is idealized; circuit I would not limit perfectly, but that need not be considered in part *a*.

(*b*) Sketch v_O versus v_I accurately (both circuits) if

$$E = 15 \text{ V dc} \qquad E_z = 5 \text{ V}$$

$$R_1 = 4 \text{ k}\Omega \qquad R_3 = 50 \text{ k}\Omega$$

$$R_2 = 1 \text{ k}\Omega \qquad R_4 = 1 \text{ k}\Omega$$

$$R_5 = 10 \text{ k}\Omega$$

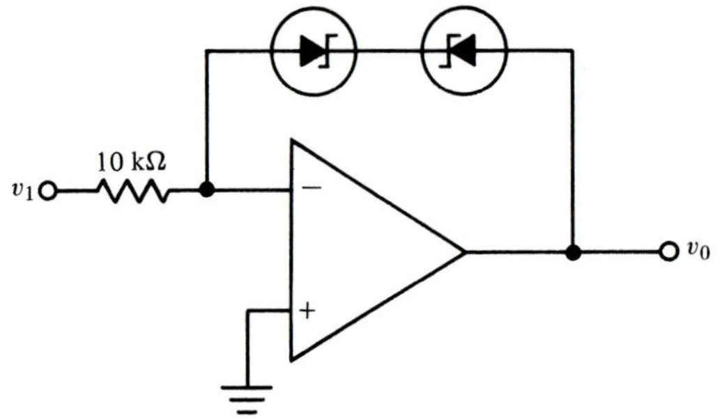

(a)

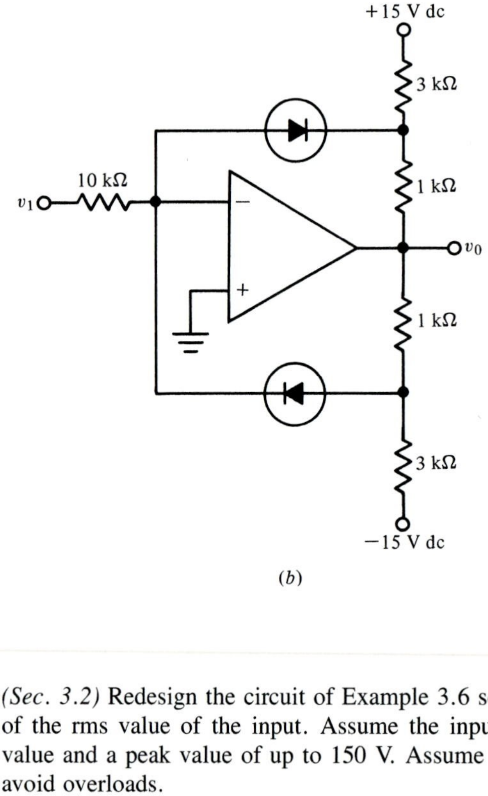

(b)

FIGURE P3.10

3.12. *(Sec. 3.2)* Redesign the circuit of Example 3.6 so that the output will be one-tenth of the rms value of the input. Assume the input is a sinusoid with zero average value and a peak value of up to 150 V. Assume ±10-V amplifiers, and be sure to avoid overloads.

3.13. *(Sec. 3.3)* Redesign the circuit for Example 3.6 to make use of the full-wave-rectifier circuit of Fig. 3.26*c*; comment on the advantages of this approach.

3.14. *(Sec. 3.3)* Design a circuit based on Fig. 3.26*c* which will implement accurately the transfer characteristic

$$v_O = \begin{cases} -4v_I & v_I < 0 \\ 3v_I & v_I > 0 \end{cases}$$

Circuit I

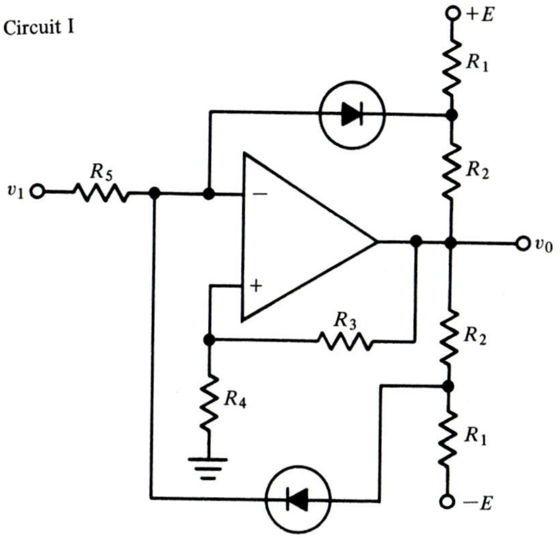

Circuit II

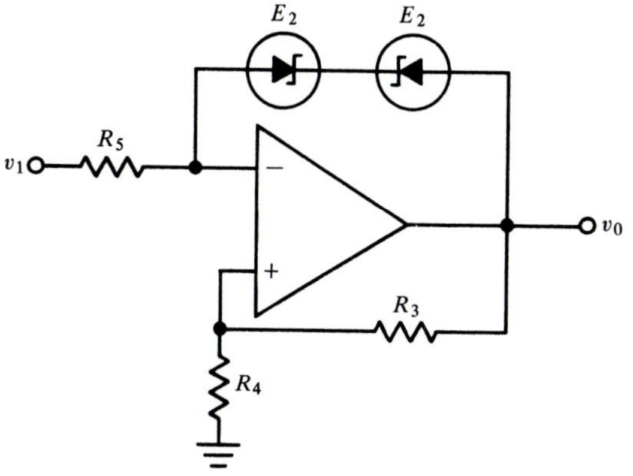

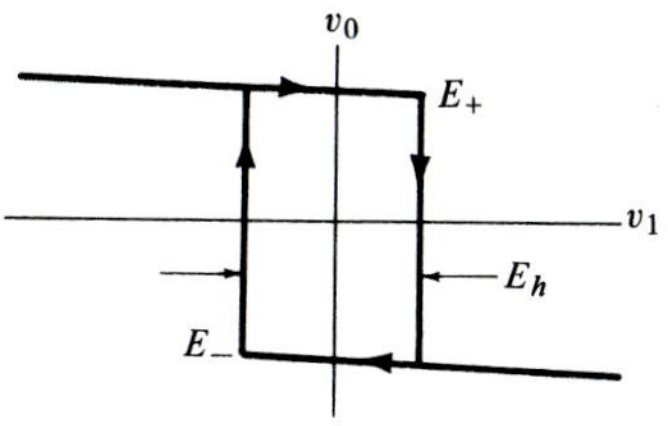

FIGURE P3.11

3.15. *(Sec. 3.3)* Adapt the absolute-value circuit of Fig. 3.26 to provide the *window comparator* characteristic of Fig. P3.15*a*. Figure P3.15*b* shows a block diagram; complete the design of the entire circuit, using two operational amplifiers, resistances in the range of 10 to 100 kΩ, switching diodes, and a Zener diode for the level clamp. Assume ΔV is set by the potentiometer, and $0 < \Delta V \leq 2$ V. The window comparator is very useful for precise measurement of limits on a signal or for signaling an overload or out-of-tolerance condition. Assume ±15 V dc is available.

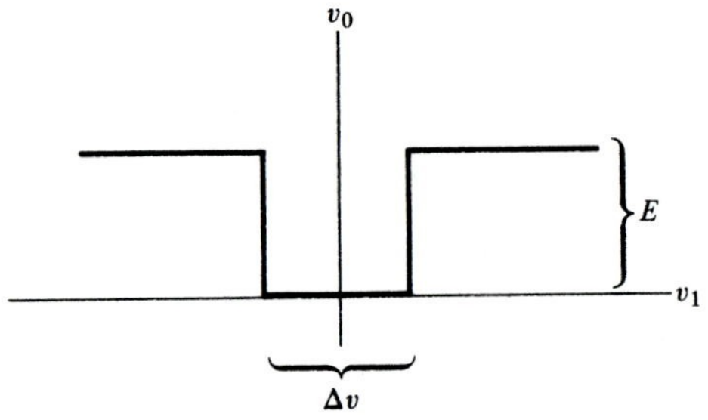

(*a*)

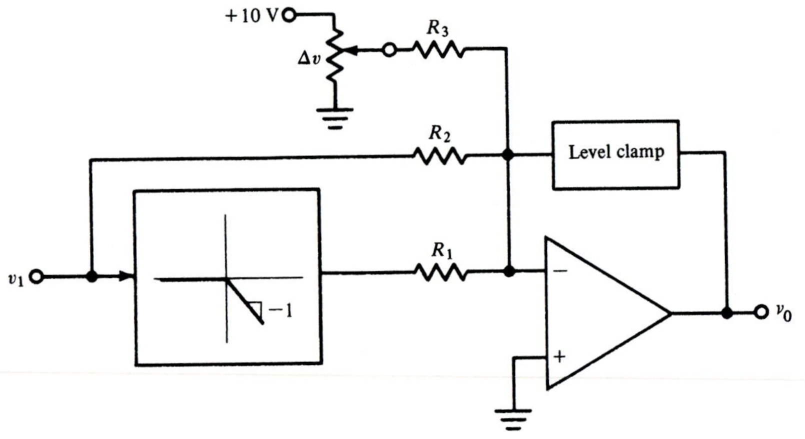

(*b*)

FIGURE P3.15

3.16. *(Sec. 3.3)* Design a circuit to accurately implement the transfer characteristic of Fig. P3.16; *D* can be adjusted by a potentiometer between 0 and 10.0 V. Assume ±10-V dc power is available; use no more than two operational amplifiers. The output impedance need not be low.

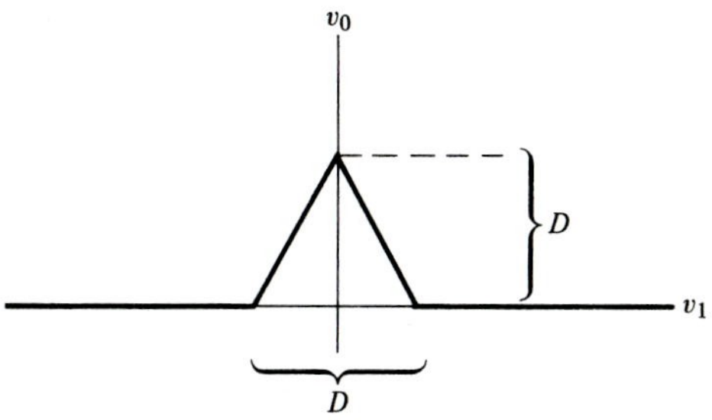

FIGURE P3.16

3.17. *(Sec. 3.3)* Design a dead-space circuit to implement the characteristic of Fig. P3.17. Assume ±15 V dc is available.

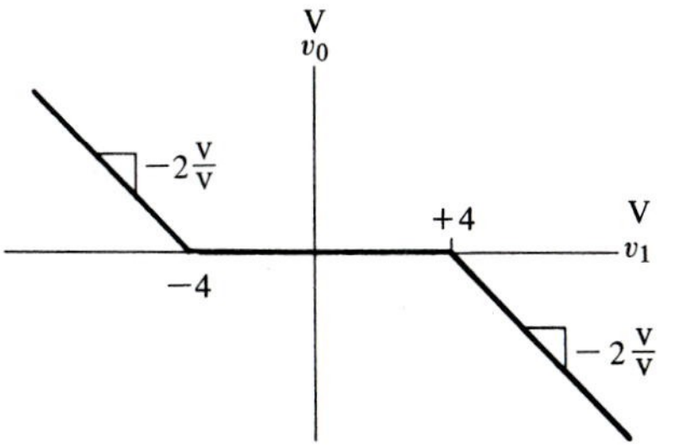

FIGURE P3.17

3.18. *(Sec. 3.3)* Sketch the transfer characteristic of the circuit in Fig. P3.18, $-10\text{ V} \le v_I \le +10\text{ V}$. Calculate the slopes of the characteristic carefully. (See p. 232.)

3.19. *(Sec. 3.3)* Refer to Fig. P3.19.

(*a*) Sketch the transfer characteristic, v_O versus v_I, if the network of Fig. P3.19*a* is used in an amplifier input path, as in Fig. P3.19*b*.

(*b*) Again sketch v_O versus v_I if the network is used in a feedback path, as in Fig. P3.19*c*. (See p. 233.)

3.20. *(Sec. 3.3)* A diode function generation network (similar to that of Fig. 3.29) is designed to implement an approximation to the sine function. When it is used in the circuit of Fig. P3.20*a*, (see p. 233),

$$v_O = -10 \sin \frac{v_I}{K_2} \qquad |v_I| \le 10\text{ V}$$

and it is designed so that $v_O = -10$ V when $v_I = +10$ V, that is, ±10 V at the input corresponds to ±90°.

(*a*) Find K_2.

(*b*) Find the *short-circuit transfer conductance* of the diode network, i.e., in Fig. P3.20*a*

$$i_{SJ} = G(v_I)$$

Be sure to specify units.

(*c*) If the same diode network is used in the circuit of Fig. P3.20*b*, find v_O as a function of v_2. Specify the allowable ranges of v_O and v_2.

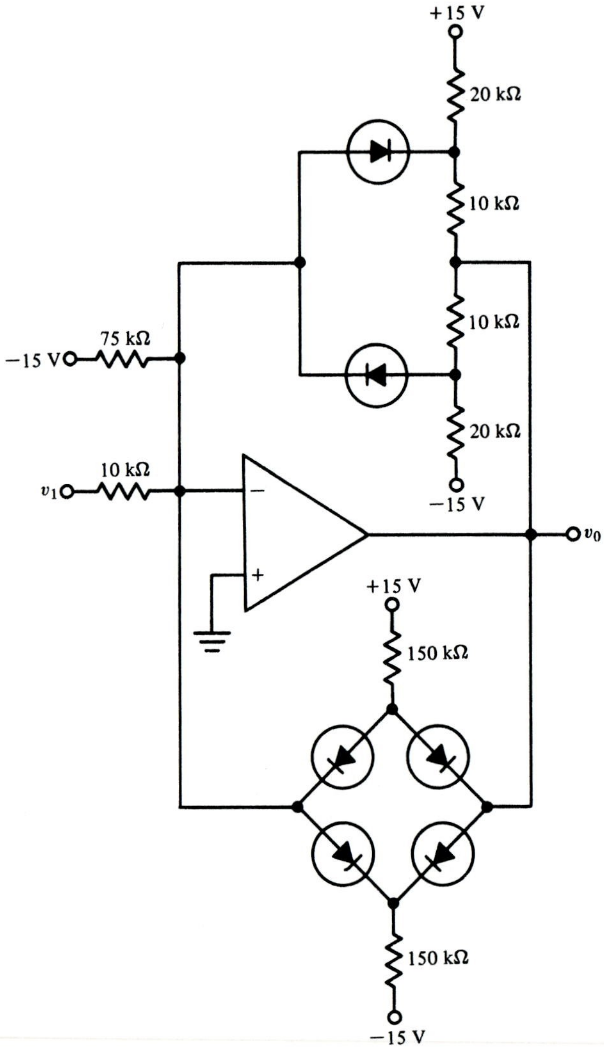

FIGURE P3.18

3.21. *(Sec. 3.3)* Using the circuit of Fig. 3.34, design a bistable circuit with the following specifications:

$$L_+ = +9\text{ V} \qquad L_- = -6\text{ V}$$

$$S_+ = +3\text{ V} \qquad S_- = 0\text{ V}$$

Hint: You may have to introduce an offset into the circuit by providing a second input to the summing junction of A_1.

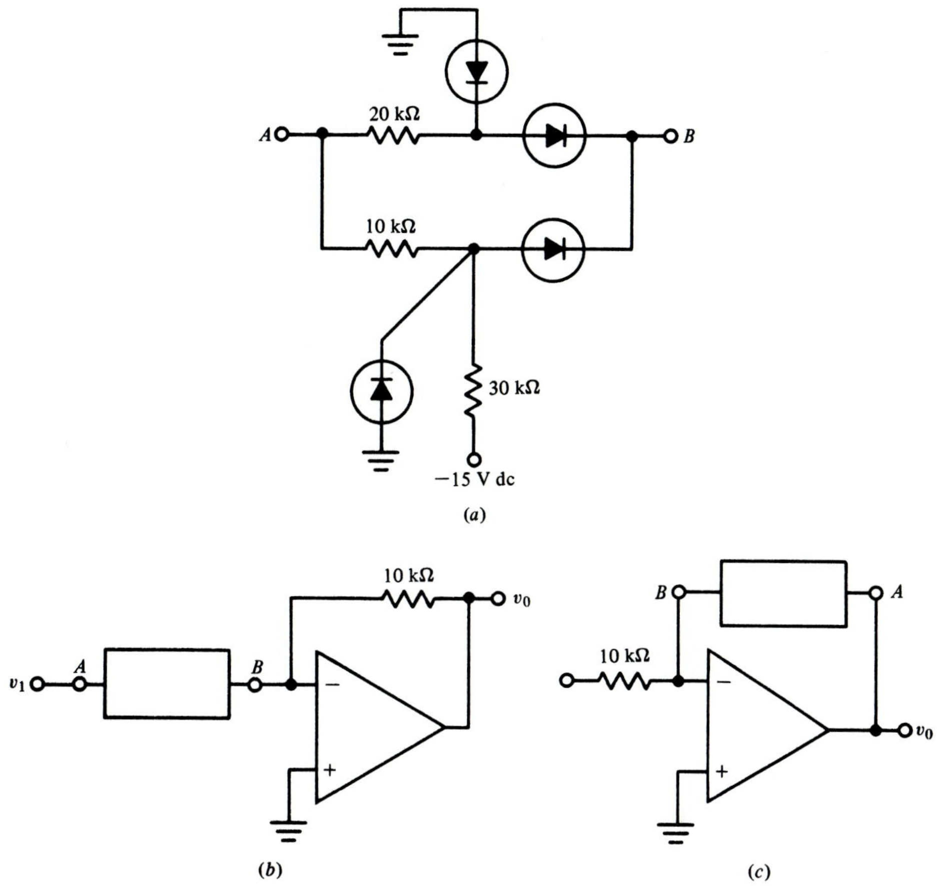

FIGURE P3.19

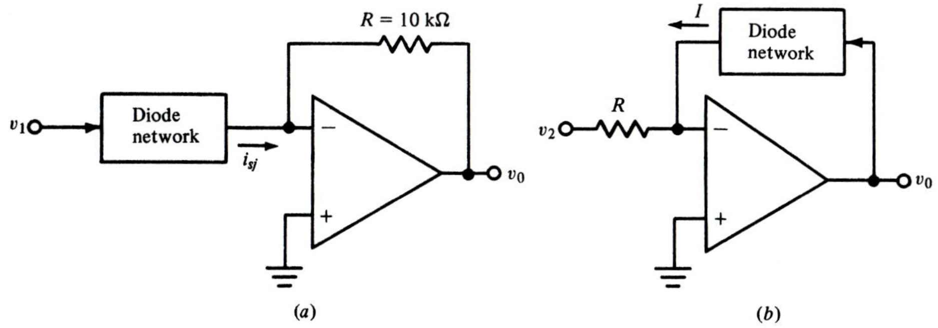

FIGURE P3.20

3.22. *(Sec. 3.3)*

(*a*) Analyze the circuit of Fig. P3.22; sketch v_0 versus v_1. What type of circuit is this?

(*b*) Sketch v_2 versus v_1.

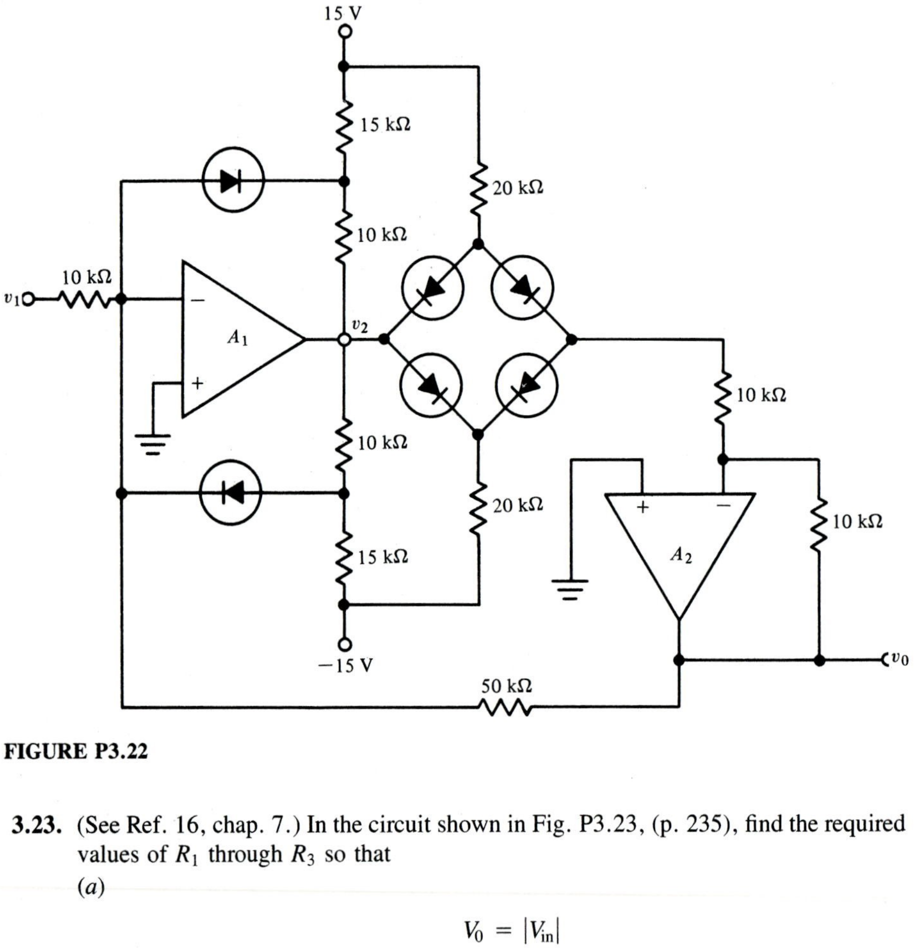

FIGURE P3.22

3.23. (See Ref. 16, chap. 7.) In the circuit shown in Fig. P3.23, (p. 235), find the required values of R_1 through R_3 so that

(*a*)

$$V_0 = |V_{in}|$$

i.e., the output is the absolute value of the input, and

(*b*)

$$V_0 = 0.1\,|V_{in}|$$

Make the smallest resistance values 10,000 Ω in both parts of this problem.

3.24. *(Sec. 3.4)* Given the circuit of Fig. 3.37, develop Eq. (3.38). Assume very high transistor current gains and ideal operational amplifiers, and use the current and voltage labels of the figure.

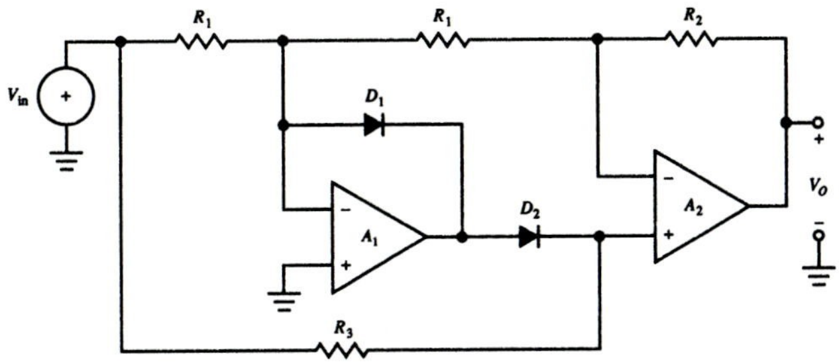

FIGURE P3.23

3.25. *(Sec. 3.4)* Redesign the circuit of Example 3.10 so that the input-output characteristic is described by

v_I	v_O
10 V	−10 V
10^{-2} V	0 V
10^{-5} V	+10 V

You may alter R_1, R_2, R_3, and R_5, but R_1 must be less than 20 kΩ.

3.26. *(Sec. 3.4)* Given the circuit of Fig. 3.39, develop Eq. (3.42). Assume very high transistor current gains and ideal operational amplifiers, and use the current and voltage labels of the figure. Why do we make $R_3 = R_7$ and $R_5 = R_6$?

3.27. *(Sec. 3.4)* The circuit of Fig. 3.39 can also be described by

$$v_O = K_1 A^{-K_4 v_I} \qquad v_I > 0$$

Find K_1 and K_4 at 27°C.

3.28. *(Sec. 3.4)* Redesign the circuit in Example 3.11 to obtain

(*a*) $v_O = 5 \times 10^{-0.1 v_I} \qquad v_I > 0$

(*b*) $v_O = 10 \times 2^{-v_I} \qquad v_I > 0$

3.29. *(Sec. 3.4)* Suppose we have a general-purpose module that can implement

$$v_4 = v_3 \left(\frac{v_1}{v_2}\right)^m$$

where $0.2 \le m \le 5.0$

$$0 < v_1, v_2, v_3, v_4 < 10 \text{ V}$$

Assume that m can be adjusted by some externally provided potentiometer to any value in the range. Also assume that the module has a 10-kΩ input impedance at each input terminal, and a very low output impedance. The terminals are numbered as shown in Fig. P3.29.

(*a*) Use the module, along with an extra operational amplifier and resistors, to implement the following approximation to sin θ (good to 0.25 percent, Ref. 10).

$$\sin\theta \cong \theta - \frac{\theta^{2.827}}{2\pi} \qquad 0 < \theta \le \frac{\pi}{2}$$

Represent θ by a signal scaled so that +10 V corresponds to 90°; implement this approximation so that the output is +10 V when the input is +10 V; be sure no output overloads.

(*b*) Repeat the above for the cosine approximation (good to 1 percent)

$$\cos\theta \cong 1 + 0.235\theta - \frac{\theta^{1.504}}{1.445}$$

Assume ±10 V dc is available; set it up so that the output is +10 V when the input is zero and a +10-V input represents $\pi/2$ rad.

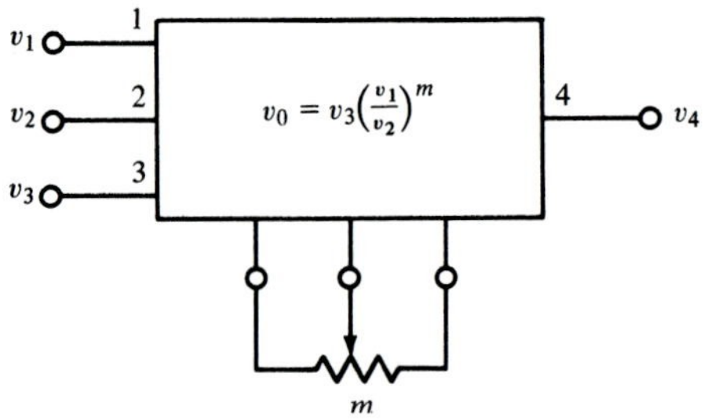

FIGURE P3.29

3.30. *(Sec. 3.5)* Show why there must be a minus sign on the right side of Eq. (3.71). *Hint:* Find the small-signal short-circuit transfer resistance in the feedback path around some nominal value of v_O.

3.31. *(Sec. 3.5)* In Fig. 3.43*b*, suppose we have an inverting multiplier, i.e.,

$$v_3 = \frac{-v_2 v_O}{10}$$

Repeat the development of the resulting circuit transfer equation, and comment on the signal polarity restrictions.

3.32. *(Sec. 3.5)* Redesign the circuit of Example 3.12 to implement vector magnitude resolution. That is, assume

$$v_1 \le 10 \qquad v_2 \le 10$$

and

$$v_1^2 + v_2^2 \le 100$$

so that the desired operation is

$$v_O = -\sqrt{v_1^2 + v_2^2}$$

Make $R_2 = 10\ \text{k}\Omega$.

3.33. *(Sec. 3.5)* Using noninverting ±10-V multipliers, implement

$$v_O = -K\sqrt{v_1^2 + v_2^2 + v_3^2}$$

so that the output never exceeds 10 V if all inputs are less than 10 V.

3.34. *(Sec. 3.5)* Refer to Fig. 3.46 and Example 3.13. Suppose we know that the input signal to be measured never exceeds 6 V in magnitude, but may have an rms value as large as 3 V. Redesign the circuit to implement

$$v_O = 3v_{1,\text{rms}}$$

Include a redesign of the input amplifier so that the input multiplier M_1 is driven over its 10-V range.

3.35. *(Sec. 3.5)* In the circuit of Example 3.13 (Fig. 3.46), suppose that the input was a 1-V rms, 200-Hz sine wave. Sketch the waveforms at points v_2, v_3, v_4, and v_O. Include an estimate of the ripple signal on v_4 and v_O.

$$(\sin \omega t)^2 = \frac{1 - \cos 2\omega t}{2}$$

3.36. *(Sec. 3.6)*

(*a*) Given the block diagram of Fig. 3.47*a*, show that (3.74) holds.

(*b*) If, in the block diagram of Fig. 3.47*b*, the noninverting integrator has the transfer function

$$\frac{V_O(s)}{V_2(s)} = \frac{\omega_0}{s - \delta}$$

show that (3.74) holds, and find ϵ in terms of δ.

(*c*) Find the location of the poles of $V_O(s)$ in the Laplace transform of (3.74), and show that a positive value of ϵ leads to regeneration.

3.37. *(Sec. 3.6)* Design a sine wave oscillator, based on Fig. 3.48*b*, that produces a 2000-Hz ±8-V waveform. Use 0.01-μF capacitors. Assume that ±15 V dc is available for the limiter circuit; the clipping level can be approximate.

3.38. *(Sec. 3.6)* Refer to Fig. 3.51*a*. Derive the conditions for oscillation of the Wien bridge oscillator (3.77). The trick is to assume that the voltage on the two amplifier input terminals is the same.

3.39. *(Sec. 3.6)* Refer to Fig. 3.51*b*.

(*a*) Derive the conditions for oscillation for the phase-shift oscillator, i.e.,

$$\omega_0 \cong \frac{1}{\sqrt{3}RC}$$

and $R_1 \cong 12R$.

(*b*) Find $v_1(t)$ in terms of $v_O(t)$.

3.40. *(Sec. 3.6)* Refer to the FM oscillator circuit block diagram of Fig. 3.49.

(*a*) Show that (3.76) gives the frequency of oscillation.

(*b*) If $|v_m| \leq$ 1-V peak, design an oscillator circuit, based on Example 3.14, that produces an FM waveform that varies ±100 Hz from a nominal center frequency of 2000 Hz. *Note:* Be sure $v_m + A \leq 10$ V at all times.

3.41. *(Sec. 3.6)*

(*a*) Using the circuit of Fig. 3.53, design an oscillator that will generate ±5-V square and triangle waves, with a frequency that may be switched between 1 and 10 kHz (manual switching of one resistor).

(*b*) Modify the design so that the circuit generates ±10-V square waves, but ±5-V triangle waves.

3.42. *(Sec. 3.6)* Refer to Fig. 3.52.

(*a*) Demonstrate that (3.78) holds for the case where the bistable has symmetric switch points and limiting levels.

(*b*) Suppose we have an asymmetric bistable. Develop a general expression for the period of the resulting square wave; also find the expression for the time intervals of the positive and negative portions of the square wave. Call T_1 the duration of the positive part and T_2 the duration of the negative part. Of course, assume $L_+ > 0$, $L_- < 0$, $S_+ > 0$, $S_- < 0$.

3.43. *(Sec. 3.6)* Design a circuit based on Figs. 3.52 to 3.54 which will generate an FM waveform. The modulating signal (to which a constant must be added) should produce the following amount of frequency change:

Value of v_m, V	Output frequency, Hz
−1	900
0	1000
+1	1100

The output waveform should be a square wave with a ±5-V peak value.

CHAPTER 4

ACTIVE RC FILTERS

In the preceding chapters we showed how operational amplifiers could be applied to the realization of many useful circuit functions such as comparators, limiters, multipliers, etc. In this chapter we continue the presentation of operational amplifier applications. The subject here, however, is a quite different one, namely, the realization of *linear* frequency-dependent circuit functions. Such functions are widely used whenever it is necessary to control or shape the time or frequency domain characteristics of signals. Circuits which perform such shaping are called *filters*. Some examples of first-order filters have already been given in Sec. 1.9. Originally, filter circuits were composed only of passive network components, namely, resistors, capacitors, and inductors. In more recent years, however, it has been found that if active elements, such as operational amplifiers, are also included, then the inductor may be eliminated without loss of generality of circuit capabilities. Since inductors, in general, are undesirable circuit elements with respect to their nonlinearity, dissipation, size, and weight, the inductorless or *active RC* filter has become the preferred choice for many applications. Another major advantage of eliminating the inductor as a circuit element is that capacitors and resistors are readily included in integrated-circuit fabrications using thick-film, thin-film, or monolithic techniques, while inductors must be wired into such circuits as discrete elements. There are many advantages and disadvantages with respect to using active *RC* filters as a replacement for their purely passive counterparts. Among the advantages are small size and weight, simplicity of synthesis and tuning, isolation of stages, and large available gain. Among the disadvantages are the limitations on voltage and current levels, inability to "float" a given realization, and parasitic effects introduced by the active element, such as dc output voltage offset, input bias currents, etc. In most applications the advantages outweigh the disadvantages. As a result, active *RC* filters are becoming increasingly important in a wide range of engineering applications. They are available both as stock items and as custom-designed ones from a number of manufacturers. In this chapter we introduce the subject of active *RC* filters. The subject is a large one, and several books have been devoted entirely to it. As a result, our treatment here must necessarily be a limited one. We do, however, present details of many of the best-known and most useful active *RC* circuits. Suggestions for additional reading are given at the end of the chapter for the reader who wishes to pursue the subject in more depth.

4.1 FILTER TYPES

Filter circuits are used where frequency-dependent processing of input signal information is needed. The general form of such a usage is shown in Fig. 4.1, where unwanted frequency components are deleted, or filtered, from an input signal. In this section we review the most commonly used types of filters, their sinusoidal steady-state frequency domain properties, and the general form of their network functions. The types are called low-pass, high-pass, bandpass, band-reject, and all-pass. In the following paragraphs we use $N(s)$ to refer to the transfer network function for these various filter types.

Low-Pass Filters

A low-pass filter has the ideal (unrealizable) response $|N(j\omega)|$ shown in Fig. 4.2*a*. The cutoff frequency is ω_0. In the passband $\omega < \omega_0$, $|N(j\omega)|$ is constant, and all input frequencies are amplified or attenuated (depending on whether $H < 1$ or $H > 1$) the same amount. In the stopband $\omega > \omega_0$, $|N(j\omega)|$ is zero and no signal appears in the output. A typical practical (realizable) low-pass filter response is shown in Fig. 4.2*b*. It is called a *Butterworth* or *maximally flat magnitude characteristic*. It is monotonic throughout the passband and the stopband. The *cutoff frequency* ω_0 is defined as the frequency at which the magnitude is 0.7071 of its dc value. On a logarithmic scale, this corresponds with an attenuation of -3.01 dB. Another practical low-pass response is shown in Fig. 4.2*c*. It is called a *Chebyshev* or *equal-ripple magnitude characteristic*. The magnitude of the passband ripple is usually given in decibels ($= 20 \log k$). The number of ripples

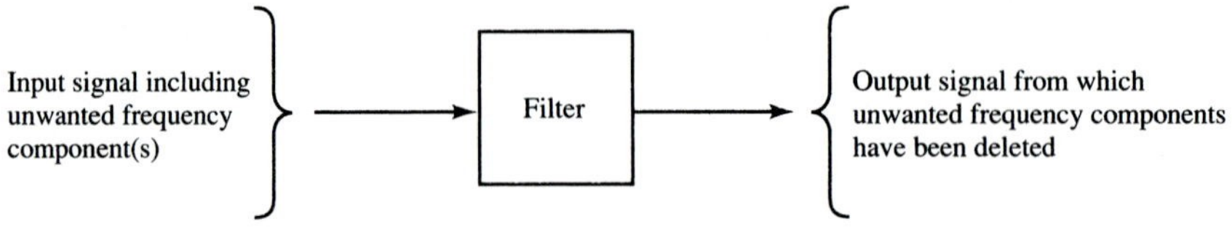

FIGURE 4.1
Frequency-dependent processing of signal information.

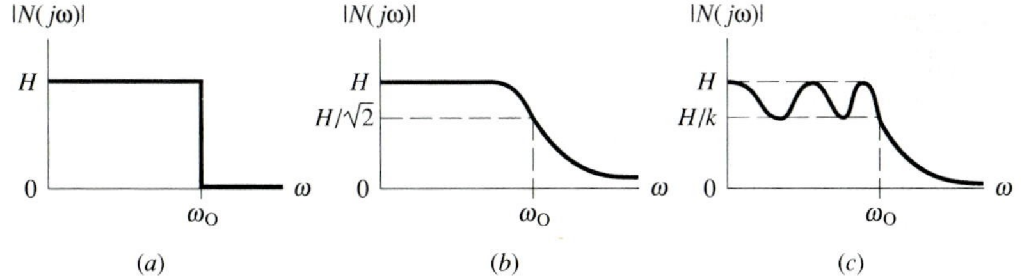

FIGURE 4.2
ow-pass filter response.

depends on the order of the filter. The cutoff frequency ω_0 is defined where the attenuation first exceeds that of the passband. For the stopband $\omega > \omega_0$, the response characteristic is monotonic.

For many signal-processing applications, a low-pass filter with a linear phase characteristic is required. The ideal (nonrealizable) magnitude and phase characteristics of such a function are shown in Fig. 4.3*a* and *b*. A practical (realizable) set of characteristics is shown in Fig. 4.3*c* and *d*. These define a *Thomson* or *linear phase function*. The properties of such a function are specified in the time domain as the delay t_0 provided by the filter to an input step function.

The network functions for the low-pass filter characteristics shown in Figs. 4.2 and 4.3 have the form

$$N(s) = \frac{Ha_0}{s^n + a_{n-1}s^{n-1} + \cdots + a_1 s + a_0} \tag{4.1}$$

Low-pass filters have many applications in signal processing. An example is the elimination of an undesirable high-frequency noise component of a signal.

Highpass Filters

A high-pass filter has the ideal (nonrealizable) response $|N(j\omega)|$ shown in Fig. 4.4*a*. The cutoff frequency is ω_0. In the passband $\omega > \omega_0$, $|N(j\omega)|$ is constant (and nonzero) and all frequencies are passed with the same gain. In the stopband $\omega < \omega_0$, $|N(j\omega)|$ is zero and there is no transmission of signals.

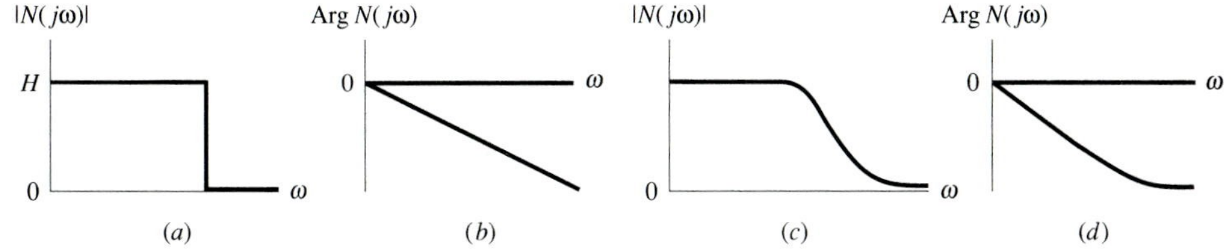

FIGURE 4.3
Low-pass filter with linear phase characteristic.

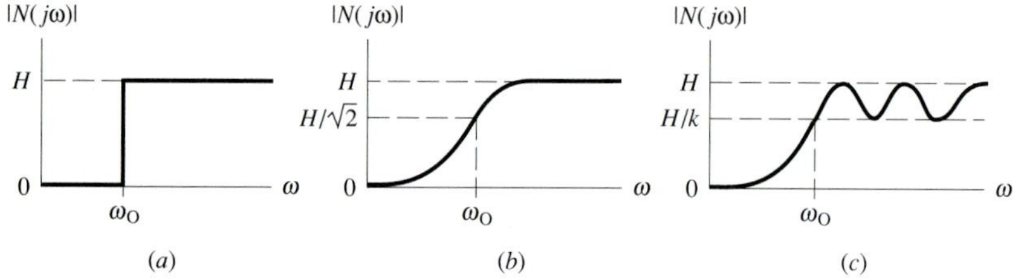

FIGURE 4.4
High-pass filter response.

A typical practical (realizable) high-pass filter response is shown in Fig. 4.4*b*. It is monotonic and is the inverse of the Butterworth low-pass characteristic shown in Fig. 4.2*b*. The cutoff frequency ω_0 is defined as the frequency at which the magnitude is 0.7071 (−3.01 dB) of its infinite frequency value. Another practical high-frequency response is shown in Fig. 4.4*c*. It is the inverse of the Chebyshev low-pass characteristic shown in Fig. 4.2*c* and is monotonic in the stopband and equal-ripple in the passband. The cutoff frequency ω_0 is defined where the attenuation first exceeds that of the passband.

The network functions for the high-pass characteristics shown in Fig. 4.4 have the form

$$N(s) = \frac{Hs^n}{s^n + a_{n-1}s^{n-1} + \cdots + a_1 s + a_0} \tag{4.2}$$

An example of an application for a high-pass filter is the elimination of low-frequency "rumble" components arising from the drive mechanism of the turntable in a high-fidelity sound system.

Bandpass Filter

A bandpass filter has the ideal (nonrealizable) response shown in Fig. 4.5*a*. There are two cutoff frequencies ω_L and ω_H. In the passband $\omega_L < \omega < \omega_H$, $|N(j\omega)|$ is constant (and nonzero) and all input frequencies are multiplied by the same factor. In the stopbands $\omega < \omega_L$ and $\omega > \omega_H$, $|N(j\omega)|$ is zero and there is no transmission of signal. A typical practical (realizable) bandpass filter response is

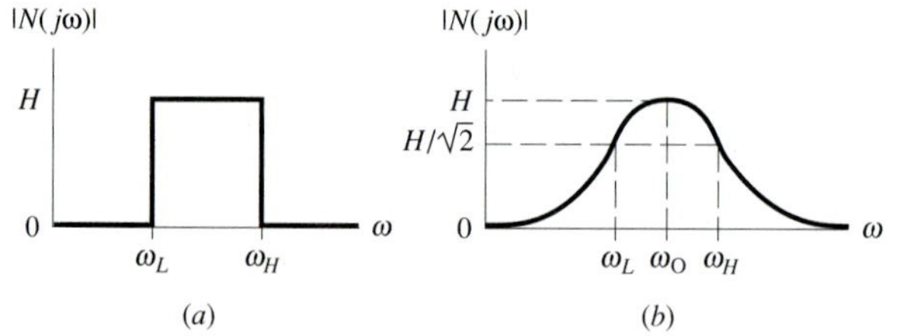

FIGURE 4.5
Bandpass filter response.

shown in Fig. 4.5*b*. It is monotonic on each side of the center frequency ω_0 where the peak of the magnitude occurs. The cutoff frequencies ω_L and ω_H are defined as the frequencies at which the magnitude is 0.7071 (−3.01 dB) of its peak value. The network function for a bandpass filter has the form

$$N(s) = \frac{Hs^{n/2}}{s^n + a_{n-1}s^{n-1} + \cdots + a_1 s + a_0} \qquad n \text{ even} \tag{4.3}$$

An example of an application for a bandpass filter is the selection of a desired frequency range from a broad spectrum of frequencies as in the tuning of radios, television sets, etc.

Band-Reject Filter

A band-reject filter has the ideal (nonrealizable) response shown in Fig. 4.6*a*. There are two cutoff frequencies ω_L and ω_H. In the passbands $\omega < \omega_L$ and $\omega > \omega_H$, $|N(j\omega)|$ is constant (and nonzero) and all input frequencies are multiplied by the same factor. In the stopband $\omega_L < \omega < \omega_H$, $|N(j\omega)|$ is zero and there is no transmission of signal. An example of a typical practical (realizable) band-reject filter characteristic is shown in Fig. 4.6*b*. It is monotonic on each side of the center frequency ω_0 at which the minimum value occurs. The cutoff frequencies ω_L and ω_H are defined as the frequencies at which the magnitude is 0.7071 (−3.01 dB) of its peak value. The network function for a band-reject filter has the form

$$N(s) = \frac{H(s^2 + \omega_0^2)^{n/2}}{s^n + a_{n-1}s^{n-1} + \cdots + a_1 s + a_0} \qquad n \text{ even} \tag{4.4}$$

An example of an application for a band-reject filter is the elimination of stray 60-Hz power line pickup from the input to a sensitive electronic circuit.

All-Pass Filter

An all-pass filter, sometimes called a *phase correction filter,* has a constant-magnitude response and a decreasing phase characteristic, as shown in Fig. 4.7.

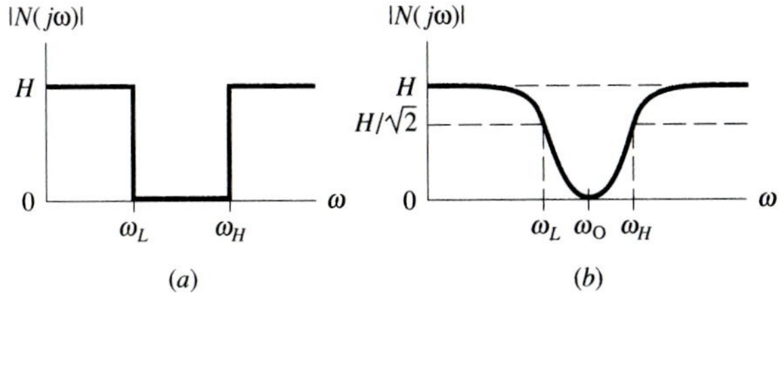

FIGURE 4.6
Band-reject filter response.

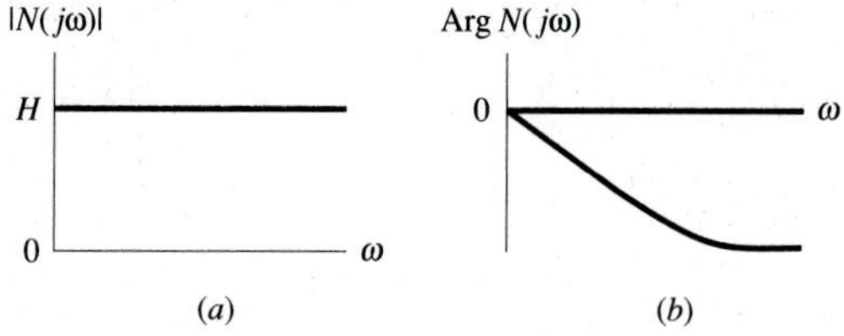

FIGURE 4.7
All-pass filter response.

The network function for an all-pass filter has the form

$$N(s) = \frac{H(s^n - a_{n-1}s^{n-1} + \cdots + a_2 s^2 - a_1 s + a_0)}{s^n + a_{n-1}s^{n-1} + \cdots + a_2 s^2 + a_1 s + a_0} \tag{4.5}$$

Note that in this function the numerator and denominator coefficients have the same magnitude, but the signs associated with the numerator coefficients alternate. An example of an application for an all-pass filter is the phase correction of the response characteristics of a transmission line.

4.2 NORMALIZED AND DENORMALIZED FILTERS

In developing and applying circuits to realize the different types of filter responses introduced in Sec. 4.1, it is convenient to use normalized values for the circuit elements. For example, consider the active *RC* filter shown in Fig. 4.8. The portion of the circuit consisting of the operational amplifier and resistors R_A and R_B forms a noninverting amplifier [an ideal *voltage controlled voltage source* (VCVS)] with gain

$$K = \frac{R_A + R_B}{R_A} \tag{4.6}$$

This amplifier was previously discussed in Sec. 1.6. If in Fig. 4.8 we choose the element values $R_1 = R_2 = 1\ \Omega$, $C_1 = C_2 = 1$ F, $R_A = 1\ \Omega$, and

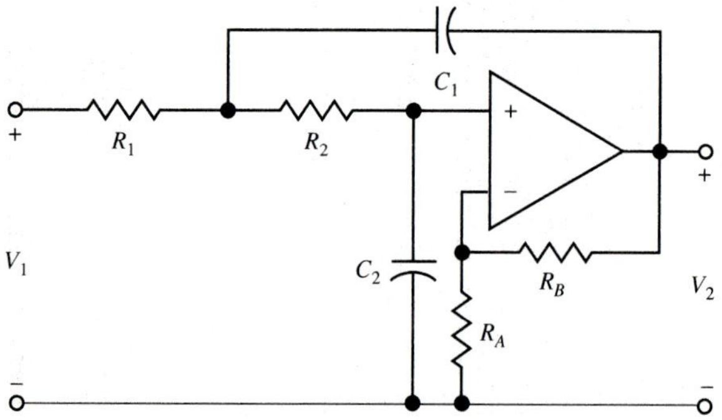

FIGURE 4.8
Active *RC* filter.

$R_B = 0.586\ \Omega (K = 1.586)$, then this active *RC* circuit realizes the voltage transfer

$$N(s) = \frac{V_2(s)}{V_1(s)} = \frac{1.586}{s^2 + 1.414s + 1} \tag{4.7}$$

This transfer function corresponds to the low-pass magnitude response shown in Fig. 4.2*b* with a dc gain $H = 1.586$ and a cutoff frequency $\omega_0 = 1$ rad/s. The circuit and response plot are shown in Fig. 4.9*a*. The results are referred to as a *normalized realization*.

Frequency Denormalization

Now let us convert the values of the filter elements shown in Fig. 4.9*a* to a more practical denormalized realization. We first consider a *frequency denormalization* (*FD*). In general, if we let s be a normalized complex frequency variable and s_{FD} be a denormalized one, then we define a frequency denormalization as

$$s_{FD} = \Omega_D s \tag{4.8}$$

where Ω_D is a frequency denormalization constant. Note that since $s = j\omega$, this equation also implies the relation

$$\omega_{FD} = \Omega_D \omega \tag{4.9}$$

Inserting the denormalization expression of (4.8) in the transfer function given in (4.7) we obtain

$$N_{FD}(s_{FD}) = \frac{1.586\ \Omega_D^2}{s_{FD}^2 + 1.414\ \Omega_D s_{FD} + \Omega_D^2} \tag{4.10}$$

The frequency denormalization given in (4.8) affects not only a network function but also any network element whose impedance is a function of s. Thus, the values of the capacitors in an active *RC* circuit are affected by a frequency denormalization while the values of the resistors and the gain of the amplifier are not. For a capacitor of impedance $Z(s) = 1/sC$, we obtain

$$Z_{FD}(s_{FD}) = s_{FD}(C/\Omega_{FD}) \tag{4.11}$$

From this equation, the value of the frequency denormalized capacitor is C/Ω_D.

As an example of the use of a frequency denormalization, assume that it is desired to modify the filter realization shown in Fig. 4.9*a* so that the cutoff frequency is 10 kHz rather than 1 rad/s. For this, we use $\Omega_D = 2\pi 10^4$. Thus, $C_1 = C_2 = 1/(2\pi 10^4)$ F. The resulting filter and its response characteristic are shown in Fig. 4.9*b*. To further improve the practicality of this circuit, we now introduce another type of denormalization.

Impedence Denormalization

A second type of denormalization used in the design of practical filters is an *impedance denormalization (ID)*. If we let $Z(s)$ be the normalized impedance

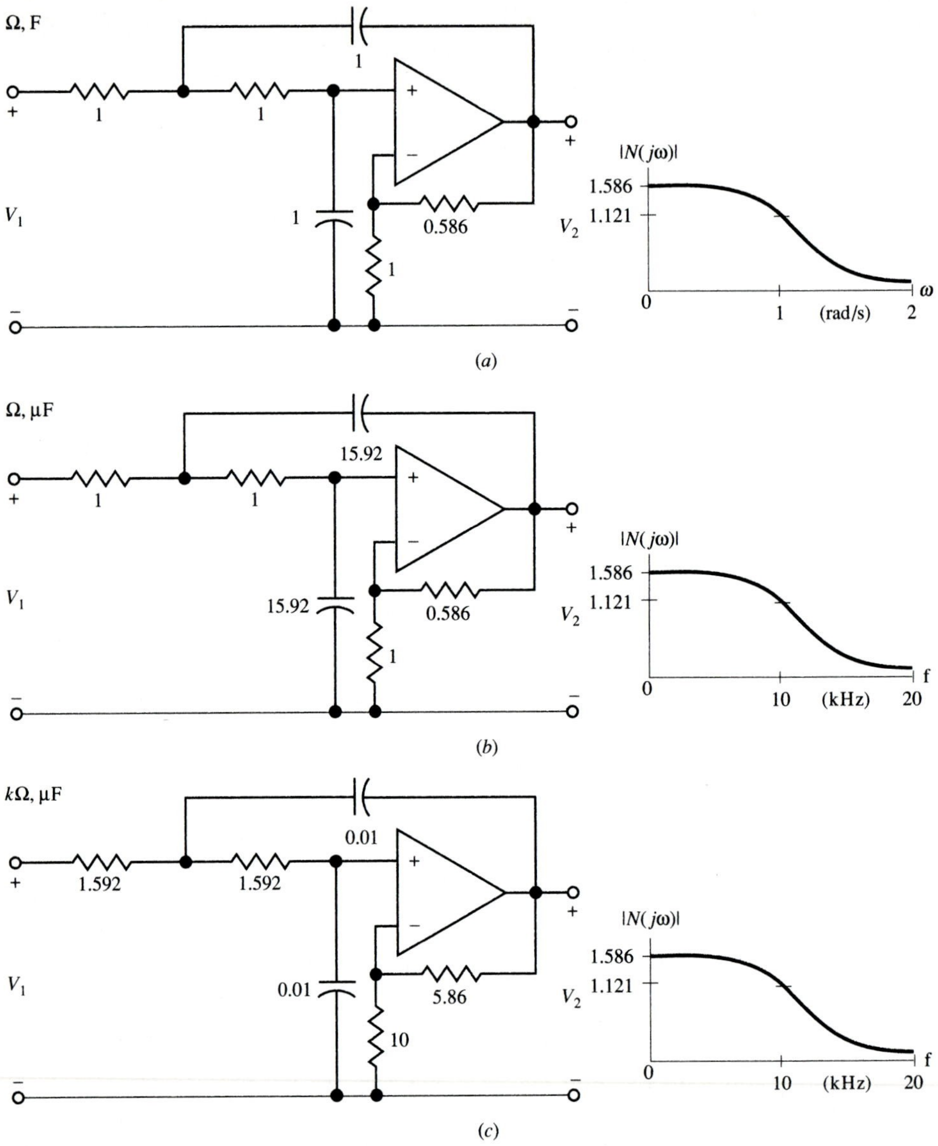

FIGURE 4.9
Frequency and impedance denormalization.

and $Z_{ID}(s)$ be the denormalized one, then we may write

$$Z_{ID}(s) = Z_D Z(s) \tag{4.12}$$

where Z_D is the impedance denormalization constant. Applying this to a resistor, we obtain

$$Z_{ID}(s) = Z_D R \tag{4.13}$$

The impedance denormalization resistor has a value $Z_D R\ \Omega$. Similarly, applying an impedance denormalization to a capacitor, we obtain

$$Z_{ID}(s) = \frac{1}{s(C/Z_D)} \tag{4.14}$$

The impedance denormalized capacitor has a value C/Z_D F. The impedance denormalization has no effect on dimensionless quantities, such as a voltage transfer function, or the value of the amplifier gain.

As an example of the use of an impedance denormalization, let us modify the filter realization shown in Fig. 4.9*b*. For convenience we choose an impedance normalization constant such that the capacitors have a value of 0.01μ F. This requires

$$10^{-8} = \frac{1/(2\pi 10^4)}{Z_D} \qquad Z_D = \frac{1}{2\pi}10^4 \tag{4.15}$$

The resulting circuit and its response characteristic are shown in Fig. 4.9*c*. Note that the impedance denormalization for resistors R_A and R_B (see Fig. 4.8) is independent of the one used for the other circuit elements. For convenience we have used $Z_D = 10^4$ for these gain-determining resistors. The network function is identical with the one shown as (4.10). Note that by using a frequency denormalization and an impedance denormalization, we have converted the impractical normalized filter realization shown in Fig. 4.9*a* to the practical one shown in Fig. 4.9*c*. We make frequent use of the denormalization techniques presented in this section in much of the following material of this chapter.

4.3 A SECOND-ORDER LOW-PASS FILTER—SENSITIVITY

In this section we present a discussion of some general design and evaluation techniques which are applicable to many active *RC* filters. We illustrate in detail their application to a specific filter, the second-order low-pass noninverting low-gain-amplifier one. In the following sections we discuss the results obtained by applying these techniques to other types of filters.

A general active *RC* filter configuration is shown in Fig. 4.10. Other circuit configurations are, of course, possible, but this one has been shown to be very practical in the realization of voltage transfer functions, one of the most commonly required filtering applications. In the figure, the box with the three numbered terminals represents the passive portion of the filter. It is comprised of resistors and capacitors. The operational amplifier and the two resistors R_A and R_B form a VCVS (voltage-controlled voltage source) with infinite input impedance, zero output impedance, and finite noninverting gain K given as

$$K = \frac{R_A + R_B}{R_A} \tag{4.16}$$

The VCVS effectively has input node 3 and output node 2.

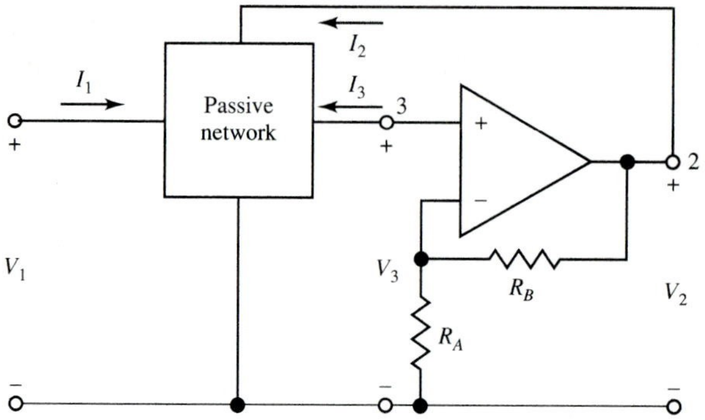

FIGURE 4.10
General active *RC* filter configuration.

To analyze the circuit shown in Fig. 4.10, we may begin by first considering only the passive network. The properties of this are readily defined using a set of *y* parameters similar to those used for two-port networks. Here, however, there are three voltage and current variables, so a 3 by 3 matrix of *y* parameters is required rather than the usual 2 by 2 matrix used in the two-port case. We now define the passive network by the relations

$$\begin{bmatrix} I_1(s) \\ I_2(s) \\ I_3(s) \end{bmatrix} = \begin{bmatrix} y_{11}(s) & y_{12}(s) & y_{13}(s) \\ y_{12}(s) & y_{22}(s) & y_{23}(s) \\ y_{13}(s) & y_{23}(s) & y_{33}(s) \end{bmatrix} \begin{bmatrix} V_1(s) \\ V_2(s) \\ V_3(s) \end{bmatrix} \tag{4.17}$$

where the quantities $V_i(s)$ and $I_i(s)$ are defined in Fig. 4.10. The square matrix in (4.17) is called the *y-parameter matrix*. Note that it is symmetric. This is the result of the reciprocal nature of the passive network. The *y* parameters are readily found using the relation

$$y_{ij}(s) = \frac{I_i(s)}{V_j(s)}\Big|_{V_k(s)=0} \qquad k = 1, 2, 3;\ k \neq j \tag{4.18}$$

The requirement that the voltage $V_k(s) = 0$ for all terminals except the one at which the excitation $V_j(s)$ is applied is, of course, satisfied by short-circuiting these terminals to ground.

Now let us consider the effect that the VCVS has on the relations of (4.17). In terms of the variables defined in Fig. 4.17, the VCVS imposes the constraints

$$V_2(s) = KV_3(s) \quad \text{and} \quad I_3(s) = 0 \tag{4.19}$$

where K is the *gain* of the VCVS as given in (4.16). Substituting these relations in (4.17), we obtain

$$\frac{V_2(s)}{V_1(s)} = \frac{-Ky_{13}(s)}{y_{33}(s) + Ky_{23}(s)} \tag{4.20}$$

as the overall voltage transfer function for the circuit of Fig. 4.10. In the following paragraphs we show how this relation forms a basis for finding active *RC* network configurations for a variety of specific filtering characteristics.

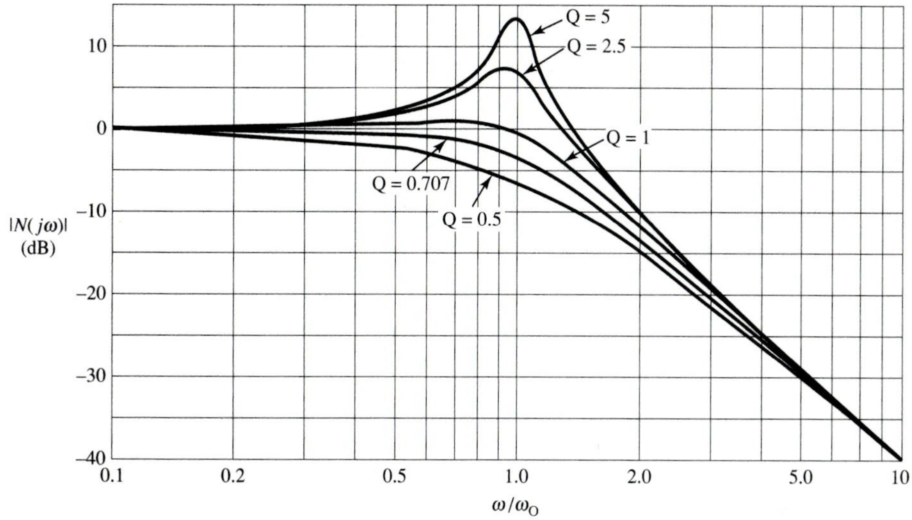

FIGURE 4.11
Low-pass response characteristics.

The first filter characteristic that we consider realizing, using the general active *RC* configuration shown in Fig. 4.10, is the second-order low-pass one. This has the form

$$\frac{V_2(s)}{V_1(s)} = \frac{H}{s^2 + a_1 s + a_0} = \frac{H_0 \omega_0^2}{s^2 + (\omega_0/Q)s + \omega_0^2} \tag{4.21}$$

where the last member is called the *standard form* of the second-order low-pass function. In this, H_0 is the dc gain, ω_0 is the cutoff frequency (in radians per second), and Q is the *quality factor* which represents the sharpness of the resonant peak in the magnitude response. Some plots of the response characteristic for $H_0 = 1$ (0 dB) and various values of Q are given in Fig. 4.11. The values of the different terms used in (4.21) are related by the expressions

$$\omega_0 = \sqrt{a_0} \qquad \text{and} \qquad Q = \frac{\sqrt{a_0}}{a_1} \tag{4.22}$$

The values of the coefficients a_0 and a_1 are usually chosen so as to achieve some specific low-pass characteristic. Several of the more common ones are summarized in Table 4.1.

Now let us consider what form the passive network must have to produce the function given in (4.21). First we note that since the function is second-order, a minimum of two capacitors is required. Second, comparing the relations given in (4.20) and (4.21), we see that the numerator of the transfer admittance $y_{13}(s)$ must be a constant.† Since any series capacitor placed between terminals 1 and 3

† In general, except when private poles are present, the denominators in all the y parameters will be the same. Thus, in (4.20), the denominators of the parameters will cancel, and the numerators may be directly equated with the polynomials in the numerator and denominator of (4.21).

TABLE 4.1
Normalized low-pass network function characteristics for (4.21)

No.	Name	Characteristic	a_1	a_0	Pole locations
1	Butterworth (maximally flat magnitude)	At 1 rad/s down 3.01 dB from zero frequency value	1.41421	1.00000	$-0.70711 \pm j0.70711$
2*a*	Chebychev (equal-ripple magnitude)	$\frac{1}{2}$-dB ripple in passband from 0–1 rad/s	1.42562	1.51620	$-0.71281 \pm j1.00404$
2*b*		1-dB ripple in passband from 0–1 rad/s	1.09773	1.10251	$-0.54887 \pm j0.89513$
3	Thomson (maximally flat delay, linear phase)	1-s delay, unity phase constant	3.00000	3.00000	$-1.50000 \pm j0.86603$

will produce a transmission zero at the origin and thus a term s in the numerator of $y_{13}(s)$, we conclude that no such series capacitors can occur. An active *RC* network configuration whose passive portion satisfies the above requirements is shown in Fig. 4.12.† The numerators $n_{ij}(s)$ of the y parameters $y_{ij}(s)$ for the passive portion of this network are

$$n_{11}(s) = \frac{1}{R_1}\left(s + \frac{1}{R_2C_1}\right) \qquad n_{12}(s) = \frac{-s}{R_1} \qquad n_{13}(s) = \frac{1}{C_1R_1R_2}$$

$$n_{22}(s) = s\frac{R_1 + R_2}{R_1R_2} \qquad n_{23}(s) = \frac{-s}{R_2} \tag{4.23}$$

$$n_{33}(s) = C_2\left[s^2 + s\left(\frac{1}{R_1C_1} + \frac{1}{R_2C_1} + \frac{1}{R_2C_2}\right) + \frac{1}{R_1R_2C_1C_2}\right]$$

The denominators $d_{ij}(s)$ of the parameters are all the same:

$$d_{ij}(s) = s + \frac{R_1 + R_2}{C_1R_1R_2} \tag{4.24}$$

Inserting these values in (4.20), we obtain the voltage transfer function for the network

$$\frac{V_2(s)}{V_1(s)} = \frac{K/R_1R_2C_1C_2}{s^2 + s[1/R_1C_1 + 1/R_2C_1 + (1-K)/R_2C_2] + 1/R_1R_2C_1C_2} \tag{4.25}$$

This obviously has the required low-pass form of (4.21).

† This and several of the other active *RC* circuits presented in this and the following section were originally described in R. P. Sallen and E. L. Key, "A Practical Method of Designing *RC* Active Filters," *IRE Transactions on Circuit Theory,* vol. CT-2, no. 1, pp. 74–85, March 1955.

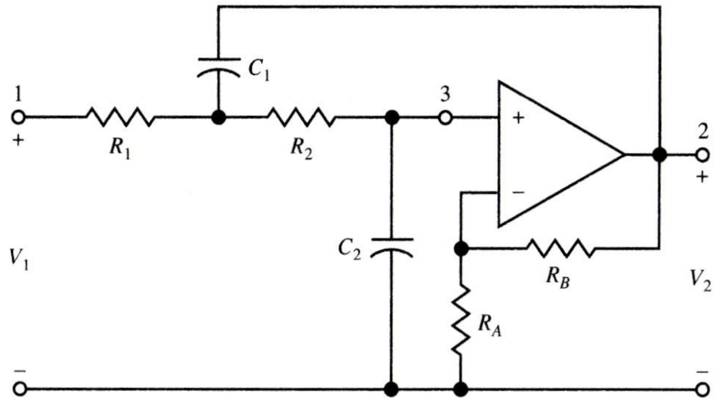

FIGURE 4.12
Sallen and Key low-pass filter.

It is interesting to note the manner in which the complex conjugate poles required by the filter characteristics defined in Table 4.1 are produced by the function given in (4.25). Passive *RC* networks, of course, can only have poles on the negative real axis. The same restriction holds for the zeros of *RC* driving-point functions such as $y_{33}(s)$. Thus the numerator of this *y* parameter must have the form $C_2(s - \sigma_1)(s - \sigma_2)$, where σ_1 and σ_2 are functions of the passive network elements R_1, R_2, C_1, and C_2. Now consider $y_{23}(s)$. From the circuit shown in Fig. 4.12, we note that one series capacitor is present in the path from terminal 2 to terminal 3. Thus, $y_{23}(s)$ must have a zero of transmission at the origin. Verifying this from (4.23), we see that its numerator is $-s/R_2$. Inserting these results in the denominator of (4.20) and comparing this with the denominator of (4.21), we obtain

$$(s - \sigma_1)(s - \sigma_2) - \frac{sK}{R_2C_2} = s^2 + a_1s + a_0 \tag{4.26}$$

When $K = 0$, the poles of the low-pass network function (the zeros of $s^2 + a_1s + a_0$) are at σ_1 and σ_2. As *K* increases, the poles follow a circular locus as indicated in Fig. 4.13. Obviously, any desired pole location can be obtained by choosing an appropriate value for *K* and, if necessary, making a frequency denormalization. It is also apparent from the locus that too high a value of *K* will produce instability, i.e., right half-plane poles.

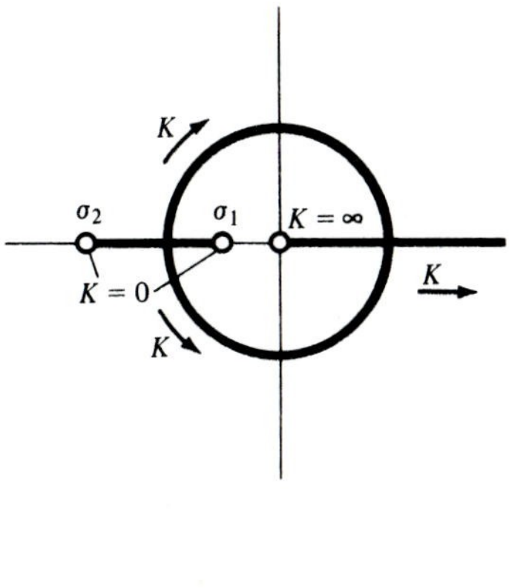

FIGURE 4.13
Locus of pole positions for the filter of Fig. 4.12.

The values of the network elements required to realize a specific low-pass characteristic can be readily determined by assigning some of the elements fixed values and solving for the values of the other ones. Since, in the network shown in Fig. 4.12, there are five element values to be selected and only three specifications to be met [the values of a_0, a_1, and H in (4.21)], obviously there are 2 degrees of freedom which may be used to arbitrarily specify element values or impose restrictions relating the values of different elements. In some cases, the values of more than two elements may be chosen. For example, choosing all passive elements to be unity, i.e., setting $R_1 = R_2 = C_1 = C_2 = 1$ (ohms and farads), (4.25) becomes

$$\frac{V_2(s)}{V_1(s)} = \frac{K}{s^2 + s(3 - K) + 1} \tag{4.27}$$

Thus a Butterworth characteristic with a bandwidth or −3-dB frequency of 1 rad/s (no. 1 in Table 4.1) is obtained by setting $3 - K = 1.414$, that is, $K = 1.586$. In this case the constant H of (4.21) is also 1.586. More generally, for this filter we may specify the following three useful designs.

Design 1: Equal-Resistor Equal-Capacitor Low-Pass Filter

The low-pass filter shown in Fig. 4.12 may be realized in an equal-resistor equal-capacitor design which is convenient since it reduces the number of different component values which must be procured. The constraints on the element values are

$$R_1 = R_2 = R \qquad \text{and} \qquad C_1 = C_2 = C \tag{4.28}$$

Using these relations in (4.25) and comparing the results with (4.21), we obtain

$$\omega_0 = \frac{1}{RC} \qquad \frac{1}{Q} = 3 - K \qquad H_0 = K \tag{4.29}$$

For specified values of Q and ω_0 the design process is as follows:

1. Choose a convenient value of C to effectively establish an impedance denormalization.
2. Use (4.29) to determine the values of R and K.
3. Use (4.16) plus an appropriate choice of impedance level to find values for R_A and R_B.

Example 4.1. A low-pass filter is to be designed with a cutoff frequency of 1 kHz ($\omega_0 = 6283$ rad/s) and a Q of 0.7071 (a Butterworth characteristic). From the procedure given above we choose $RC = 1/6283 = 1.592 \times 10^{-4}$ and $K = 1.586$. Suitable choices would be $C = 0.1\ \mu$F and $R = 1.592$ kΩ.

Design 2: Unity Gain Low-Pass Filter

The low-pass filter shown in Fig. 4.12 may be realized by using the unity gain follower shown in Fig. 1.21*b*. Such a design eliminates resistors R_A (= infinity) and R_B (= 0) but usually results in unequal resistor and capacitor values. The constraints are

$$K = 1 \qquad R = R_1 = \frac{R_2}{n} \qquad C = C_1 = \frac{C_2}{m} \tag{4.30}$$

Using these relations in (4.25) and comparing the results with (4.21), we obtain

$$\omega_0 = \frac{1}{\sqrt{mn}RC} \qquad \frac{1}{Q} = (n+1)\sqrt{\frac{m}{n}} \tag{4.31}$$

For specific values of Q and ω_0, the design procedure is:

1. Choose a convenient value of C to effectively establish an impedance denormalization.
2. Choose a value of m compatible with standard capacitor value ratios such that

$$m \leq \frac{1}{4Q^2} \tag{4.32}$$

   and use (4.30) to find $C_2 = mC_1$.
3. Calculate n, using the following relation (either the plus or the minus sign may be used, the resulting values are reciprocal):

$$n = \frac{1}{2mQ^2} - 1 \pm \frac{1}{2mQ^2}\sqrt{1 - 4mQ^2} \tag{4.33}$$

4. Use (4.31) to find $R = R_1$ and (4.30) to find $R_2 = nR_1$.

Example 4.2. A low-pass filter is to be designed with the same specifications given in Example 4.1 ($Q = 0.7071$, $\omega_0 = 6283$ rad/s), but using design 2. From the procedure given above, we choose $C = C_1 = 0.1\ \mu$F. From $m \leq 1/4Q^2 = 0.5$ we choose $m = 0.5$ and $C_2 = mC_1 = 0.05\ \mu$F. For n we substitute $2mQ^2 = 0.5$ in (4.33) to obtain $n = 1$. From (4.31) and (4.30) we find $R = R_1 = R_2/n = R_2 = 1/\omega_0 C\sqrt{mn} = 2251\ \Omega$.

Design 3: Equal-Capacitor Low-Pass Filter

The low-pass filter shown in Fig. 4.12 may also be realized by choosing $R_A = R_B = R$. In this case, from (4.16), $K = 2$. Such a choice allows the use of identical precision-value resistors to provide a highly accurate and stable value of the VCVS gain. Such a design also permits the use of equal-valued capacitors. The constraints are

$$R_A = R_B = R \qquad K = 2 \qquad C = C_1 = C_2 \tag{4.34}$$

Using these relations in (4.25) and comparing the results with (4.21), we obtain

$$R_1 = \frac{Q}{\omega_0 C} \qquad R_2 = \frac{R_1}{Q^2} \tag{4.35}$$

For specific values of Q and ω_0 the design procedure is:

1. Choose a convenient value of C to effectively establish an impedance denormalization for the circuit.
2. Choose an appropriate value for $R = R_A = R_B$ to establish an impedance denormalization for the gain-determining resistors of the VCVS.
3. Use (4.35) to find values for R_1 and R_2.

Example 4.3. A low-pass filter is to be designed with the same specifications used in the two preceding examples ($Q = 0.7071, \omega_0 = 6283$ rad/s). From the procedure given above we choose $C = C_1 = C_2 = 0.1\ \mu$F, and $R = R_A = R_B = 10$ kΩ. We find $R_1 = Q/\omega_0 C = 1125\ \Omega$ and $R_2 = R_1/Q^2 = 2250\ \Omega$.

Sensitivity

One of the most important criteria used in evaluating active RC filter designs is *sensitivity*. This is a measure of the amount that some characteristic of the filter changes as a result of a change in some circuit element. The filter characteristics most frequently of interest are Q and ω_0 (the cutoff frequency) of the filter. For these quantities we define

$$S_x^Q = \frac{\delta Q}{\delta x}\frac{x}{Q} \qquad S_x^{\omega_0} = \frac{\delta \omega_0}{\delta x}\frac{x}{\omega_0} \tag{4.36}$$

as the Q sensitivity and ω_0 sensitivity of the filter with respect to any element x. For the low-pass filter shown in Fig. 4.12, x may be any of the following: K, R_A, R_B, R_1, R_2, C_1, C_2. Note that the sensitivity measure of (4.36) is a "golflike" quantity; i.e., *lower is better*. For this filter we obtain the following expressions for sensitivity:

$$\begin{aligned}
S_{R_1}^Q &= -\frac{1}{2} + Q\sqrt{\frac{R_2C_2}{R_1C_1}} = -S_{R_2}^Q \\
S_{C_1}^Q &= -\frac{1}{2} + Q\left(\sqrt{\frac{R_1C_2}{R_2C_1}} + \sqrt{\frac{R_2C_2}{R_1C_1}}\right) = -S_{C_2}^Q \\
S_K^Q &= QK\sqrt{\frac{R_1C_1}{R_2C_2}} \\
S_{R_A}^Q &= -Q(K-1)\sqrt{\frac{R_1C_1}{R_2C_2}} = -S_{R_B}^Q \\
S_{R_1,R_2,C_1,C_2}^{\omega_n} &= -\frac{1}{2} \\
S_{K,R_A,R_B}^{\omega_n} &= 0
\end{aligned} \tag{4.37}$$

TABLE 4.2
Sensitivities for three designs of Sallen and Key low-pass filters

Sensitivity	Design 1	Design 2	Design 3
$S_{R_1}^{Q} = -S_{R_2}^{Q}$	$-\frac{1}{2} + Q$	$-\frac{1}{2} + Q\sqrt{mn}$	$\frac{1}{2}$
$S_{C_1}^{Q} = -S_{C_2}^{Q}$	$-\frac{1}{2} + 2Q$	$-\frac{1}{2} + Q\left(\sqrt{\frac{m}{n}} + \sqrt{mn}\right)$	$\frac{1}{2} + Q^2$
S_{K}^{Q}	$3Q - 1$	$\frac{Q}{\sqrt{mn}}$	$2Q^2$
$S_{R_A}^{Q} = -S_{R_B}^{Q}$	$1 - 2Q$	0	$-Q^2$
$S_{R_1,R_2,C_1,C_2}^{\omega_0}$	$-\frac{1}{2}$	$-\frac{1}{2}$	$-\frac{1}{2}$
$S_{K,R_A,R_B}^{\omega_0}$	0	0	0

The values of these sensitivities for designs 1, 2, and 3 are given in Table 4.2. This table shows that the Q sensitivities vary widely for the different designs while the ω_0 sensitivities are all the same. Since the Q sensitivities are readily seen to be proportional to either Q or (even worse) Q^2, the use of this filter is only practical for low values of Q such as $Q < 0.5$.

An important part of the actual fabrication of any active *RC* filter is a final tuning to compensate for inaccuracies in the network element values, parasitic effects, etc. In the low-pass low-gain-amplifier filter, tuning is readily accomplished since the adjustment of the Q and ω_0 of (4.21) may be made separately without interaction occurring. Specifically, equal percentage changes in the values of R_1 and R_2 will vary ω_0 without affecting Q, and changes in K will vary Q without affecting ω_0.

In this section we have presented our first active *RC* circuit configuration and showed how to evaluate its performance by determining its sensitivities. In the following section we introduce some other similar active *RC* filters which realize different transfer characteristics.

4.4 OTHER SECOND-ORDER VCVS FILTERS

In the preceding section we discussed a general approach to the design and evaluation of active *RC* filters using a specific low-pass network configuration with a VCVS as the active element. In this section we apply the same technique to realize several other types of second-order network functions, namely, high-pass, bandpass, and band-reject.

High-Pass Filter

The first second-order filter function that we consider in this section is the high-pass voltage transfer function having the form

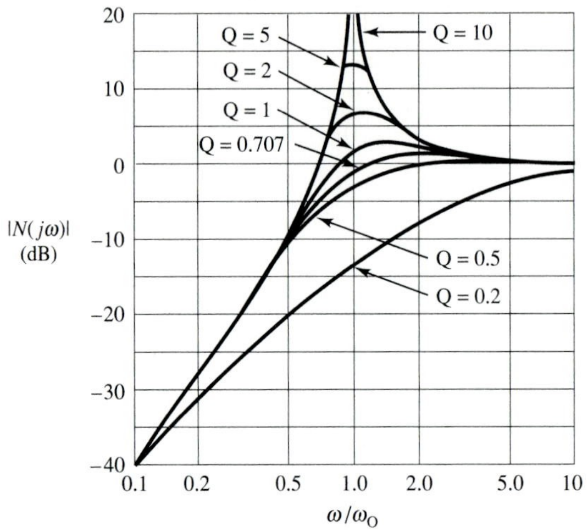

FIGURE 4.14
High-pass response characteristics.

$$\frac{V_2(s)}{V_1(s)} = \frac{Hs^2}{s^2 + a_1 s + a_0} = \frac{H_0 s^2}{s^2 + (\omega_0/Q)s + \omega_0^2} \tag{4.38}$$

where the last member is called the *standard form* of the second-order high-pass function. In this, H_0 is the infinite frequency gain, ω_0 is the cutoff frequency (in radians per second), and Q is the quality factor. Some plots of the response characteristic for $H_0 = 1$ (0 dB) and various values of Q are given in Fig. 4.14. The values of the coefficients a_0 and a_1 are usually chosen so as to achieve some specific high-pass characteristic. Some of the more common ones are summarized in Table 4.3. In general, high-pass functions may be determined directly from low-pass ones by substituting $1/s$ for the complex frequency variable s. Thus, if a network function $N(s)$ is a low-pass function, the function $N(1/s)$ will be a high-pass one. Such a substitution inverts the magnitude characteristic of the low-pass function around the frequency 1 rad/s. For example, a low-pass function having a 1-dB ripple in its passband from 0 to 1 rad/s would be transformed to a high-pass function having a 1-dB ripple in its passband from 1 to infinite

TABLE 4.3
Normalized high-pass network function characteristics for (4.38)

No.	Name	Characteristic	a_1	a_0	Pole locations
1	Butterworth (maximally flat magnitude)	At 1 rad/s down 3.01 dB from infinite frequency value	1.41421	1.00000	-0.70711 ± 0.70711
2*a*	Chebychev (equal-ripple magnitude)	$\frac{1}{2}$-dB ripple in passband from 1 rad/s to infinity	0.94026	0.65954	$-0.47013 \pm j0.66221$
2*b*		1-db ripple in passband from 1 rad/s to infinity	0.99566	0.90702	$-0.49783 \pm j0.81191$

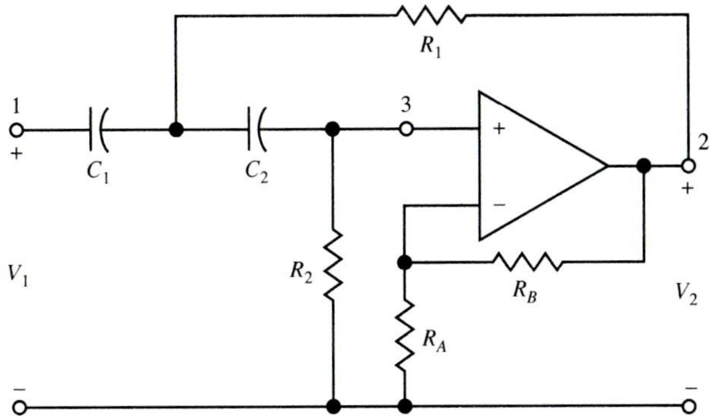

FIGURE 4.15
Sallen and Key high-pass filter.

rad/s, etc. This means that the low-pass characteristics given in Table 4.1 may be directly related to corresponding high-pass characteristics. The results for cases 1 and 2 are shown in Table 4.3.

To find an active *RC* configuration capable of realizing a voltage transfer function having the form of (4.38), we note from (4.20) that, for the general active *RC* configuration shown in Fig. 4.10, the numerator of the transfer admittance $y_{13}(s)$ must be some multiple of s^2; that is, it must have a second-order transmission zero at the origin, which means that there must be two independent series capacitors in the path between terminals 1 and 3. Such a condition is satisfied by the network configuration shown in Fig. 4.15. By direct analysis we find that the voltage transfer function for this circuit is

$$\frac{V_2(s)}{V_1(s)} = \frac{Ks^2}{s^2 + s[1/R_2C_1 + 1/R_2C_2 + (1/R_1C_1)(1-K)] + 1/R_1R_2C_1C_2} \tag{4.39}$$

This obviously is of the required form. The decomposition of the denominator follows a form similar to that given in (4.26). Thus, K is required to be positive, the pole locus has the form shown in Fig. 4.13, and the comments on sensitivity given in Sec. 4.3 are directly applicable.

The determination of the values of the elements of the network shown in Fig. 4.15 is similar to the process used in the preceding section. For example, setting $R_1 = R_2 = C_1 = C_2 = 1$ (ohms and farads) in (4.39) gives

$$\frac{V_2(s)}{V_1(s)} = \frac{Ks^2}{s^2 + s(3-K) + 1} \tag{4.40}$$

More generally, for this filter we may specify the following designs.

Design 1: Equal-Resistor Equal-Capacitor High-Pass Filter

The high-pass filter shown in Fig. 4.15 may be realized in an equal-resistor equal-capacitor design which is convenient since it reduces the number of different component values that must be procured. The constraints on the element values are

$$R_1 = R_2 = R \qquad \text{and} \qquad C_1 = C_2 = C \tag{4.41}$$

Using these relations in (4.39) and comparing the results with (4.38), we obtain

$$\omega_0 = \frac{1}{RC} \qquad \frac{1}{Q} = 3 - K \qquad H_0 = K \tag{4.42}$$

For specified values of Q and ω_0 the design process is:

1. Choose a convenient value of C to effectively establish an impedance denormalization.
2. Use (4.42) to determine the values of R and K.
3. Use (4.16) plus an appropriate choice of impedance level to find values for R_A and R_B.

Example 4.4. A high-pass filter is to be designed with a cutoff frequency of 1 kHz ($\omega_0 = 6283$ rad/s) and a Q of 0.7071 (a Butterworth characteristic). From the procedure given above we obtain $RC = 1/6283 = 1.592 \times 10^{-4}$ and $K = 1.586$. Suitable choices would be $C = 0.1\ \mu$F and $R = 1.592$ kΩ. If we choose $R_A =$ 10 kΩ, then $R_B = 5.86$ kΩ.

Design 2: Unity Gain High-Pass Filter

The high-pass filter shown in Fig. 4.15 may be realized by using the unity gain follower shown in Fig. 1.21*b*. Such a design eliminates resistors R_A (= infinity) and R_B (= 0) but usually results in unequal resistor and capacitor values. The constraints are

$$K = 1 \qquad R = R_1 = \frac{R_2}{n} \qquad C = C_1 = \frac{C_2}{m} \tag{4.43}$$

Using these relations in (4.39) and comparing the results with (4.38), we obtain

$$\omega_0 = \frac{1}{\sqrt{mn}RC} \qquad \frac{1}{Q} = \frac{m+1}{\sqrt{mn}} \tag{4.44}$$

For these, for any given value of n, the minimum value of $1/Q$ will occur when $m = 1$. With this value of m, Eq. (4.44) becomes

$$\omega_0 = \frac{1}{\sqrt{n}RC} \qquad Q = \frac{\sqrt{n}}{2} \tag{4.45}$$

For specific values of Q and ω_0, the design procedure is:

1. Choose a convenient value of C to effectively establish an impedance denormalization, and let $C = C_1 = C_2$.
2. Determine the value of n from (4.45).
3. Find the value of R, using (4.45), and let $R = R_1 = R_2/n$.

Example 4.5. A high-pass filter is to be designed with the same specifications given in Example 4.4 ($Q = 0.7071$, $\omega_0 = 6283$ rad/s), but using design 2. From the procedure given above, we choose $C = C_1 = C_2 = 0.1\ \mu$F. From (4.45) we find $n = (2Q)^2 = 2, R = R_1 = 1/\omega_0 RC\sqrt{n} = 1.125$ kΩ, and $R_2 = nR_1 = 2.25$ kΩ.

Sensitivity considerations for this filter are similar to those covered for the low-pass one in the preceding section and are left for the reader as an exercise.

Bandpass Filter

The next second-order filter function that we consider is the bandpass voltage transfer function having the form

$$\frac{V_2(s)}{V_1(s)} = \frac{Hs}{s^2 + a_1 s + a_0} = \frac{H_0(\omega_0/Q)s}{s^2 + (\omega_0/Q)s + \omega_0^2} \tag{4.46}$$

where, in the standard form given in the right member, H_0 is the gain at resonance, ω_0 is the resonant frequency, and Q is the quality factor. Some plots of the response characteristic for $H_0 = 1$ (0 dB) are given in Fig. 4.16. An active *RC* network capable of realizing such a function may be found by noting from (4.20) that, for the general active *RC* configuration shown in Fig. 4.10, the numerator of the transfer admittance $y_{13}(s)$ must be some multiple of s, that is, it must have a single transmission zero at the origin and thus a single series capacitor is required in the path between terminals 1 and 3. Such a condition is satisfied by the configuration shown in Fig. 4.17.† By direct analysis we find that the voltage transfer function

† Many other bandpass configurations are possible. This one is due to R.P. Sallen and E.L. Key, *op. cit.*, and has been shown to be of practical value.

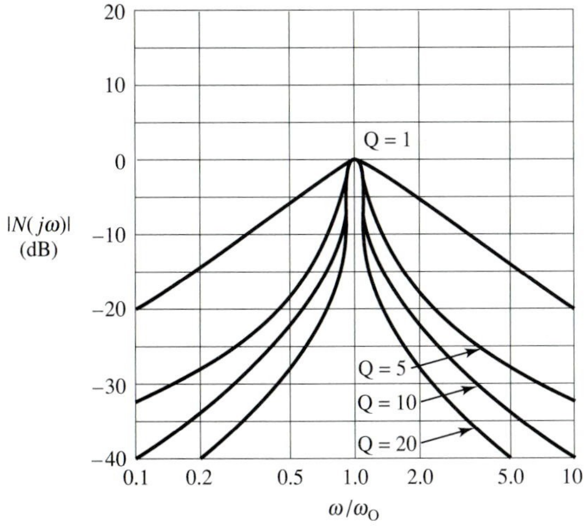

FIGURE 4.16
Bandpass response characteristics.

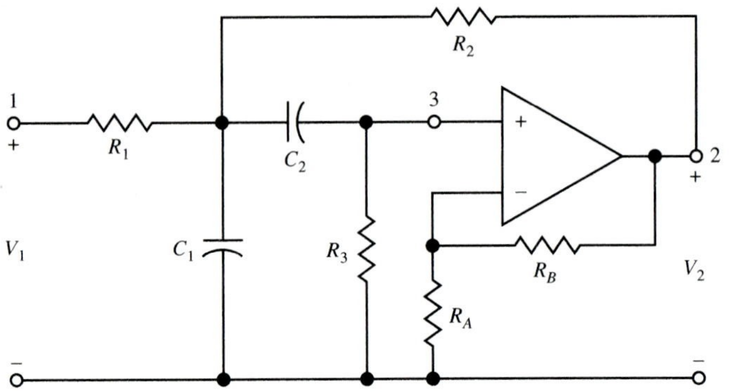

FIGURE 4.17
Sallen and Key bandpass filter.

for this circuit is

$$\frac{V_2(s)}{V_1(s)} = \frac{sK/R_1C_1}{s^2 + s[1/R_1C_1 + 1/R_3C_1 + 1/R_3C_2 + (1-K)/R_2C_1] + (1/R_1 + 1/R_2)/R_3C_1C_2} \tag{4.47}$$

The comments on the denominator decomposition are identical with those given for the high-pass network described above. For this filter we may specify the following designs.

Design 1: Equal-Resistor Equal-Capacitor Bandpass Filter

The bandpass filter shown in Fig. 4.17 may be realized in an equal-resistor equal-capacitor design which is convenient since it reduces the number of different component values that must be procured. The constraints on the element values are

$$R_1 = R_2 = R_3 = R \qquad \text{and} \qquad C_1 = C_2 = C \tag{4.48}$$

Note that for K to be positive, Q must be greater than $\sqrt{2}/4$. Using these relations in (4.47) and comparing the results with (4.46), we obtain

$$\omega_0 = \frac{\sqrt{2}}{RC} \qquad Q = \frac{\sqrt{2}}{4-K} \qquad H_0 = \frac{K}{4-K} \tag{4.49}$$

Solving for RC and K, we find

$$RC = \frac{\sqrt{2}}{\omega_0} \qquad K = 4 - \frac{\sqrt{2}}{Q} \tag{4.50}$$

For specified values of Q and ω_0 the design process is:

1. Choose a convenient value of C to effectively establish an impedance denormalization.
2. Use (4.50) to determine the values of R and K.
3. Use (4.16) plus an appropriate choice of impedance level to find values for R_A and R_B.

Example 4.6. A bandpass filter is to be designed with a resonant frequency of 1 kHz ($\omega_0 = 6283$ rad/s) and a Q of 5. From the procedure given above we obtain $RC = \sqrt{2}/6283 = 2.251 \times 10^{-4}$ and $K = 3.717$. Suitable choices would be $C = 0.1\ \mu\text{F}$ and $R = 2.251\ \text{k}\Omega$. To realize the VCVS gain, we may choose $R_A = 10\ \text{k}\Omega$ and $R_B = 27.17\ \text{k}\Omega$.

Design 2: Equal-Capacitor Bandpass Filter

The bandpass filter shown in Fig. 4.17 may also be realized by choosing $R_A = R_B = R$. In this case, from (4.16), $K = 2$. Like the low-pass design 3, this choice allows the use of identical precision resistors to obtain a stable and precise value of VCVS gain and equal capacitor values. It is convenient to work with a normalized design. For this, the constraints are

$$R_1 = 1 \qquad C_1 = C_2 = 1 \qquad K = 2 \tag{4.51}$$

Using these relations in (4.47) and comparing the results with (4.46), we obtain

$$1 = \left(1 + \frac{1}{R_2}\right)\frac{1}{R_3} \qquad \frac{1}{Q} = 1 + \frac{2}{R_3} - \frac{1}{R_2} \tag{4.52}$$

For specified values of Q and ω_0 the design process is:

1. Solve the simultaneous equations of (4.52) to obtain normalized values for R_2 and R_3.
2. Frequency-denormalize the capacitor values to obtain the desired resonant frequency.
3. Impedance-denormalize the elements to obtain convenient values for the capacitors.

Example 4.7. A bandpass filter is to be designed with the same specifications given in Example 4.6 ($Q = 5$ and $\omega_0 = 6283$ rad/s) but using design 2. From the procedure given above we obtain $R_2 = 0.6345\ \Omega$ and $R_3 = 2.576\ \Omega$. Applying the frequency denormalization $\Omega_D = 2\pi \times 10^3$, we obtain $C = 0.1\ \mu\text{F}$. Impedance-denormalizing by $Z_D = 1/2\pi \times 10^4$, we obtain $C = 0.1\ \mu\text{F}$, $R_1 = 1.592\ \text{k}\Omega$, $R_2 = 1.010\ \text{k}\Omega$, and $R_3 = 4.100\ \text{k}\Omega$. If higher values of the resistors are desired, additional impedance denormalizations may be applied.

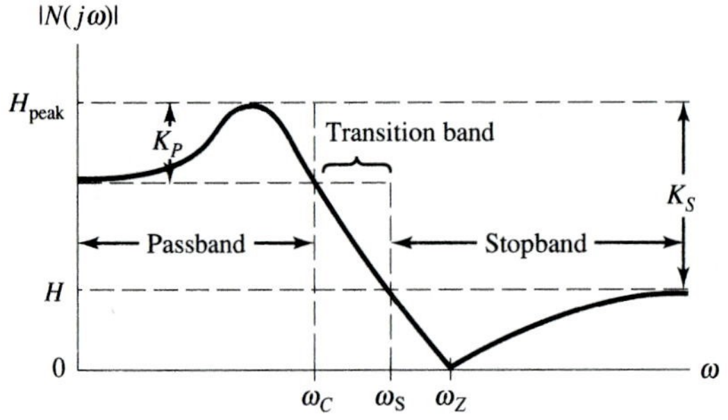

FIGURE 4.18
Low-pass band-reject response.

Band-Reject Filter

Frequently, filtering applications require the realization of a transfer function which has transmission zeros on the $j\omega$ axis of the complex frequency plane. For the second-order case, a network function of this type has the voltage transfer function

$$\frac{V_2(s)}{V_1(s)} = \frac{H(s^2 + b_0)}{s^2 + a_1 s + a_0} = \frac{H(s^2 + \omega_z^2)}{s^2 + (\omega_0/Q)s + \omega_0^2} \tag{4.53}$$

In the last member, ω_z is the null frequency where no transmission occurs. For the case where $a_0 < b_0$, this is considered as a low-pass function, and it has a magnitude characteristic having the general form shown in Fig. 4.18. In the figure K_p determines the ripple in the passband $0 \leq \omega \leq \omega_c$, and K_s is the minimum attenuation in the stopband $\omega_s \leq \infty$. The frequency range $\omega_c \leq \omega \leq \omega_s$ is called the *transition band*. If $a_0 > b_0$, a high-pass characteristic occurs which is similar to Fig. 4.18 but in which the frequency axis has been inverted, as shown in Fig. 4.19. If $a_0 = b_0$ $(\omega_0 = \omega_z)$, the result is called a *notch characteristic*. Plots of the magnitude for various values of Q for this case are shown in Fig. 4.20. Comparing (4.53) with (4.20) for the network shown in Fig. 4.10, we see that in the passive portion of an active *RC* network capable of realizing (4.53), the numerator of its parameter $y_{13}(s)$ must be a multiple of $s^2 + b_0$. Thus, there must be a twin-tee, parallel ladder or other similar $j\omega$ axis zero-producing network connected between terminals 1 and 3. Such a condition is

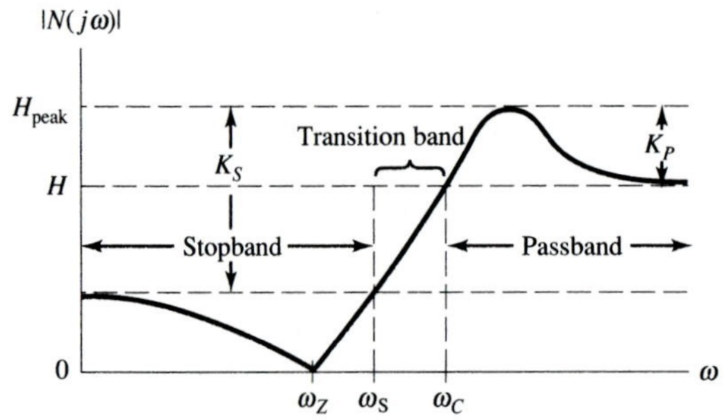

FIGURE 4.19
High-pass band-reject response.

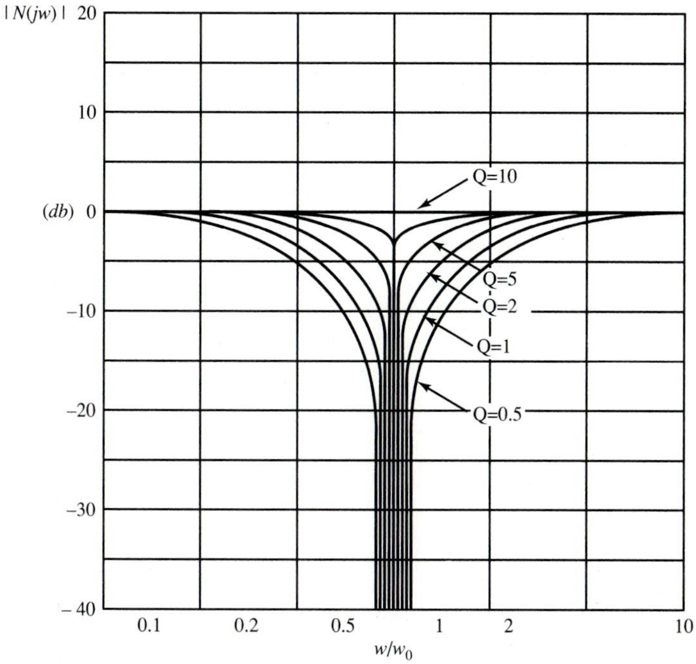

FIGURE 4.20
Notch response characteristics.

met by the normalized network configuration shown in Fig. 4.21.† For the case in which $a_0 > b_0$, $Y = 1/R$ and the transfer function is

$$\frac{V_2(s)}{V_1(s)} = \frac{K(s^2 + 1/a^2)}{s^2 + s[(K+1)/a][1/R + (2-K)/k] + [1 + (k+1)/R]/a^2} \tag{4.54}$$

In this relation, the constant k may be chosen as unity or may be used to obtain a particular value of K or to reduce sensitivity. In the latter cases, however, a wider spread of element values results. For this network the following design relations may be used:

$$a = \sqrt{\frac{1}{b_0}} \qquad R = \frac{k+1}{a_0/b_0 - 1} \qquad H = K$$
$$K = 2 + \frac{k}{k+1}\left(\frac{a_0}{b_0} - 1 - \frac{a_1}{\sqrt{b_0}}\right) \tag{4.55}$$

† W. J. Kerwin and L. P. Huelsman, "The Design of High-Performance Active *RC* Bandpass Filters," *Proceedings of the IEEE International Convention Record,* part 10, pp. 74–80, March 1966; and W. J. Kerwin, "An Active *RC* Elliptic Function Filter," *IEEE Region Six Conference Record,* vol. 2, pp. 647–654, April 1966.

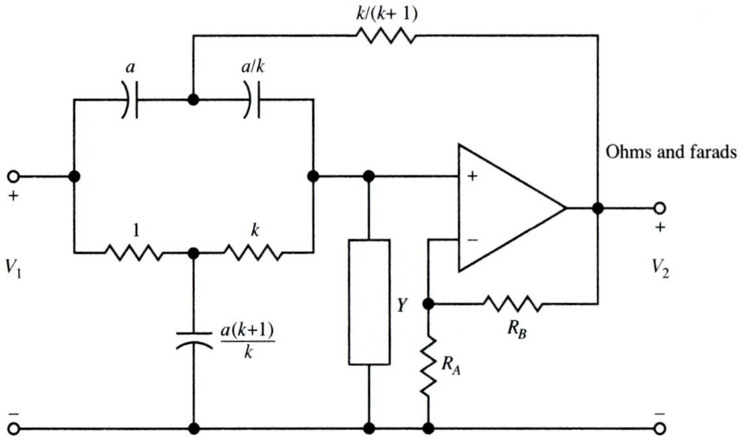

FIGURE 4.21
Band-reject filter.

For the case where $b_0 > a_0$ in (4.53), $Y = aCs$ and the transfer function is

$$\frac{V_2(s)}{V_1(s)} = \frac{K/[(k+1)C+1](s^2+1/a^2)}{s^2+s(k+1)[C+(2-K)/k]/a[(k+1)C+1]+1/a^2[(k+1)C+1]} \tag{4.56}$$

The following design equations may be used:

$$a = \sqrt{\frac{1}{b_0}} \qquad C = \frac{b_0/a_0 - 1}{k+1} \qquad H = \frac{a_0}{b_0}K$$
$$K = 2 + \frac{k}{k+1}\left(\frac{b_0}{a_0} - 1 - \frac{a_1\sqrt{b_0}}{a_0}\right) \tag{4.57}$$

For the case where $a_0 = b_0$, it is convenient to consider the normalized situation $\omega_z = \omega_0 = a_0 = b_0 = 1$. In this case the shunt element labeled Y in Fig. 4.21 is not used, and we obtain

$$a = 1 \qquad H = K = 2 - \frac{k}{k+1}\left(\frac{1}{Q}\right) \tag{4.58}$$

Example 4.8. A notch filter is to be designed with a notch frequency of 1 kHz ($\omega_z = 6283$ rad/s) and a Q of 5. From the relations of (4.58), choosing $k = 1$, we find $K = 2 - \frac{1}{10} = 1.9$. Denormalizing frequency and impedance, we obtain the circuit shown in Fig. 4.22.

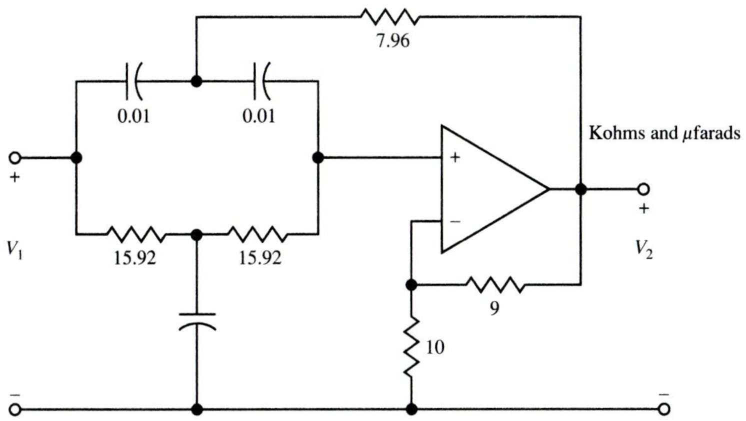

FIGURE 4.22
Filter for Example 4.8.

4.5 SECOND-ORDER MULTIPLE-FEEDBACK FILTERS

The active *RC* filters discussed in the preceding sections of this chapter used low-gain amplifiers as their active element. In this section we present a different approach to the design of active *RC* filters, namely, the use of the entire passive *RC* network to provide the feedback around the operational amplifier. A general configuration for the second-order case is shown in Fig. 4.23. It is called a *multiple-feedback filter*. The resulting transfer function is an inverting one; i.e., the dc gain is negative. The voltage transfer function for the circuit is

$$\frac{V_2(s)}{V_1(s)} = \frac{-Y_1Y_3}{Y_5(Y_1 + Y_2 + Y_3 + Y_4) + Y_3Y_4 - \dfrac{1}{A}[(Y_3 + Y_5)(Y_1 + Y_2 + Y_4) + Y_3Y_5]} \tag{4.59}$$

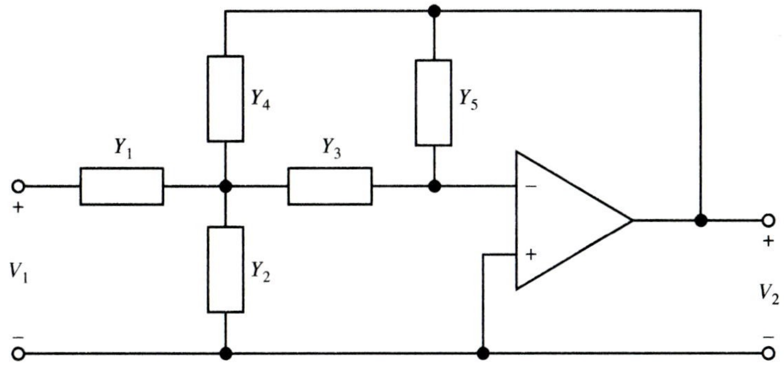

FIGURE 4.23
General multiple-feedback filter.

where A is the open-loop gain of the operational amplifier. In general, this gain is very large, thus (4.59) may be approximated as

$$\frac{V_2(s)}{V_1(s)} = \frac{-Y_1Y_3}{Y_5(Y_1 + Y_2 + Y_3 + Y_4) + Y_3Y_4} \tag{4.60}$$

If the individual admittances Y_i are suitably chosen as either resistors or capacitors, this network function will yield a low-pass, high-pass, or bandpass characteristic. The details are given in the following paragraphs.

Low-Pass Multiple-Feedback Filter

The circuit configuration shown in Fig. 4.23 may be used to realize a second-order low-pass function having the general form given in (4.21) by choosing Y_1, Y_3, and Y_4 as resistors and Y_2 and Y_5 as capacitors. The resulting network configuration is shown in Fig. 4.24. The voltage transfer function (for large values of the operational amplifier open-loop gain) is

$$\frac{V_2(s)}{V_1(s)} = \frac{-1/R_1R_3C_2C_5}{s^2 + (s/C_2)(1/R_1 + 1/R_3 + 1/R_4) + 1/R_3R_4C_2C_5} \tag{4.61}$$

The dc gain is $-R_4/R_1$. For specified values of Q, ω_0, and H_0 ($H_0 < 0$) as given in (4.21), the design process is:

1. Choose a convenient value of $C = C_2$ to effectively establish an impedance denormalization.
2. Calculate $C_5 = mC$ where m is a convenient ratio chosen subject to the restriction

$$m \leq \frac{1}{4Q^2(1 + |H_0|)} \tag{4.62}$$

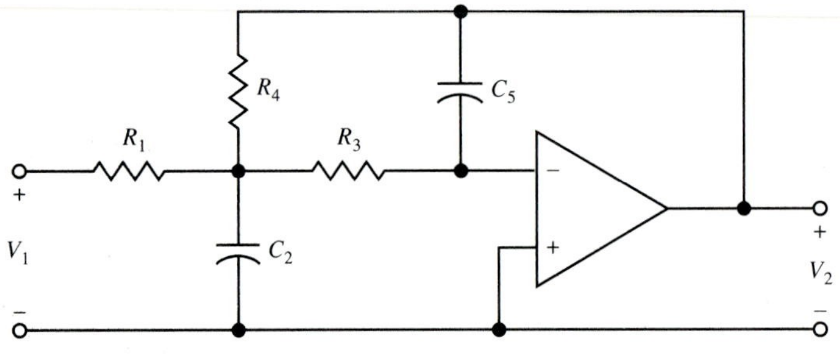

FIGURE 4.24
Low-pass multiple-feedback filter.

3. Calculate

$$R_4 = \frac{1}{2\omega_0 C m Q}[1 \pm \sqrt{1 - 4mQ^2(1 + |H_0|)}]$$

$$R_1 = \frac{R_4}{|H_0|} \qquad (4.63)$$

$$R_3 = \frac{1}{\omega_0^2 C^2 R_4 m}$$

Example 4.9. A low-pass filter is to be designed with the same specifications used in Example 4.1 (a cutoff frequency of 1 kHz and a Q of 0.7071) but with a dc gain $H_0 = -8$. From the procedure given above we choose $C = C_2 = 0.1\ \mu\text{F}$. From (4.62) we find $m = 0.0556$. For the choice $m = 0.05$ we obtain $C_5 = 0.05\ \mu\text{F}$. From (4.63) we find $R_4 = 24.76\,\text{k}\Omega$, $R_1 = 3.095\ \text{k}\Omega$, and $R_3 = 2.046\ \text{k}\Omega$.

A comparison of this realization with the one given in Example 4.1 illustrates a disadvantage of the multiple-feedback filter; namely, it is not usually possible to have elements of the same type with the same value. In general, as the values of Q and $|H_0|$ increase, so does the spread of element values, and this is one of the limitations on the use of this filter.

The multiple-feedback filter has several advantages compared with the low-gain (VCVS-based) filter configuration. One is its ability to exactly realize a specified dc gain. Another more significant advantage is its low sensitivity. The sensitivity to gain changes of the operational amplifier is practically zero. In addition, the sensitivities to changes in the passive elements are usually lower than the ones encountered in comparable low-gain filters.

High-Pass Multiple-Feedback Filter

The circuit configuration shown in Fig. 4.23 may be used to realize a second-order high-pass function having the general form given in (4.38) by choosing Y_2 and Y_5 as resistors and Y_1, Y_3, and Y_4 as capacitors. The resulting network configuration is shown in Fig. 4.25. The voltage transfer function (for large values of operational

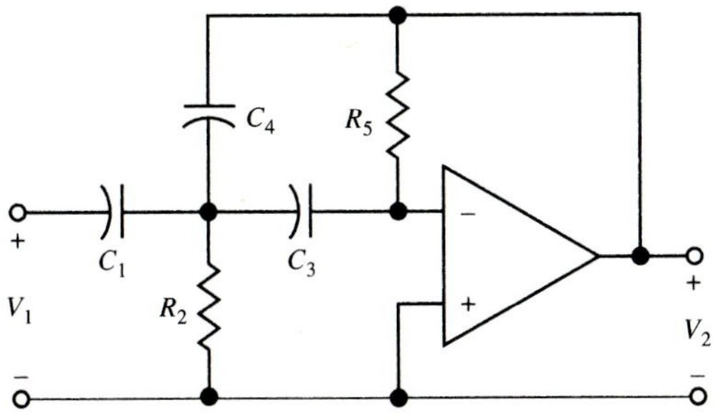

FIGURE 4.25
High-pass multiple-feedback filter.

amplifier open-loop gain) is

$$\frac{V_2(s)}{V_1(s)} = \frac{-s^2C_1/C_4}{s^2 + (s/R_5)(C_1/C_3C_4 + 1/C_3 + 1/C_4) + 1/R_2R_5C_3C_4} \tag{4.64}$$

The gain at infinite frequencies is $-C_1/C_4$. For specified values of Q, ω_0, and H_0 ($H_0 < 0$) as given in (4.38), the design process is:

1. Choose a convenient value of $C = C_1 = C_3$ to effectively establish an impedance denormalization.
2. Calculate

$$R_2 = \frac{|H_0|}{\omega_0 QC(2|H_0| + 1)}$$
$$R_5 = \frac{(2|H_0| + 1)Q}{\omega_0 C} \tag{4.65}$$
$$C_4 = \frac{C}{|H_0|}$$

Example 4.10. A high-pass filter is to be designed with the same specifications used in Example 4.4 (a cutoff frequency of 1 kHz and a Q of 0.7071) but with an infinite frequency gain $H_0 = -8$. From the procedure given above we choose $C = C_1 = C_3 = 0.1\ \mu$F. From (4.65) we find $R_2 = 1.059$ kΩ, $R_5 = 19.13$ kΩ, and $C_4 = 0.0125\ \mu$F.

Comments similar to those made for the low-pass multiple-feedback filter with respect to sensitivity and spread of element values also apply to this filter.

Bandpass Multiple-Feedback Filter

There are several different choices of elements which may be used in the configuration shown in Fig. 4.23 to realize a second-order bandpass function having the general form given in (4.46). One of the most practical configurations is obtained by choosing Y_1, Y_2, and Y_5 as resistors and Y_3 and Y_4 as capacitors. The resulting network configuration is shown in Fig. 4.26. The voltage transfer function (for high values of open-loop operational amplifier gain) is

$$\frac{V_2(s)}{V_1(s)} = \frac{-s/R_1C_4}{s^2 + (s/R_5)(1/C_3 + 1/C_4) + (1/R_5C_3C_4)(1/R_1 + 1/R_2)} \tag{4.66}$$

The gain at resonance is $-R_5C_3/[R_1(C_3 + C_4)]$. For specified values of Q, ω_0, and $|H_0|$ ($H_0 < 0$) as given in (4.46), the design procedure is:

1. Choose a convenient value of $C = C_3 = C_4$ to effectively establish an impedance denormalization.

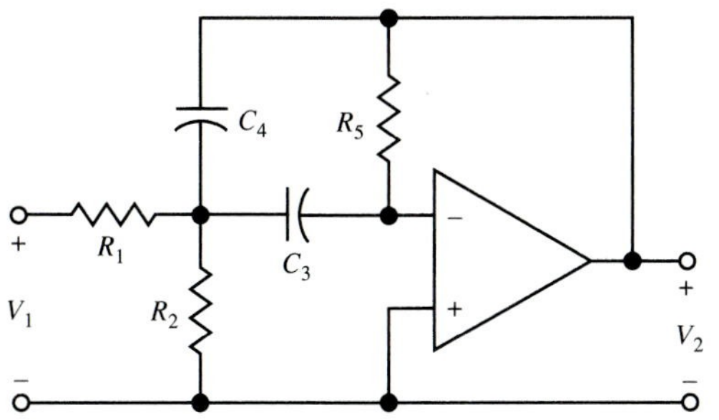

FIGURE 4.26
Bandpass multiple-feedback filter.

2. Calculate the resistor values as follows:

$$R_1 = \frac{Q}{\omega_0 C|H_0|}$$

$$R_2 = \frac{Q}{(2Q^2 - |H_0|)\omega_0 C} \tag{4.67}$$

$$R_5 = \frac{2Q}{\omega_0 C}$$

Example 4.11. A bandpass filter is to be designed with the same specifications used in Example 4.6 (a resonant frequency of 1 kHz and a Q of 5) but with a resonant gain $H_0 = -10$. From the procedure given above we choose $C = C_3 = C_4 = 0.01\ \mu\text{F}$. From (4.67) we find $R_1 = 7.958\ \text{k}\Omega$, $R_2 = 1.989\,\text{k}\Omega$, and $R_5 = 159.2\ \text{k}\Omega$.

The element spread and sensitivities are similar to those of the low-pass and high-pass realizations.

Bandpass Filter with Positive Feedback

A modification of the bandpass circuit discussed above can be used to reduce the spread of element values for high-Q realizations. This is accomplished by providing positive feedback to the "+" input terminal of the operational amplifier in addition to the negative (multiple-loop) feedback already present at the "−" terminal. A circuit with this provision is shown in Fig. 4.27.† The positive feedback is provided by the voltage divider consisting of the two resistors R_A and R_B. To simplify the equations describing the circuit, we may represent the effect of these two resistors by defining a constant K, where

$$K = \frac{R_A}{R_B} \tag{4.68}$$

† T. Deliyannis, "High Q-Factor Circuit with Reduced Sensitivity," *Electronics Letters*, vol. 4, p. 577, December 1968.

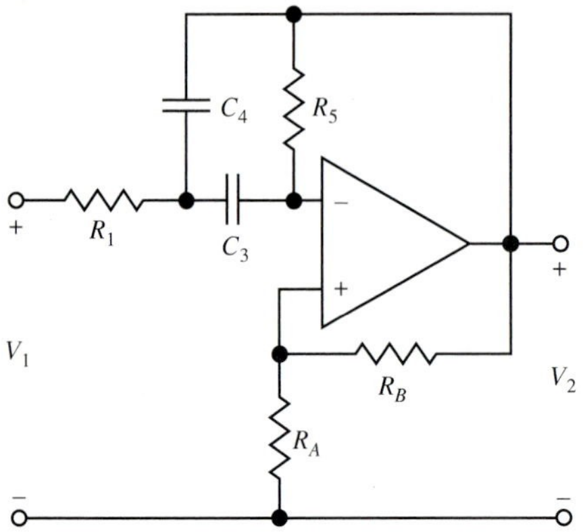

FIGURE 4.27
Bandpass multiple-feedback filter with positive feedback.

The voltage transfer function for the circuit is

$$\frac{V_2(s)}{V_1(s)} = \frac{-s(K+1)/R_1C_4}{s^2 + s(1/R_5C_4 + 1/R_5C_3 - K/R_1C_4) + 1/R_1R_5C_3C_4} \tag{4.69}$$

Using the quantities Q, and ω_0 defined for the second-order bandpass function in (4.26), we obtain the relations

$$\begin{aligned} \omega_0 &= \frac{1}{\sqrt{R_1R_5C_3C_4}} \\ \frac{1}{Q} &= \sqrt{\frac{R_1}{R_5}}\left(\sqrt{\frac{C_3}{C_4}} + \sqrt{\frac{C_4}{C_3}} - K\frac{R_5}{R_1}\sqrt{\frac{C_4}{C_3}}\right) \\ |H_0| &= \frac{(K+1)/R_1C_3}{1/R_5C_4 + 1/R_5C_3 - K/R_1C_3} \end{aligned} \tag{4.70}$$

The design procedure is:

1. Choose a convenient value of $C = C_3 = C_4$ to effectively establish an impedance normalization.
2. Determine the resistance ratio parameter m_0 that would be required if there were no positive feedback:

$$m_0 = \frac{1}{4Q^2} \tag{4.71}$$

3. Select the desired resistor ratio m in the range $m_0 < m < 1$, and use it to determine the positive feedback constant K of (4.68):

$$K = 2m - \frac{\sqrt{m}}{Q} \tag{4.72}$$

Choose convenient values for R_A and R_B to satisfy (4.68).

4. Determine

$$R_5 = \frac{1}{\omega_0 C \sqrt{m}} \qquad R_1 = mR_2 \tag{4.73}$$

Example 4.12. A bandpass filter is to be designed with a resonant frequency of 1.592 kHz ($\omega_0 = 10^4$ rad/s), a Q of 10, and a maximum resistance value spread of 100. From the procedure given above we choose $C = C_3 = C_4 = 0.01\ \mu$F. Since $m_0 = 0.0025$, we may choose $m = 0.01$. From (4.72) we find $K = 0.01$, so for $R_A = 1\ \text{k}\Omega, R_B = 100\,\text{k}\Omega$. From (4.73) we find $R_5 = 100\,\text{k}\Omega$, and $R_1 = 1\ \text{k}\Omega$.

Although, due to the positive feedback, this circuit has slightly larger sensitivities than the previously described bandpass multiple-feedback one, it still provides satisfactory performance for the realization of high-Q functions. Note that the gain at resonance H_0 is not available as a design parameter. In some cases the value of $|H_0|$ that results from the procedure given may be too large to be practical. If it is desired to reduce this value, it may be done by adding a resistor R_0 to the circuit as shown in Fig. 4.28. In this figure the designation for the input resistor has been changed from R_1 to R_1'. If $|H_0'|$ is the desired gain (where $|H_0'|$ is less than $|H_0|$), the values of R_1' and R_0 are found as follows.

$$\begin{aligned} R_1' &= R_1 \frac{|H_0|}{|H_0'|} \\ R_0 &= \frac{R_1}{1 - |H_0'|/|H_0|} \end{aligned} \tag{4.74}$$

Note that the addition of the resistor R_0 to the circuit effectively creates a voltage divider, which lowers the level of the input to the rest of the circuit and thus reduces its overall gain. The adjustment to the gain provided by this step is only

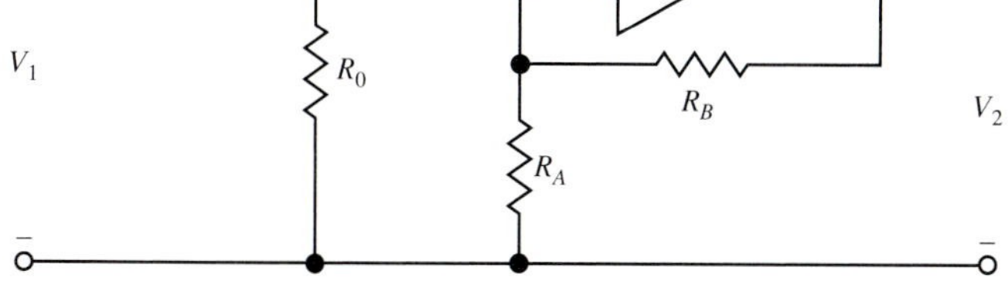

FIGURE 4.28
Bandpass filter with reduced gain.

approximate, since the calculations assume that the current flowing through the resistors R_1' and R_0 is much greater than the current flowing out of the junction of the resistors. For small values of m, the error produced by this assumption may usually be ignored.

The positive feedback circuit described in this section provides a compromise between the low sensitivities (and high element-value spreads) of the multiple-loop (negative) feedback circuit and the low element-value spread (and high sensitivity), positive-feedback Sallen-Key circuits described in Secs. 4.3 and 4.4. Since it allows the designer to specify the amount of element-value spread as part of the design process, this technique produces a versatile practical circuit that is well adapted to the realization of filters with Q's in the general range of 5 to 20.

4.6 HIGH-Q FILTERS

The filters described in the preceding sections of this chapter are, in general, not suited for the realization of high-Q network functions. In this section, however, we describe two active *RC* filters which are well suited to realizing functions with Q's whose values may go into the hundreds. The first such circuit is called a *state-variable filter*.† To see how such a filter works, we modify the second-order bandpass function given in (4.46) by multiplying numerator and denominator by the quotient $X(s)/s^2$, where $X(s)$ is an arbitrary function. Thus, assuming an inverting function, we obtain

$$\frac{V_2(s)}{V_1(s)} = \frac{-|H|s}{s^2 + a_1 s + a_0} = \frac{-|H|X(s)/s}{X(s) + a_1 X(s)/s + a_0 X(s)/s^2} \tag{4.75}$$

This relation may now be written as the two separate equations

$$X(s) = V_1(s) - \frac{a_1 X(s)}{s} - \frac{a_0 X(s)}{s^2} \qquad V_2(s) = -|H|\frac{X(s)}{s} \tag{4.76}$$

Since multiplying by $1/s$ in the (transformed) complex frequency domain corresponds to integration in the time domain, the equations of (4.76) are readily implemented by the analog-computer flowchart shown in Fig. 4.29. In this figure, the quantities $x(t) = L^{-1}[X(s)]$, $\int x(t)dt$, and $\int\int x(t)dt\,dt$ are called the state variables. The operations of integration and summation called for in the flowchart are readily implemented by the operational amplifier circuits shown in Figs. 1.12 and 1.28, respectively. Using these, we obtain in Fig. 4.30 a circuit realization of the flowchart. For the case where the open-loop gain of the operational amplifiers

---

† W. J. Kerwin, L. P. Huelsman, and R. W. Newcomb, "State-Variable Synthesis for Insensitive Integrated Circuit Transfer Functions." *IEEE Journal of Solid-State Circuits,* vol. SC-2, no. 3, pp. 87–92, September 1967.

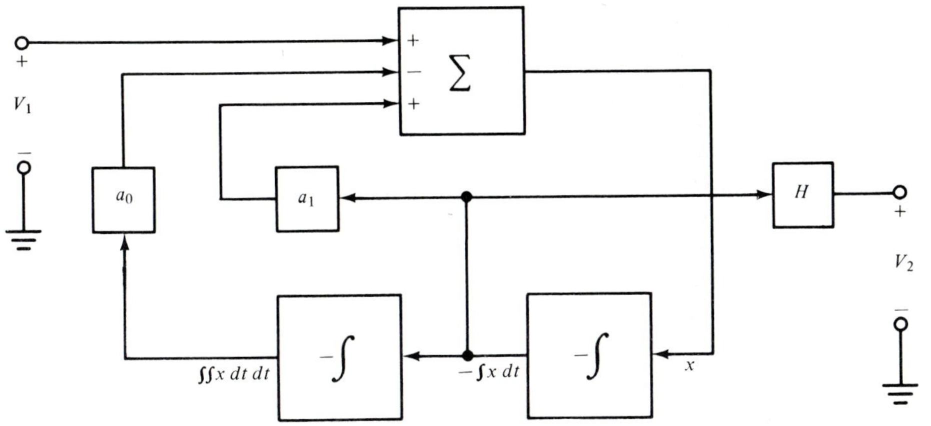

FIGURE 4.29
State-variable flowchart.

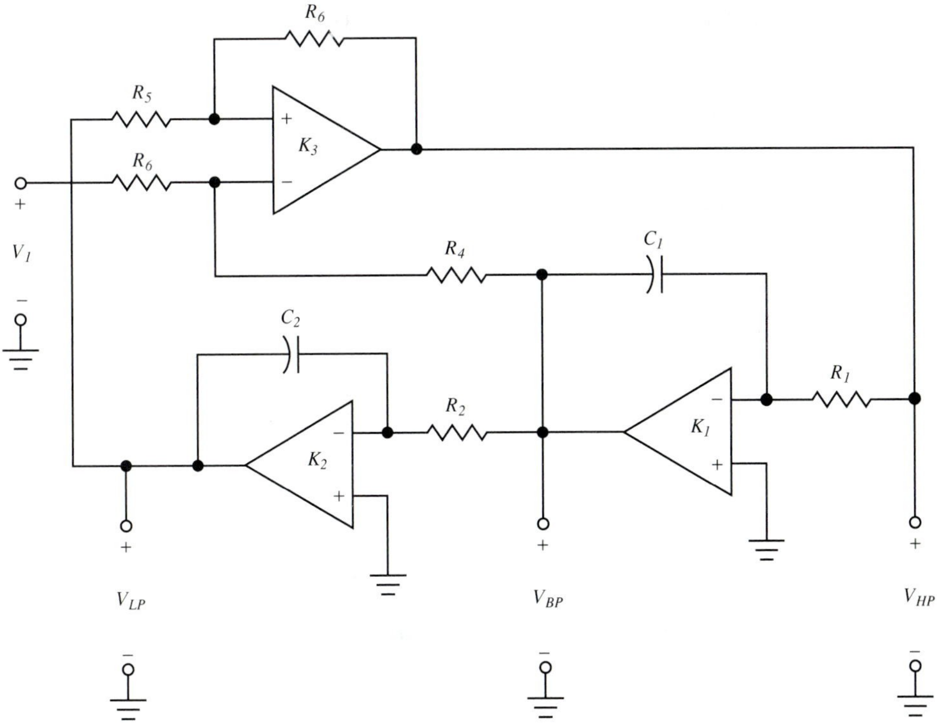

FIGURE 4.30
State-variable filter.

is considered infinite, the transfer function is

$$\frac{V_{BP}(s)}{V_1(s)} = \frac{N(s)}{D(s)}$$

$$= \frac{-s(1/R_1C_1)[(1 + R_6/R_5)/(1 + R_3/R_4)]}{s^2 + s(1/R_1C_1)[(1 + R_6/R_5)/(1 + R_4/R_3)] + (R_6/R_5)(1/R_1R_2C_1C_2)} \tag{4.77}$$

This circuit may also be used to produce a low-pass or a high-pass function by taking the output from the points indicated in Fig. 4.30. For these outputs,

$$\frac{V_{LP}(s)}{V_1(s)} = \frac{(1/R_1R_2C_1C_2)[(1 + R_6/R_5)/(1 + R_3/R_4)]}{D(s)}$$
$$\frac{V_{HP}(s)}{V_1(s)} = \frac{s^2[(1 + R_6/R_5)/(1 + R_3/R_4)]}{D(s)} \tag{4.78}$$

where $D(s)$ is defined in (4.77). Other biquadratic functions are easily produced by appropriate summing of the three indicated outputs.†

For specified values of Q and ω_0 as given in (4.46), the design process is:

1. Choose a convenient value for $C = C_1 = C_2$ to effectively establish an impedance denormalization.
2. Calculate

$$R_1 = R_2 = \frac{1}{\omega_0 C} \tag{4.79}$$

3. Select a convenient value for $R_3 = R_5 = R_6$ and calculate

$$R_4 = (2Q - 1)R_3 \tag{4.80}$$

The resulting values of H_0 are

$$\text{Low-pass and high-pass:} \qquad H_0 = \frac{2Q - 1}{Q}$$
$$\text{Bandpass:} \qquad H_0 = 1 - 2Q \tag{4.81}$$

Note that the bandpass output is inverting while the low-pass and high-pass ones are noninverting.

Example 4.13. A bandpass filter is to be designed with a center frequency of 1 kHz and $Q = 20$. From the procedure given above we choose $C_1 = C_2 = 0.01\mu$F

---

† Biquadratic functions may also be obtained by a slight modification of the original circuit. See P. E. Fleischer and J. Tow, "Design Formulas for Biquad Active Filters Using Three Operational Amplifiers," *Proc. of the IEEE,* vol. 61, no. 5, pp. 662–663, May 1973.

and $R_3 = R_5 = R_6 = 10\ \text{k}\Omega$. We find $R_1 = R_2 = 15.9\ \text{k}\Omega$ and $R_4 = 390\ \text{k}\Omega$. The gain at resonance H_0 is –39. If this gain is too large, the input resistor R_3 may be replaced by a voltage divider following the procedure given in connection with Fig. 4.28.

One of the reasons for the popularity of the state-variable filter is the low sensitivities that characterize its performance. Using the definitions of Sec. 4.3, we find

$$
\begin{aligned}
S^Q_{R_1,C_1} &= -S^Q_{R_2,C_2} = -S^{\omega_n}_{R_1,R_2,R_5,C_1,C_2} = +S^{\omega_n}_{R_6} = \tfrac{1}{2} \\
S^Q_{R_3} &= \frac{-1}{1+R_3/R_4} = -S^Q_{R_4} \\
S^Q_{R_5} &= -S^Q_{R_6} = -\frac{Q}{2}\left(\frac{R_5-R_6}{1+R_4/R_3}\right)\sqrt{\frac{R_2C_2}{R_5R_6R_1C_1}}
\end{aligned}
\tag{4.82}
$$

An examination of these relations shows that the Q sensitivity for R_5 and R_6 can be set to zero by choosing $R_5 = R_6$. The remaining sensitivities are all less than unity in magnitude, and most have a magnitude of $\frac{1}{2}$. The Q sensitivities of the realization to the gains of the operational amplifiers are even smaller, being of the order of Q/K_0, where K_0 is the open-loop operational amplifier gain. As a result of its low sensitivities, the state-variable filter has been successfully used to realize functions with Q's in the hundreds. Many commercial products based on the state-variable filter are available. An example is the UAF41 Universal Active Filter.†

A second multiamplifier filter which may be used to realize high-Q functions is the resonator active filter shown in Fig 4.31. It provides a low-pass and a band-pass output and, like the state-variable filter, has excellent sensitivity properties.

† Burr-Brown Corporation, International Airport Industrial Park, PO Box 11400, Tucson, Ariz. 85734.

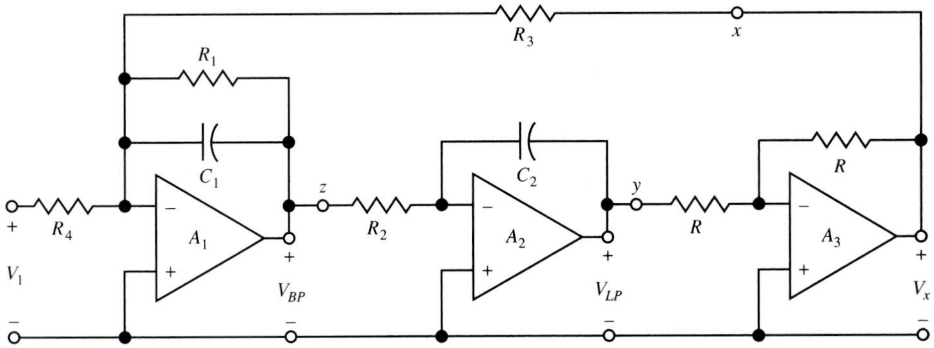

FIGURE 4.31
Resonator filter.

The transfer functions are

$$\frac{V_{BP}(s)}{V_1(s)} = \frac{-s/R_4C_1}{s^2 + s/R_1C_1 + 1/R_2R_3C_1C_2}$$
$$\frac{V_{LP}(s)}{V_1(s)} = \frac{1/R_2R_4C_1C_2}{s^2 + s/R_1C_1 + 1/R_2R_3C_1C_2} \tag{4.83}$$

For specified values of Q and ω_0 as given in (4.46), the design process is:

1. Choose a convenient value for $C = C_1 = C_2$ to effectively establish an impedance denormalization.
2. Calculate $R = R_2 = R_3$ from

$$R = \frac{1}{\omega_0 C} \tag{4.84}$$

3. Calculate

$$R_1 = QR \tag{4.85}$$

4. For a desired value of H_0 choose

$$\text{Low-pass:}\quad R_4 = \frac{R}{H_0} \qquad H_0 > 0$$
$$\text{Bandpass:}\quad R_4 = \frac{R_1}{|H_0|} \qquad H_0 < 0 \tag{4.86}$$

Example 4.14. A bandpass filter is to be designed with a 1-kHz resonant frequency and a Q of 20. From the procedure given above we choose $C = C_1 = C_2 = 0.01\ \mu\text{F}$. We find $R = 15.915\ \text{k}\Omega$ and $R_1 = 318.31\ \text{k}\Omega$. For a gain at resonance of $H_0 = -1$, we require $R_4 = 318.31\ \text{k}\Omega$.

4.7 CASCADE REALIZATION OF HIGH-ORDER FILTERS—TUNING

In the preceding sections of this chapter we considered the realization of second-order filters using various finite- and infinite-gain-amplifier circuits. Many filtering applications, however, require filters of higher than second order, either to provide greater stopband attenuation and sharper cutoff at the edge of the passband in the low-pass or high-pass case or to provide a broad passband with some special transmission characteristic in the bandpass case. In this and the following sections we discuss some means of obtaining these high-order filters.

One of the simplest approaches to the realization of high-order filters is to factor the specified network function into quadratic factors and realize each by a separate circuit, using the second-order realizations given in Secs. 4.3 to 4.6. Since these circuits all have an operational amplifier as an output element, their output impedance is low (ideally, zero), and thus a simple cascade of such second-order realizations may be made without interaction occurring between the individual stages. As a result, the overall voltage transfer function is simply the product of the individual transfer functions. An advantage of this approach is that each of the filters may be tuned separately, a point of considerable practical importance when high-order network functions are to be realized. The success of the cascade method, of course, depends on the use of operational amplifiers which have as low an output impedance as possible. The use of high impedance denormalization levels for the passive elements, which minimizes loading on the interior operational amplifier output stages, is also significant in reducing interaction. Some details of cascade realizations for various types of high-order filters follow.

In the nth-order case, the general low-pass voltage transfer function is

$$\frac{V_2(s)}{V_1(s)} = \frac{H_0 a_0}{s^n + a_{n-1}s^{n-1} + \cdots + a_1 s + a_0} \tag{4.87}$$

Such a function has a magnitude characteristic which decreases at the rate of $20n$ dB/decade outside the passband. If n is even, (4.87) may be written in the form

$$\frac{V_2(s)}{V_1(s)} = \prod_{i=1}^{n/2} \frac{H_i}{s^2 + a_{i1}s + a_{i0}} \tag{4.88}$$

and the individual quadratic functions $H_i/(s^2 + a_{i1}s + a_{i0})$ may be synthesized by any of the second-order low-pass realizations given in the preceding sections. The realizations may be individually impedance-normalized to provide convenient element values. The constants H_i are, of course, arbitrary, unless we want to realize H_0 exactly, in which case the product of the H_i must equal $H_0 a_0$. If n is odd, (4.87) may be written

$$\frac{V_2(s)}{V_1(s)} = \frac{1}{s - \sigma_0} \prod_{i=1}^{(n-1)/2} \frac{H_i}{s^2 + a_{i1}s + a_{i0}} \tag{4.89}$$

The quadratic factors are realized as before, and the first-order factor $1/(s - \sigma_0)$ may be realized (within a multiplicative constant) by the circuit shown in Fig. 4.32, which has the voltage transfer function

$$\frac{V_2(s)}{V_1(s)} = \frac{1/RC}{s + 1/RC} \tag{4.90}$$

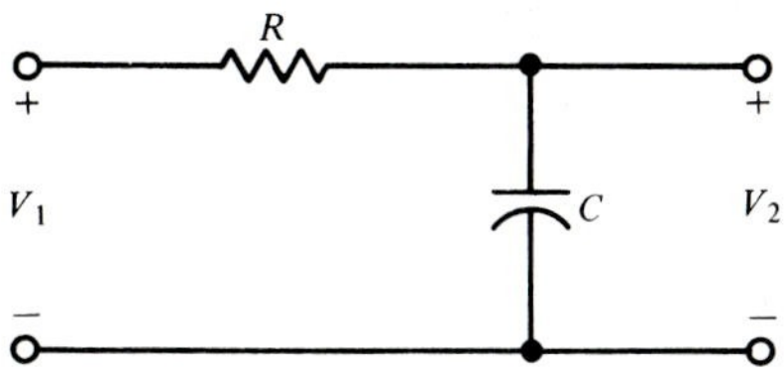

FIGURE 4.32
Realization of a low-pass first-order transfer function.

The first-order network must, of course, be isolated so as not to interact with the other second-order realizations.

The determination of the coefficients a_i of (4.87), and thus of the coefficients a_{i1} and a_{i0} of (4.88), is done according to the specific magnitude or phase characteristic desired. Some frequently used choices are given in Table 4.4, which is an extension of Table 4.1. As an example of the cascade technique, consider the realization of a fifth-order frequency-normalized low-pass filter having a maximally flat magnitude (Butterworth) characteristic which is 3 dB down at the passband edge of 1 rad/s. From Table 4.4 we see that the network function has the form

$$\frac{V_2(s)}{V_1(s)} = \frac{H}{(s+1)(s^2+0.618s+1)(s^2+1.618s+1)} \tag{4.91}$$

The resulting realization using design 1 of Sec. 4.3 is shown in Fig. 4.33.

High-Pass Filters

In the nth-order case, the general high-pass voltage transfer function has the form

$$\frac{V_2(s)}{V_1(s)} = \frac{H_0 s^n}{s^n + a_{n-1}s^{n-1} + \cdots + a_1 s + a_0} \tag{4.92}$$

TABLE 4.4*a*

Denominator polynomials for normalized low-pass network functions of the form $s^n + a_{n-1}s^{n-1} + \cdots + a_1 s + a_0$

Case (Table 4.1)	Order	a_0	a_1	a_2	a_3	a_4	a_5
1	3	1.00000	2.00000	2.00000			
	4	1.00000	2.61313	3.41421	2.61313		
	5	1.00000	3.23607	5.23607	5.23607	3.23607	
	6	1.00000	3.86370	7.46410	9.14162	7.46410	3.86370
2*a*	3	0.71569	1.53489	1.25291			
	4	0.37905	1.02545	1.71687	1.19739		
	5	0.17892	0.75252	1.30957	1.93737	1.17249	
	6	0.09476	0.43237	1.17186	1.58976	2.17184	1.15918
2*b*	3	0.49131	1.23841	0.98834			
	4	0.27563	0.74262	1.45392	0.95281		
	5	0.12283	0.58053	0.97440	1.68882	0.93682	
	6	0.06891	0.30708	0.93934	1.20214	1.93083	0.92825
3	3	15	15	6			
	4	105	105	45	10		
	5	945	945	420	105	15	
	6	10,395	10,395	4725	1260	210	21

TABLE 4.4*b*

Roots of denominator polynomials for normalized low-pass network functions

Case (Table 4.1)	Order	Roots		
1	3	$-0.50000 \pm j0.86603$	-1.00000	
	4	$-0.38268 \pm j0.92388$	$-0.92388 \pm j0.38268$	
	5	$-0.30902 \pm j0.95106$	$-0.80902 \pm j0.58778$	-1.00000
	6	$-0.25882 \pm j0.96593$	$-0.70711 \pm j0.70711$	$-0.96593 \pm j0.25882$
2*a*	3	$-0.31323 \pm j1.02193$	-0.62646	
	4	$-0.17535 \pm j1.01626$	$-0.42334 \pm j0.42095$	
	5	$-0.11196 \pm j1.01156$	$-0.29312 \pm j0.62518$	-0.36232
	6	$-0.07765 \pm j1.00846$	$-0.21214 \pm j0.73824$	$-0.28979 \pm j0.27022$
2*b*	3	$-0.24708 \pm j0.96600$	-0.49417	
	4	$-0.13954 \pm j0.98338$	$-0.33687 \pm j0.40733$	
	5	$-0.08946 \pm j0.99011$	$-0.23420 \pm j0.61192$	-0.28951
	6	$-0.06218 \pm j0.99341$	$-0.16988 \pm j0.72723$	$-0.23206 \pm j0.26618$
3	3	$-1.83891 \pm j1.75438$	-2.32218	
	4	$-2.89621 \pm j0.86723$	$-2.10379 \pm j2.65742$	
	5	$-3.35196 \pm j1.74266$	$-2.32467 \pm j3.57102$	-3.63674
	6	$-4.24836 \pm j0.86751$	$-3.73571 \pm j2.62627$	$-2.51593 \pm j4.49267$

TABLE 4.4*c*

Factors of denominator polynomials for normalized low-pass network functions of the form $(s + a)(s^2 + a_{11}s + a_{10})(s^2 + a_{21}s + a_{20})(s^2 + a_{31}s + a_{30})$

Case (Table 4.1)	Order	a	a_{11}	a_{10}	a_{21}	a_{20}	a_{31}	a_{30}
1	3	1.00000	1.00000	1.00000				
	4		1.84777	1.00000	0.76536	1.00000		
	5	1.00000	0.61803	1.00000	1.61804	1.00000		
	6		0.51764	1.00000	1.93185	1.00000	1.41421	1.00000
2*a*	3	0.62646	0.62646	1.14244				
	4		0.84667	0.35641	0.35072	1.06351		
	5	0.36231	0.58625	0.47677	0.22393	1.03578		
	6		0.57959	0.15699	0.42429	0.59004	0.15530	1.02300
2*b*	3	0.49417	0.49417	0.99421				
	4		0.67374	0.27940	0.27907	0.98650		
	5	0.28951	0.46839	0.42929	0.17893	0.98833		
	6		0.46413	0.12472	0.33973	0.55769	0.12438	0.99075
3	3	2.32218	3.67781	6.45943				
	4		4.20758	11.48780	5.79242	9.14014		
	5	3.64674	4.64935	18.15631	6.70391	14.27248		
	6		5.03186	26.51402	7.47142	20.85282	8.49672	18.80113

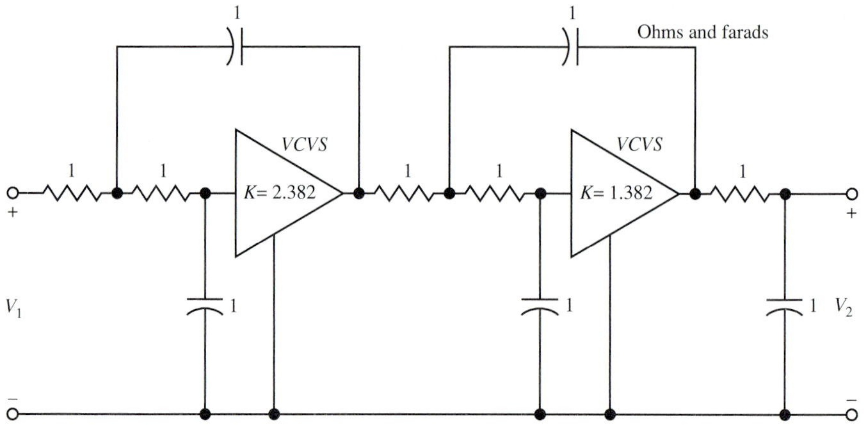

FIGURE 4.33
Fifth-order low-pass cascade filter realization.

If n is even, this may be put in the form

$$\frac{V_2(s)}{V_1(s)} = \prod_{i=1}^{n/2} \frac{H_i s^2}{s^2 + a_{i1}s + a_{i0}} \tag{4.93}$$

The separate quadratic factors are easily realized by using any of the second-order high-pass circuits presented in the preceding sections. For n odd, (4.92) may be factored as

$$\frac{V_2(s)}{V_1(s)} = \frac{s}{s - \sigma_0} \prod_{i=1}^{(n-1)/2} \frac{H_i s^2}{s^2 + a_{i1}s + a_{i0}} \tag{4.94}$$

The quadratic factors are realized as before, and the first-order factor $s/(s - \sigma_0)$ is realized by the circuit shown in Fig. 4.34, which has the transfer function

$$\frac{V_2(s)}{V_1(s)} = \frac{s}{s + 1/RC} \tag{4.95}$$

Some frequently used cases defining the values of the coefficients of the high-order high-pass functions are given in Table 4.5, which is an extension of Table 4.3.

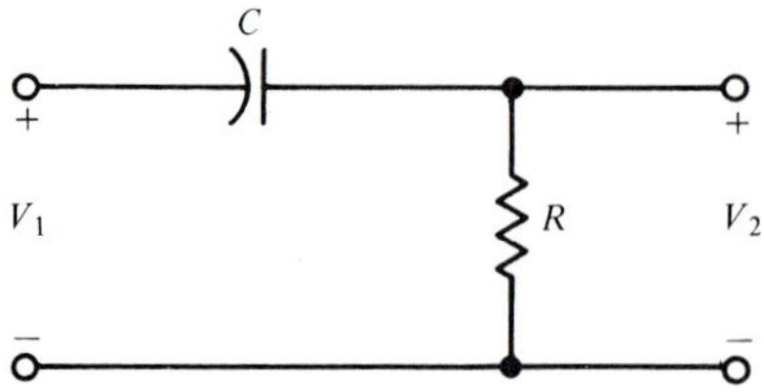

FIGURE 4.34
Realization of the first-order factor of (4.86).

TABLE 4.5*a*
Denominator polynomials for normalized high-pass network functions of the form $s^n + a_{n-1}s^{n-1} + \cdots + a_1 s + a_0$

Case (Table 4.3)	Order	a_0	a_1	a_2	a_3	a_4	a_5
1	3–6	Same as low-pass (Table 4.4a)					
2*a*	3	1.39725	1.75063	2.14463			
	4	2.63717	3.15892	4.52940	2.70531		
	5	5.58899	6.55303	10.82793	7.31919	4.20580	
	6	10.55268	12.23240	22.91877	16.77626	12.36628	4.56263
2*b*	3	2.03537	2.01164	2.52063			
	4	3.62805	3.45684	5.27494	2.69426		
	5	8.14153	7.62716	13.74958	7.93310	4.72645	
	6	14.51233	13.47109	28.02078	17.44587	13.63210	4.72645

Bandpass Filters

In the *n*th-order case, the general bandpass voltage transfer function has the form

$$\frac{V_2(s)}{V_1(s)} = \frac{H s^{n/2}}{s^n + a_{n-1}s^{n-1} + \cdots + a_1 s + a_0} \tag{4.96}$$

where n can only be even. The factored form of this function is

$$\frac{V_2(s)}{V_1(s)} = \prod_{i=1}^{n/2} \frac{H_i s}{s^2 + a_{i1}s + a_{i0}} \tag{4.97}$$

The separate quadratic factors are easily realized by any of the second-order bandpass circuits presented in the preceding sections. An alternative decomposition of

TABLE 4.5*b*
Roots of denominator polynomials for normalized high-pass network functions

Case (Table 4.3)	Order	Roots		
1	3–6	Same as low-pass (Table 4.4*b*)		
2*a*	3	$-0.27418 \pm j0.89451$	-1.59628	
	4	$-1.18777 \pm j1.18108$	$-0.16489 \pm j0.95556$	
	5	$-0.61484 \pm j1.31128$	$-0.10809 \pm j0.97660$	-2.76005
	6	$-1.84584 \pm j1.72114$	$-0.35955 \pm j1.25121$	$-0.07590 \pm j0.98578$
2*b*	3	$-0.24852 \pm j0.97163$	-2.02358	
	4	$-1.20569 \pm j1.45785$	$-0.14144 \pm j0.99684$	
	5	$-0.54554 \pm j1.42542$	$-0.09052 \pm j1.00181$	-3.45427
	6	$-1.86087 \pm j2.13442$	$-0.30458 \pm j1.30397$	$-0.06277 \pm j1.00269$

TABLE 4.5c
Factors of denominator polynomials for normalized high-pass network functions of the form $(s + a)(s^2 + a_{11}s + a_{10})(s^2 + a_{21}s + a_{20})(s^2 + a_{31}s + a_{30})$

Case (Table 4.3)	Order	a	a_{11}	a_{10}	a_{21}	a_{20}	a_{31}	a_{30}
1	3–6	Same as low-pass (Table 4.4c)						
2a	3	1.59628	0.54334	0.87531				
	4		2.37553	2.80573	0.32977	0.94028		
	5	2.76004	1.22968	2.09749	0.21618	0.96544		
	6		3.69168	6.36953	0.71910	1.69491	0.15180	0.97752
2b	3	2.02358	0.49705	1.00582				
	4		2.41138	3.57912	0.28288	1.01370		
	5	3.45427	1.09108	2.32938	0.18104	1.01181		
	6		3.72173	8.01881	0.60917	1.79300	0.12555	1.00933

a bandpass network function into a product of low-pass and high-pass functions is also possible; for example, for $n = 4$, (4.97) could be written

$$\frac{V_2(s)}{V_1(s)} = \frac{H_1}{s^2 + a_{11}s + a_{10}} \frac{H_2 s^2}{s^2 + a_{21}s + a_{20}} \tag{4.98}$$

which could be realized as a cascade connection of second-order low-pass and high-pass filters.

The determination of the coefficients of high-order bandpass functions is most easily done by using a transformation of the complex frequency variable. If we let p be the new complex frequency variable, then the transformation is defined as

$$s = \frac{p^2 + 1}{p} \tag{4.99}$$

where s is the original complex frequency variable. Such a transformation, when applied to a low-pass network function, produces a bandpass network function with a center frequency of unity. The bandwidth of the low-pass function becomes the bandwidth of the bandpass function, and the band-edge frequencies of the bandpass function have 1 rad/s as their geometric mean. Thus, applying (4.99) (with $s = j\omega$) to the low-pass characteristic shown in Fig. 4.35*a* produces the bandpass characteristic shown in Fig. 4.35*b*, in which the indicated frequencies satisfy the relations

$$\omega_2 - \omega_1 = \omega_0 \qquad \omega_1\omega_2 = 1 \tag{4.100}$$

Some transformation of the second- and third-order low-pass characteristics defined in Tables 4.1 and 4.5 are given in Table 4.6. If the transformation defined by (4.99) is applied to a high-pass function, a band elimination characteristic (notch filter) results. Relations similar to those given in (4.100) also apply to this case.

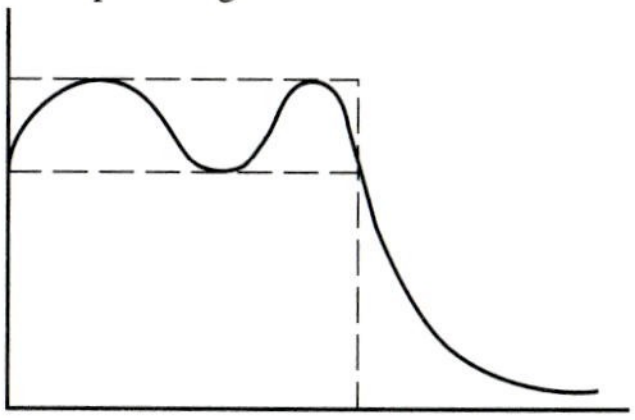

(a)

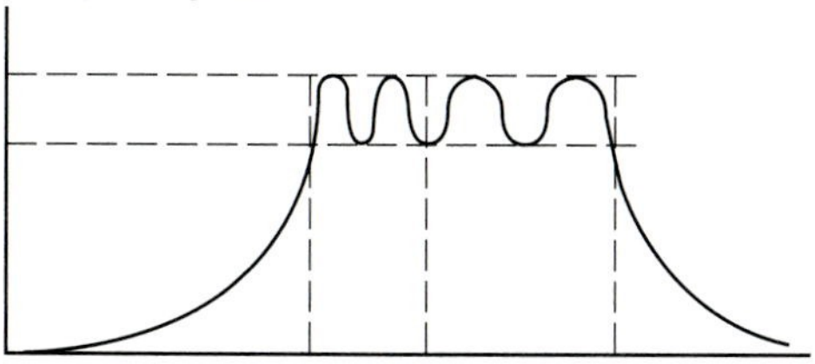

(b)

FIGURE 4.35
Low-pass-to-bandpass transformation.

TABLE 4.6*a*
Denominator polynomials of normalized bandpass network functions of the form $s_4 + a_1 s^3 + a_2 s^2 + a_1 s + 1 = (s^2 + a_{11} s + a_{10})(s^2 + a_{21} s + a_{20})$

Low-pass (Table 4.1)	BW	a_2	a_1	a_{11}	a_{10}	a_{21}	a_{20}
1	0.5	2.25000	0.70711	0.41602	1.42922	0.29108	0.69968
	0.2	2.04000	0.28285	0.15142	1.15218	0.13142	0.86792
	0.1	2.01000	0.14121	0.07321	1.07330	0.06821	0.93170
	0.05	2.00250	0.07071	0.03598	1.03599	0.03473	0.96528
2*a*	0.5	2.37905	0.71281	0.44442	1.65585	0.26839	0.60392
	0.2	2.06065	0.28512	0.15684	1.22260	0.12828	0.81793
	0.1	2.01516	0.14256	0.07485	1.10564	0.06770	0.90445
	0.05	2.00379	0.07128	0.03654	1.05148	0.03475	0.95103
2*b*	0.5	2.27563	0.54887	0.33488	1.56498	0.21398	0.63898
	0.2	2.04410	0.21954	0.11957	1.19609	0.09997	0.83606
	0.1	2.01103	0.10977	0.05734	1.09364	0.05243	0.91437
	0.05	2.00276	0.05489	0.02806	1.04577	0.02683	0.95623

TABLE 4.6*b*
Denominator polynomials of normalized bandpass network functions of the form $s^6 + a_1 s^5 + a_2 s^4 + a_3 s^3 + a_2 s^2 + a_1 s + 1 = (s^2 + a_{11} s + a_{10})(s^2 + a_{21} s + a_{20})(s^2 + a_{31} s + a_{30})$

Low-pass (Table 4.1)	BW	a_3	a_2	a_1	a_{11}	a_{10}	a_{21}	a_{20}	a_{31}	a_{30}
1	0.5	2.12500	3.50000	1.00000	0.30328	1.54171	0.50000	1.00000	0.19672	0.64863
	0.2	0.80800	3.08000	0.40000	0.10864	1.18911	0.20000	1.00000	0.09136	0.84097
	0.1	0.40100	3.02000	0.20000	0.05216	1.09046	0.10000	1.00000	0.04783	0.91704
	0.05	0.20013	3.00500	0.10000	0.02554	1.04425	0.05000	1.00000	0.02446	0.95762
2*a*	0.5	1.34237	3.38372	0.62645	0.19548	1.66029	0.31323	1.00000	0.11774	0.60230
	0.2	0.50689	3.06139	0.25058	0.06902	1.22646	0.12529	1.00000	0.05627	0.81536
	0.1	0.25129	3.01534	0.12529	0.03292	1.10756	0.06264	1.00000	0.02972	0.90288
	0.05	0.12538	3.00383	0.06264	0.01606	1.05242	0.03132	1.00000	0.01526	0.95019
2*b*	0.5	1.04975	3.30960	0.49417	0.15259	1.61489	0.24708	1.00000	0.09449	0.61923
	0.2	0.39927	3.04953	0.19767	0.05417	1.21283	0.09883	1.00000	0.04466	0.82451
	0.1	0.19816	3.01238	0.09883	0.02590	1.10138	0.04942	1.00000	0.02352	0.90795
	0.05	0.09889	3.00309	0.04942	0.01265	1.04948	0.02471	1.00000	0.01206	0.95285

Elliptic Filters

Filters with transmission zeros on the $j\omega$ axis, originally introduced in Sec. 4.4, may also be required to be of greater than second order. In the general case, for n even, the transfer function of such a filter has the factored form

$$\frac{V_2(s)}{V_1(s)} = \prod_{i=1}^{n/2} \frac{H_i(s^2 + b_i)}{s^2 + a_{i1}s + a_{i0}} \tag{4.101}$$

For n odd, the function has the form

$$\frac{V_2(s)}{V_1(s)} = \frac{\prod_{i=1}^{(n-1)/2} H_i(s^2 + b_i)}{(s + a) \prod_{i=1}^{(n-1)/2} (s^2 + a_{i1}s + a_{i0})} \tag{4.102}$$

Specifically, the third-order elliptic filter will have the form

$$\frac{V_2(s)}{V_1(s)} = \frac{H(s^2 + b_i)}{(s + a)(s^2 + a_{11}s + a_{10})} \tag{4.103}$$

while the fourth-order filter has the form

$$\frac{V_2(s)}{V_1(s)} = \frac{H(s^2 + b_1)(s^2 + b_2)}{(s^2 + a_{11}s + a_{10})(s^2 + a_{21}s + a_{20})} \tag{4.104}$$

Values of these coefficients for various values of the constants K_p, K_s, and ω_s defined in Fig. 4.18, for the normalized case where $\omega_c = 1$ rad/s, are given in Table 4.7.

Example 4.15. A normalized elliptic network function is to be found which has 1.0-dB ripple in the passband from 0 to 1 rad/s, a minimum of 39 dB at all frequencies

TABLE 4.7*a*
Coefficients of third-order low-pass elliptic network functions having the form of (4.103)†

K_p, dB	ω_s	K_s, dB	b_1	a	a_{11}	a_{10}
0.1	1.1	3.374	1.3703	2.2408	0.1708	1.2658
0.1	1.5	14.848	2.8060	1.2982	0.5793	1.5539
0.1	2.0	24.010	5.1532	1.1168	0.7637	1.6291
1.0	1.1	11.480	1.3703	0.8162	0.1953	1.0424
1.0	1.5	25.176	2.8060	0.5910	0.3754	1.0237
1.0	2.0	34.454	5.1532	0.5400	0.4341	1.0106

† This table was computed using a program from David Jose Miguel Baezlopez, "Determination of Elliptic Network Functions," M.S. thesis, University of Arizona, Tucson, 1977.

TABLE 4.7*b*
Coefficients of fourth-order low-pass elliptic network functions having the form of (4.104)†

K_p, dB	ω_s	K_s, dB	b_1	b_2	a_{11}	a_{10}	a_{21}	a_{20}
0.1	1.1	10.721	1.2909	4.3499	1.4076	1.4489	0.1335	1.1411
0.1	1.5	29.064	2.5356	12.0993	1.3975	0.8689	0.3473	1.2581
0.1	2.0	41.447	4.5933	24.2272	1.3409	0.7364	0.4325	1.2941
1.0	1.1	20.832	1.2909	4.3499	0.7985	0.5670	0.1090	1.0097
1.0	1.5	39.518	2.5356	12.0993	0.7280	0.3643	0.2088	0.9988
1.0	2.0	51.906	4.5933	24.2272	0.7025	0.3192	0.2430	0.9932

† This table was computed using a program from David Jose Miguel Baezlopez, "Determination of Elliptic Network Functions," M.S. thesis, University of Arizona, Tucson, 1977.

greater than 1.5 rad/s, and a maximum passband gain of 0 dB. From Table 4.7 we find

$$N(s) = \frac{H(s^2 + 2.5356)(s^2 + 12.0993)}{(s^2 + 0.7280s + 0.3643)(s^2 + 0.2088s + 0.9988)} \tag{4.105}$$

At $s = 0$, $N(s)$ must have a value of 0.89125 (1-dB attenuation). Evaluating (4.105) at $s = 0$, we find $0.89125 = 84.3147H$ and $H = 0.010571$. A realization of this function (within a multiplicative constant) can be made by using a cascade of two of the circuits shown in Fig. 4.21. The details are left to the reader as an exercise.

Tuning

In practice, to complete the actual construction of an active *RC* filter, a final tuning or adjustment of the element values and the gain may be required to compensate for element tolerance and parasitic effects. This becomes especially important in high-order cascaded realizations and in the high-Q case. First let us consider the tuning of a low-pass second-order filter having a network function of the form given in (4.21). Plots of the magnitude as a function of normalized frequency ω/ω_0 readily show that the magnitude at low frequencies, say $\omega \leq 0.1\omega_0$, is constant, independent of Q, and equal to H_0. Similarly, at the high frequency of $10\omega_0$, the magnitude characteristic is independent of Q and is –40 dB below the values of H_0. Thus, the adjustment of the element values which determine H_0 and ω_0 is most easily done at these values. The tuning to correctly realize Q is most easily done at the ω_Q frequency, defined as

$$\omega_Q = \omega_0\sqrt{1 - \frac{1}{2Q^2}} \tag{4.106}$$

at which the peak magnitude occurs (for $1/Q < \sqrt{2}$). Some iterations between the three adjustments described above may be required, depending on the particular filter configuration being tuned. The same three steps may be used to tune high-pass second-order filters. In this case, the frequencies are the reciprocals of those referred to above.

The other basic second-order filter function to be considered here is the bandpass one having the form given in (4.46). Here the peak response occurs at ω_0, and thus the adjustment of element values to provide this frequency is one of the tuning steps. For the remaining steps, we note that the bandwidth (BW) is the difference between the frequencies at which the magnitude characteristic is down 3 dB from its peak value. These frequencies are

$$f_1 = \frac{f_0}{2Q}\left(-1 + \sqrt{1+4Q^2}\right) \qquad f_2 = \frac{f_0}{2Q}\left(1 + \sqrt{1+4Q^2}\right) \tag{4.107}$$

where $Q = f_0/\mathrm{BW} = f_0/(f_2 - f_1)$. As described in connection with the low-pass-to-bandpass transformation earlier in this section, the relation $f_1 f_2 = f_0^2$ must hold. As in the tuning procedure for the low-pass case, some iteration between the tuning steps described above may be required. In the high-Q case, the relations of (4.107) may be approximated as

$$f_1 = f_0 - \frac{\mathrm{BW}}{2} \qquad f_2 = f_0 + \frac{\mathrm{BW}}{2} \tag{4.108}$$

4.8 GENERALIZED IMPEDANCE CONVERTER FILTERS

Before active *RC* filters became popular, practically all filtering was done using passive *RLC* filters. The realization of such filters is well documented, and extensive sets of tables which cover various applications may be found in the literature. Some examples are given in Table 4.8. Passive *RLC* realizations have certain advantages over active *RC* ones, the most important being their low sensitivity. Their main disadvantage, of course, is that they require the use of inductors, which, as was pointed out at the beginning of this chapter, are a rather undesirable element. In this section, we present a method of designing active *RC* filters in which the low sensitivity of the passive *RLC* realization is retained, but the inductors are eliminated. To understand how the method works, consider the low-pass second-order filter shown in Fig. 4.36*a*. The elements of this are specified in terms of their impedances $Z(s)$. The voltage transfer function for this filter is

$$\frac{V_2(s)}{V_1(s)} = \frac{1}{s^2LC + s(L+C) + 2} \tag{4.109}$$

Now let us make an impedance denormalization of the elements of this filter. Since the transfer function is dimensionless, such a denormalization will not affect it. Specifically we will choose $1/s$ as the denormalization. The individual passive impedances transform as

$$R \rightarrow \frac{R}{s} \qquad sL \rightarrow L \qquad \frac{1}{sC} \rightarrow \frac{1}{s^2C} \tag{4.110}$$

TABLE 4.8
Double-*R* terminated normalized low-pass passive *LC* ladder realizations

Case (Table 4.1)	Order	C_1/L_1'	L_2/C_2'	C_3/L_3'	L_4/C_4'	C_5/L_5'	L_6/C_6'
1	2	1.4142	1.4142				
$R = 1$	3	1.0000	2.0000	1.0000			
$R' = 1$	4	0.7654	1.8478	1.8478	0.7654		
	5	0.6180	1.6180	2.0000	1.6180	0.6180	
	6	0.5176	1.4142	1.9319	1.9319	1.4142	0.5176
2*a*	2	1.5132	0.6538				
$R = 2$	3	2.9431	0.6503	2.1903			
$R' = \frac{1}{2}$	4	1.8158	1.1328	2.4881	0.7732		
	5	3.2228	0.7645	4.1228	0.7116	2.3197	
	6	1.8786	1.1884	2.7589	1.2403	2.5976	0.7976
2*b*	2	2.5721	0.4702				
$R = 3$	3	4.9893	0.4286	3.8075			
$R' = \frac{1}{3}$	4	3.0355	0.7929	3.7589	0.5347		
	5	5.3830	0.4915	6.6673	0.4622	3.9944	
	6	3.1307	0.8287	4.1451	0.8467	3.8812	0.5475
3	2	1.5774	0.4226				
$R = 1$	3	1.2550	0.5528	0.1922			
$R' = 1$	4	1.0598	0.5116	0.3181	0.1104		
	5	0.9303	0.4577	0.3312	0.2090	0.0718	
	6	0.8377	0.4116	0.3158	0.2364	0.1480	0.0505

Unprimed Network

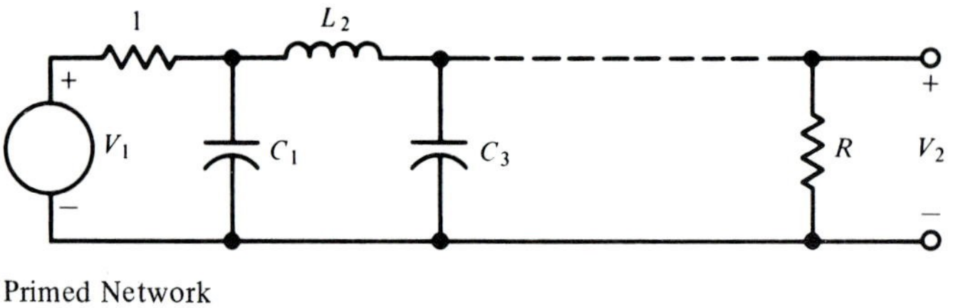

Primed Network

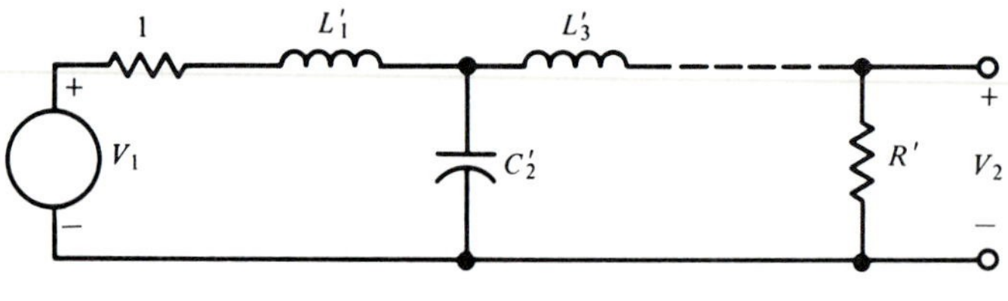

NOTES:

1 Either of the above configurations may be used to realize the indicated network function.

2 Normalized *high-pass* networks for cases 1, 2*a*, and 2*b* of Table 4.3 may be derived from the above network configurations (primed or unprimed) by making the following transformations of elements (values are given in henrys and farads).

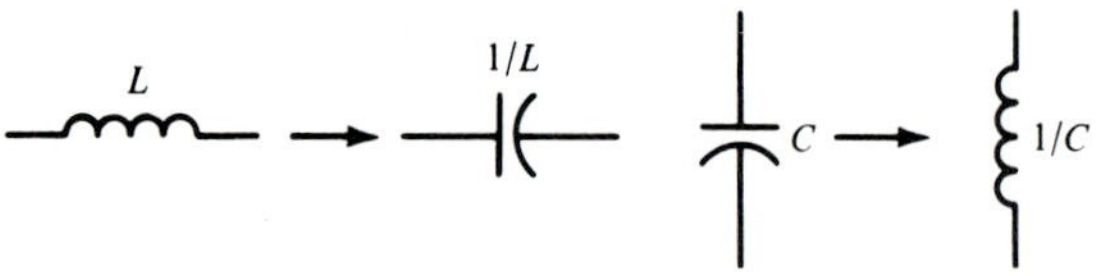

3 Normalized *bandpass* (center frequency of 1 rad/s) networks may be derived from the above network configurations (primed and unprimed) by making the following transformations of elements (values are given in henrys and farads, BW is the desired bandwidth in radians per second).

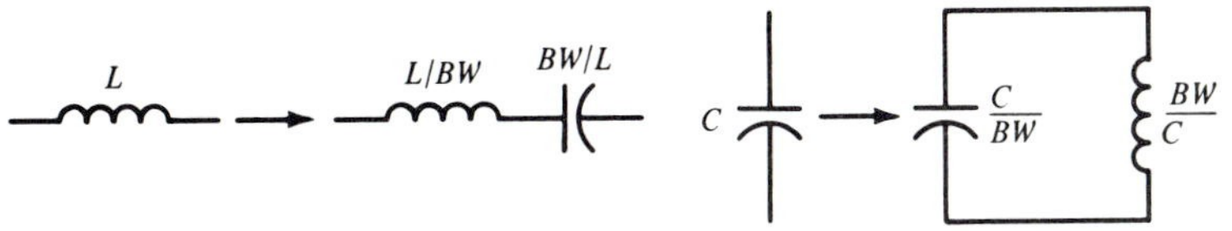

Thus, resistors are transformed to capacitors, inductors to resistors, and capacitors to elements which have impedances proportional to $1/s^2$ and are symbolized by four bars, as shown in the transformed network in Fig. 4.36*b*. This latter element is called a *frequency-dependent negative resistor* (FDNR), since its impedance $Z_{\mathrm{FDNR}}(s) = 1/s^2C$ has the property

$$Z_{\mathrm{FDNR}}(s)\Big|_{s=j\omega} = \frac{1}{s^2C}\Big|_{s=j\omega} = \frac{-1}{\omega^2C} \tag{4.111}$$

which is purely real (and thus resistive), negative; and obviously a function of frequency, hence the name.† The voltage transfer function for the circuit shown in Fig. 4.36*b* is given by (4.109), just as that for the circuit of Fig. 4.36*a* is. As a result of the impedance transformation, however, the floating inductor of Fig. 4.36*a* has been removed, leaving in its place a resistor which is, of course, easy to realize. The FDNR is now the only active element to be considered.

† L.T. Bruton, "Network Transfer Functions Using the Concept of Frequency-Dependent Negative Resistance," *IEEE Transactions on Circuit Theory,* vol. CT-16, no. 3, pp. 406–408, August 1969.

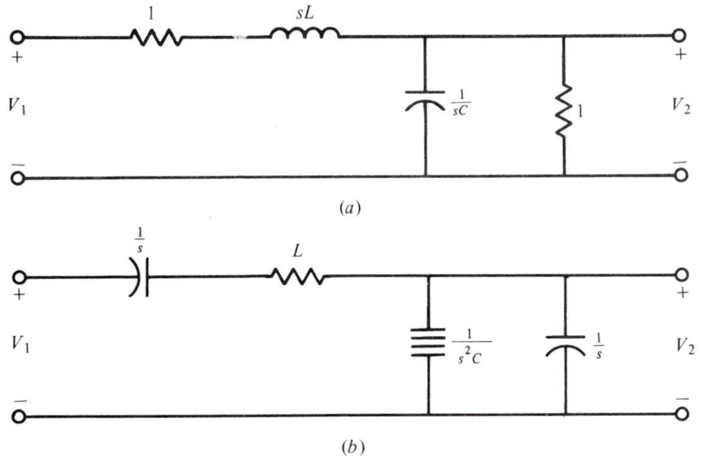

FIGURE 4.36
Impedance transformation by $1/s$ of a low-pass second-order filter.

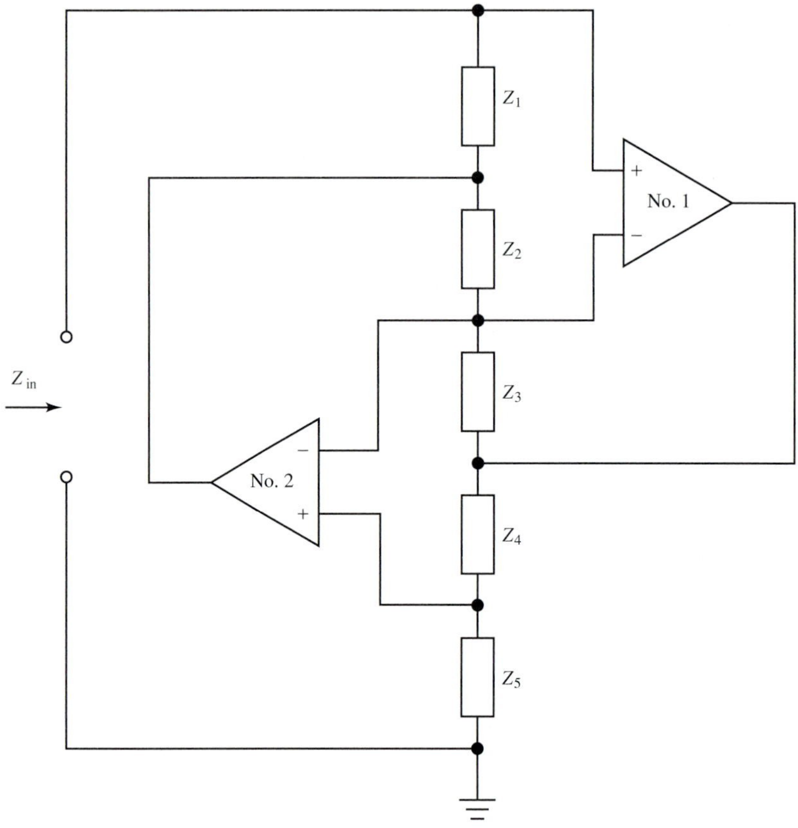

FIGURE 4.37
Generalized impedance converter.

To realize a grounded FDNR, we may use a *generalized impedance converter* (GIC). Such a circuit is shown in Fig. 4.37. For this configuration, the input impedance $Z_{in}(s)$ is

$$Z_{in}(s) = \frac{Z_1(s)Z_3(s)Z_5(s)}{Z_2(s)Z_4(s)} \tag{4.112}$$

If any two of impedances $Z_1(s)$, $Z_3(s)$, and $Z_5(s)$ are chosen as capacitors, and the remaining ones as resistors, then an FDNR is realized.† The circuit of Fig. 4.37 is unconditionally stable (assuming ideal operational amplifiers) for all combinations of $Z_1(s)$ through $Z_5(s)$ which contain not more than two capacitors.‡

† Similarly, if $Z_2(s)$ or $Z_4(s)$ is chosen as a capacitor, and the remaining elements as resistors, an inductor is realized.

‡ L. T. Bruton, "Nonideal Performance of Two-Amplifier Positive-Impedance Converters," *IEEE Transactions on Circuit Theory,* vol. CT-17, no. 4, pp. 541–549, November 1970.

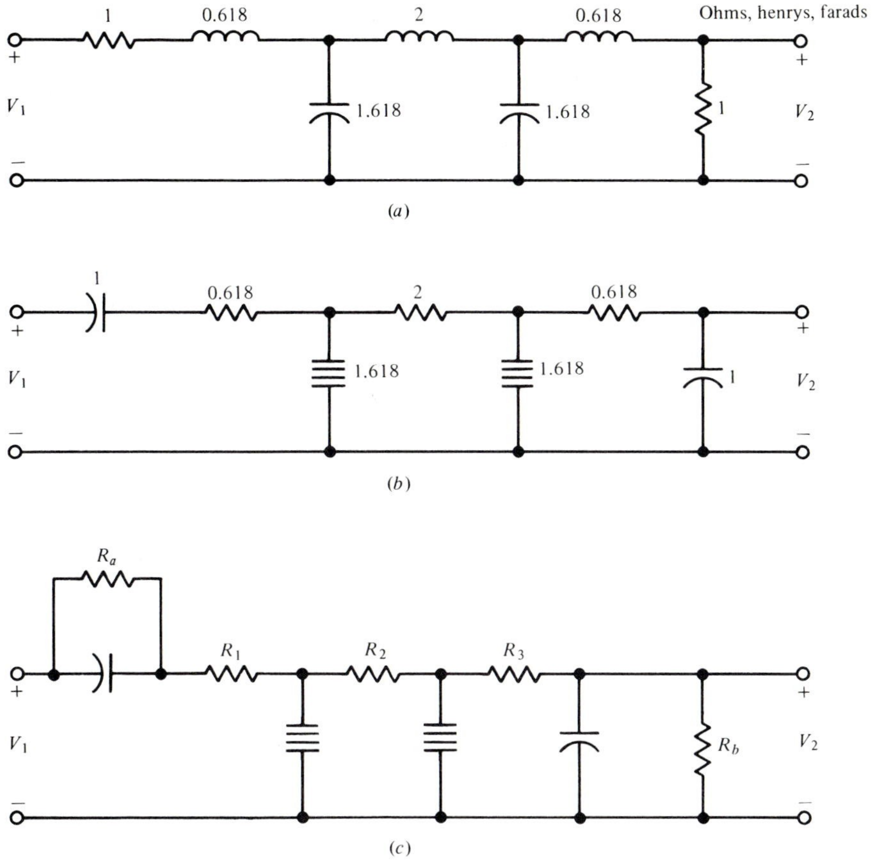

FIGURE 4.38
Steps in the realization of a fifth-order low-pass filter.

As an example of the application of FDNR elements, consider the realization of a fifth-order maximally flat magnitude low-pass filter with equal-resistance terminations. From Table 4.8*a* we obtain the realization shown in Fig. 4.38*a*. After an impedance transformation of $1/s$, the circuit has the form shown in Fig. 4.38*b*, which requires two FDNRs. This network must be slightly modified to provide a resistive path from the FDNRs to ground, since, as shown in Fig. 4.37, with $Z_1(s)$ and $Z_3(s)$ chosen as capacitors, such a path is required to provide input bias current to the noninverting terminal of operational amplifier 2. In theory such a resistive path could be provided by placing a large shunt resistor to ground from the input terminals of either of the FDNR realizations in Fig. 4.38*b*. Such an element, however, corresponds with a shunt inductor to ground in the *RLC* prototype filter, and this would produce zero transmission at zero frequency. An alternative modification which avoids this problem is to use two resistors R_a and R_b as shown in Fig. 4.38*c*. To select the values of these components, we note that the dc gain of the filter shown in Fig. 4.38*a* is 0.5. Thus, for the circuit shown in Fig. 4.38*c*,

$$\lim_{\omega\to 0} \frac{V_2}{V_1}(j\omega) = \frac{R_b}{R_a + R_b + R_1 + R_2 + R_3} = \frac{R_b}{R_a + R_b + 3.236} \tag{4.113}$$

By choosing R_a and R_b large with respect to the other resistors and by requiring that $R_b = R_a + 3.236$, the desired criteria are met. For example, $R_a = 10$ and $R_b = 13.236$ works well. The complete filter, frequency-denormalized to provide a cutoff frequency of 1 kHz, is shown in Fig. 4.39. An impedance denormalization of 1000 has also been used. Note that the GIC elements have been chosen so as to obtain a nonunity conversion ratio, thus permitting the use of all equal-valued capacitors in the realization.

The use of FDNRs works well when all the capacitors are grounded, as in the low-pass case. An extension of the FDNR approach may be used, however, when floating FDNRs are to be realized, as in the case of bandpass filters. To understand how this extension works, consider the GIC configuration shown in Fig. 4.40, in which a passive network defined by the transmission parameters

$$\begin{bmatrix} A(s) & B(s) \\ C(s) & D(s) \end{bmatrix} \tag{4.114}$$

is connected in cascade with two identical GICs. The overall transmission parameters for the cascade are

$$\begin{bmatrix} 1 & 0 \\ 0 & \dfrac{Z_2(s)Z_4(s)}{Z_1(s)Z_3(s)} \end{bmatrix} \begin{bmatrix} A(s) & B(s) \\ C(s) & D(s) \end{bmatrix} \begin{bmatrix} 1 & 0 \\ 0 & \dfrac{Z_1(s)Z_3(s)}{Z_2(s)Z_4(s)} \end{bmatrix}$$

$$= \begin{bmatrix} A(s) & B(s)\dfrac{Z_1(s)Z_3(s)}{Z_2(s)Z_4(s)}i \\ C(s)\dfrac{Z_2(s)Z_4(s)}{Z_1(s)Z_3(s)} & D(s) \end{bmatrix} \tag{4.115}$$

This set of parameters is identical to those of a network derived from the original network by making an impedance transformation of $Z_1(s)Z_3(s)/Z_2(s)Z_4(s)$ on each element. In practice, rather than using the cascade shown in Fig. 4.40 to separately realize each floating FDNR, a combination of such elements can be simultaneously realized, thus reducing the total number of GICs required.

As an example of the application of GIC techniques in the general case, consider the fourteenth-order resistance-terminated bandpass filter configuration shown in Fig. 4.41*a*.† After all the elements have been impedance-transformed by $1/s$, it has the form shown in Fig. 4.41*b*. Now let us consider the realization of some of the separate elements of this circuit. The series FDNR-capacitor

† L. T. Bruton and D. Treleaven, "Active Filter Design Using Generalized Impedance Converters," *EDN*, pp. 68–75, Feb. 5, 1973.

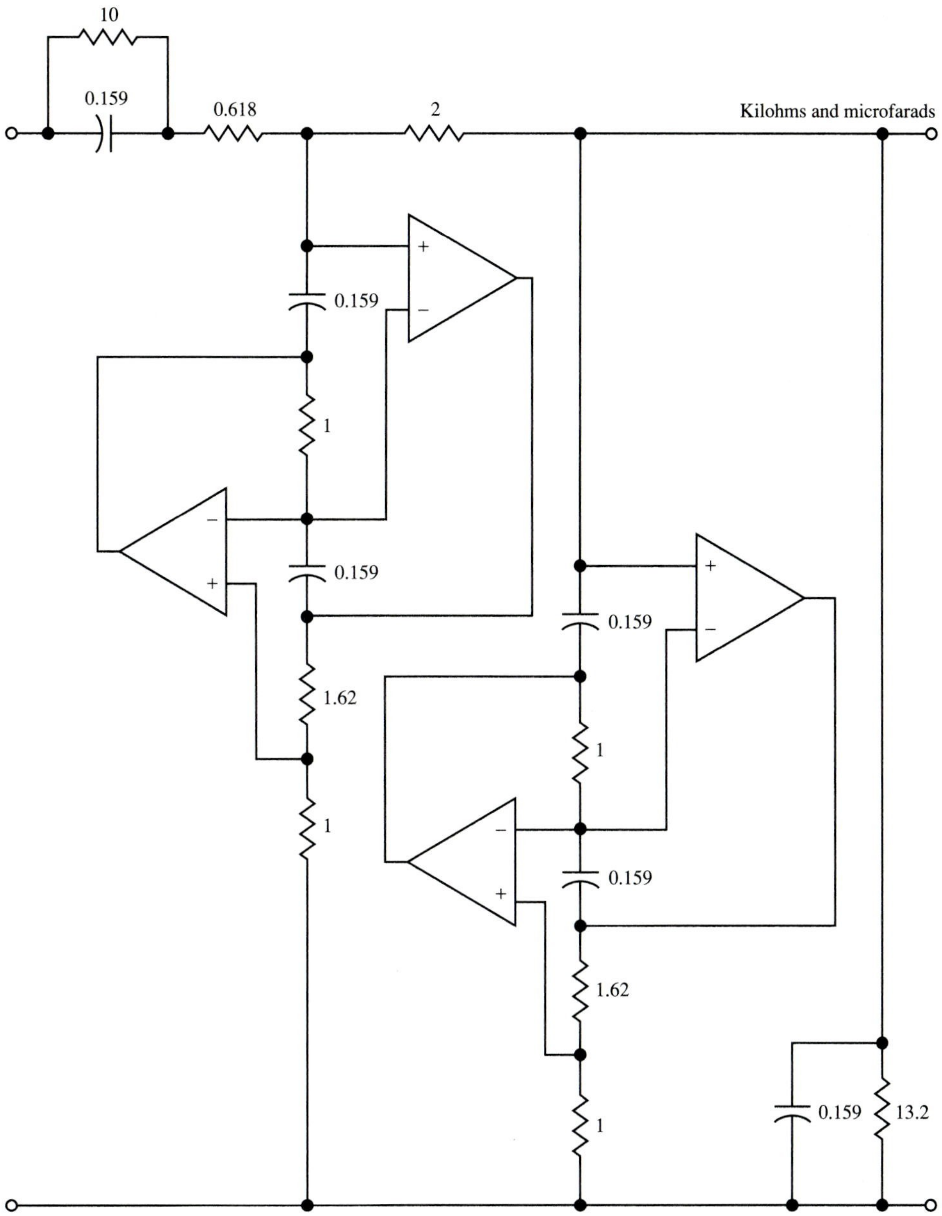

FIGURE 4.39
Circuit diagram for fifth-order FDNR filter.

combinations at the ends of the filters are most easily realized by using the GIC circuit shown in Fig. 4.37, in which $Z_1(s) = 1/s, Z_2(s) = Z_4(s) = Z_5(s) = 1$, and $Z_3(s) = R + 1/sC_1$. Inserting these values in (4.112), we obtain

$$Z_{\text{in}}(s) = \frac{1}{s^2 C_1} + \frac{R_s}{s} \tag{4.116}$$

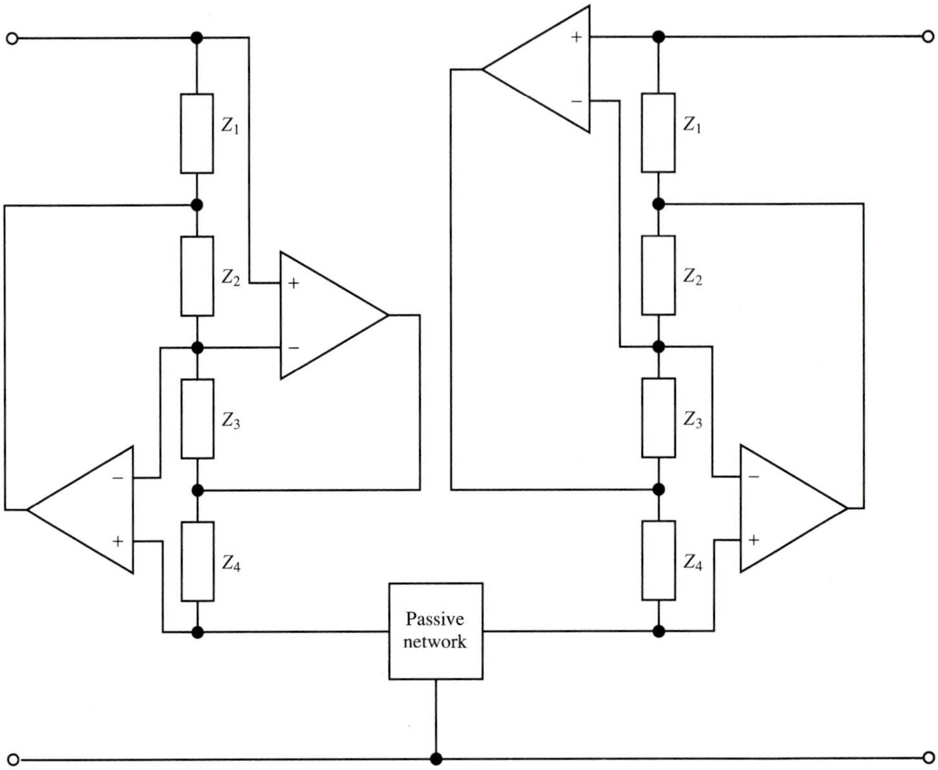

FIGURE 4.40
Realization of a floating FDNR.

which provides the input configuration for GIC A shown in Fig. 4.41b. Similarly, choosing $Z_1(s) = 1/s, Z_2(s) = Z_3(s) = Z_4(s) = 1$, and $Z_5(s) = R_L + 1/sC_7$ provides the output configuration of GIC G. The grounded FDNRs of value $1/C_2$ and $1/C_6$ shown in GICs B and F are realized by setting $Z_1(s) = Z_3(s) = 1/s, Z_2(s) = Z_4(s) = 1$, and $Z_5(s)$ to $1/C_2$ and $1/C_6$, respectively. The "T" configuration of two floating and one grounded FDNRs of values $1/C_3, 1/C_4$, and $1/C_5$ and the grounded resistor of value L_4 shown in the center of the network is realized by impedance-transforming each of these elements by s^2, then setting $Z_1(s) = Z_3(s) = 1$, and $Z_2(s) = Z_4(s) = 1/s$ in the circuit shown in Fig. 4.40 for GICs C and D. Thus the T of FDNRs is realized by a T of resistors of values $1/C_3, 1/C_4$, and $1/C_5$. The above procedure transforms the resistor of value L_4 to an element with impedance s^2L_4 connected to ground. This is easily realized by the GIC of Fig. 4.37, in which $Z_1(s) = Z_3(s) = 1, Z_2(s) = Z_4(s) = 1/s$, and $Z_5(s) = L_4$. The overall circuit is shown in Fig. 4.42. In this figure the GICs are indicated schematically by showing their impedances $Z_i(s)$. The realization of the fourteenth-order bandpass filter thus requires 14 capacitors and 7 GICs. Note that the direct realization of each of the floating FDNRs in Fig. 4.41b by a separate

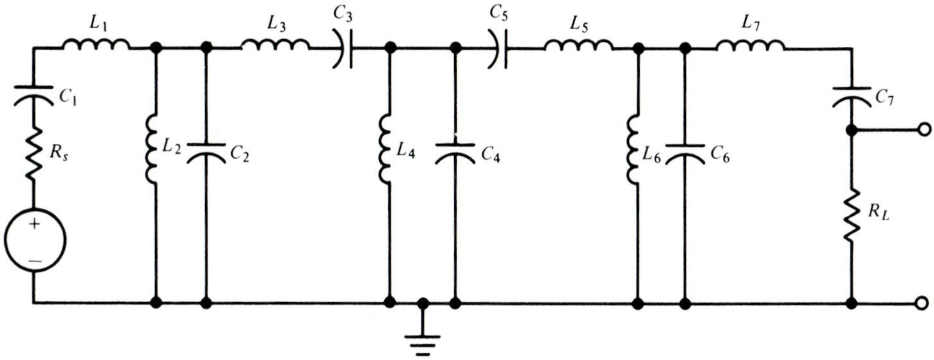

(a)

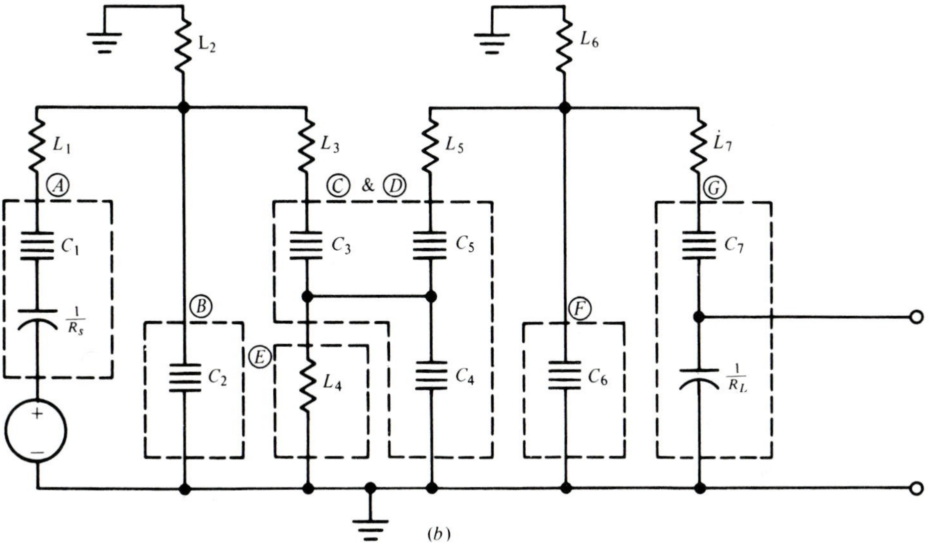

(b)

FIGURE 4.41
Steps in the realization of a fourteenth-order bandpass filter.

cascade of two GICs and a resistor would be considerably less efficient, resulting in a circuit containing 24 capacitors and 11 GICs.

4.9 LEAPFROG FILTERS

In the last section we introduced an active *RC* filter design method based on the use of a passive *RLC* prototype. A GIC (generalized impedance converter) was used to directly synthesize impedance-transformed network elements. In this section we introduce another active *RC* design method which is based on a passive network. In the low-pass case it uses damped and undamped integrators to simulate the

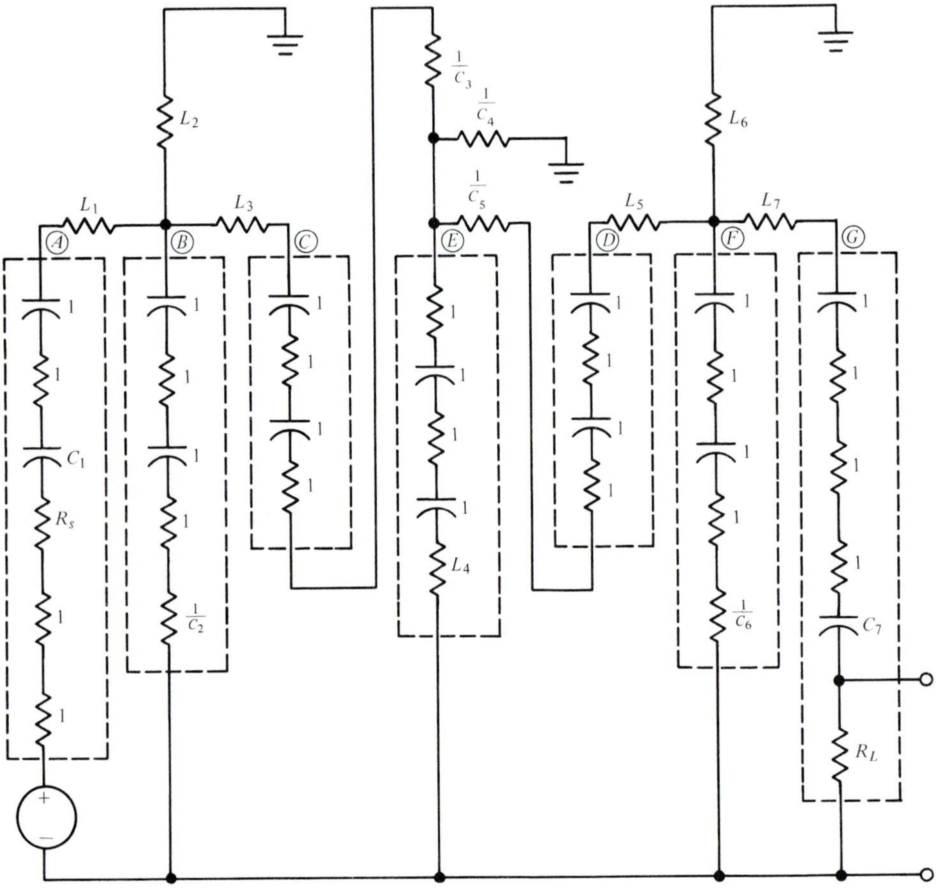

FIGURE 4.42
Final realization for the fourteenth-order bandpass filter.

voltage and current variables of the prototype. Like the GIC method, it retains the excellent sensitivity properties of the passive realization. It is called the *leapfrog filter*.

To see how a leapfrog filter is designed, consider the ladder network shown in Fig. 4.43. The branch relations may be written as

$$\begin{aligned} I_1 &= (V_1 - V_2)Y_1 \\ V_2 &= (I_1 - I_3)Z_2 \\ I_3 &= (V_2 - V_4)Y_3 \\ V_4 &= (I_3 - I_5)Z_4 \\ I_5 &= (V_4 - V_6)Y_5 \\ &\cdots \end{aligned} \tag{4.117}$$

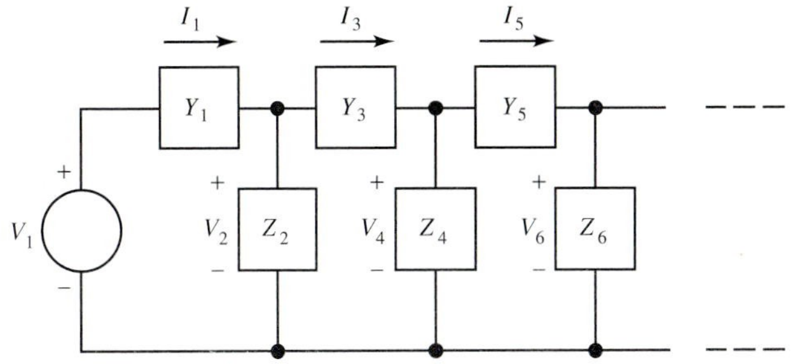

FIGURE 4.43
Ladder network.

where all the quantities V_i, I_i, Y_i, and Z_i are functions of s. If we let each of the currents I_1, I_3, I_5, etc., be simulated by voltages V_{i1}, V_{i3}, V_{i5}, etc., the equations of (4.117) become

$$\begin{aligned} V_{i1} &= (V_1 - V_2)Y_1 \\ V_2 &= (V_{i1} - V_{i3})Z_2 \\ V_{i3} &= (V_2 - V_4)Y_3 \\ V_4 &= (V_{i3} - V_{i5})Z_4 \\ V_{i5} &= (V_4 - V_6)Y_5 \\ &\cdots \end{aligned} \tag{4.118}$$

By introducing negative signs in both members of the second, third, and fourth equations, we obtain

$$\begin{aligned} V_{i1} &= (V_1 - V_2)Y_1 \\ -V_2 &= (V_{i1} - V_{i3})(-Z_2) \\ -V_{i3} &= (-V_2 + V_4)Y_3 \\ V_4 &= (-V_{i3} + V_{i5})(-Z_4) \\ V_{i5} &= (V_4 - V_6)Y_5 \\ &\cdots \end{aligned} \tag{4.119}$$

A block diagram of a network realization of these equations is shown in Fig. 4.44. The square boxes represent voltage transfer functions, and the circles represent summers. The alternation of the forward and backward signal flow paths is what gives the configuration its "leapfrog" epithet.

Low-Pass Leapfrog Filter

Now let us consider the application of the leapfrog structure to the realization of a low-pass filter. In this case, the series elements Y_i of Fig. 4.43 will be inductors

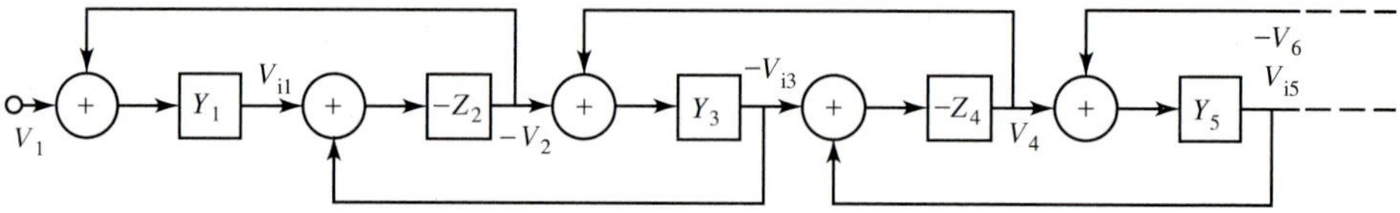

FIGURE 4.44
Block diagram simulation of a ladder network.

(or a series connection of an inductor and a resistor), and the shunt elements Z_i will be capacitors (or a shunt connection of a capacitor and a resistor). As an example, consider a fifth-order low-pass filter which is driven by a voltage source and realizes a voltage transfer function. The configuration is shown in Fig. 4.45. In this case the voltage transfer functions of the blocks shown in Fig. 4.44 are

$$\begin{aligned} T_1(s) &= Y_1(s) = \frac{1/L_1}{s + R_s/L_1} \\ T_2(s) &= -Z_2(s) = -\frac{1}{sC_2} \\ T_3(s) &= Y_3(s) = \frac{1}{sL_3} \\ T_4(s) &= -Z_4(s) = -\frac{1}{sC_4} \\ T_5(s) &= Y_5(s) = \frac{1/L_5}{s + R_L/L_5} \end{aligned} \tag{4.120}$$

Of these voltage transfer functions, $T_2(s)$ and $T_4(s)$ may be realized by conventional inverting operational amplifier integrators, $T_3(s)$ may be realized by an inverting integrator in series with a unity gain inverter, and $T_1(s)$ and $T_5(s)$ may

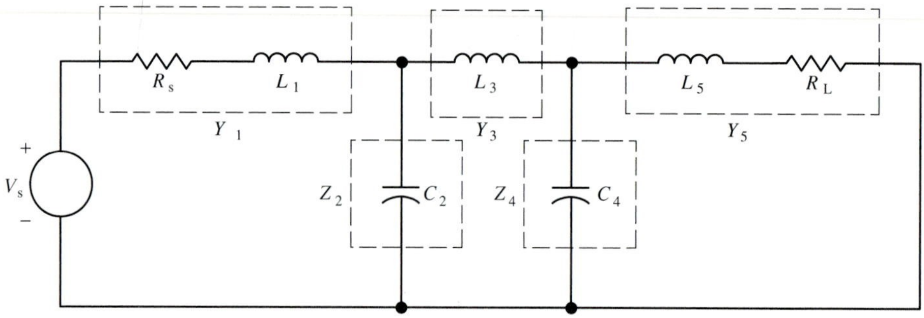

FIGURE 4.45
Fifth-order low-pass filter.

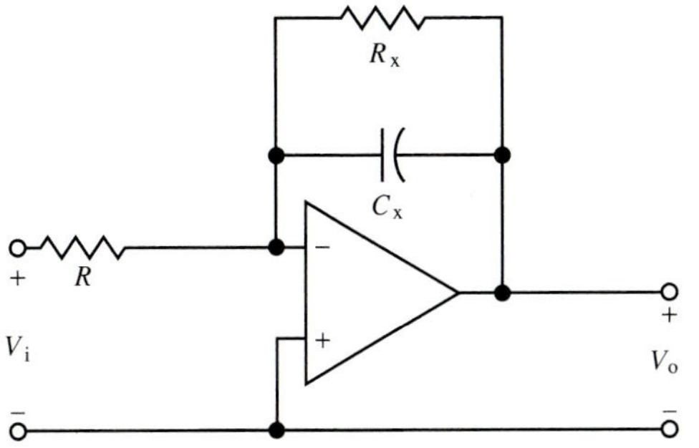

FIGURE 4.46
Damped inverting integrator.

be realized by an inverting damped integrator in series with an inverter. A realization for the damped integrator is shown in Fig. 4.46. Its voltage transfer function is

$$\frac{V_0(s)}{V_i(s)} = \frac{-1/RC_x}{s + 1/R_x C_x} \tag{4.121}$$

The resulting leapfrog realization of the filter is shown in Fig. 4.47. All the resistor values are shown as functions of an arbitrary resistance parameter R which may be chosen to achieve a suitable impedance denormalization. The values of the capacitors used in the integrators are given in terms of R and the element values in Fig. 4.45.

Example 4.16. It is desired to realize a third-order low-pass filter with a Butterworth characteristic (case 1 in Table 4.4) with a cutoff frequency $\omega_0 = 1$ krad/s and equal source and load resistors. From Table 4.8 we obtain the circuit shown in Fig. 4.48. For the normalized value $R = 1$ we obtain the leapfrog circuit shown in Fig. 4.49. To reduce the number of inverters required in the circuit, we have based this realization on a modified block diagram arrived at by multiplying the equations of (4.119) by minus signs. By choosing a frequency denormalization of 10^3 and an impedance denormalization of 10^4, all resistors would have a value of 10 kΩ and all the capacitor values would be multiplied by 10^{-7}.

Bandpass Leapfrog Filter

The leapfrog technique described above for low-pass filters is also applicable to bandpass ones. In this case the prototype network is derived by applying the low-pass-to-bandpass transformation to a low-pass prototype network. The general form of the resulting series and shunt branches is shown in Fig. 4.50. In all except the first series branch R_i in Fig. 4.50*a* will be zero, and in all but the last shunt branch R_j in Fig. 4.50*b* will be infinity. The admittance of Fig. 4.50*a* is

$$Y_i(s) = \frac{(1/L_i)s}{s^2 + (R_i/L_i)s + 1/L_iC_i} \tag{4.122}$$

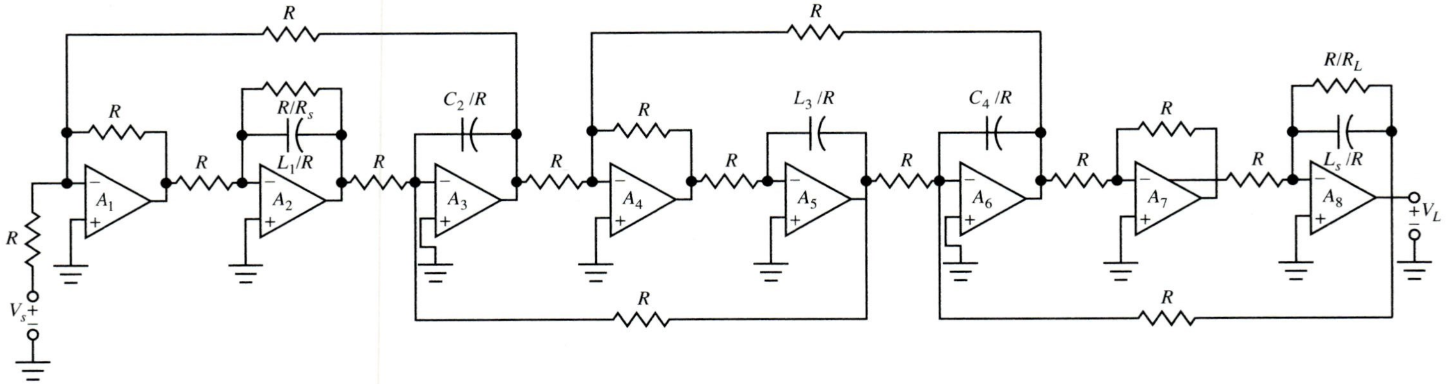

FIGURE 4.47
Leapfrog realization of a fifth-order low-pass filter.

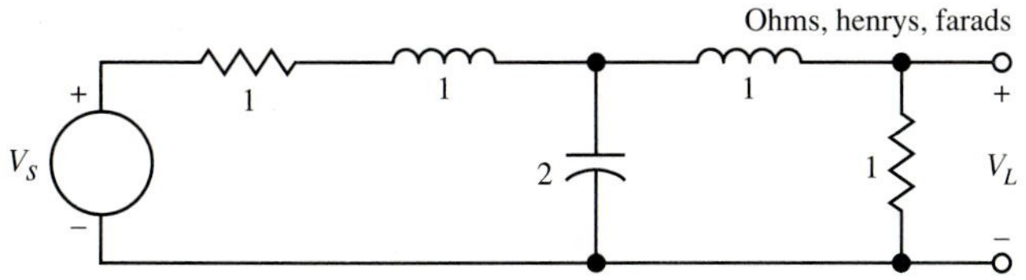

FIGURE 4.48
Third-order low-pass filter.

The impedance of Fig. 4.50*b* is

$$Z_j(s) = \frac{(1/C_j)s}{s^2 + (1/R_jC_j)s + 1/L_jC_j} \tag{4.123}$$

For the bandpass case we see that the required voltage transfer functions are second-order bandpass ones. Thus, the integrators used in the low-pass leapfrog realization are replaced by one of the second-order active *RC* bandpass realizations presented in this chapter. One of the most useful circuits for this is obtained by modifying the resonator filter presented in Sec. 4.6 so that it has two inputs (and thus may be used to provide the required summing function) and both inverting and noninverting bandpass outputs. A modified resonator circuit with these properties is shown in Fig. 4.51. The input and output variables are related by

$$V_{o2}(s) = -V_{o1}(s) = \left(\frac{s/R_4C_1}{s^2 + s/R_1C_1 + 1/R_2R_3C_1C_2}\right)(V_{i1} + V_{i2}) \tag{4.124}$$

Equating the coefficients of the transfer function given in (4.124) with those of the functions given in (4.122) and (4.123) provides the relationships needed to obtain the realization.

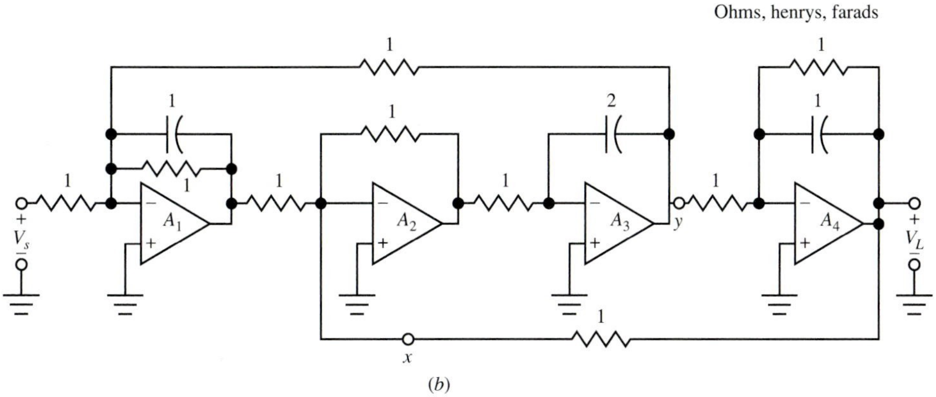

FIGURE 4.49
Leapfrog realization of a third-order low-pass filter.

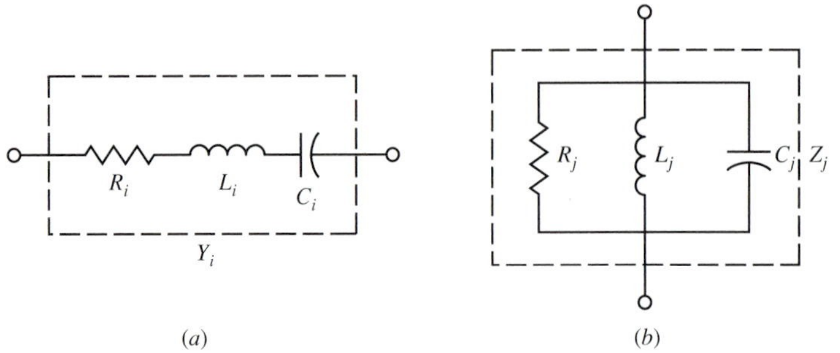

FIGURE 4.50
Series and shunt branches of a passive bandpass filter.

Example 4.17. It is desired to use the leapfrog method to design a sixth-order broadband filter with a Butterworth characteristic (case 1 of Table 4.6). A normalized design is to be prepared in which $\omega_0 = 1$ rad/s and the bandwidth BW $= 1/\sqrt{2}$ rad/s. The third-order low-pass Butterworth prototype with a bandwidth of 1 rad/s is shown in Fig. 4.52*a*. Applying the low-pass-to-bandpass transformation to this, we obtain the bandpass filter shown in Fig. 4.52*b*. The realization is shown in Fig. 4.52*c*. Each of the blocks in this figure represents one of the modified resonator circuits shown in Fig. 4.51 with the indicated element values. For a 1-kHz center frequency and a $1/\sqrt{2}$ kHz bandwidth, a frequency denormalization of $2\pi \times 10^3$ would be made on all capacitor values. An impedance denormalization would then be applied to obtain practical resistor (and capacitor) values.

The leapfrog method is probably one of the best methods available for realizing bandpass filters since it makes repeated use of a second-order filter as a "building block" yet provides sensitivities very close to those of the passive network from which it is derived. The second-order functions can be individually tuned to provide for the proper realization of high-order network functions.

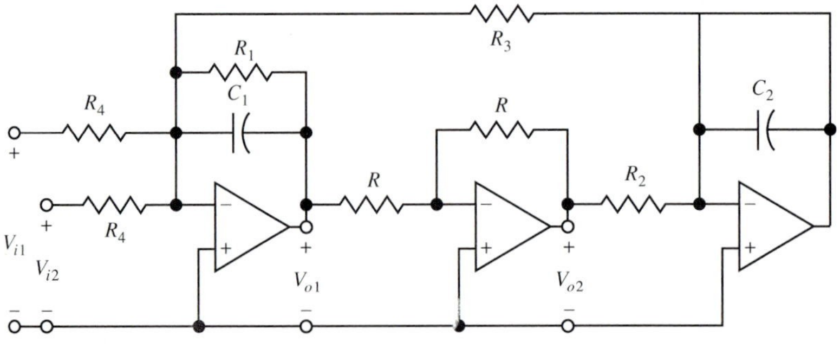

FIGURE 4.51
Modification of the resonator filter.

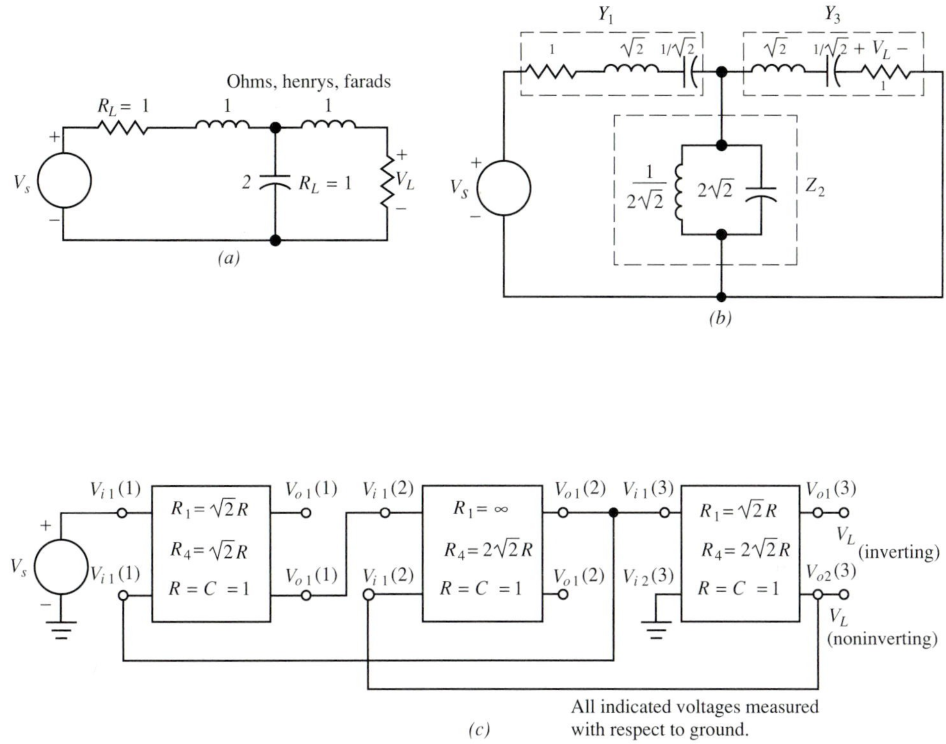

FIGURE 4.52
Leapfrog realization of a sixth-order bandpass filter.

4.10 SWITCHED CAPACITOR FILTERS

In the last two sections of this chapter we introduced filter realizations based on the use of a passive *RLC* prototype. In this section we present another design. It provides an alternate technique for physically realizing the leapfrog filter described in the preceding section. It can also be used for realizing second-order filter sections. It uses a device called a *switched capacitor* and is especially suited to very large-scale integrated-circuit (VLSI) implementations.

The Switched Capacitor

The basic premise in the development of *switched-capacitor* (SC) filters is that a resistor may be simulated by using periodically operated MOSFET switches and a capacitor. Such filters are especially attractive for realization using VLSI technologies in which the basic elements are MOS transistors and picofarad-range capacitors. SC filters have three main advantages:

1. Resistors are eliminated. In VLSI realizations, resistors have the disadvantage of requiring large amounts of die area. In addition, their tolerance levels and drift parameters are usually too large to be acceptable.

2. The frequency properties of the filter are determined by the *ratio* of capacitor values. These ratios usually can be controlled to 0.1 percent accuracy and maintain excellent tracking with respect to time and temperature variations. Active *RC* filters, on the other hand, have frequency characteristics determined by *RC* products which provide dramatically inferior performance in VLSI implementations.
3. The frequency performance of the filters is proportional to a clock frequency which can be controlled to provide programmable filter characteristics.

A switched-capacitor circuit realization of a resistor is shown in Fig. 4.53*a*. It consists of two MOSFET switches driven by the nonoverlapping clock waveforms shown in Fig. 4.53*b*. When the gate voltage applied to these switches is high, the typical enhancement *n*MOS transistor has a channel resistance of the order of $10^3\ \Omega$. When its gate voltage is low, the channel resistance is $10^{12}\ \Omega$ or more. As a result of the large ratio of these resistance values, the transistor functions effectively as an on-off switch. The two transistors shown in Fig. 4.53*a* can thus be modeled by an SPDT switch, as shown in Fig. 4.54*a*. With the switch in the left position, as shown in the figure, capacitor C charges to voltage V_1. When the switch is moved to the right, C discharges to V_2. The net charge transfer

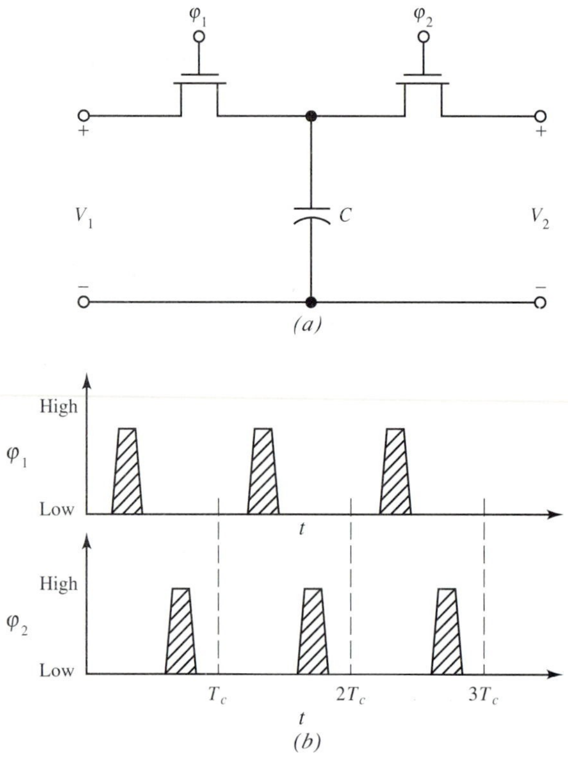

FIGURE 4.53
SC realization of a resistor.

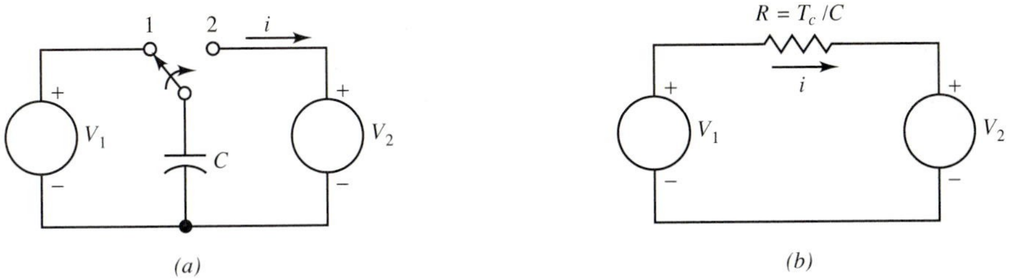

FIGURE 4.54
SC circuit symbol and equivalent resistor.

per cycle of the switch is

$$\Delta Q = C\ \Delta V = C(V_1 - V_2) \tag{4.125}$$

If the switch cycles back and forth at $f_c = 1/T_c$ Hz, the average current (for constant values of V_1 and V_2) is

$$I_{\text{av}} = \frac{\Delta Q}{\Delta T} = Cf_c(V_1 - V_2) \tag{4.126}$$

If f_c is much higher than the highest frequency component in V_1 and V_2, then the charge transfer process may be treated as a continuous one and the SC circuit shown in Fig. 4.53*a* may be modeled as an equivalent resistor, as shown in Fig. 4.54*b*, where

$$R_{\text{eq}} = \frac{V_1 - V_2}{I_{\text{av}}} = \frac{1}{Cf_c} \tag{4.127}$$

Such equivalent resistors can be combined with other capacitors and with operational amplifiers to obtain many of the filter functions normally realized with active *RC* filters.

The Basic SC Integrator

As an example of the use of SC circuits, consider the inverting integrator shown in Fig. 4.55*a*. The voltage transfer function is given as

$$\frac{V_{\text{out}}}{V_{\text{in}}} = \frac{-1}{sR_1C_2} \tag{4.128}$$

If we represent R_1 by the SC-resistor implementation shown in Fig. 4.53*a*, we obtain the SC inverting integrator circuit shown in Fig. 4.55*b*. Assuming that the highest signal frequency component is much lower than f_c, the voltage transfer function for this circuit is

$$\frac{V_{\text{out}}}{V_{\text{in}}} = \frac{-C_1 f_c}{sC_s} \tag{4.129}$$

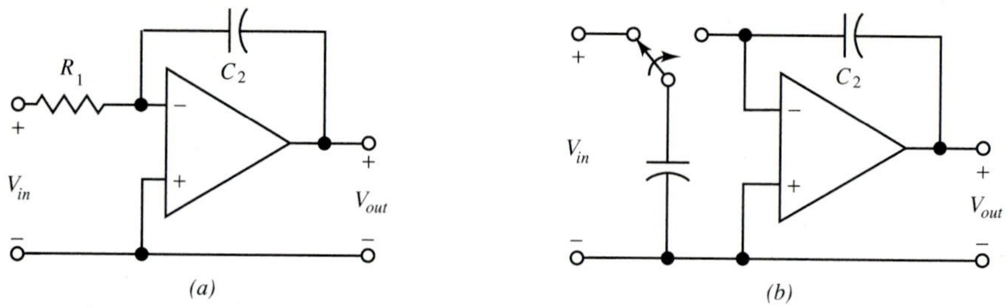

FIGURE 4.55
Inverting integrator and SC realization.

The SC integrator circuit shown in Fig. 4.55*b* is of limited use due to stray capacitances in the circuit. These occur in the MOSFET switches, the interconnections, and the capacitor plates and may significantly alter the frequency response of the circuit. To avoid these effects, it is common practice to use dual switches to replace each of the MOSFETs shown in Fig. 4.53*a*. The additional switches effectively ground the stray capacitances. The resulting realization is called a *stray-insensitive* one. A dual-switch version of the inverting integrator is shown in Fig. 4.56. The schematic representation of this circuit is shown in Fig. 4.57*a*. Another advantage of dual switches is that by inverting the phase of one of the switches, the network function which is realized is multiplied by a minus sign. This effect is illustrated in Fig. 4.57*b*. The resulting circuit realizes a noninverting integrator.

Other SC Integrators

The basic SC integrator circuit described above is easily modified to provide other useful integrator configurations. Four such circuits are the summing integrator, the differential integrator, the integrator/summer, and the damped integrator. A schematic of a *summing integrator* is shown in Fig. 4.58. The output voltage is given as

$$V_{\text{out}} = -\frac{C_1 f_c}{sC_3} V_1 - \frac{C_2 f_c}{sC_3} V_2 \tag{4.130}$$

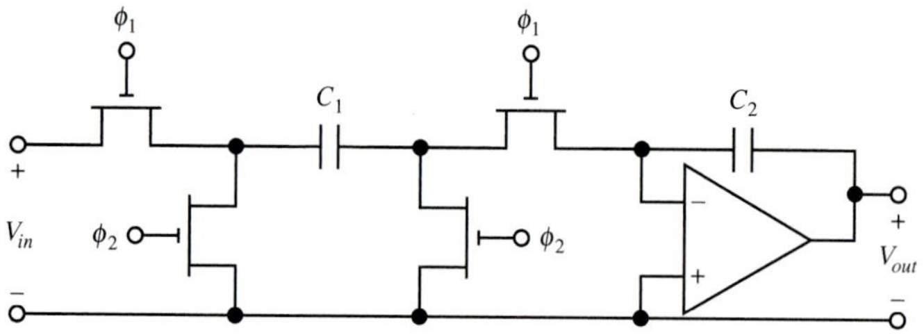

FIGURE 4.56
Stray-insensitive dual switches.

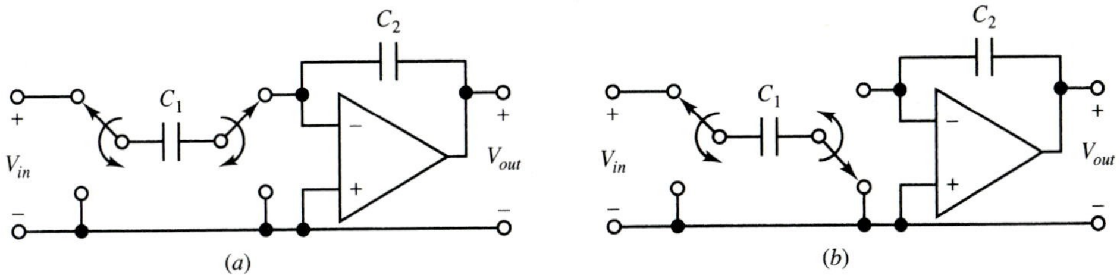

FIGURE 4.57
Stray-insensitive inverting and noninverting integrators.

In most applications, the SPDT switches would need to be implemented by dual-switch configurations to obtain stray-insensitive performance. A schematic of a *differential integrator* is shown in Fig. 4.59. The output voltage is given as

$$V_{\text{out}} = \frac{C_1 f_c}{sC_2}(V_1 - V_2) \tag{4.131}$$

This circuit is a stray-insensitive one. A schematic of an *integrator/summer* is shown in Fig. 4.60. The output voltage is

$$V_{\text{out}} = -\frac{C_1 f_c}{sC_3}V_1 - \frac{C_2}{C_3}V_2 \tag{4.132}$$

Dual-switch configurations would usually be required for this circuit. A schematic of a *lossy integrator* is shown in Fig. 4.61. The voltage transfer function is given as

$$\frac{V_{\text{out}}}{V_{\text{in}}} = -\frac{C_1 f_c / C_2}{s + C_3 f_c / C_2} \tag{4.133}$$

As in the summing integrator and integrator/summer configurations, dual-switch configurations are usually required for stray-insensitive performance of this circuit.

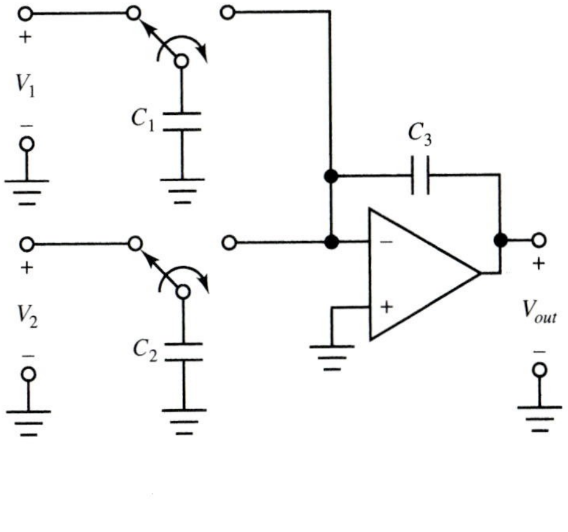

FIGURE 4.58
SC summing integrator.

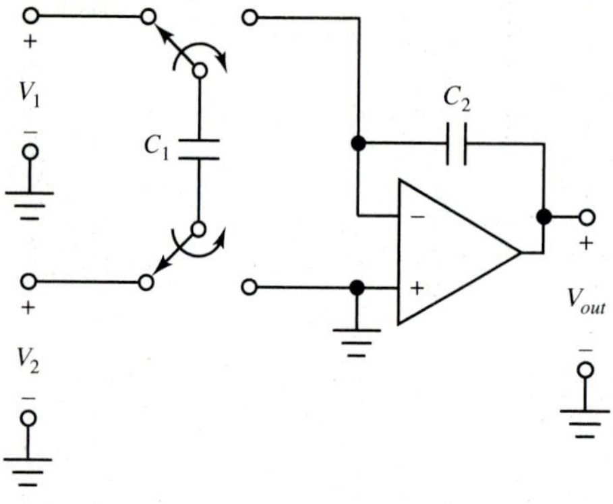

FIGURE 4.59
SC differential integrator.

SC Resonator Circuit

The circuits described above can be used as building blocks to realize many filter functions. As an example, consider the resonator filter described in Sec. 4.6 and shown in Fig. 4.31. It consists of a lossy (or damped) integrator, an inverting integrator, and an inverter. In an SC realization, the functions of inverting integrator and inverter can be combined; thus only two circuits are required—one to provide lossy integration and the other for noninverting integration. An SC realization of the resonator is shown in Fig. 4.62. For this the voltage transfer functions are

$$\frac{V_{\rm BP}}{V_{\rm in}} = \frac{sC_1f_c/C_2}{s^2 + s(C_3f_3/C_2) + (C_1^2f_c^2/C_2^2)}$$
$$\frac{V_{\rm LP}}{V_{\rm in}} = \frac{(C_1f_c/C_2)^2}{s^2 + s(C_3f_c/C_2) + (C_1f_3/C_2)^2} \qquad (4.134)$$

SC Leapfrog Filter

The leapfrog filter introduced in Sec. 4.9 is readily implemented by using SC techniques. Consider the fifth-order low-pass filter shown in Fig. 4.45. The SC circuits

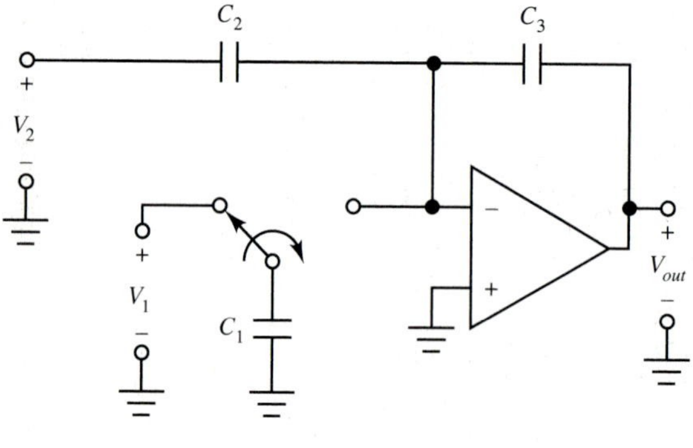

FIGURE 4.60
SC integrator/summer.

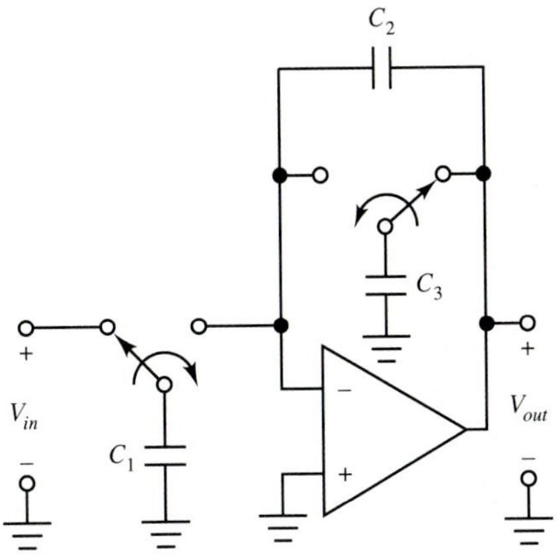

FIGURE 4.61
SC lossy integrator.

of choice for use in the realization are the differential input damped and undamped inverting integrators. To implement the realization, the equations of (4.118) may be modified so that all the transfer functions are inverting. The new equations (for the fifth-order case) become

$$\begin{aligned} -V_{i1} &= [V_1 - (+V_2)](-Y_1) \\ V_2 &= [-V_{i1} - (-V_{i3})](-Z_2) \\ -V_{i3} &= [V_2 - (+V_4)](-Y_3) \\ V_4 &= [-V_{i3} - (-V_{i5})](-Z_4) \\ -V_{i5} &= V_4(-Y_5) \end{aligned} \tag{4.135}$$

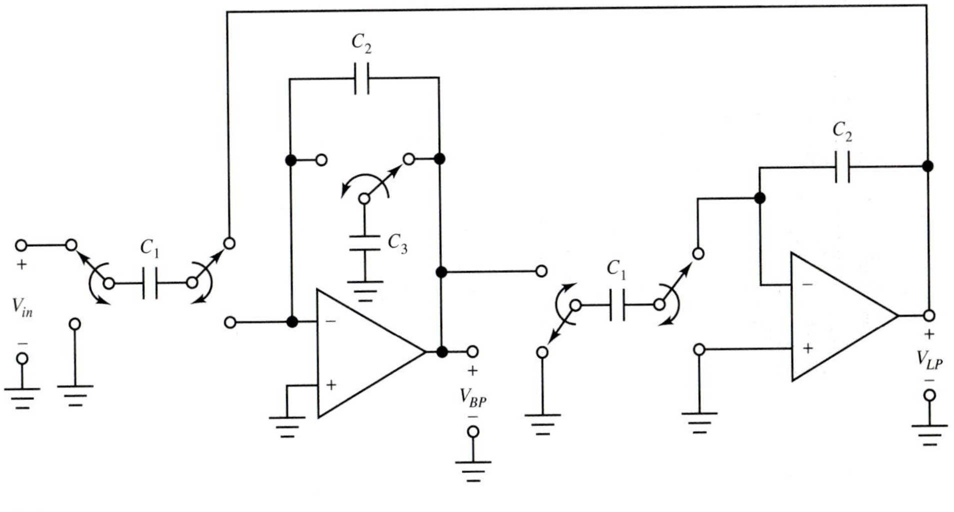

FIGURE 4.62
SC resonator.

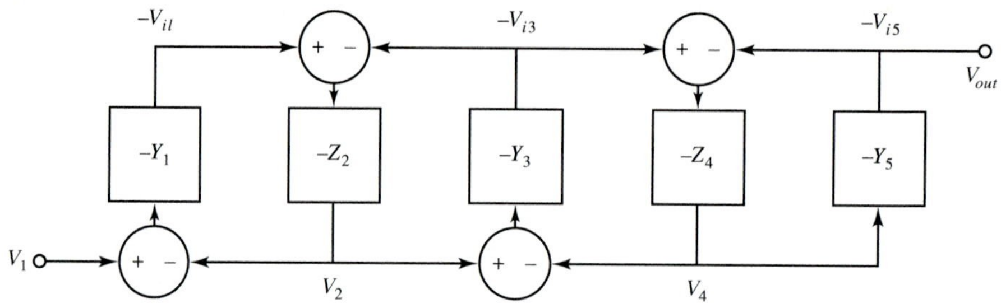

FIGURE 4.63
Block-diagram realization of leapfrog equations.

A block-diagram realization of these equations is shown in Fig. 4.63. The voltage transfer functions realized as $Y_1(s), Z_2(s), Y_3(s), Z_4(s)$, and $Y_5(s)$ are defined by the relations given in (4.120). An SC realization of the block diagram is shown in Fig. 4.64. The values of the capacitors used to replace resistors are

$$C_0 = \frac{1}{Rf_c} \qquad C_{R_s} = \frac{R_S}{Rf_c} \qquad C_{R_L} = \frac{R_L}{Rf_c} \tag{4.136}$$

where R is an impedance scaling constant. The values of the capacitors used in the integrators are shown in the figure in terms of the element values given in the prototype filter shown in Fig. 4.45.

Example 4.18. Consider the use of SC techniques to realize a fifth-order Butterworth low-pass filter with a bandwidth of 1 kHz using a clock frequency $f_c = 100$ kHz. From Table 4.8 we find $L_1 = L_5 = 0.618$ H, $C_2 = C_4 = 1.618$ F, $L_3 = 2$ H for the normalized filter. If we choose $C_0 = 1/Rf_c = 1$ pF, we obtain $R = 10^7\ \Omega$. Using this value and frequency-denormalizing by $2\pi \times 10^3$, we obtain $C_{L_1} = C_{L_5} = 9.836$ pF, $C_{C_2} = C_{C_4} = 25.751$ pF, and $C_{L_3} = 31.831$ pF.

SC techniques may also be used to realize bandpass leapfrog filters by using the SC resonator filter shown in Fig. 4.62 as a building block and following the leapfrog construction for bandpass filters given in Sec. 4.9.

Commercial implementations of SC filters are available from several manufacturers. Like the Universal Active Filter described in Sec. 4.6, these implementations allow the user to add external resistors to obtain a wide range of filter characteristics. An example of such a product is the Universal SC Filter which provides two dual-integrator-loop sections, each of which independently realizes the state-variable filter described in Sec. 4.6. It is available as the MF10 (National Semiconductor) and the LTC1060 (Linear Technology).

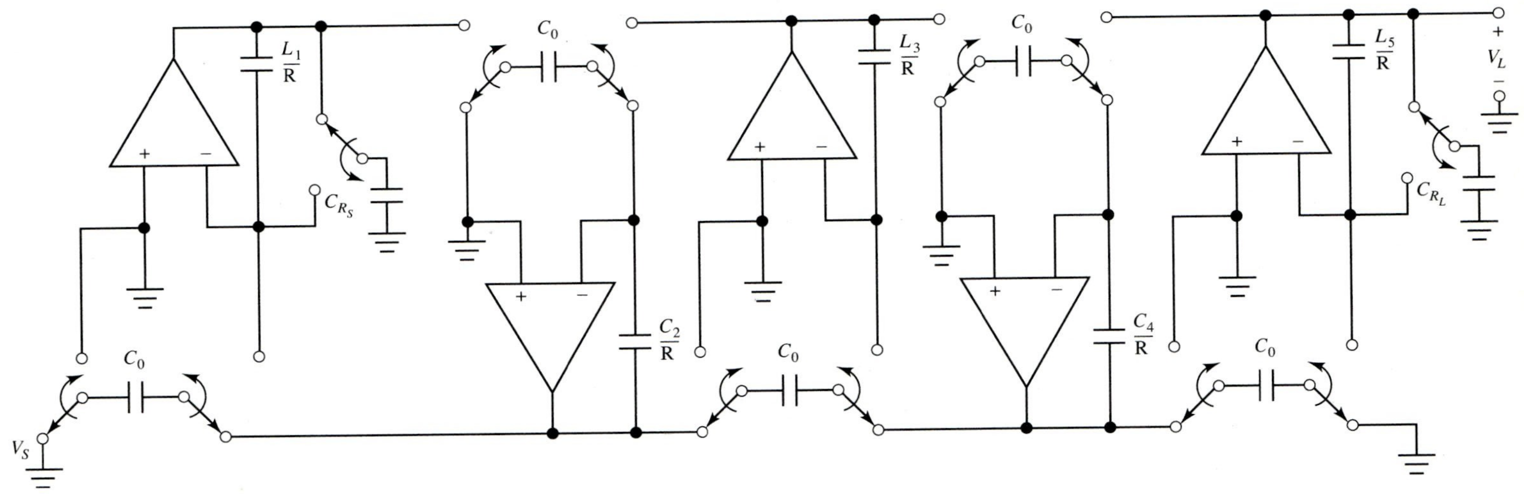

FIGURE 4.64
SC realization of fifth-order low-pass filter.

PROBLEMS

4.1. (*Sec. 4.1*) Construct a plot of the magnitude of each of the following network functions, and compare them with the waveshapes shown in Fig. 4.2.

$$N_a(s) = \frac{H}{s^2 + \sqrt{2}s + 1}$$

$$N_b(s) = \frac{H}{s^4 + 2.6131s^3 + 3.4142s^2 + 2.6131s + 1}$$

$$N_c(s) = \frac{H}{s^2 + 1.0977s + 1.1025}$$

$$N_d(s) = \frac{H}{s^4 + 0.9528s^3 + 1.4539s^2 + 0.7426s + 0.2756}$$

4.2. (*Sec. 4.1*) Construct a plot of the magnitude of each of the following network functions, and compare them with the waveshapes shown in Fig. 4.4.

$$N_a(s) = \frac{Hs^2}{s^2 + \sqrt{2}s + 1}$$

$$N_b(s) = \frac{Hs^4}{s^4 + 2.6131s^3 + 3.4142s^2 + 2.6131s + 1}$$

$$N_c(s) = \frac{Hs^2}{s^2 + 0.9957s + 0.9070}$$

$$N_d(s) = \frac{Hs^4}{s^4 + 3.6986s^3 + 5.6440s^2 + 2.8828s + 3.8820}$$

4.3. (*Sec. 4.1*) Construct a plot of the magnitude of the following network function, and compare it with the waveshape shown in Fig. 4.5.

$$N(s) = \frac{Hs^2}{s^4 + 0.7071s^3 + 2.25s^2 + 0.7071s + 1}$$

4.4. (*Sec. 4.1*) Construct a plot of the magnitude of the following network function, and compare it with the waveshape shown in Fig. 4.6.

$$N(s) = \frac{H(s^2 + 2s + 1)}{s^4 + \sqrt{2}s^3 + 3s^2 + \sqrt{2}s + 1}$$

4.5. (*Sec. 4.1*) Construct plots of the magnitude and phase of the following network function, and compare them with the waveshapes shown in Fig. 4.7.

$$N(s) = \frac{s^2 - s + 1}{s^2 + s + 1}$$

4.6. (*Sec. 4.1*) Convert the filter circuit shown in Fig. 4.8*a* to one with a 20-kHz bandwidth and with resistance values $R_1 = R_2 = 10\ \text{k}\Omega$.

4.7. (*Sec. 4.2*) Change the element values of the filter realization in Example 4.4 (Sec. 4.4) so that its cutoff frequency is 2 kHz and the resistor values shown in Fig. 4.15 are $R_1 = R_2 = 10\ \text{k}\Omega$.

4.8. (*Sec. 4.3*) Derive the y parameters given in (4.23) and (4.24) for the low-pass second-order active *RC* circuit shown in Fig. 4.12.

4.9. (*Sec. 4.3*) Prove that the locus shown in Fig. 4.13, as defined by the roots of the denominator polynomial in (4.25), is a circle.

4.10. (*Sec. 4.3*) Repeat the design of Example 4.1 for a Chebychev $\frac{1}{2}$-dB ripple realization (case 2*a* of Table 4.1) with a cutoff frequency of 1 kHz. Choose $C = 0.05\ \mu$F. Find the Q and ω_0 sensitivities.

4.11. (*Sec. 4.4*) Repeat the design of Example 4.2 for a Chebychev 1-dB ripple realization (case 2*b* of Table 4.1) with a cutoff frequency of 1 kHz. Choose $C = 0.05\ \mu$F. Find the Q and ω_0 sensitivities.

4.12. (*Sec. 4.4*) Repeat the design of Example 4.3 for a Thomson realization (case 3 of Table 4.1) with a 0.1-ms delay. Choose $C = 0.05\ \mu$F. Find Q and ω_0 sensitivities.

4.13. (*Sec. 4.4*) Find expressions for the Q and ω_0 sensitivities [in a form similar to that given in (4.37)] for the second-order high-pass filter shown in Fig. 4.15.

4.14. (*Sec. 4.4*) Use the expressions found in Problem 4.13 to construct a table similar to Table 4.2 for the sensitivities of the high-pass filter shown in Fig. 4.15.

4.15. (*Sec. 4.4*) Repeat the design of Example 4.4 for a Chebychev $\frac{1}{2}$-dB ripple realization (case 2*a* of Table 4.3) with a cutoff frequency of 1 kHz. Choose $C = 0.05\ \mu$F.

4.16. (*Sec. 4.4*) Repeat the design of Example 4.5 for a Chebychev 1-dB ripple realization (case 2*a* of Table 4.3) with a cutoff frequency of 1 kHz. Choose $C = 0.05\ \mu$F.

4.17. (*Sec. 4.4*) Find the voltage transfer function for the bandpass filter shown in Fig. 4.17 for the case where elements R_1 and C_2 are interchanged. Develop a design procedure for the resulting filter.

4.18. (*Sec. 4.4*) Repeat the design given in Example 4.6 for a bandpass filter with a Q of 2 and a resonant frequency of 5 kHz. Choose $C = 0.01\ \mu$F.

4.19. (*Sec. 4.4*) Repeat the design given in Example 4.7 for the specifications given in Problem 4.18.

4.20. (*Sec. 4.4*) Repeat the design given in Example 4.8 for a notch filter, using a design parameter $k = 0.8$. Repeat for $k = 1.2$.

4.21. (*Sec. 4.4*) Design a band-reject filter with poles at $-0.15 \pm j1.1$ and zeros at $\pm j1.12$. Choose a multiplicative constant so that the infinite frequency transmission is 0.1.

4.22. (*Sec. 4.4*) Design a band-reject filter with poles at $-0.15 \pm j0.9$ and zeros at $\pm j0.88$. Choose a multiplicative constant so that the dc transmission is 0.1.

4.23. (*Sec. 4.4*) Plot the magnitude characteristic of the cascade connection of the two filters realized in Problems 4.21 and 4.22.

4.24. (*Sec. 4.5*) The multiple-feedback filter shown in Fig. 4.24 is used to realize a normalized maximally flat magnitude low-pass characteristic (case 1 of Table 4.1) with a dc gain of 1.586. Find the passive element values and their sensitivities. Compare the sensitivities with those obtained by using design 1 for the same specifications and the filter shown in Fig. 4.12.

4.25. (*Sec. 4.5*) (*a*) Determine the actual network function obtained by using the element values given in Example 4.9 if the open-loop gain of the operational amplifier is 100 kHz. (*b*) Repeat for an open-loop gain of 10 kHz.

4.26. (*Sec. 4.5*) Repeat the design of Example 4.9 for a Chebychev $\frac{1}{2}$-dB ripple realization (case 2*a* of Table 4.1) with a cutoff frequency of 1 kHz. Choose $C = 0.05\ \mu$F and a dc gain of -5. Find the Q and ω_0 sensitivities.

4.27. (*Sec. 4.5*) Repeat the design of Example 4.9 for a Chebychev 1-dB ripple realization (case 2*b* of Table 4.1) with a cutoff frequency of 1 kHz. Choose $C = 0.05\ \mu F$ and a dc gain of -5. Find the Q and ω_0 sensitivities.

4.28. (*Sec. 4.5*) Repeat the design of Example 4.9 for a Thomson realization (case 3 of Table 4.1) with a 0.1-ms delay and a dc gain of -5. Choose $C = 0.05\ \mu F$. Find the Q and ω_0 sensitivities.

4.29. (*Sec. 4.5*) Repeat the design of Example 4.10 for a Chebychev $\frac{1}{2}$-dB ripple realization (case 2*a* of Table 4.3) with a cutoff frequency of 1 kHz. Choose $C = 0.05\ \mu F$, and use an infinite frequency gain of -5.

4.30. (*Sec. 4.5*) Repeat the design of Example 4.10 for a Chebychev 1-dB ripple realization (case 2*b* of Table 4.3) with a cutoff frequency of 1 kHz. Choose $C = 0.05\ \mu F$, and use an infinite frequency gain of -5.

4.31. (*Sec. 4.5*) Repeat the design given in Example 4.11 for a bandpass filter, but use a Q of 2, a resonant frequency of 2 kHz, and a gain at resonance of -5. Choose $C = 0.005\ \mu F$.

4.32. (*Sec. 4.5*) Repeat the design given in Example 4.12 for a maximum resistance spread m of 0.1.

4.33. (*Sec. 4.6*) Modify the design given in Example 4.13 so that the bandpass gain at resonance is -10.

4.34. (*Sec. 4.6*) Repeat the design given in Example 4.13 for a Q of 50.

4.35. (*Sec. 4.6*) Repeat the design given in Example 4.14 for a Q of 50. Choose a gain at resonance of -10.

4.36. (*Sec. 4.7*) Design a sixth-order Butterworth low-pass filter with a cutoff frequency of 5 kHz using a cascade connection of the Sallen and Key filter described in Sec. 4.3. Use design 1 and choose $C = 0.01\ \mu F$.

4.37. (*Sec. 4.7*) Design a sixth-order Butterworth low-pass filter with a cutoff frequency of 5 kHz using a cascade connection of the multiple-feedback filter described in Sec. 4.5. Choose $C = 0.01\ \mu F$, and let each stage of the realization have a dc gain of -2.

4.38. (*Sec. 4.7*) Design a sixth-order Chebyshev 1-dB ripple low-pass filter with a cutoff frequency of 5 kHz using a cascade connection of the multiple-feedback filter described in Sec. 4.5. Choose $C = 0.01\ \mu F$, and let each stage have a gain of -2 at direct current.

4.39. (*Sec. 4.7*) Design a sixth-order Butterworth bandpass filter having a center frequency of 5 kHz and a Q of 10. Use the Sallen and Key filter configuration described in Sec. 4.4, and choose $C = 0.01\ \mu F$. Find the gain at resonance of the resulting filter.

4.40. (*Sec. 4.7*) Find a denormalized filter realization of the normalized low-pass elliptic network function derived in Example 4.15. The cutoff frequency should be 1 kHz.

4.41. (*Sec. 4.7*) Find a normalized filter realization for a sixth-order elliptic bandpass filter with a center frequency of 1 rad/s and a bandwidth of 1 rad/s. The ripple in the passband should be 1 dB, and the attenuation at 0.414 and 2.414 rad/s should be at least 34 dB.

4.42. (*Sec. 4.8*) Find a realization for a sixth-order Butterworth low-pass leapfrog filter with a cutoff frequency of 1 kHz using FDNRs. Choose an impedance denormalization constant and GIC gain constants such that all the capacitors have a value of 0.01 μF.

4.43. (*Sec. 4.8*) Find a GIC realization of a fourth-order Butterworth bandpass equal-resistance terminated filter with a center frequency of 1 kHz and a bandwidth of 0.5 kHz. Choose an impedance denormalization constant and GIC gain constants such that all the capacitors have a value of 0.01 μF.

4.44. (*Sec. 4.9*) Find a way of introducing negative signs into the equations of (4.118) so that three of the transfer functions given in (4.120) will be inverting. Draw a block diagram (similar to the one shown in Fig. 4.44) and a realization (similar to the one shown in Fig. 4.47) for the resulting set of equations.

4.45. (*Sec. 4.9*) Realize a sixth-order Butterworth low-pass filter with a cutoff frequency of 1 kHz using the leapfrog technique. Let all the resistors have a value of 10 kΩ.

4.46. (*Sec. 4.9*) Use the leapfrog filter technique to design a fourth-order bandpass Butterworth filter with a center frequency of 1 kHz and a bandwidth of 1 kHz. Let all the resistors have a value of 10 kΩ.

4.47. (*Sec. 4.10*) Design a second-order Butterworth SC filter using the resonator circuit shown in Fig. 4.62. Let the clock frequency be 100 kHz and the cutoff frequency be 1 kHz. Choose $C_1 = 1$ pF.

4.48. (*Sec. 4.10*) Repeat the design given in Example 4.18 for the case where the filter is to have a Chebyshev 1-dB ripple characteristic rather than a Butterworth one.

4.49. (*Sec. 4.10*) Design a fourth-order Butterworth SC low-pass filter with a cutoff frequency of 1 kHz and a clock frequency of 1 kHz. Choose $C_0 = 1$ pF.

BIBLIOGRAPHY

Chen, W.: *Passive and Active Filters*, John Wiley & Sons, New York, 1986.

Huelsman, L. P., and P. E. Allen: *Introduction to the Theory and Design of Active Filters*, McGraw-Hill, New York, 1980.

Johnson, D. E., J. R. Johnson, and H. P. Moore: *A Handbook of Active Filters*, Prentice-Hall, Inc., Englewood Cliffs, N.J., 1980.

Moschytz, G. S., and P. E. Horn: *Active Filter Design Handbook*, John Wiley & Sons, New York, 1981.

Stephenson, F. W. (Ed.): *RC Active Filter Design Handbook*, John Wiley & Sons, New York, 1985.

Van Valkenburg, M. E.: *Analog Filter Design*, Holt, Rinehart, and Winston, New York, 1982.

CHAPTER 5

ELECTRONIC SWITCHING CIRCUITS

5.1 INTRODUCTION

We refer to continuously variable quantities in instrumentation, computing, control, and communication systems as *analog variables*; the term is historically derived from "analogies" between mechanical and electrical quantities in early analog computers. *Digital variables*, on the other hand, take countable sets of *discrete* states (0 or 1; ON or OFF; 1, 2, 3, 4, . . . , etc.). *Analog comparators* (Sec. 3.2) are the basic *analog-to-digital* transducers, producing a digital (discrete-state) output as a function of analog inputs. Similarly, *switches* are the basis of all *digital-to-analog* transducers, which produce different continuously variable outputs given different discrete (digital) inputs. The many possible kinds of switching circuits (e.g., single-pole/single-throw, double-pole/double-throw, star switches connecting one point to one of several different points, and crossbar switches for matrix interconnections) can all be constructed from simple *on-off switches*, each controlled by a basic digital ON/OFF (0 or 1) decision. Accurate analog switches are needed for the following system design purposes:

1. *General-purpose analog-data interconnections* in instrumentation, control, computer, and communication circuits
2. *Multiplexing* (timesharing) of devices such as analog-to-digital converters
3. *Timed sampling* of continuously changing variables, e.g., for analog-to-digital conversion
4. *Digital-to-analog conversion*; programming of test signals (Chap. 6)

5. *Integrator switching* in analog computers
6. *Accurate modulation and demodulation*; analog multiplication

5.2 IDEAL SWITCHES AND REAL SWITCHING CIRCUITS

Switches in automatic systems are most frequently operated by *electric signals from digital logic circuits*, typically 0 V for digital 0 and +4 V for digital 1. The earliest automatic switches, still widely used, were *relays*, which afford good low-resistance contacts but switch far too slowly for many purposes; many relays also tend to bounce a few times between ON and OFF when they are turned on. An *ideal electronic switch* would, in fact, have the following desirable properties:

1. *Zero forward impedance* Z_{on} when the switch is on.
2. *Infinite back impedance* Z_{off} when the switch is off.
3. *Zero switching time* between ON and OFF conditions.
4. No *offset voltages or currents* introduced by the switch into the signal circuit.
5. *Zero cross talk* from the switch-control circuit into the signal circuit.
6. These properties should be obtained *without critical circuit adjustments*.

Electronic switching circuits approximating these desirable properties are not easy to design. In general, compromises will be necessary.

To turn an electric signal circuit on or off, one can insert an electronic switch *in series with the signal current*, or the switch may *shunt the signal current to ground* (Fig. 5.1*a*). Real electronic switches do have a finite forward resistance R_{on} when the switch is on and a finite back impedance (leakage impedance) Z_{off} representing both resistive and capacitive leakage when the switch is off (Fig. 5.1*b*). The leakage resistance R_{off} is quite large (greater than 50 MΩ) for most electronic switches, but even a few picofarads of leakage capacitance C_{off} can produce serious leakage at high frequencies. A very effective method of minimizing such leakage effects is to *combine shunt and series switches* into tee-network transfer impedances (Fig. 5.1*c*). Given the forward impedance $Z_{\text{on}} = R_{\text{on}}$ of the closed shunt switch and the leakage impedance

$$Z_{\text{off}} = \frac{R_{\text{off}}}{j\omega R_{\text{off}} C_{\text{off}} + 1} \tag{5.1}$$

the tee-network short-circuit transfer impedance (Sec. 1.5) of the switching network in Fig. 5.1*c* is

$$Z = R + Z_{\text{off}} + \frac{RZ_{\text{off}}}{Z_{\text{on}}} = R + \frac{R_{\text{off}}}{j\omega R_{\text{off}} C_{\text{off}} + 1}\left(1 + \frac{R}{R_{\text{on}}}\right) \tag{5.2}$$

We see that capacitive as well as resistive leakage is effectively shunted to ground. For $R = 10\ \text{k}\Omega$, $R_{\text{on}} = 50\ \Omega$, $R_{\text{off}} = 50\ \text{M}\Omega$, $C_{\text{off}} = 5\ \text{pF}$, $|Z|$ exceeds 10^4 MΩ at direct current and 60 MΩ at 100 kHz. To switch even higher

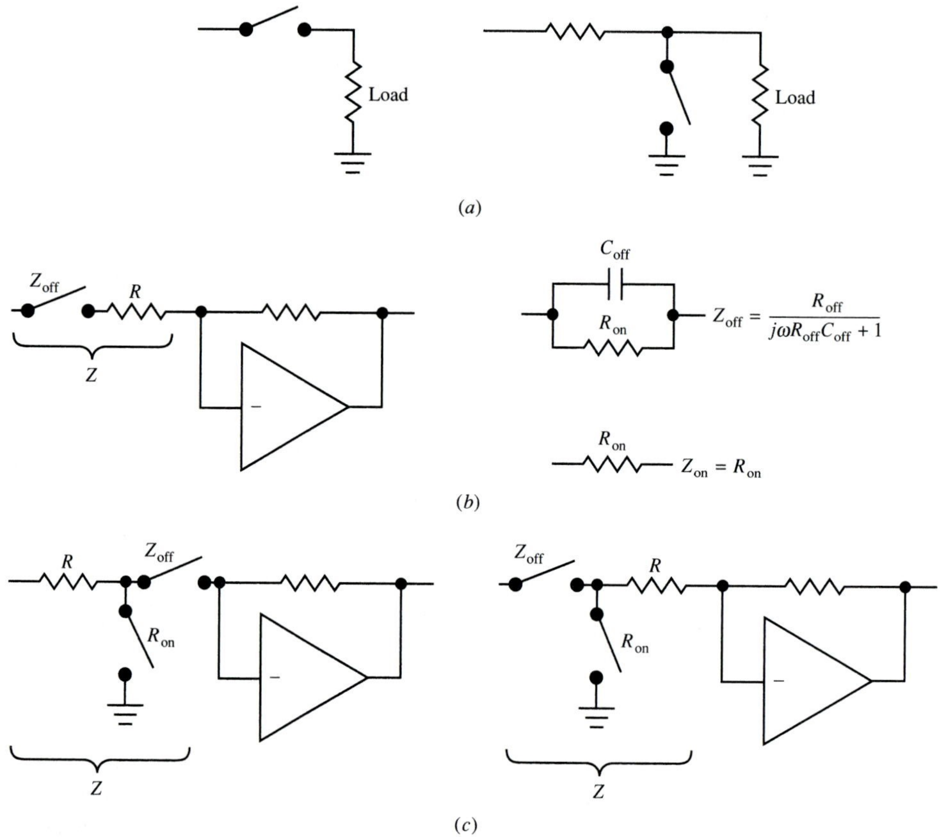

FIGURE 5.1
(*a*) Series and shunt switches; (*b*) a simple series switch; (*c*) the large tee-network transfer impedance formed by combination series-shunt switching.

frequencies, the resistance R can be replaced by a third switch, so that

$$Z = 2Z_{off} + \frac{Z_{off}^2}{Z_{on}} \tag{5.3}$$

(see also Fig. 5.3*b* and *d* and Ref. 1).

Electronic switching devices such as diodes and bipolar transistors can introduce not only spurious ON and OFF switch impedances, but also *offset voltages and currents* (Secs. 5.3 and 5.4).

If the switch-control circuit and the switched (signal) circuit have common dc impedances, a portion of the switching waveform may appear in the signal circuit, where switching steps give rise to control-voltage–dependent voltage offsets or *pedestals*.

For fast switching, switch-control voltages are usually fast-rising steps or pulses derived from digital logic circuits. Even small coupling capacitances C_C between the control circuit and the switched (signal) circuit *differentiate* these steep-sided waveforms and create *switching spikes* and/or steps in the signal cir-

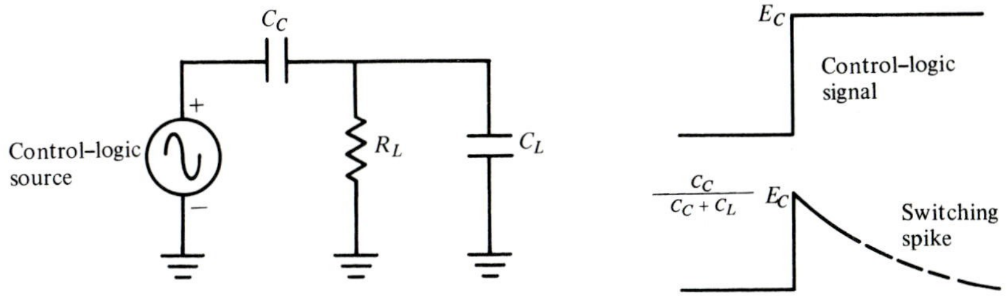

FIGURE 5.2
Turn-on spike (grounded control-signal source, grounded load) due to a switch capacitance C_C. The dashed lines indicate that the spike will be modified when the switch actually turns on. Until then (neglecting nonlinear circuit effects), the spike has the form of a decaying exponential with initial amplitude $E_C C_C/(C_C + C_L)$ and time constant $R_L(C_C + C_L)$. For a purely capacitive load ($R_L = \infty$), the exponential becomes a step (based on Ref. 1).

cuits (Fig. 5.2). Switching spikes may have the full control-voltage step amplitude and constitute very annoying interference which can trip signal-circuit comparators and may even overload amplifiers in low-voltage signal circuits. In capacitive signal-load circuits, switching spikes are integrated into steps (Sec. 5.8).

Reducing the coupling capacitance C_C will make switching spikes shorter but not smaller. Spike amplitudes can be reduced by increasing the signal load capacitance or by otherwise reducing the frequency response of the signal circuit, if this is compatible with signal-bandwidth requirements.

5.3 DIODE SWITCHING CIRCUITS

Figure 5.3 shows a variety of *diode switches* (*diode-bridge switches*). Figure 5.4 presents simple diode equivalent circuits in the ON and OFF conditions. Diode switches are perhaps the simplest electronic switches and can be quite fast. It is best to employ modern fast-recovery, low-leakage silicon-junction diodes in matched monolithic sets. The leakage capacitance C_{off} in Fig. 5.4 can be less than 2 pF, R_{off} is greater than 100 MΩ, and R_{on} will be 25 to 50 Ω. Unfortunately, diode characteristics (see also Chap. 4) imply an *offset voltage* of about 0.6 V when the diode is on and an *offset current* I_0 when the diode is off. The offset voltage rises with temperature by 2 to 3 mV/°C and can be canceled in matched diode-pair circuits. The small leakage current, typically 0.1 to 2 nA at room temperature, doubles every 7 to 10 °C.
To turn each diode switch in Fig. 5.3 fully off (i.e., to turn *all* series diodes off) requires control voltages *exceeding the largest signal-voltage excursion in absolute value*. Again, to keep a diode-bridge switch on, the control circuits must supply a *current* at least *as large as the largest signal current required to flow through the switch*. Otherwise, one of the series diodes could turn off, and the bridge would act as a voltage limiter in the manner discussed in Chap. 3. The resulting control-current requirement can be quite large if the switch must pass high-frequency currents charging capacitors in sample-hold circuits (Sec. 5.8). The six-diode switches of Fig. 5.3*b* and *d* combine shunt and series switching

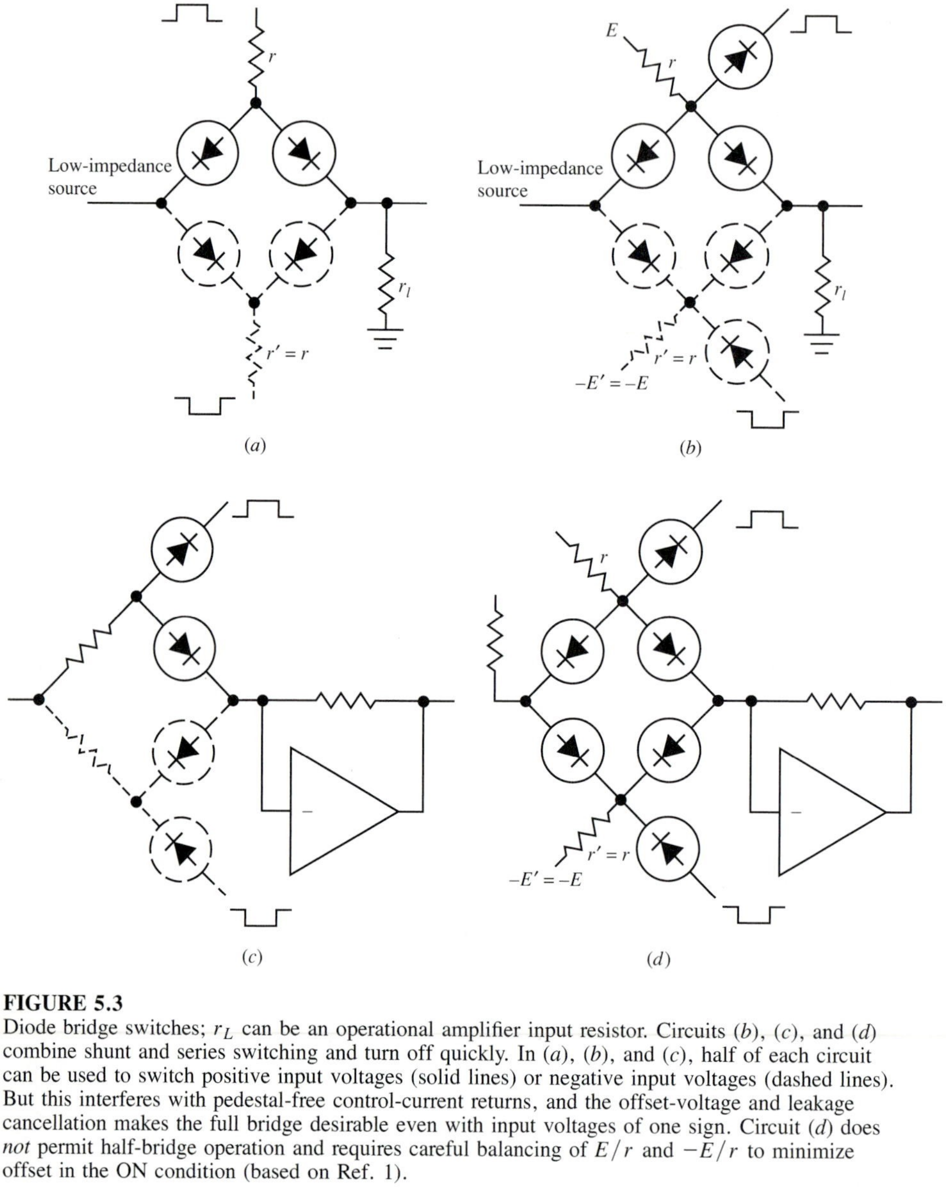

FIGURE 5.3
Diode bridge switches; r_L can be an operational amplifier input resistor. Circuits (*b*), (*c*), and (*d*) combine shunt and series switching and turn off quickly. In (*a*), (*b*), and (*c*), half of each circuit can be used to switch positive input voltages (solid lines) or negative input voltages (dashed lines). But this interferes with pedestal-free control-current returns, and the offset-voltage and leakage cancellation makes the full bridge desirable even with input voltages of one sign. Circuit (*d*) does *not* permit half-bridge operation and requires careful balancing of E/r and $-E/r$ to minimize offset in the ON condition (based on Ref. 1).

in the manner discussed in Sec. 5.2. The symmetric bridge circuits in Fig. 5.3, moreover, can reduce the effective diode leakage current by a factor of at least 3 if matched bridge diodes are fabricated on the same monolithic chip. Unfortunately, switching spikes from the push-pull control circuit will not cancel accurately because of the nonlinear nature of the diode capacitances. One may try to use trimmer capacitors to vary the control-step rise and fall times on one side of the bridge.

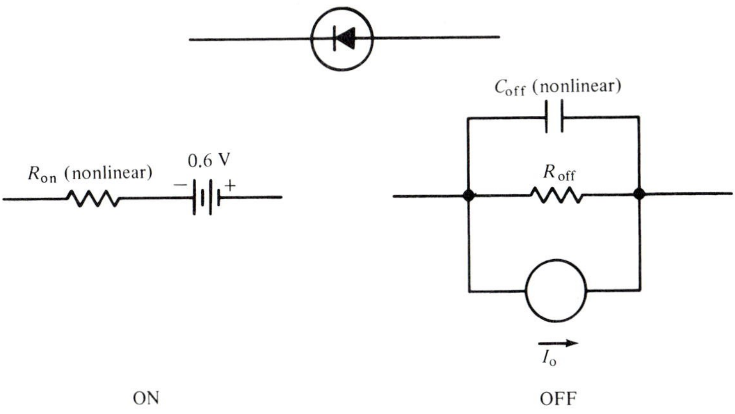

FIGURE 5.4
Diode models in the OFF and ON states.

The *turnoff time* of a diode switch will depend on the time constant of the circuit which charges the OFF-state diode capacitance or capacitances until each diode is forward-biased. In the case of the six-diode switch of Fig. 5.3*b*, the turnoff time constant is of the order of $C_{off}R_{on}$, so that six-diode switches can turn off very quickly. The *turn-on time* will depend on the resistance in the signal circuit.

5.4 BIPOLAR-TRANSISTOR SWITCHES (REF. 1)

Switch-Transistor Characteristics; Inverted Connection

The collector characteristics of a bipolar transistor (Fig. 5.5) indicate that a relatively small change of the transistor base current will switch either *pnp* or *npn* transistors between *saturation* (ON condition) and *cutoff* (OFF condition); note that a saturated transistor will pass current in both directions. In the *npn* circuit of Fig. 5.5, a sufficiently *positive* base current i_B will cause saturation, and the transistor is cut off for zero or negative i_B.

Actual transistor collector characteristics show that (Ref. 1)

1. The reciprocal slopes $\Delta e_{CE}/\Delta i_C = R$ of the saturation and cutoff characteristics, respectively, yield *forward resistances* $R = R_{on}$ between 10 and 100 Ω and *back resistances* $R = R_{off}$ between 1 and 500 MΩ, much as in the case of junction-diode switches.
2. The saturation and cutoff characteristics in Fig. 5.5*b* do not pass exactly through the coordinate origin, as ideal switch characteristics would. The resulting offset errors are interpreted as follows: The saturation characteristic is displaced horizontally by a positive *saturation offset voltage* $i_B r_E + V_0(i_B)$ of 10 to 500 mV, and the cutoff characteristic is displaced vertically by a positive *leakage current* I_0 between 0.02 and 50 nA at 25° C.

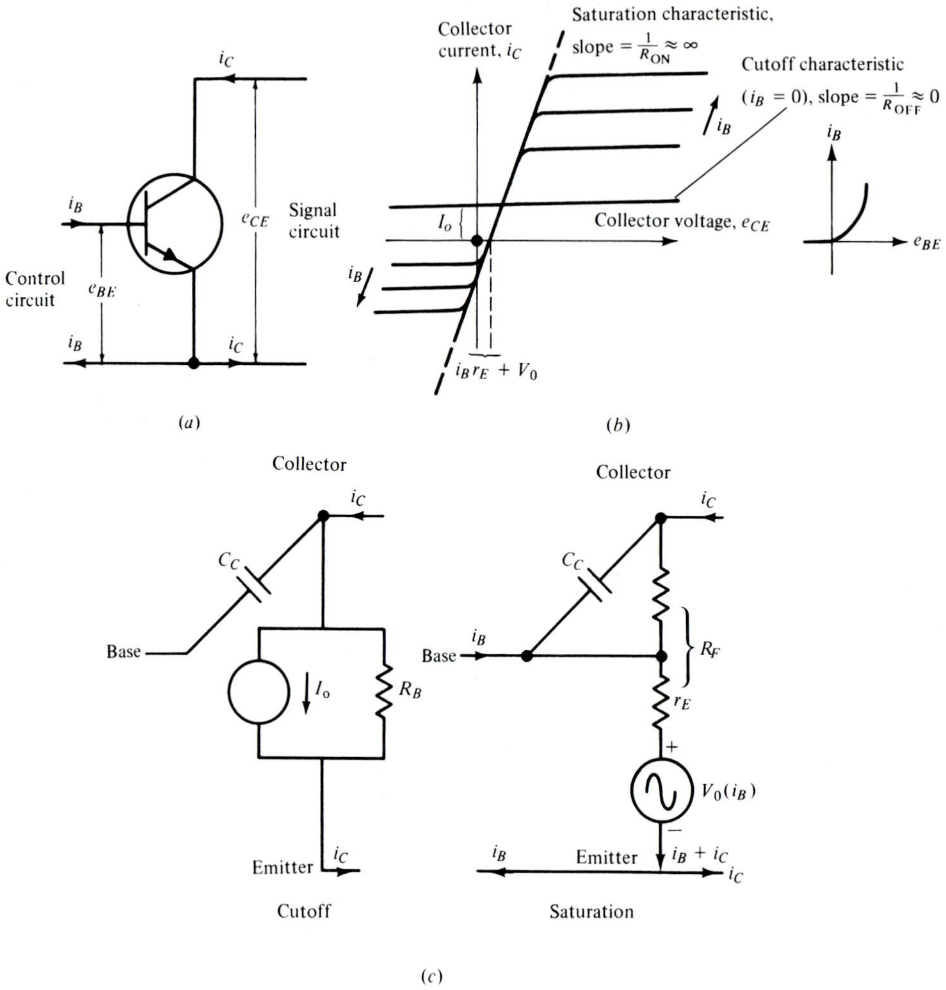

FIGURE 5.5
(*a*) An *npn* transistor switch; (*b*) transistor characteristics; (*c*) simplified equivalent circuits (based on Ref. 1).

This situation is summarized by the simplified *equivalent circuits* of Fig. 5.5*c*. If an *npn* or *pnp* junction transistor is *inverted*, i.e., if one interchanges the roles of emitter and collector, the device still operates as a transistor whose characteristics are shaped like those in Fig. 5.5*b*, but they are *compressed vertically*. More specifically, experimental and theoretical investigation of the normal and inverted transistor characteristics reveals that

1. The current amplification factor $\beta = \partial i_C / \partial i_B$ is decreased by a factor of 5 to 100 in the inverted connection, and $-\ln \alpha = -\ln(\partial i_C / \partial i_E)$ is decreased by about a factor of 10. The inverted connection is, therefore, not suitable for amplification or for power switching with small control currents.

2. The switch forward resistance (reciprocal slope of the saturation characteristic) is only slightly higher in the inverted connection.
3. *The inverted transistor connection typically decreases the offset voltage* $V_0 + i_B r_E$ *by a factor of at least 10.*

Note: In earlier (germanium, silicon-alloy) switching transistors, the leakage-current offset in the OFF state (as well as the offset voltage in the ON state) was substantially reduced by the inverted connection. This is *not* true for modern silicon switching transistors, where what little leakage is left is due mainly to surface leakage and charge generation rather than to diffusion in the junction.

Transistors suitable for analog switching are silicon epitaxial-planar types combining low saturation resistance and leakage with relatively small junctions for high switching speed. For such typical silicon transistors (e.g., Motorola 2N2220, 2N2369, and MPS6512) used in the inverted connection at 25°C, we find

Inverted-saturation voltage $i_B r_E + V_0 i_B < 1$ mV ($i_B = 0.5$ mA)
Inverted leakage current $I_0 < 0.1$ nA ($i_B \approx 0$)
Inverted-saturation resistance ("dynamic resistance")

$$R_{\text{on}} < 25\ \Omega \quad (i_B = 0.5 \text{ mA})$$

Back resistance $R_{\text{off}} > 1$ to 100 MΩ

Since the emitter-base and collector-base junctions of a junction transistor behave essentially as junction diodes, a transistor switch exhibits nonlinear emitter-base and collector-base capacitances (typically 5 to 15 pF for 1-V reverse bias) and turnoff storage time (below 0.1 μs for switching transistors). Transistor capacitances and storage times are, approximately, inversely proportional to the quoted alpha-cutoff frequency f_α, which is, therefore, a useful measure of the attainable switching speed. Since the collector-base junction is larger than the emitter-base junction, an inverted transistor has a higher input capacitance than a normal transistor, but typical switching times are still well below 1 μs for switching transistors with $f_\alpha > 100$ MHz.

Circuit Design

To turn a transistor switch off, the base-control circuit must supply a voltage at least large enough to reverse-bias *both* the emitter-base junction and the collector-base junction. To turn the transistor switch on, the base current must exceed the signal current divided by β_N (normal connection) or β_I (inverted connection).

Bipolar-transistor switching circuits (like diode switches but unlike FET switches) are controlled by current rather than voltage, so that *the designer must guard against impedances common to this control current and the signal circuit* if pedestal-type offsets in the signal circuit are to be avoided.

This isolation between control and signal circuits is achieved in the *transistor*

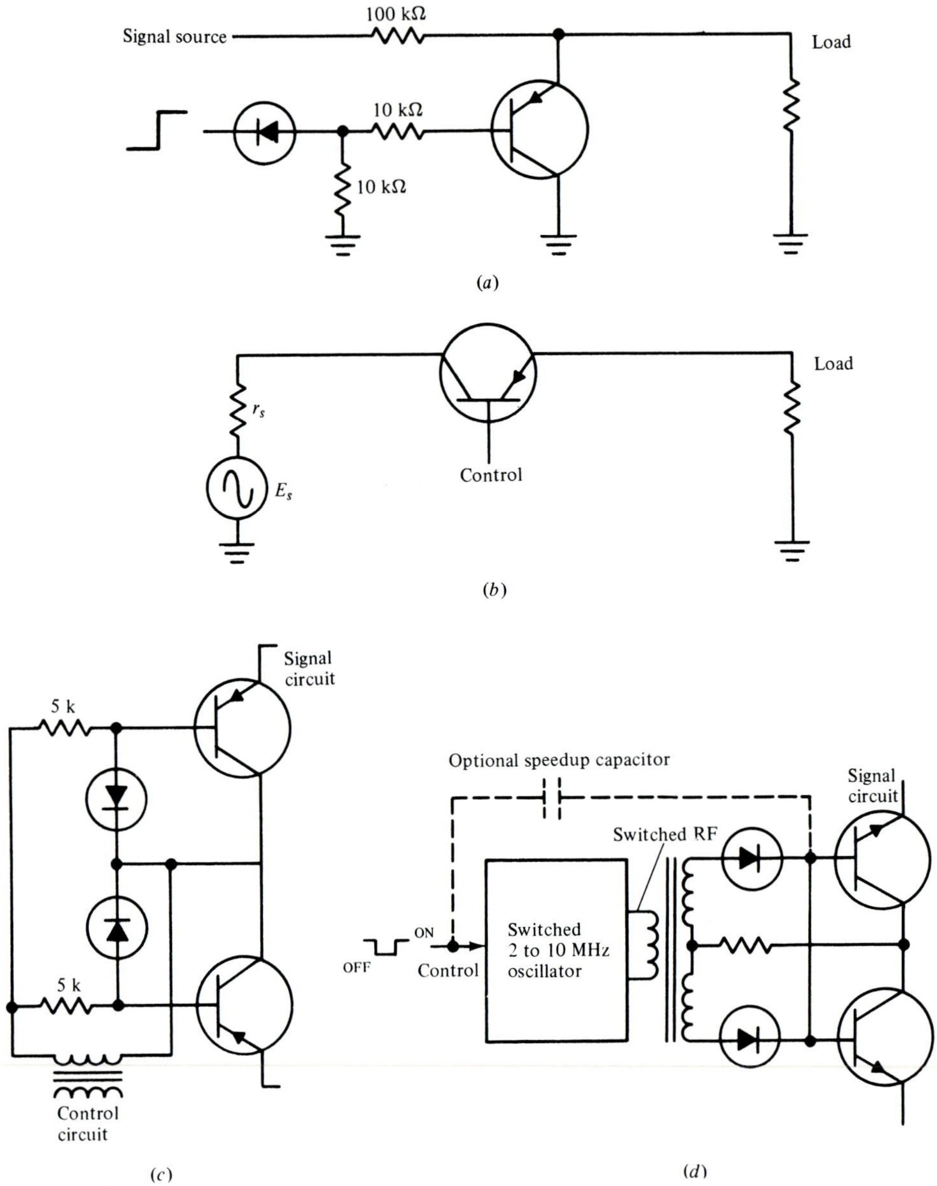

FIGURE 5.6
(*a*) Inverted-transistor shunt switch; (*b*) series switch; (*c*) matched inverted-transistor pair with floating control circuit; (*d*) dc transformer control for a transistor pair.

shunt switch (simple bipolar-transistor chopper) of Fig. 5.6*a*, where the switching transistor returns the control current to ground and causes only a relatively small saturation offset. But control-current return poses a real difficulty for the *series-transistor switch* of Fig. 5.6*b*, *unless the signal-source impedance* R_S *is quite small compared to the load.*

The best way to return the series-transistor-switch control current is with a *transformer-floated circuit* as in Fig. 5.6*c*; the circuit shown employs two matched inverted transistors to cancel much of the saturation offset voltage. But such a transformer-coupled circuit cannot be turned on for an indefinite time (since the transformer will not pass direct current) and is therefore mainly used for chopper operation. The *dc transformer circuit* of Fig. 5.6*d* solves this problem at the expense of some circuit complication.

Nonsaturating Transistor Switches

To avoid the extra time ("storage time") required to sweep carriers out of a saturated-transistor base-collector junction when the switch is turned off, two types of bipolar-transistor switches avoid saturation and utilize the "working" region of the transistor characteristics in Fig. 5.5*b* (i.e., the transistor works like an amplifier in the ON condition of the switch). The first such switch is the *switched current source* described in Sec. 6.3, and the second method involves *switching an amplifier stage* (Secs. 5.8 and 5.9). Such switches turn off within 10 to 30 ns even when they must supply large currents (10 to 100 mA) in the ON condition.

5.5 JUNCTION FIELD-EFFECT TRANSISTOR SWITCHES

The *junction field-effect transistors* (JFETs) sketched in Fig. 5.7 contain "channels" of diffused *p*-type or *n*-type impurities which act, for *zero* gate-to-source voltage e_{GS}, *like a passive silicon resistor*. In this ON state, the JFET passes current in either direction, and *there is no voltage offset*. JFET forward resistances range between 3 and 100 Ω and increase slightly with temperature (0.1 to 0.7 percent/°C) and with current (about 2 percent/mA).

Increasingly positive (for *p*-channel JFETs) or increasingly negative (for *n*-channel JFETs) e_{GS} will "pinch" the flow of carriers in the channel (electrostatic field effect) and decrease the channel conductance g_{DS} (Fig. 5.7) until the current ceases at the *pinchoff voltage*. In this OFF condition, there are small leakage currents (typically less than 0.01 to 0.3 nA at 25°C, doubling every 10°C) across the gate-source and gate-drain junctions; these junctions also act as nonlinear capacitances. Figure 5.8 shows simple equivalent circuits in the ON and OFF states.

Forward resistances of *n*-channel JFETs are slightly lower than those of *p*-channel JFETs of similar geometry. Unfortunately, JFET junction capacitances increase with decreasing R_F (larger junctions). High-priced units may have $R_F = 4\ \Omega$ at 25° C with 20-pF junction capacitances, but $R_F = 30\ \Omega$ is more typical.

Figure 5.9 shows practical *n*-channel JFET switching circuits (reverse voltages and junctions for *p*-channel). To turn each switch in Fig. 5.9 off, *the negative driver-amplifier output must exceed the largest signal excursion at S by 8 to 12 V in absolute value*. To turn the JFET on, *we would like to keep the gate-to-source voltage at 0 V*. We must not make the gate so positive that the gate-to-channel

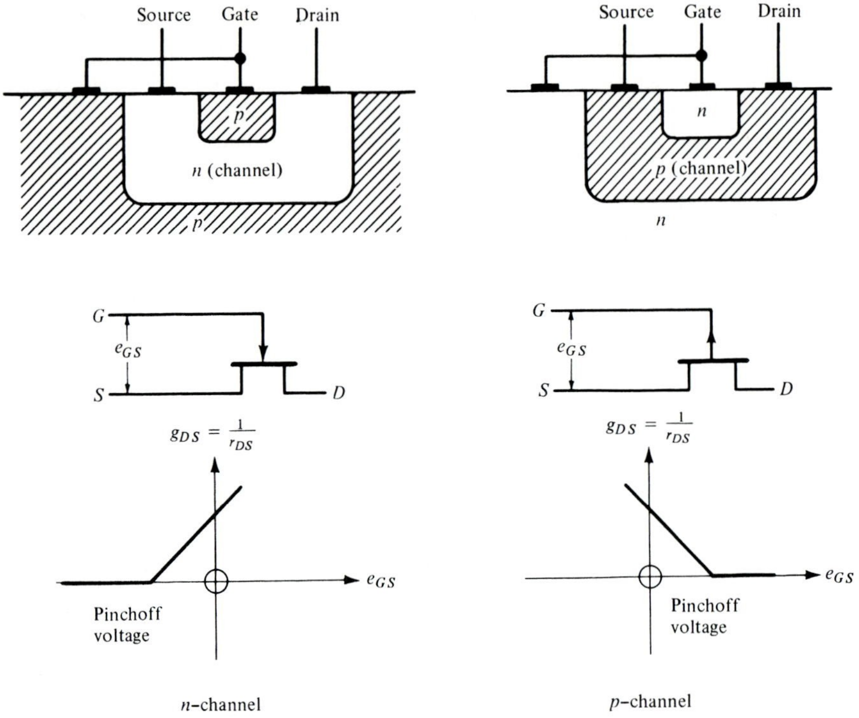

FIGURE 5.7
Junction field-effect transistors and their dc switching characteristics. Note that switch forward resistances will depend on the control voltage e_{GS}.

junction (Fig. 5.7) is forward-biased; any changes of e_{GS} from 0 V will, moreover, "modulate" the switch conductance and disturb resistance matching in accurate computing circuits.

In Fig. 5.9*a*, the ON value of the driver-amplifier output exceeds the largest signal excursion by at least 0.6 V to turn the coupling diode off. The gate voltage

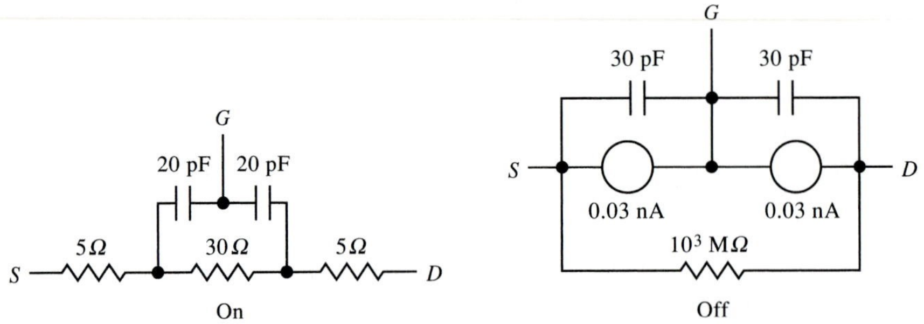

FIGURE 5.8
Simple ON and OFF equivalent circuits for a typical JFET switch. Note that gate-junction capacitances are nonlinear (highest at low junction voltages).

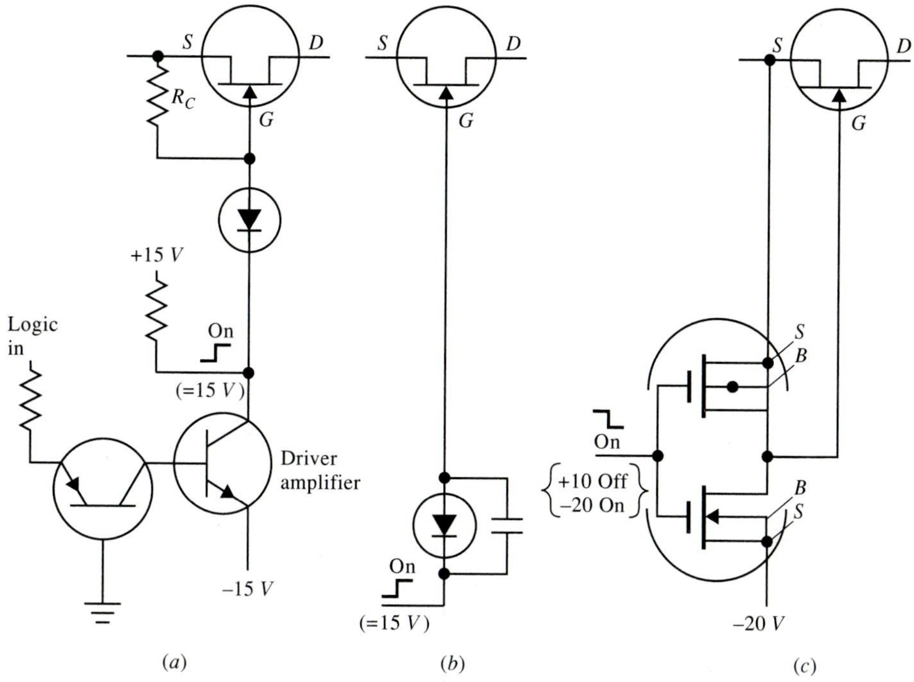

FIGURE 5.9
n-channel JFET switching circuits.

will then follow the signal voltage at the source, unless signal excursions are fast compared to the time constant determined by R_C (typically 20 to 100 kΩ) and the circuit capacitances at G, so that e_{GS} is near zero at least for low signal frequencies. Unfortunately, lower R_C values load the signal circuit at S when the switch is on, but low R_C also speeds up the turn-on time of the switch.

The circuit of Fig. 5.9*b* serves to switch a voltage on for a short time, during which the signal remains more or less constant (as in a multiplexer sampling successive signal sources, Sec. 5.7). Here R_C is omitted; e_{GS} is reduced to (approximately) zero by the positive charge transmitted by the coupling-diode capacitance as the driver output swings more positive than the diode turnoff voltage. One may employ a varactor diode (Ref. 5) or an external 10- to 100-pF capacitor to get the correct turn-on charge. But the gate will not follow the signal when the switch is on; large positive signal excursions may even pinch the FET off and get clipped.

In the improved circuit of Fig. 5.9*c*, the output stage of a CMOS switch driver shorts the JFET gate to the source through a few hundred ohms of MOSFET resistance (see also Sec. 5.6) when the JFET is on, so that e_{GS} is kept neatly near zero.

Note that JFET switches will normally turn on when the control-voltage supply is turned off or fails. If this creates problems, special precautions may be needed.

5.6 MOSFET AND CMOS INTEGRATED CIRCUITS

The *enhancement-mode* insulated-gate FETs illustrated in Fig. 5.10 are off for zero gate-to-source voltage e_{GS}. The MOSFET conductance increases to turn on as e_{GS} is made more positive (*n*-channel) or more negative (*p*-channel). MOSFETs with *depletion* characteristics similar to those in Fig. 5.7 also exist but are rarely used as switches.

Figure 5.11 shows simple equivalent circuits. With the gate electrode insulated by a thin silicon oxide layer, gate leakage is essentially negligible, and gate-source and gate-drain capacitances are very low (<2 pF, Fig. 5.12). On the

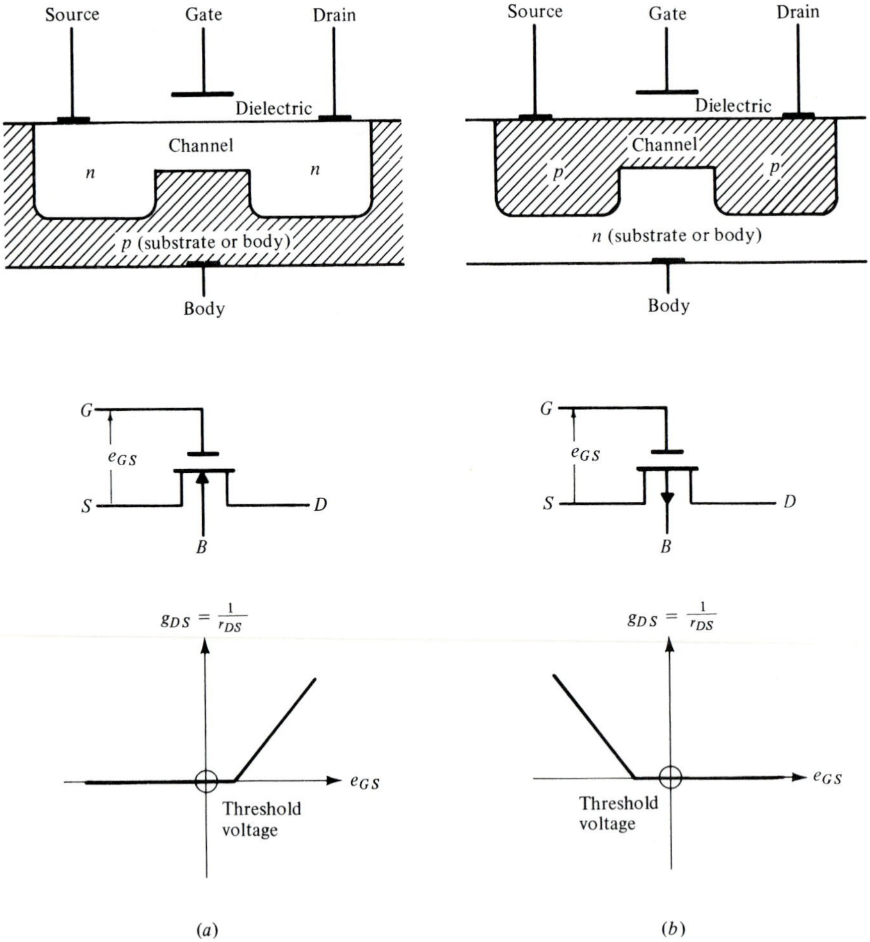

FIGURE 5.10
Enhancement-type MOSFETs and their dc switching characteristics.

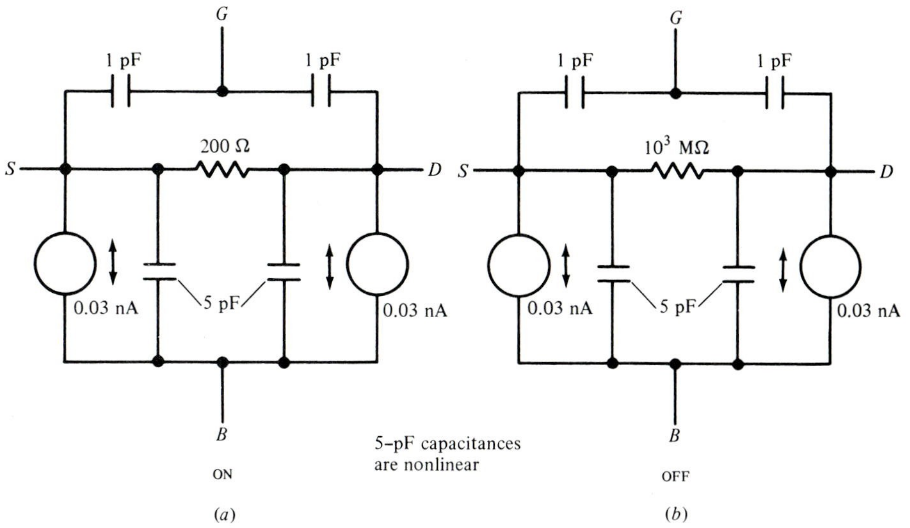

FIGURE 5.11
Simple MOSFET switch equivalent circuits. The substrate-junction capacitances are nonlinear (highest for low junction voltages).

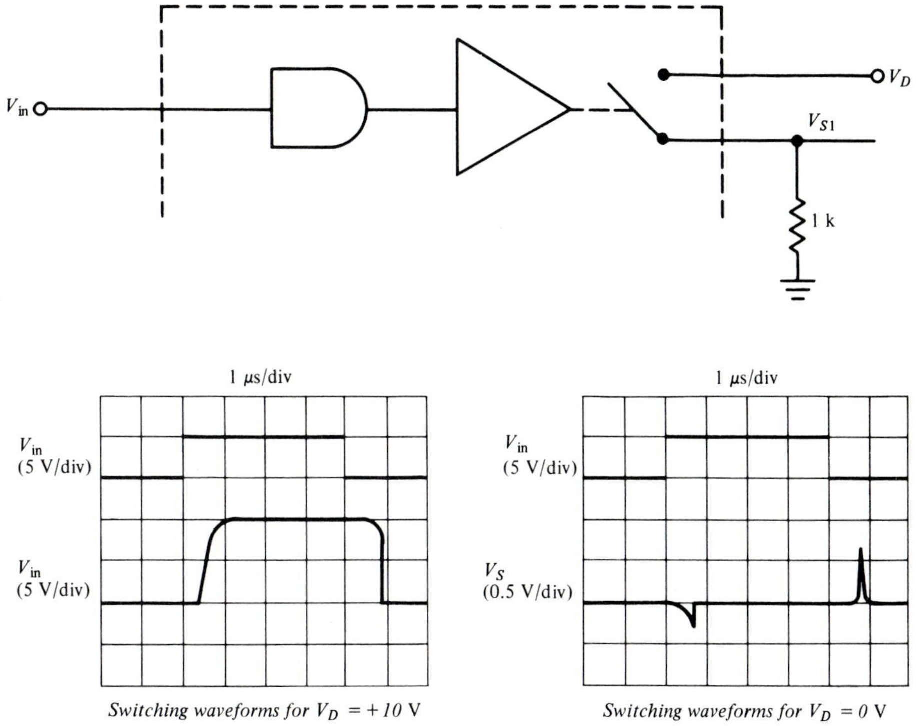

FIGURE 5.12
Switching performance of an integrated-circuit MOSFET switch/switch driver combination, showing the low switching delay and short switching spikes (*Analog Devices, Inc.*).

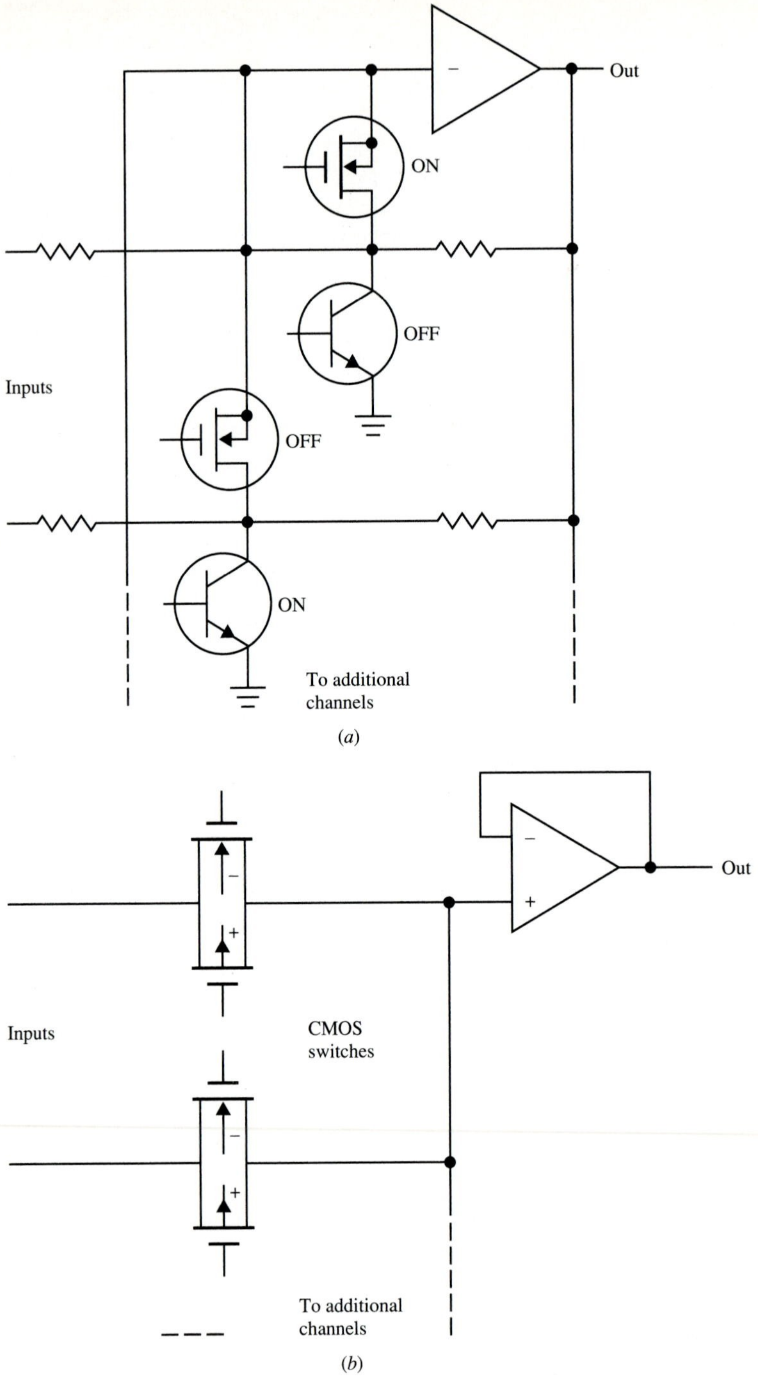

FIGURE 5.13
Multiplexer circuits designed to minimize the effects of FET switch forward resistances.

other hand, MOSFET forward resistances (30 to 800 Ω) tend to be larger than JFET resistances.

In MOSFET switches, the *substrate-to-channel junction must be back-biased for all signal-voltage excursions* to avoid surprises depending on the circuit design (the circuit may "latch up" with a forward-biased substrate junction until all power supplies are momentarily turned off). For signals below 0.6 V in absolute value, it may be possible to simply short substrate and source. Some MOSFETs come with an extra insulating diode in their substrate connection to prevent latch-up. Note also that the thin gate-insulation layer of many MOSFETs is easily damaged by electrostatic charges during circuit construction.

If their higher forward resistances can be tolerated, the design of MOSFET switches is much simpler than that of JFET switches because the insulated MOSFET gate can simply be overdriven in the ON and OFF conditions. Multiple MOSFET switches and MOSFET switch drivers are easily constructed as monolithic integrated circuits. In particular, complementary pairs of *n*-channel and *p*-channel MOSFETs (*CMOS integrated circuits*) are very useful as switches (Figure. 5.13) and as switch-driving amplifiers (Fig. 5.9*c*).

Integrated circuits incorporating MOSFET switches are usually laser-trimmed at the wafer stage to allow for FET forward resistances as well as circuit-resistance tolerances (Chap. 6). Since the MOSFET forward resistance, like that of a JFET, changes slightly with current (Sec. 5.5), accurate MOSFET multiplexers either drive a high-impedance follower or use sufficient series resistance to minimize FET resistance-change effects (Fig. 5.13). Figure 5.14 shows circuit examples using complementary MOSFETs.

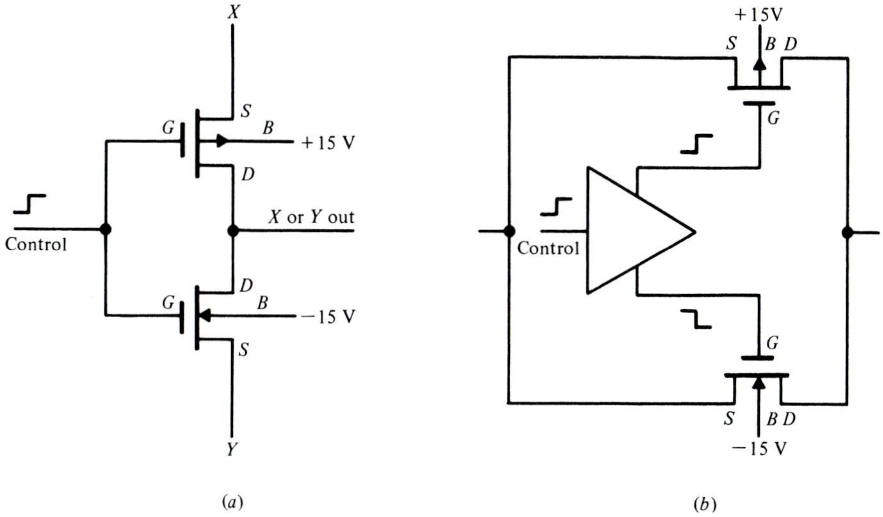

FIGURE 5.14
(*a*) A CMOS SPDT switch; (*b*) a CMOS SPST switch with complementary gate-control signals from a CMOS push-pull switch driver.

5.7 DESIGN DECISIONS AND IMPROVED SWITCHES

Sections 5.3 to 5.6 describe properties of different electronic switches. Designers must select the best switching device and circuit for a specific application. Often switches are components of larger integrated circuits, whose fabrication will affect device selection. The following are general selection guidelines (see also Chap. 6):

1. *Diode-bridge switches* (Sec. 5.3) serve in modulators and sample-hold circuits (Sec. 5.8) if they operate at high switching rates (1 to 100 MHz) but require only moderate gain and dc accuracy (0.2 to 3 percent error).
2. *Saturated bipolar-transistor switches* (Sec. 5.4) serve mainly as modulators (choppers), to turn off (shunt) transistor amplifiers, and as auxiliary shunt switches (see below). They are used less and less, except where their device characteristics fit an overall integrated-circuit design. *Nonsaturating bipolar-transistor switches* (Sec. 5.4) serve widely as fast and accurate switched current sources in nonmultiplying digital-to-analog converters (Chap. 6).
3. *CMOS FET switches* (Sec. 5.6) are now used for most general-purpose switching (Ref. 3). They permit accurate switching and can switch (less accurately) at low megahertz rates. Although *junction-FET* switches (Sec. 5.5) may have somewhat lower forward resistances, driving circuits are less convenient to fabricate than CMOS switch/driver combinations.

At relatively low signal frequences (below 10 kHz), *combinations of shunt switches, series switches, and operational amplifier feedback* permits instrumentation-quality switching with gain and dc accuracies approaching those of linear operational amplifier circuits, with the exception of switching spikes. The latter can often be eliminated with "deglitching" sample-hold circuits (Sec. 5.8 and Chap. 6).

Figure 5.15 illustrates a technique for switching operational amplifier inputs with FET switches. Each circuit employs complementary shunt and series switching of the input resistor to ground potential (ground or summing junction), which yields significant advantages:

1. The voltage across each switch is practically zero in both ON and OFF states. This simplifies gate-drive requirements and minimizes forward resistance changes.
2. Combined series-shunt switching minimizes signal leakage through the FET junction capacitances (Sec. 5.2).

In addition, the impedances seen by the signal source in the ON and OFF states are similar and can be accurately matched in Fig. 5.15*b* (e.g., by laser trimming). This is important in ADC ladder circuits (Fig. 6.7). The MOSFET switches in Fig. 5.15*b* could be replaced with CMOS switches.

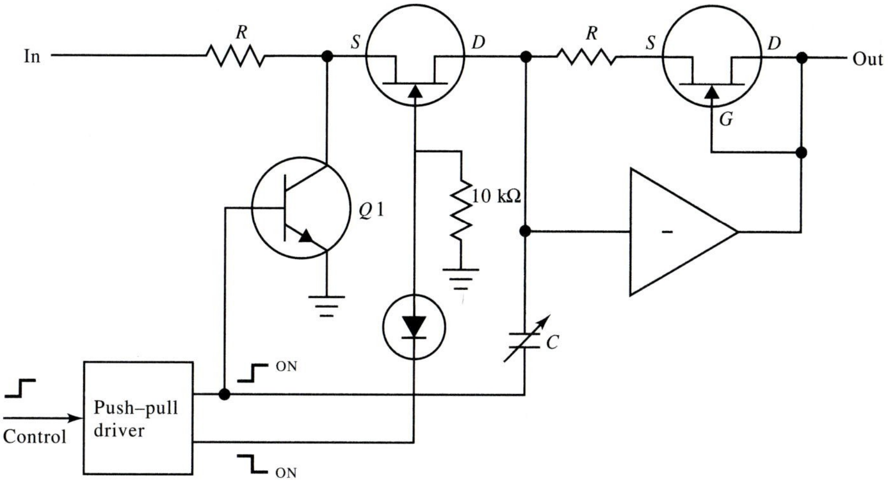

(*a*)

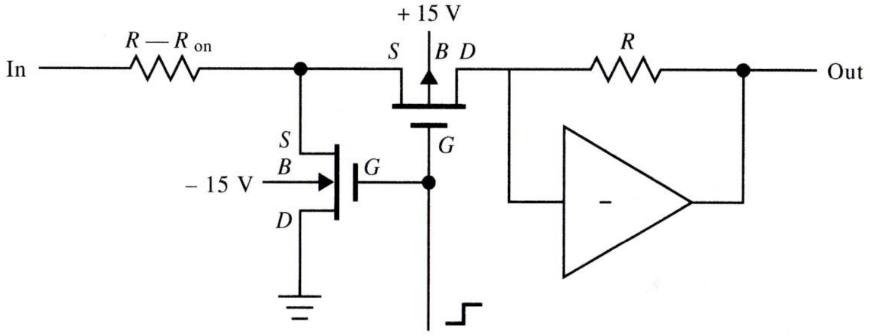

(*b*)

FIGURE 5.15
Summing-junction switching with FETs. In (*a*), variations of R_{ON} with temperature and current are compensated by a matched FET in the feedback circuit. And *C* is a trimmer for partial switching-spike cancellation. The shunt switch is sometimes replaced by a pair of shunt limiter diodes limiting FET source excursions to $\pm 0.6V$. In (*b*), the signal source (e.g., the ladder circuit of Fig. 6.7) sees roughly equal switch impedances to ground in the ON and OFF states (see also Fig. 6.6*b*).

Finally, the accurate shunt switch shown in Figure 5.16 has its FET switch inside a high-gain operational amplifier feedback loop, so that the FET forward resistance is effectively divided by the loop gain (Ref. 1).

5.8 SAMPLE-HOLD CIRCUITS

Sample-hold (or *track-hold*) *circuits* (see Ref. 1) hold a timed sample of an analog input for analog-to-digital conversion or analog storage (delay). The output X_0 of the basic capacitor-storage track-hold circuit shown in Fig. 5.17*a* tracks the input X_1 while the switch is on (*TRACK mode*). A HOLD command opens

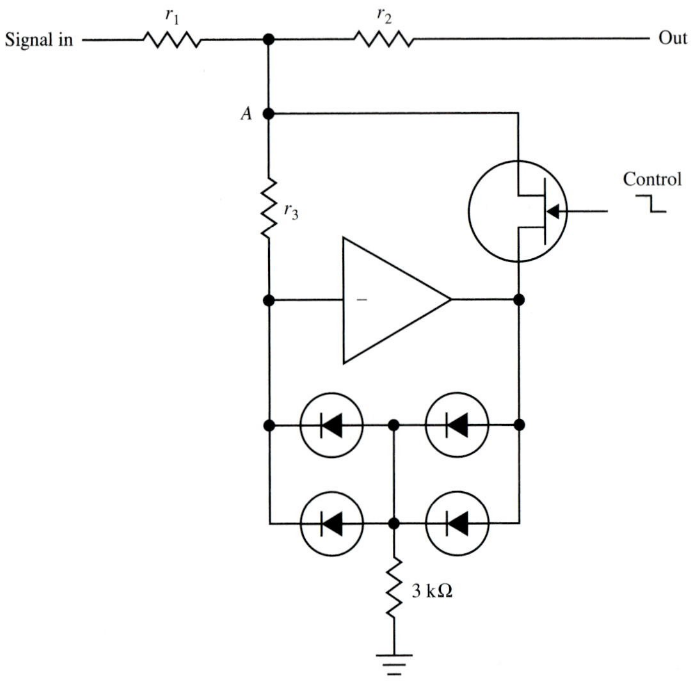

FIGURE 5.16
Tee network with an active shunt switch. Point A is returned to ground potential either through r_3 or through the FET resistance divided by the operational amplifier loop gain.

the switch, so that the capacitor stores the last input value, and the output ideally remains constant (*HOLD mode*). The storage circuit can drive external loads without discharging the holding capacitor.

An ideal track-hold circuit would follow its input instantaneously and exactly in TRACK and would hold instantaneously and exactly for an indefinite time after receiving a HOLD command. Figure 5.17*b* illustrates the response of an actual track-hold circuit. The *current capacity of amplifiers and switches* not only limits the maximum tracking rate, but also places a lower limit on the TRACK period required to catch up with specified input-signal excursions. For accurate tracking of high-frequency signal components too small to cause rate limiting, we require low phase shift in TRACK, i.e., sufficient bandwidth in the capacitor-charging input circuit or amplifier. The effect of the capacitor-charging time constant on small-signal phase shift can be compensated with suitable equalizing networks in the input amplifier. The charging time constant must, in any case, be less than one-seventh of the shortest contemplated TRACK period for 0.1 percent accuracy (about one-tenth for 0.01 percent) to allow for exponential step response.

The HOLD output, ideally constant, drifts because of capacitor leakage and dielectric absorption, switch and amplifier-input leakage, and amplifier voltage

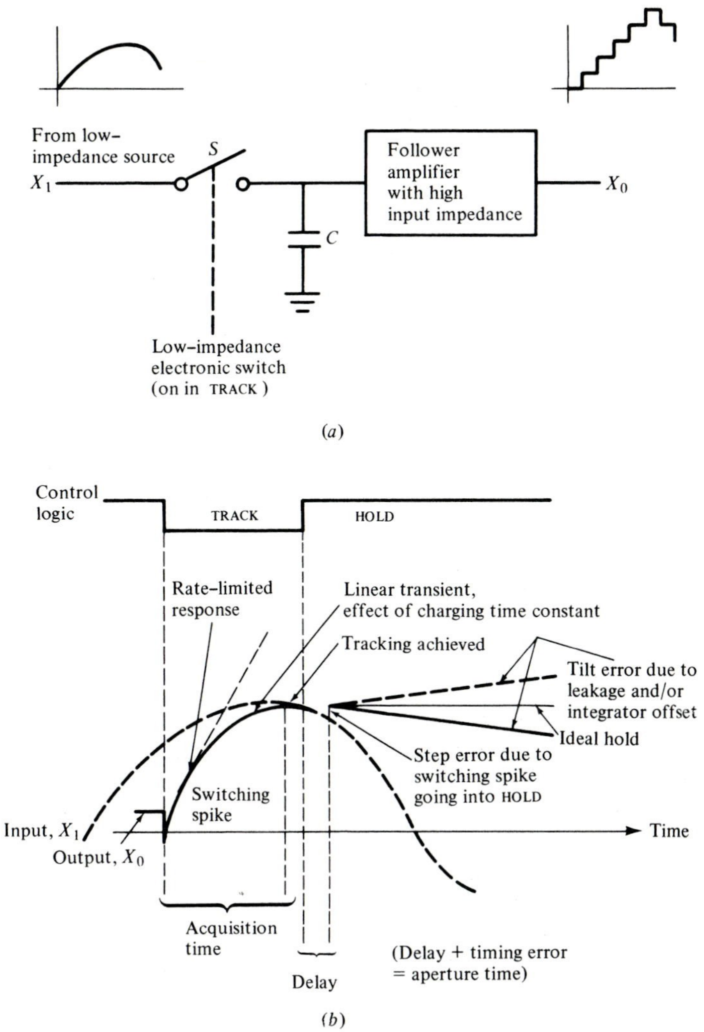

FIGURE 5.17
(*a*) Switched-capacitor sample-hold circuit; (*b*) typical response, showing error effects. Rate limits are determined by current capacities of switches and/or amplifiers and may differ for positive and negative rates. Switching spikes can be positive or negative. Timing error and switching spikes going into TRACK will not affect HOLD output. Errors due to dc offset in TRACK are not shown (based on Ref. 1).

offset. A total error current i_E contributes the absolute drift error

$$e_H = \frac{|i_E|}{C}\tau_H \tag{5.4}$$

where C is the holding capacitance and τ_H is the elapsed time in HOLD. An amplifier offset of e_D V referred to the input will be reproduced at the follower-amplifier output.

With electronic switching, the *switching spike* caused by differentiation of the HOLD control step in the switch capacitances charges the holding capacitor and contributes a spurious voltage step (Fig. 5.17*b*) whose amplitude is approximately

$$e_s = a_S E_C \frac{C_s}{C} \tag{5.5}$$

where E_C is the control-step amplitude and C_s is the effective switch capacitance. And a_S is a proportionality factor, which we may attempt to reduce through push-pull spike cancellation in symmetric switching circuits, and/or by adding inverted spikes through a trimmer capacitor. With well-designed circuits, switching-spike errors are not serious in ordinary track-hold operation, but may accumulate in certain computations (Ref. 1).

Finally, switch-response delays in executing the HOLD command result in possibly serious *timing errors*; i.e., the sample-hold circuit actually holds a later input voltage value than desired.

The holding errors in (5.4) and (5.5) are seen to improve with *increasing* holding capacitance *C*, while *low* values of *C* require less current and *favor fast tracking*. The required compromise is best resolved with high-current solid-state amplifiers and switches, which permit the use of relatively large holding capacitances. In practice, i_D and e_D can be kept below 1 to 10 nA and 50 to 500 μV, respectively.

If one must track fast signals accurately *and* one requires relatively long holding periods, it is best to employ a fast (low-*C*) sample-hold to track the signal and to sample the output of this fast sample-hold with a second, accurately holding (high-*C*) sample-hold after the input sample-hold has just settled into HOLD. Such *track-hold pairs* are, in any case, needed in applications requiring true zero-order-hold output (i.e., output of successive sample levels not separated by noisy TRACK periods, as in simple sample-holds).

Figure 5.18 shows an improved sample-hold circuit (see also Sec. 5.8), and Table 5.1 shows typical specifications.

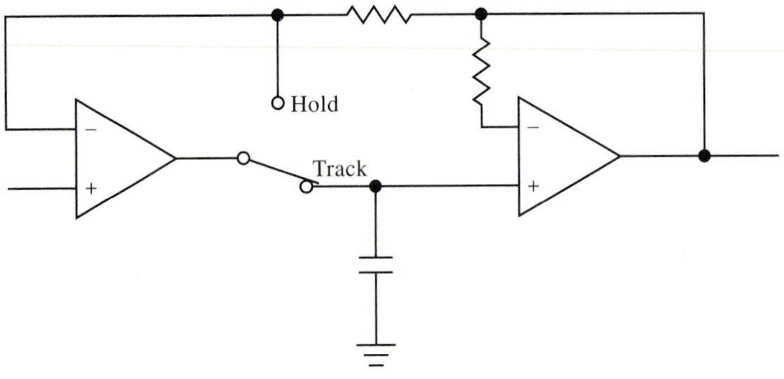

FIGURE 5.18
This commonly used sample-hold circuit improves accuracy and frequency response in the TRACK mode by enclosing the entire circuit in a feedback loop.

TABLE 5.1
Typical specifications of 10-V sample-hold circuits

(*a*) TRACK Mode

Gain nonlinearity	0.005 to 0.1%	
Gain drift (after zero adjustment)	0.002 to 0.005%/°C	
Noise	1 mV peak (output)	
Full-signal bandwidth	10 to 500 kHz	1 to 100 MHz
Small-signal dynamic sine wave error	0.1 to 0.5% of half-scale to full-scale bandwidth	
Acquisition time (settling time for 10-V step): Within 10 mV Within 1 mV	— 1 to 20 μs	10 to 1000 ns 100 to 2000 ns

(*b*) HOLD Mode

Aperture time (time to switch into HOLD)	100 to 600 ns	10 to 100 ns
rms aperture uncertainty ("jitter")	0.5 to 20 ns	5 to 100 ps
HOLD drift ("droop," 25°C)	0.1 to 2 mV/ms (roughly doubles every 10°C)	0.1 to 2 mV/μs
Switching step (pedestal)	1 to 1 mV	2 to 8 mV
Input/output cross talk	0.01 to 0.04% of half-scale	
HOLD capacitance	50 to 2000 pF	1 to 25 pF
HOLD capacitor	IC capacitor, Teflon or polystyrene capacitor	
Dielectric absorption error (following 1-s 10-V pulse)	2 to 5 mV	

Note: The faster sample-hold circuits (greater bandwidth, shorter acquisition time and aperture time) necessarily exhibit larger dc errors and drift. Their output can be resampled with a slower, more accurate sample-hold if necessary.

5.9 INTEGRATOR MODE CONTROL AND RELATED SWITCHING CIRCUITS

The simplest scheme for resetting an integrator merely resets it to *zero* by short-circuiting the integrating capacitor (say with a JFET switch); the desired initial output is then added to the output of the integrator proper. In spite of its simplicity, this resetting method is rarely used because the two voltages to be added must be scaled down and thus lose accuracy.

Figure 5.19 shows a *three-mode integrator control circuit* used for analog/hybrid computation (Refs. 1 and 11). In the RESET mode, the amplifier output follows the phase-inverted initial-condition (IC) input and charges

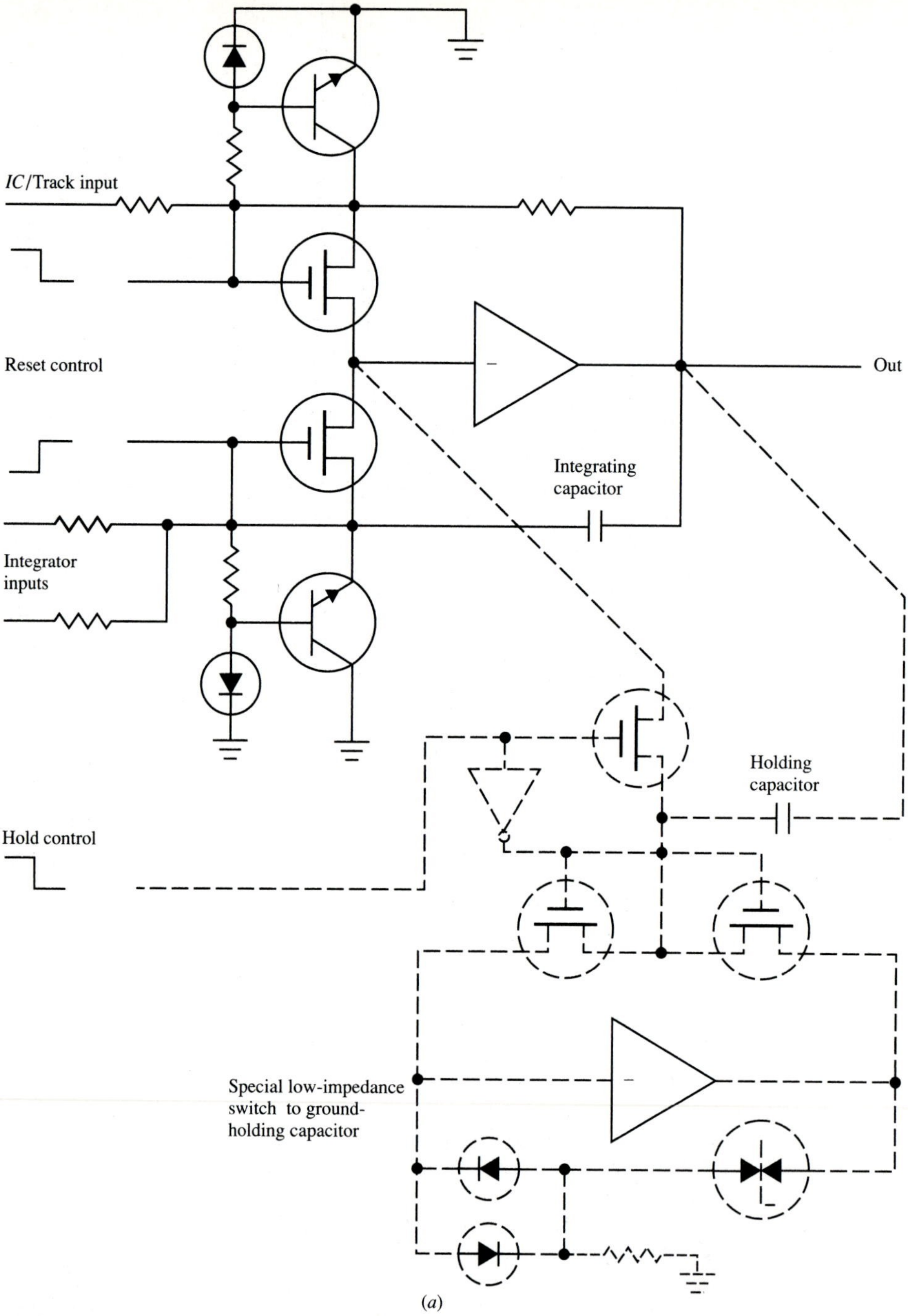

FIGURE 5.19
Three-mode integrator-control circuit. Typical integrators of this type can hold within 0.01 percent of half-scale (10 or 100 V) for at least 200 s/μF and switch into COMPUTE or HOLD within 200 to 400 ns, with at most 50 to 100 ns between integrators. An "active" switch uses operational amplifier feedback to ground the holding capacitor through a very small impedance (0.01 to 1 Ω); this minimizes tracking phase shift (Ref. 11). (*continued on next page*)

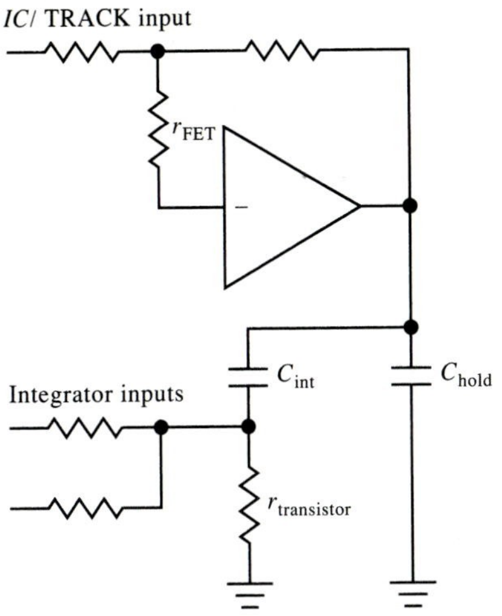

(*b*) RESET/TRACK mode.

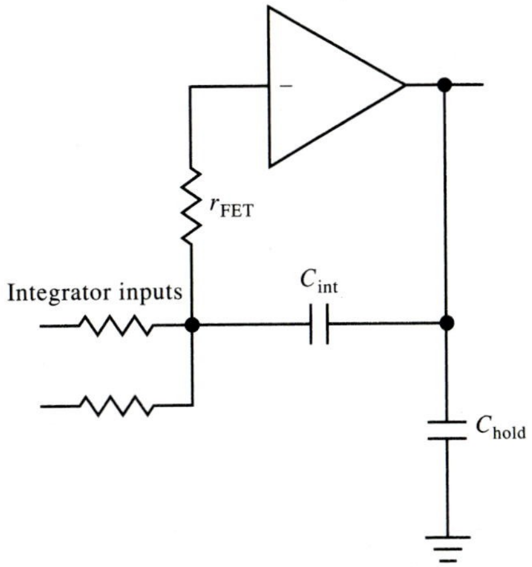

(*c*) INTEGRATE (COMPUTE) mode.

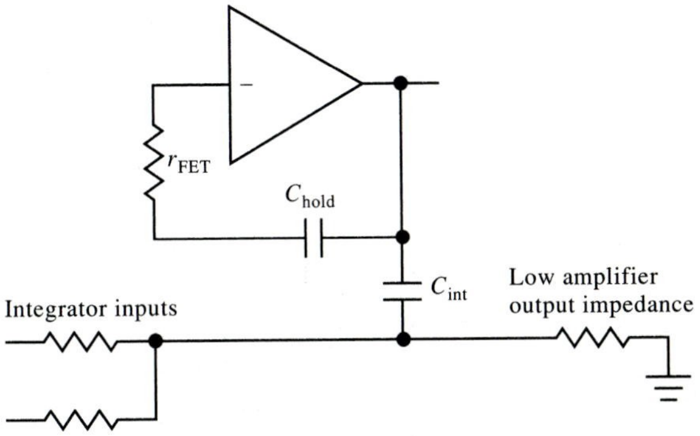

(*d*) HOLD mode.

FIGURE 5.19 (continued)

the integrating and holding capacitors through the low feedback-amplifier output impedance. In the COMPUTE mode, the integrator inputs are integrated, and the holding-capacitor charge follows. In the HOLD mode, the circuit acts as an integrator with zero input and holds its last output voltage. We see that *switched integrators can also be used as sample-hold circuits.*

Figure 5.20 shows a *two-mode integrator* designed for very fast analog computation (Refs. 1, 9, and 12); without the integrand input, the circuit is useful as a sample-hold. In the RESET/TRACK mode, the switched unity gain amplifier acts as an impedance transformer, reducing the effective RESET/TRACK-network resistances to 0.2 Ω, so that the effect of the 1- or 10-kΩ integrator inputs becomes

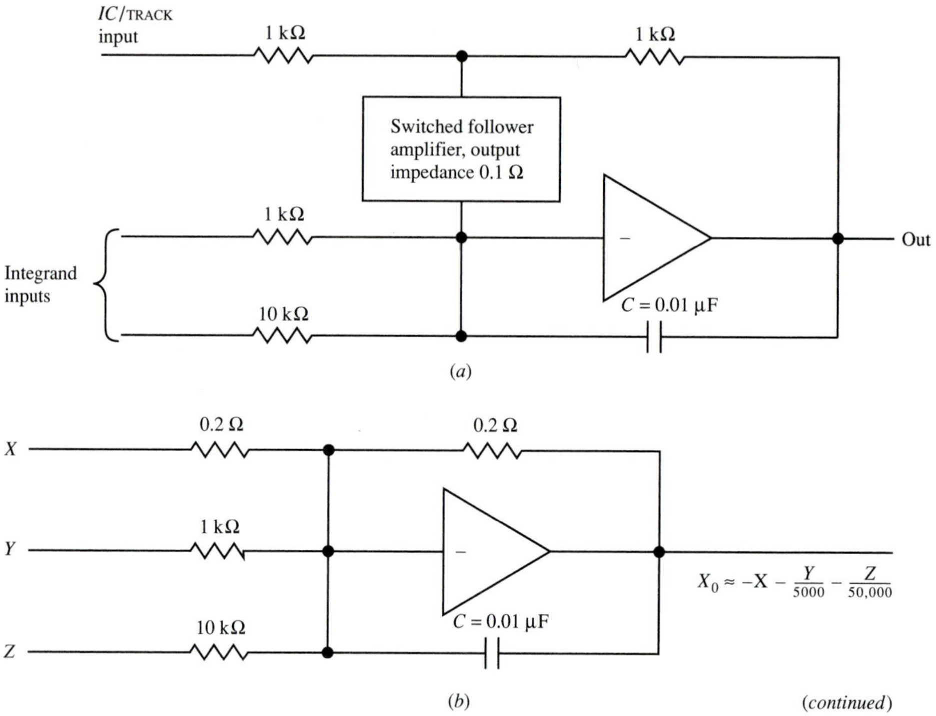

(continued)

FIGURE 5.20
(*a*) Two-mode integrator or sample-hold circuit; (*b*) equivalent circuit in the TRACK/IC mode; (*continued on next page*)

negligible. Typical integrators of this type can hold within 0.1 percent of half-scale (10 V) for at least 100 s/μF and switch into COMPUTE or HOLD within 20 to 40 ns, with at most 5 to 10 ns between integrators. The impedance transformation also reduces phase shift in RESET/TRACK.

Switched Operational Amplifiers (Ref. 9)

Substitution of a resistance for the integrator capacitance C in Fig. 5.20 produces an *electronic two-input SPDT switch* with sign-inverted low-impedance output (switched phase inverter or summer). Figure 5.21 shows a more general type of switched operational amplifier implementing

$$X_0 = -\frac{Z_0(P)}{Z_1(P)}X_1 \qquad \text{or} \qquad X_0 = -\frac{Z_0'(P)}{Z_1'(P)}X_1' \qquad \left(P = \frac{d}{dt}\right) \tag{5.6}$$

In addition to switching-spike and switch-leakage errors, switched operational amplifiers will have transfer function errors due to finite amplifier gain

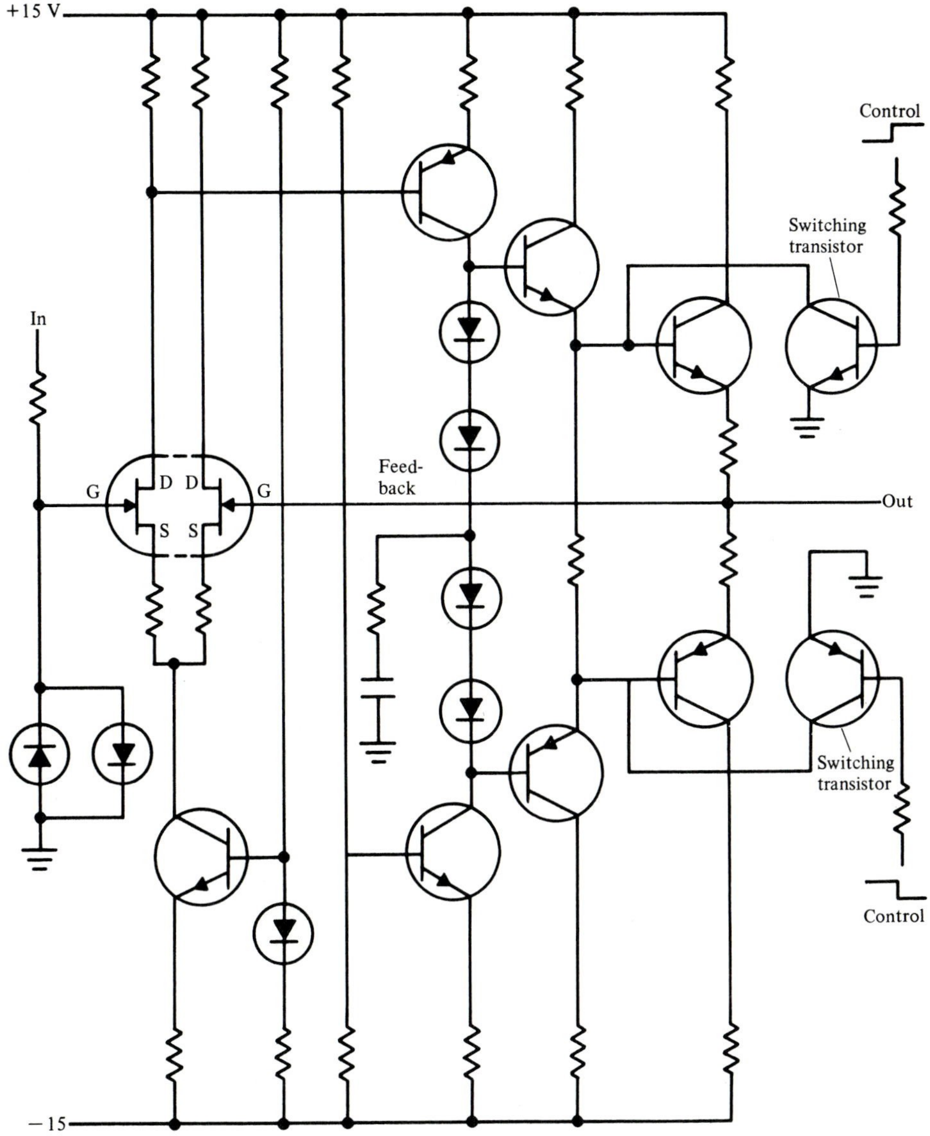

(c)

FIGURE 5.20 (continued)
(*c*) switched follower amplifier. The switched follower not only has very low output impedance, but employs a switched emitter follower output stage to avoid saturation of the switching transistors, which pass 30 mA when on. Output impedance is 0.1 Ω at 100 Hz, and output current is 30 mA. The integrator switches into COMPUTE/HOLD within 40 ns, with ±10-ns maximum difference between two integrators. Output impedance is 0.1 Ω at 100 Hz; output current is 30 mA. COMPUTE/HOLD drift in 10 ms (0.01 μF) is 2 mV ± 0.05 mV/°C. In RESET/TRACK with 0.01 μF, the integrator tracks ±10-V sine waves up to 40 kHz and follows a 10-V step within 0.1 percent in 10 μs; small-signal dynamic error is within 0.1 percent up to 10 kHz (no load), and temperature drift is ±100 μV/°C (*University of Arizona and Burr-Brown Research Corporation, Tucson, Ariz.*; based on Ref. 1).

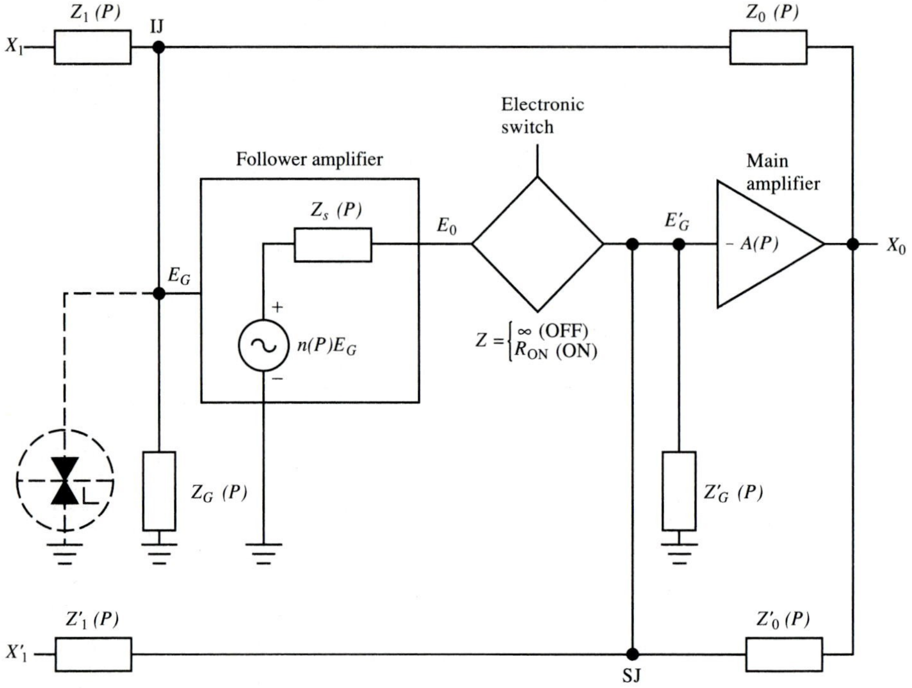

FIGURE 5.21
Switched operational amplifier (Ref. 9).

and follower and/or switch impedances. Such errors are readily calculated by the methods of Chap. 2. The circuit of Fig. 5.21, for example, is an ordinary operational amplifier with the switch off; with the switch on, $X_0 = -(Z_0/Z_1)X_1 + e$. The error $e = e(t)$ is given by

$$e \approx -\left(\frac{1}{A_{\text{on}}\beta_{\text{on}}}\frac{Z_0}{Z_1}X_1 + \frac{\beta_1}{\alpha\beta_{\text{on}}}\frac{Z_0}{Z_0'}X_0'\right) \tag{5.7}$$

$$A_{\text{on}} = \frac{\alpha A}{1 - A\beta_1} \qquad \beta_{\text{on}} = \left(1 + \frac{Z_0}{Z_1} + \frac{Z_0}{Z_G}\right)^{-1} \qquad \beta_1 \approx \frac{Z_S + R_{\text{on}}}{Z_0'}$$

at frequencies where $|A_{\text{on}}\beta_{\text{on}}| \gg 1$ and $|Z_S + R_{\text{on}}| \ll |Z_0|, |Z_1'|, |Z_G'|$. Note that the second error term makes this one-switch circuit less practical than multiswitch circuits if low computing impedances are to be used.

If the switch in Fig. 5.21 is enclosed in the follower-amplifier feedback loop in the manner of Fig. 5.20, then $Z_S + R_{\text{on}}$ in Eq. (5.7) is simply replaced by Z_S.

REFERENCES AND BIBLIOGRAPHY

1. G. A. Korn and T. M. Korn: *Electronic Analog and Hybrid Computers*, 2d ed., McGraw-Hill, New York, 1972.
2. P. G. Jespers: "Pulsed Circuits," Sec. 16, in D. G. Fink and D. Christiansen, *Electronics Engineer's Handbook*, 3d ed., McGraw-Hill, New York, 1989.
3. P. Burton: "CMOS DAC Application Guide," Analog Devices, Inc., Norwood, Mass., 1984.
4. J. G. Graeme, G. E. Tobey, and L. P. Huelsman: *Operational Amplifiers*, McGraw-Hill, New York, 1971.
5. J. Cohen: "Solid-State Signal Switching," *EDN*, Nov. 15, 1972.
6. ———:"Sample-and-Hold Circuits Using FET Analog Gates," *EEE*, January 1971.
7. A. D. Evans: "The Basics of Using FETs for Analog-Signal Switching," *EDN*, May 20, 1973.
8. L. Altman: "CMOS Forges Monolithic Analog Gate," *Electronics*, Apr. 26, 1973.
9. G. A. Korn: "Performance of Operational Amplifiers with Electronic Mode Switching," *IEEETEC*, June 1963.
10. C. S. Keay and J. A. Kennewell: "D/A Converter Switches Inputs with TTL Gates," *Electronics*, Dec. 8, 1969.
11. S. Marjanovic: "Analog Section of Hybrid Computing System HRS-100," *Proceedings of the 7th AICA Conference*, Prague, 1973.
12. J. Dostal: "Three-Mode Integrator," *Proceedings of the 7th AICA Conference*, Prague, 1973.

CHAPTER 6

DIGITAL-TO-ANALOG AND ANALOG-TO-DIGITAL CONVERTERS†

6.1 DIGITAL REPRESENTATION OF VOLTAGES; SIMPLE CONVERSION CIRCUITS

Digital computing devices represent each variable or parameter by the discrete states of switching devices. As an example, the numbers 0, 1, 2, ..., 9 could be represented by the 10 possible states (positions) of a 10-position star switch or by 9 of the 16 possible combined states of 4 on-off or SPDT switches.

The latter representation turns out to be practical and convenient. We therefore commonly represent a nonnegative voltage as *a binary fraction*

$$Y_D = \frac{1}{2}a_1 + \frac{1}{4}a_2 + \cdots + \frac{1}{2^n}a_n \tag{6.1}$$

($0 \leq Y_D < 1$) of a *reference voltage* X_1 (for example, 10 or 100 V), where a_1, a_2, ..., a_n are *binary digits (bits)*, each capable of taking the two values 0 and 1. The value (6.1) of the n-bit binary fraction is then given by the states of n binary switching devices (for example, n toggle switches or a flip-flop register); thus, 011 stands for the fraction $\frac{3}{8}$, 100 stands for $\frac{4}{8} = \frac{1}{2}$, etc. Many other binary *codes* relating the sequence $(a_1, a_2, \ldots, a_n)$ to a fraction are possible, but (6.1) is the most frequently employed coding scheme.

---

† Portions of Chap. 6 are based on material taken from Chap. 5 of G. A. Korn and T. M. Korn, *Electronic Analog and Hybrid Computers*, 2d ed., McGraw-Hill, 1972, by permission of the authors and publisher.

TABLE 6.1
Binary codes representing real fractions X by n-bit word: $(a_0, a_1, a_2, \ldots, a_{n-1})$

Each binary digit a_k is either 0 or 1; a_0 is the most significant bit (MSB), and a_{n-1} is the least significant bit (LSB).

1. **Nonnegative fractions:** $X = \frac{1}{2}a_0 + \frac{1}{2^2}a_1 + \cdots + \frac{1}{2^n}a_{n-1}$, where $0 \le X \le 1 - \frac{1}{2^n}$.

EXAMPLE (3 bits):

Decimal	Binary	Decimal	Binary
0.000	000	$\frac{4}{8} = 0.500$	100
$\frac{1}{8} = 0.125$	001	$\frac{5}{8} = 0.625$	101
$\frac{2}{8} = 0.250$	010	$\frac{6}{8} = 0.750$	110
$\frac{3}{8} = 0.375$	011	$\frac{7}{8} = 0.875$	111

2. **Signed fraction (positive, negative, or zero).** The sign bit a_0 is 0 for $X \le 0$ and 1 for $X < 0$.

(*a*) **Sign-and-magnitude code:** $X = (-1)^{a_0}\left(\frac{1}{2}a_1 + \frac{1}{2^2}a_2 + \cdots + \frac{1}{2^{n-1}}a_{n-1}\right)$, where $\frac{1}{2^{n-1}} - 1 \le X \le 1 - \frac{1}{2^{n-1}}$. There are *two* binary representations of 0: 000 . . . and 100. . . . This may cause complications, e.g., in statistical work.

(*b*) **One's-complement code:**

$$X = \left(\frac{1}{2^{n-1}} - 1\right)a_0 + \frac{1}{2^2}a_2 + \cdots + \frac{1}{2^{n-1}}a_{n-1}$$

where $\frac{1}{2^{n-1}} - 1 \le X \le 1 - \frac{1}{2^{n-1}}$. There are *two* binary representations of 0: 000 . . . and 111. . . . *One obtains the code for $-X$ very simply by complementing each bit.* This code is used in some arithmetic units.

(*continued*)

If we want to represent *negative as well as positive fractions* Y_D, we can add a *sign bit* a_0 and use the n-bit coding scheme defined by

$$Y_D = (1 - 2a_0)\left(\frac{1}{2}a_1 + \frac{1}{4}a_2 + \cdots + \frac{1}{2^{n-1}}a_{n-1}\right) \tag{6.2}$$

which is similar to (6.1), except that a sign bit $a_0 = 1$ acts as a minus sign and reverses the sign of our fraction (*sign-and-magnitude code*). Unfortunately, we now have *two* codes for $Y_D = 0$ (000 . . . and 100 . . .).

A binary code more useful for digital counting and arithmetic is the *two's-complement code*, which has a unique zero 000 . . . :

$$Y_D = -a_0 + \frac{1}{2}a_1 + \frac{1}{4}a_2 + \cdots + \frac{1}{2^{n-1}}a_{n-1} \tag{6.3}$$

Table 6.1 further illustrates binary-fraction codes. Many other codes exist (see also Sec. 6.2).

TABLE 6.1
(continued)

(*c*) **Two's-complement code:**

$$X = -a_0 + \frac{1}{2}a_1 + \frac{1}{2^2}a_2 + \cdots + \frac{1}{2^{n-1}}a_{n-1}$$

where $-1 \leq X \leq 1 - \frac{1}{2^{n-1}}$. It has a unique 0 and simple arithmetic. *To obtain code for* $-X$, *complement every bit and add 1 LSB*. This code is used in almost all minicomputers.

EXAMPLES (4-bit codes):

Decimal	Sign and magnitude	One's-complement	Two's-complement
$+\frac{7}{8} = +0.875$	0 1 1 1	0 1 1 1	0 1 1 1
$+\frac{6}{8} = +0.750$	0 1 1 0	0 1 1 0	0 1 1 0
...	...	...	...
$+\frac{3}{8} = +0.375$	0 0 1 1	0 0 1 1	0 0 1 1
$+\frac{2}{8} = +0.250$	0 0 1 0	0 0 1 0	0 0 1 0
$+\frac{1}{8} = +0.125$	0 0 0 1	0 0 0 1	0 0 0 1
+0	0 0 0 0	0 0 0 0	0 0 0 0
−0	1 0 0 0	1 1 1 1	
$-\frac{1}{8} = -0.125$	1 0 0 1	1 1 1 0	1 1 1 1
$-\frac{2}{8} = -0.250$	1 0 1 0	1 1 0 1	1 1 1 0
$-\frac{3}{8} = -0.375$	1 0 1 1	1 1 0 0	1 1 0 1
...	...	...	...
$-\frac{6}{8} = -0.750$	1 1 1 0	1 0 0 1	1 0 1 0
$-\frac{7}{8} = -0.875$	1 1 1 1	1 0 0 0	1 0 0 1
−1			1 0 0 0

Source: From G. A. Korn, *Minicomputers for Engineers and Scientists*, McGraw-Hill, New York, 1973.

A *digital-to-analog converter (DAC)* is a switched resistance network (possibly also including amplifiers) which produces the output voltage $X_1 Y_D$, where X_1 is an analog input *(reference voltage)* and Y_D is a fraction determined by the state of a set of switches, usually binary (on-off or SPDT) switches. Thus, the simple binary-code DAC of Fig. 6.1 multiplies its analog input voltage X_1 by the n-bit nonnegative fraction

$$Y_D = \frac{1}{2}a_1 + \frac{1}{4}a_2 + \cdots + \frac{1}{2^n}a_n \tag{6.4}$$

(*see also* Table 6.2)

The kth switch is on if the corresponding binary digit a_k is 1, so that $0 \leq Y_D < 1$. a_1 is the *most significant bit* (MSB) of Y_D, and a_n is the *least significant bit* (LSB). The analog input X_1 may be a fixed reference voltage *(fixed-reference DAC)* or a variable (*multiplying DAC,* or MDAC). MDACs are more difficult to design and more expensive because switch-resistance effects, and possibly switch offset voltages, will vary with X_1.

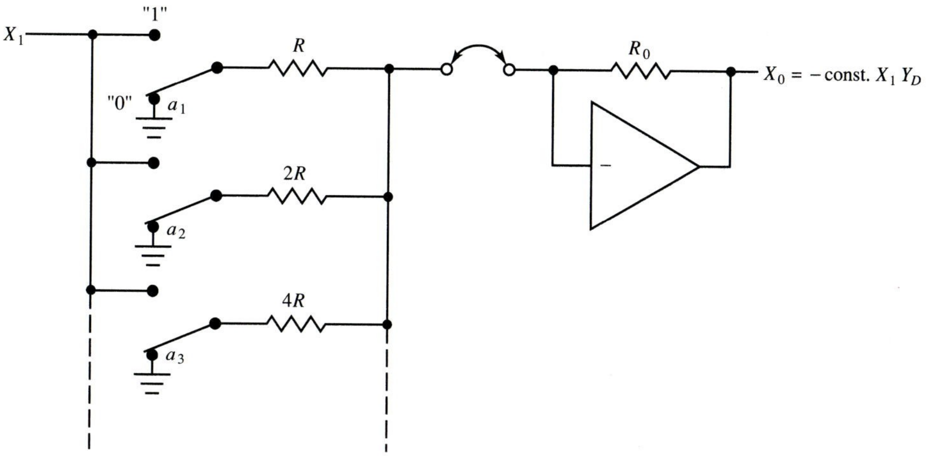

FIGURE 6.1
Simple digital-to-analog converter (DAC) for n-bit nonnegative binary numbers

$$Y_D = \frac{1}{2}a_1 + \frac{1}{4}a_2 + \cdots + \frac{1}{2^n}a_n$$

where $a_k = 0$ or $1 (k = 1, 2, \ldots, n)$. The kth switch is on if the corresponding binary digit a_k is 1. The circuit has *constant source impedance*, so that it could drive a simple load resistance r_L instead of an operational amplifier summing point.

Most frequently, a flip-flop register, which can be set by parallel bit lines from a digital computer, stores the bit values a_0, a_1, a_2, ... and produces bit voltages (for example, 0 V for $a_i = 0$ and +4 V for $a_i = 1$) which keep the correct switches turned off and on. In Fig. 6.2*a*, n-bit lines from a digital computer set a DAC register so that the output voltage X_0 is proportional to a desired "digital input" Y_D.

Figure 6.2*b* illustrates *double-buffered* DAC operation. It is often desirable to "update" two or more DAC outputs *simultaneously*, i.e., to switch these voltages simultaneously to new values. But most digital computers can output *only one number at a time*. In Fig. 6.2*b*, therefore, *buffer registers* no. 1 and no. 2 are loaded *successively* by the computer. Afterward, a common transfer pulse transfers the bit values in each buffer register to the corresponding DAC register, so that both DACs are updated together.

6.2 PRACTICAL DIGITAL-TO-ANALOG CONVERTERS; VOLTAGE-LADDER CIRCUITS

Conversion of Positive and Negative Numbers

The simple digital-to-analog converter of Fig. 6.1 was designed to convert *nonnegative* binary numbers. Such circuits will, however, correctly convert both negative and positive numbers expressed in the *sign-and-magnitude code* (Sec. 6.1) if we

TABLE 6.2
Binary-number representation of voltages

Unsigned octal numbers†	Voltage
0000	0
0001	0.00244140625
0002	0.0048828125
0004	0.009765625
0010	0.01953125
0020	0.0390625
0040	0.078125
0100	0.15625
0200	0.3125
0400	0.625
1000	1.25
2000	2.5
4000	5
6000	7.5
7000	8.75
7400	9.375
7600	9.6875
7700	9.84375
7740	9.921875
7760	9.9609375
7770	9.98046875
7774	9.990234375
7776	9.9951171875
7777	9.99755859375
(10000)	(10)

† Each octal digit may be regarded as *a shorthand notation for three binary digits*, thus:

Octal	Binary	Octal	Binary
0	000	4	100
1	001	5	101
2	010	6	110
3	011	7	111

add a *sign-bit switch* which switches either the analog input X_1 or the output X_0 through a unity gain inverter whenever the *sign bit* a_0 is 1.

Figure 6.3 illustrates the conversion of a *positive or negative two's-complement–coded n-bit binary fraction*

$$Y_D = -a_0 + \frac{1}{2}a_1 + \frac{1}{4}a_2 + \cdots + \frac{1}{2^{n-1}}a_{n-1}$$

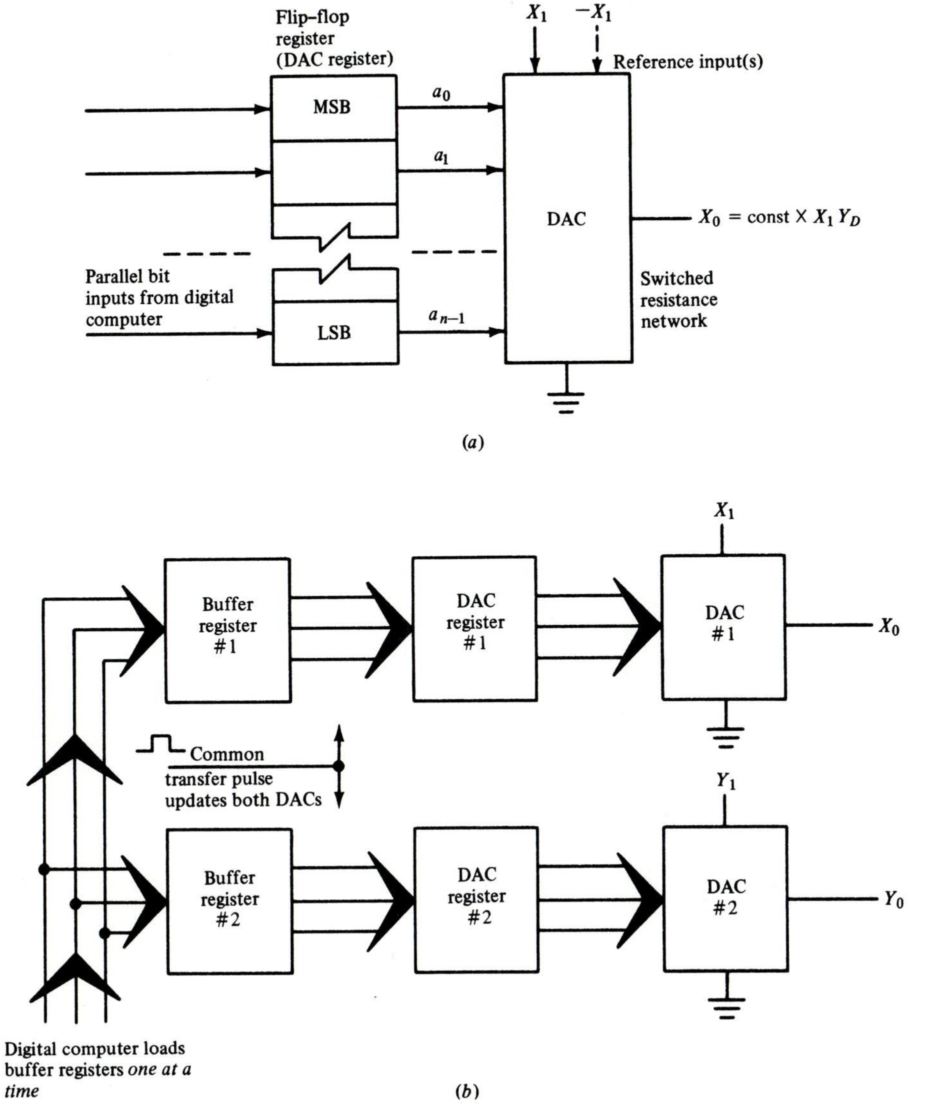

FIGURE 6.2
(*a*) DAC with flip-flop register; (*b*) two double-buffered DACs updated by the same transfer pulse. The common transfer pulse in *b* updates both DACs simultaneously.

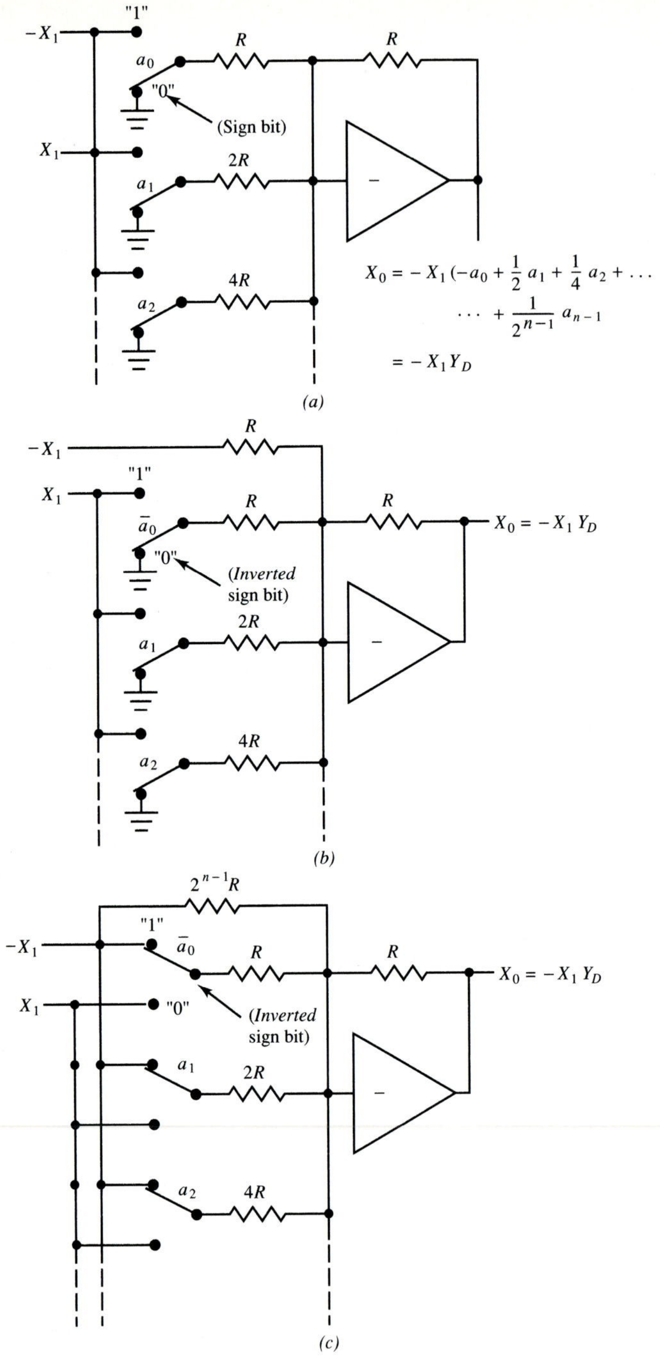

FIGURE 6.3
Three different methods for converting two's-complement–coded positive or negative binary fractions

$$Y_D = -a_0 + \frac{1}{2}a_1 + \frac{1}{4}a_2 + \cdots + \frac{1}{2^{n-1}}a_{n-1}$$

If X_1 is a fixed reference voltage, the scheme of Fig. 6.3*b* requires only switches passing current in one direction. All three schemes work just as well with ladder networks.

Note, first, that four-quadrant multiplication of the two's-complement–coded digital input Y_D and the analog input X_1 requires both $-X_1$ and X_1 as input voltages. The most significant bit will be the sign bit a_0, and the least significant bit is a_{n-1}. Note also that $-1 \le Y_D < 1$. Starting at -1, Y_D would increase as follows:

$$100\cdots00 = -1$$

$$100\cdots01 = -(1 - 2^{1-n})$$

$$100\cdots10$$

$$\cdots\cdots\cdots$$

$$111\cdots11 = -2^{1-n}$$

(Two's-complement code)

$$000\cdots00 = 0$$

$$000\cdots01 = 2^{1-n}$$

$$000\cdots10$$

$$\cdots\cdots\cdots$$

$$011\cdots11 = 1 - 2^{1-n}$$

The use of the inverted sign bit in Fig. 6.3*b* and *c* ($\overline{a}_0 = 0$ if $a_0 = 1$ and vice versa) is sometimes referred to as an *offset-binary code*. The scheme of Fig. 6.3*c* is used in some multiplying digital-to-analog converters. Note that the bit switches no longer multiply by $\overline{a}_0, a_1, a_2, \ldots,$ but by $2\overline{a}_0 - 1, 2a_1 - 1, 2a_2 - 1, \ldots,$ so that Y_D is in the form

$$Y_D = \frac{1}{2}\left[-(2a_0 - 1) + \frac{1}{2}(2a_1 - 1) + \frac{1}{4}(2a_2 - 1) + \cdots + \frac{1}{2^{n-1}}(2a_{n-1} - 1) - \frac{1}{2^{n-1}}\right] \tag{6.5}$$

Tee-Network Digital-to-Analog Converters

If more than 5 or 6 bits are needed, the resistance $2^k R$ in the simple current-summing networks of Figs. 6.2 and 6.3 becomes uncomfortably large. One way to reduce the network impedance levels when an operational amplifier does the current summing (as in Figs. 6.2 and 6.3) is to replace each resistance $2^k R$ by a corresponding *tee-network transfer impedance* (Sec. 1.5). Figure 5.16 shows a tee-network bit-switching circuit designed for fast, calibration-free MDACs (Ref. 1).

While the requirement of a given static accuracy at low cost favors *high* DAC-network resistances to make switch resistances less critical, each DAC must settle to within the specified accuracy after switching, and the required settling time will be proportional to some circuit resistance (Fig. 6.12). The sum of this settling time and the (usually smaller) switching time constitutes the conversion time, whose reciprocal is the maximum conversion rate still permitting the specified accuracy. High conversion rates, and also low MDAC phase shift, favor *low* DAC network resistances. Hence DAC design will require a compromise between accuracy, conversion rate and/or phase shift, and switch costs (see also Sec. 6.5).

Digital-to-analog converters employing *switched capacitance networks* to decode digital data arriving in serial form (i.e., bit by bit) are described in Refs. 2 and 3.

Voltage-Ladder Digital-to-Analog Converters

The basic *binary voltage-ladder network* illustrated in Fig. 6.4 is equivalent to a voltage source of internal impedance R and source voltage

$$\frac{1}{2}a_1 + \frac{1}{2^2}a_2 + \cdots + \frac{1}{2^n}a_n$$

at the terminal Q, where each bit value a_k is either 0 or 1. This fact is easily proved by the successive Thévenin-equivalent circuits shown in Fig. 6.4*b*, *c*, and *d*. Figure 6.5 illustrates how binary voltage ladders can be connected for conversion of two's-complement–coded *positive or negative* numbers; the two circuits shown are, respectively, analogous to Fig. 6.3*a* and *b*. The design of a ladder circuit analogous to Fig. 6.3*c* is left as a problem.

Ladder network DACs necessarily require SPDT switches but need fewer resistors than tee-network DACs; above all, binary ladders require *only two different resistance values of the same order of magnitude* and can, in fact, be made up of resistance R or $2R$ alone. This greatly simplifies close matching of film-resistor temperature coefficients; indeed, the entire ladder network can be deposited on a substrate by thick-film or thin-film techniques.

Figure 6.6 shows practical electronic *bit switches* (SPDT switches) suitable for voltage-ladder switching. The on resistances of such switches (5 to 15 Ω, Sec. 5.4) are within 0.01 percent of 200-kΩ ladder resistances $2R$. With smaller ladder resistances, tighter accuracy specifications, and/or FET bit switches (Sec. 5.5), the $2R$ ladder resistances must be trimmed to account for the switch on resistances. A common technique is to specify $2R$ resistances 30 Ω too low and then to select switching FETs with 30 Ω ± 3 Ω on resistances.

The "inverted" binary ladder circuit of Fig. 6.7 permits low-level FET switching in the manner of Fig. 5.9. Note the compensation of FET on resistance changes with temperature by a FET in the output-amplifier feedback circuit.

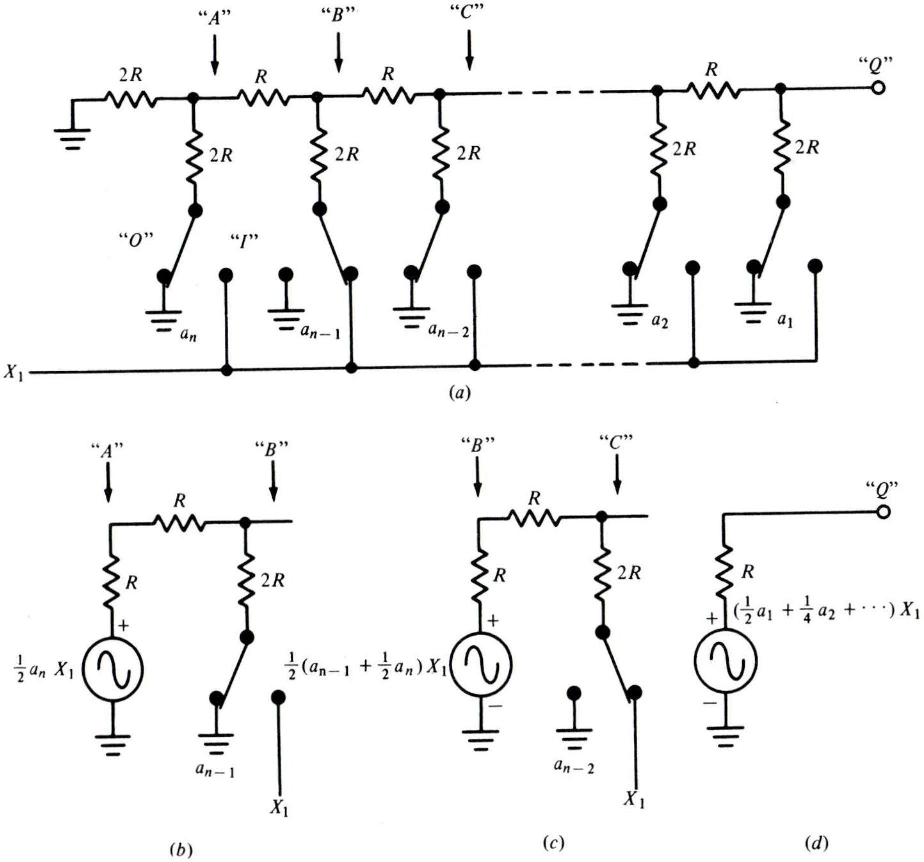

FIGURE 6.4
(*a*) A binary ladder network; (*b*) to (*d*) Thévenin-equivalent circuits for successive sets of ladder sections, showing how successive inputs are effectively divided by powers of 2. *Circuit c is the equivalent-voltage-source circuit for the complete ladder*. If such a ladder network is loaded with $2R$, the impedance between each ladder switch and ground is $3R$.

Figure 6.8 is an example of a voltage-ladder DAC converting a *binary-coded decimal (BCD) code*, with

$$Y_D = \tfrac{1}{2}a_1 + \tfrac{1}{4}a_2 + \tfrac{1}{8}a_3 + \tfrac{1}{16}a_4 + \tfrac{1}{10}\left(\tfrac{1}{2}b_1 + \tfrac{1}{4}b_2 + \tfrac{1}{8}b_3 + \tfrac{1}{16}b_4\right)$$
$$+ \tfrac{1}{100}\left(\tfrac{1}{2}c_1 + \tfrac{1}{4}c_2 + \tfrac{1}{8}c_3 + \tfrac{1}{16}c_4\right) \qquad (6.6)$$

In such a BCD code, each *decimal* digit of Y_D is separately represented in binary form. Each group of 4 bits can, for instance, operate one decimal digit of a decimal-number display. The operation of such networks is readily analyzed with the aid of Thévenin's theorem in the manner of Fig. 6.4 (Prob. 6.6).

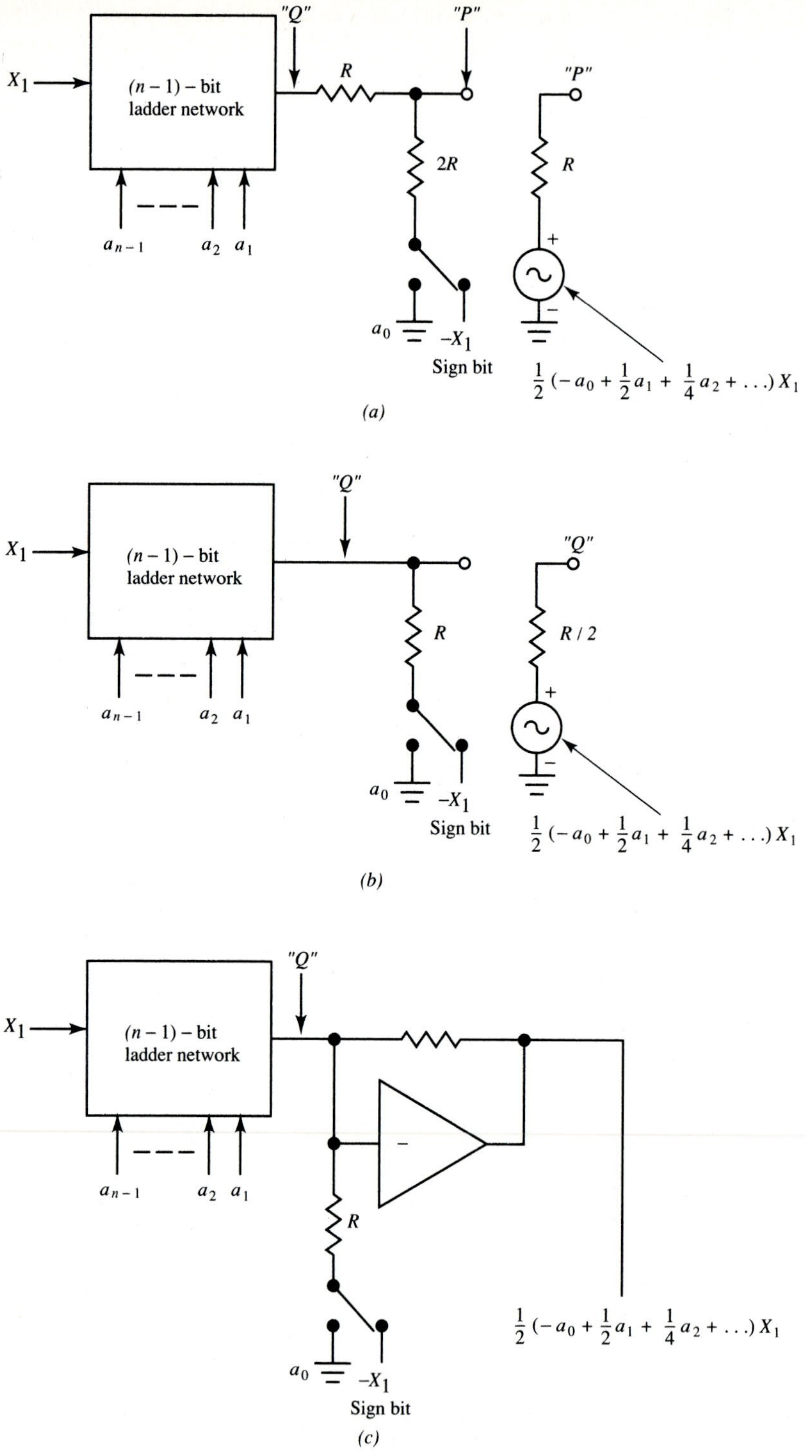

FIGURE 6.5
Ladder circuits for converting two's-complement–coded positive or negative quantities. Equivalent-voltage-source circuits are shown for (*a*) and (*b*).

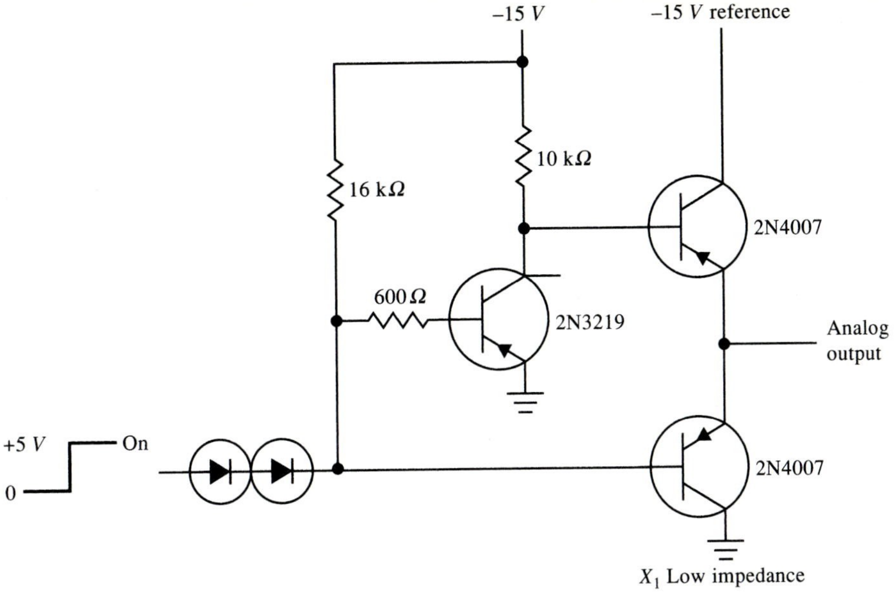

(*a*)

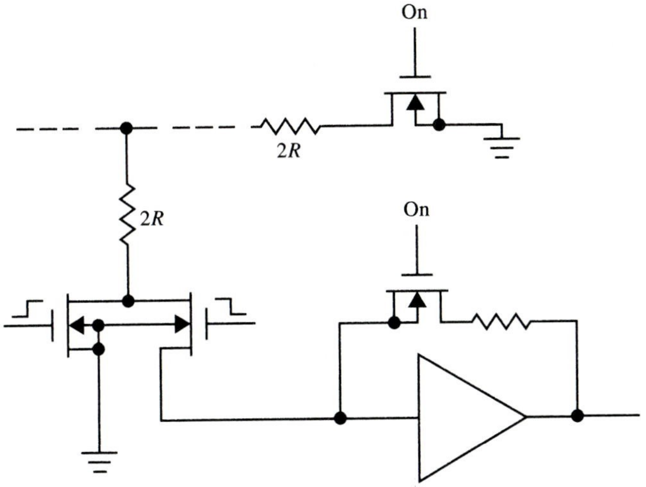

(*b*)

FIGURE 6.6
(*a*) Simple transistor switch for the fixed-reference-voltage ladder of Fig. 6.1 with $X_1 = -15$ V. DAC calibration must allow for the transistor saturation voltage. (*b*) Multiplying digital-to-analog converters typically use the "inverted" binary ladder of Fig. 6.7 with this type of MOSFET switch, driven by CMOS logic. Resistances to ground in the ON and OFF states match even better than in Fig. 5.13*b*. IC bit switches are "scaled" to contribute, say, 25, 50, 100, . . . Ω to successive ladder resistances; the last 2*R* "leg" of the ladder in Fig. 6.7 has a permanently on series MOSFET. All MOSFET voltages are then small and roughly equal in both ON and OFF states, so that forward resistance changes with current are similar and small. An extra MOSFET in the amplifier feedback loop compensates for temperature variations. The feedback capacitor C, typically between 30 and 100 pF, is adjusted to compensate for the DAC network capacitances.

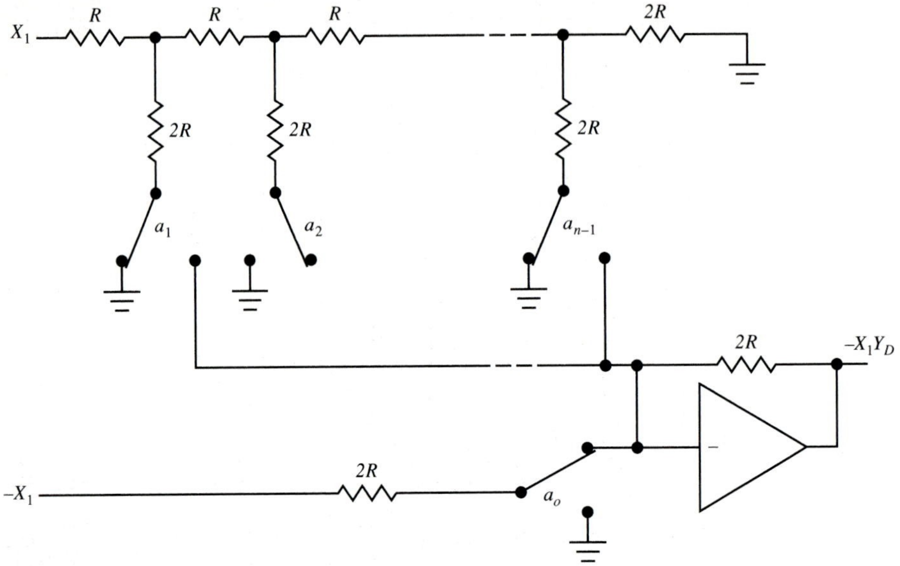

FIGURE 6.7
This "inverted" binary ladder works nicely with the FET/bipolar SPDT switch of Fig. 5.14. The 2*R* resistances in series with FET switches can be trimmed to account for the FET ON resistance, and the 2*R* feedback resistance can include a FET for temperature compensation as in Fig. 5.14a.

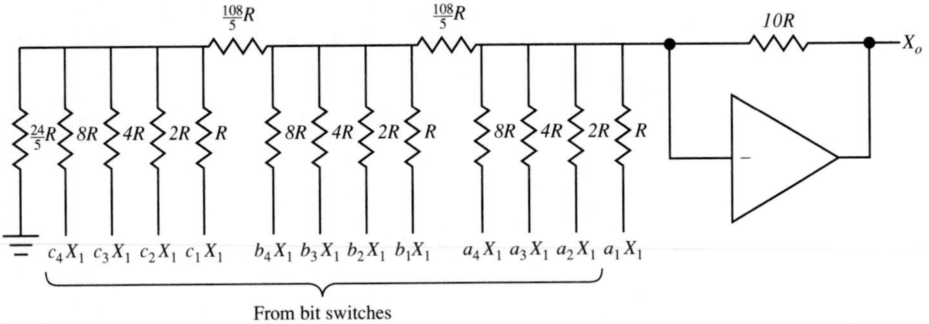

FIGURE 6.8
DAC circuit for converting binary-coded decimal (BCD) number. Many other schemes exist.

6.3 CURRENT-SWITCHING DACs

A *bipolar-transistor current switch* like that shown in Fig. 6.9 does not depend on transistor saturation to turn on. Such a switch operates as a feedback-controlled current source in the "working region" of its characteristics when it is on. To turn the switch off, the logic waveform opens an alternate current path through a diode or transistor ("current stealing") and cuts the current-source transistor off. Compared to voltage switching, current switching has two significant advantages:

1. With reasonably high current gain in the current-source feedback circuit, the output current does not depend on switch-transistor forward resistance.
2. Current switches turn off quickly because there is no need to get a transistor out of saturation.

Switched current sources can be used as operational amplifier inputs in DAC circuits similar to Figs. 6.2 and 6.3, or in current-ladder circuits like that shown in Fig. 6.11*a* which, like binary voltage ladders, employ only two different resistance values. Current switches are employed mainly in fixed-reference DACs, because the inherent high impedances of current sources tend to cause analog-input phase shift.

The output of the simple switched current sources in Fig. 6.9 depends, unfortunately, on transistor V_{BE} and thus on temperature, so that accurate current-switch

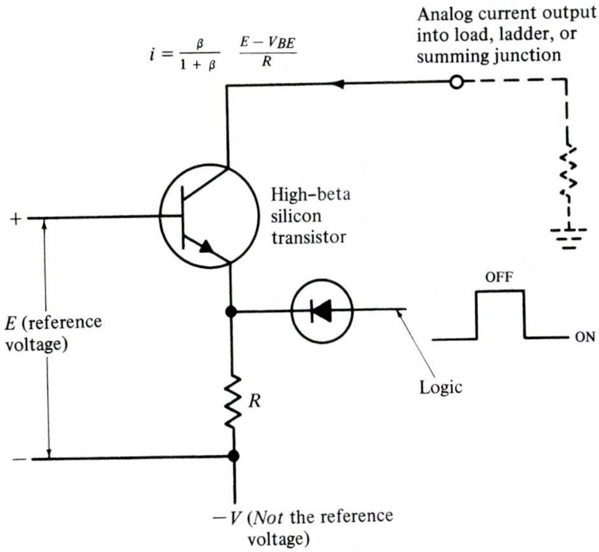

FIGURE 6.9
A simple switched current source suitable for fixed-reference DACs. Monolithic integrated-circuit construction is possible. The V_{BE} temperature dependence should be compensated in the source of the reference voltage E, or through a reference current source in the manner of Fig. 6.10. Reverse the junctions and bias for negative current sources.

DACs require some form of temperature compensation. In the current-switch network of Fig. 6.10, the on current level in five closely matched transistors is determined by an operational amplifier circuit. And V_{BE} matching is close, since all transistors are deposited on the same monolithic integrated-circuit chip. Transistors designed to pass successively larger bit currents of $\frac{1}{8}$, $\frac{1}{4}$, $\frac{1}{2}$, and 1 mA are scaled to have successively doubled emitter areas, so that good current-gain (beta) tracking is obtained. Figure 6.11 shows two more current-switching DAC circuits.

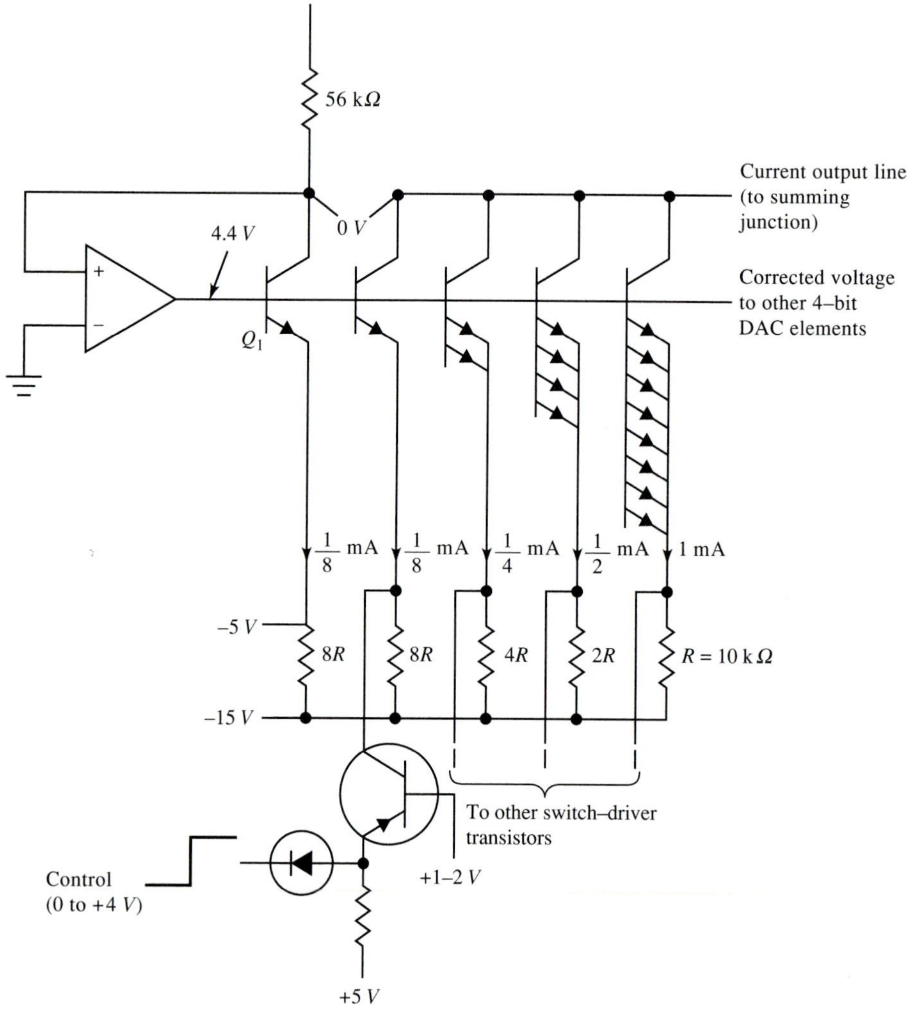

FIGURE 6.10
Trimmer-free four-bit current-switching DAC element for a 12-bit fixed-reference DAC (Fig. 6.11*b*). A feedback loop including the extra current source Q_1 sets all current-source base voltages to compensate for V_{BE} and β changes with temperature and power supply. The four current sources shown correspond to the most significant bits and are deposited on the same substrate as Q_1 for V_{BE} tracking within 500 μV between -50 and $+120$°C. Similarly good β tracking is achieved by doubling the emitter areas of successive current sources for equal current densities. This integrated circuit, using *n-p-n* current-source transistors, permits faster switching than earlier monolithic circuits using *p-n-p*'s (*Analog Devices, Inc.*).

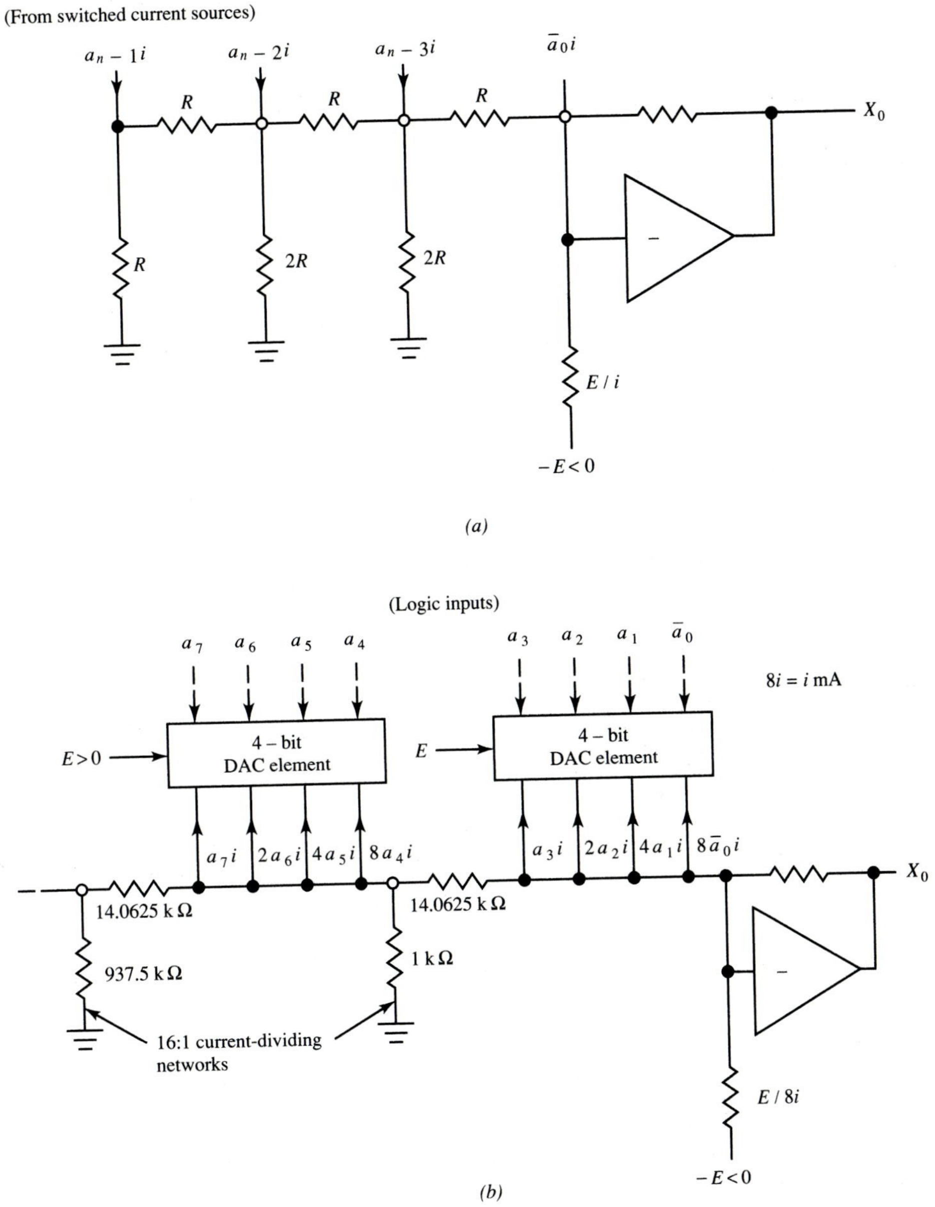

FIGURE 6.11
Fixed-reference DACs employing switched current sources. The sign-changing scheme of Fig. 6.11*b* permits the use of similar one-directional current sources. Each a_k is either 0 or 1, and E is a fixed reference voltage.

The current-switch diodes in Figs. 6.9 and 6.10 are often replaced by bipolar transistors matched to the switching transistors.

Cascaded DACs and Correction DACs

Note that the circuit of Fig. 6.11*b* *cascades* two 4-bit DACs into an 8-bit DAC. Cascading is not restricted to current-switching DACs; there are many other circuits cascading two or more DACs (see, e.g., Ref. 3). In general, a summing/attenuation network or operational amplifier adds the outputs of, say, two DACs

$$V = V_{MS} + \alpha V_{LS}$$

The attenuation factor $\alpha < 1$ attenuates the DAC output supplying the less significant bits. The more significant DAC may have only a few bits but must switch accurately enough for the full range of all bits in both DACs. For this reason, 4- and 6-bit DACs with 12-bit accuracy specifications are commercially available. Errors in the less significant DAC, though, are attenuated by α.

In a related scheme, the first DAC is a high-resolution DAC, and the second, attenuated DAC has a few bits used to *correct* the main DAC. A calibration procedure loads a memory lookup table to feed the second DAC suitable correction bits for every digital input value. The lookup table may be user-programmable or set permanently in a read-only memory (ROM) chip. Such techniques permit the design of DACs with as many as 18 bits.

6.4 STATIC ERROR SPECIFICATIONS

DAC *resolution* is given by the number of bits. Ideal conversion accuracy ought to be within the voltage limits determined by one-half of the least significant bit ($^1/_2$ LSB). Even if this is not true, DAC analog output must always increase (or always decrease) with the digital input (*monotonicity*). Figure 6.12 illustrates ideal and nonideal static performance. The DAC *transfer characteristic* is the locus of the midpoints of successive voltage steps. The ideal transfer characteristic would be a straight line through zero. Since *offset and gain errors* in the DAC transfer characteristic can, in principle, be "calibrated out" before an accuracy measurement, one usually specifies *total nonlinearity* as the deviation from a straight line fitted to the nonideal transfer characteristic. *Differential nonlinearity* is the deviation from the ideal step size after calibration. In high-resolution DACs, circuit noise as well as static DAC characteristics can contribute to differential nonlinearity (Ref. 24).

DAC static errors are due to *network and switch-resistance tolerances, switch and amplifier offsets, and reference-voltage errors*. Errors ought to be specified on a worst-case basis for all digital inputs, working temperatures, and power supply voltage limits. In general, the more significant bits (including the sign bit, if any) contribute relatively larger switch resistance and switch offset errors than the less significant bits (see also Prob. 6.5).

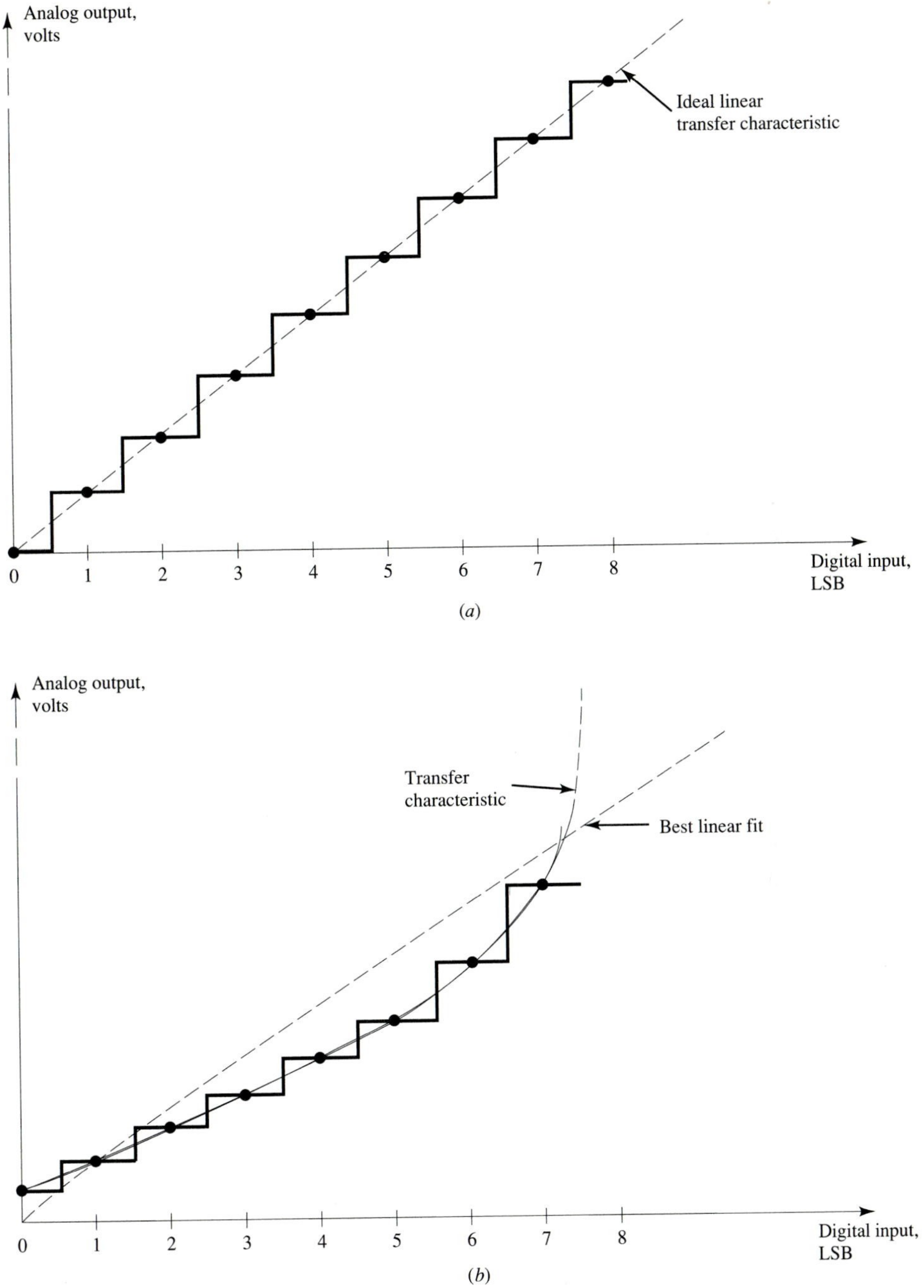

FIGURE 6.12
Transfer characteristic (analog output vs. digital input) of (*a*) an ideal digital-to-analog converter, and (*b*) a non-ideal characteristic. Offset and gain errors in (*b*) can often be calibrated out so that, after calibration, one is left with deviations from the best linear fit to the transfer characteristic.

6.5 DAC DYNAMIC PERFORMANCE: SETTLING TIME, PHASE SHIFT, SPIKES, AND GLITCHES

To permit a specified minimum time interval between samples, DACs must settle to within a specified error (say within 1 LSB, or within some percentage of full scale) in a specified time (*settling time*) after a specified input step. A full-scale step is usually specified; note, though, that some DACs need to follow only continuous signals. Switched resistance networks involve distributed circuit capacitances between 1 and 30 pF and exhibit exponential step responses approximated by that of the simple network in Fig. 6.13. Clearly, better accuracy requires a relatively longer settling time. Low network resistances produce faster settling but are more sensitive to switch resistance errors. Fast, high-resolution DACs typically use current switches.

Analog phase-shift errors in multiplying digital-to-analog converters (MDACs) are also improved by low network resistances and can be lowered with equalizing capacitors in amplifier feedback networks (Fig. 6.6*b*). Phase-shift errors within 0.1 percent up to 25 kHz are possible with special care, but phase-shift errors are rarely included in manufacturers' specifications. MDAC accuracy can also be affected by input-output ac *feedthrough*. This effect can be minimized by proper circuit layout (e.g., circuit isolation by grounded conductors).

Amplifier Settling Time

Most DACs involve an output operational amplifier or voltage follower whose response affects DAC settling but is not always included in the DAC specifications

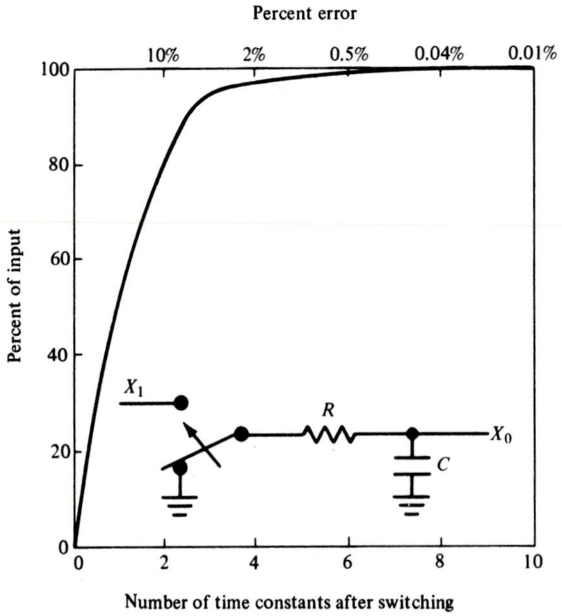

FIGURE 6.13
The step response

$$X_0 = X_1\left(1 - e^{-t/RC}\right)$$

of a switched *RC* circuit (time constant *RC*) indicates the settling time needed for a given accuracy.

proper. Correct overall DAC settling-time specifications must, then, refer to the time required for DAC *and* the amplifier to settle within a specified accuracy after large or small steps. Amplifier step response involves a short *propagation time* and, at least after large steps, a *slewing period* in which the essentially overloaded amplifier supplies the maximum possible current to load and circuit capacitances. This is followed by a *linear transient* (often with a small overshoot), which must be controlled by careful amplifier equalization. Typical DAC/amplifier settling times range from 30 ns (0.1 to 0.01 percent of half-scale) to slower and more accurate DACs settling to within 0.01 percent of half-scale within 1 to 4 μs.

Spikes and Glitches

All digital-to-analog converters will exhibit *switching spikes* due to switch capacitances (Sec. 5.9). Spike amplitude will be reduced by slow output amplifiers, but fast DACs can exhibit spike amplitudes much larger than 1 LSB. More serious transient disturbances than capacitive spikes are the *glitches*, which arise from the fact that many DAC voltage changes involve operations of multiple bit switches with unequal turn-on and turn-off times. Since one bit switch may turn off more quickly than another turns on, the DAC output can momentarily take a completely false value during the switching interval. Such multiple switch operations take

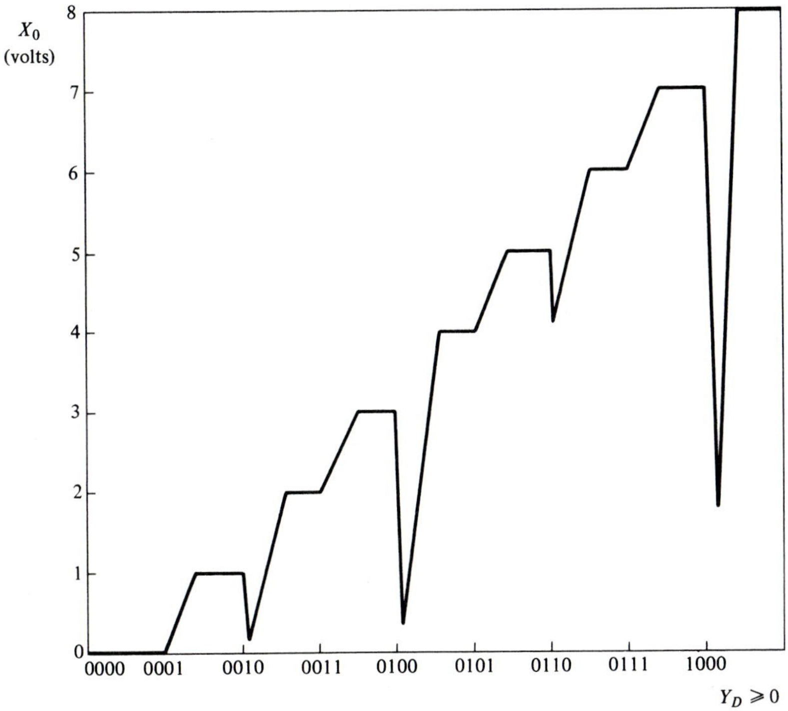

FIGURE 6.14
Output of a counter-driven fast digital-to-analog converter showing glitches associated with each counter carry. No capacitive switching spikes are shown (*Analog Devices,* based on Ref. 16).

place, for instance, on each carry when a DAC is driven by a counter (Fig. 6.14). Glitches, like spikes, are reduced by faster switches, but glitches cannot be completely eliminated in fast DACs. In applications where a smooth DAC output waveform is crucial, as in DACs driving cathode-ray tube displays, it is necessary to follow the digital-to-analog converter with a sample-hold circuit ("deglitcher").

6.6 CHOICE OF A DIGITAL-TO-ANALOG CONVERTER: DESIGN CONSIDERATIONS

Table 6.3 shows typical DAC specifications. Thin-film resistors with feedback matching contribute little circuit noise and keep gain errors below 5 ppm/°C. They can be matched within 0.1 percent and are then laser-trimmed to allow for MOSFET switch resistances.

MDACS now use MOSFET switches almost exclusively. Bipolar-transistor current switching in fixed-reference DACs permits settling times well below 100 ns. This permits, e.g., high-resolution sampling at 50- to 300-kHz rates for audio applications. Compact-disk player DACs need such sampling rates (for oversampling and stereo-channel multiplexing) together with high resolution (16 to 18 bits) and excellent linearity. Overall gain and offset accuracy is less important. Such DACs are mass-produced at relatively low cost (Ref. 25).

Either CMOS or bipolar logic works with TTL digital control levels. IC chips can combine single or multiple DACs with DAC and buffer registers, output

TABLE 6.3
Typical digital-to-analog converters for 5- to 10-V operation

Most commercial units provide current output into an operational amplifier not part of the following specifications. *Differential nonlinearity* is typically $\pm 1/2$ LSB, and *gain error before calibration* (if any) is 0.1 to 0.2 percent. *Sine wave phase-shift error of multiplying DACs* depends on the output amplifier as well as on the DAC and is rarely properly specified.

1. *Low-cost:* 12 bits, 10 to 500 kΩ, total nonlinearity 0.01 to 0.04 percent of full scale (0 to 50°C, more expensive for −50 to 150°C); settling time 0.5 to 4 μs within 1 LSB. Mostly CMOS, can multiply.
2. *Slow, high-quality, for instrumentation, digital attenuators:* 14 to 18 bits, total nonlinearity < 0.01 percent of full scale (0 to 50°C); settling time 5 to 50 μs within 1 LSB. May use correction DACs controlled by digital logic or ROMs. Can multiply.
3. *Fast, high-quality, general-purpose:* 12 bits, total nonlinearity 0.01 to 0.05 percent of full scale (0 to 50°C, more expensive for −50 to 150°C); settling time 30 to 600 ns to 1 LSB. CMOS types can multiply.
4. *Very fast, low-resolution* (display and video applications): 4 to 12 bits, resistance 1 to 10 kΩ, total nonlinearity 0.02 to 0.1 percent of full scale (0 to 50°C); settling time 30 to 100 ns within 1 LSB. Except for some display converters, most do not multiply.
5. *Very fast, high-resolution* (for audio applications): 14 to 16 bits, total nonlinearity < 0.01 percent of full scale (0 to 50°C), offset and gain errors less important; settling time 30 to 50 ns within 1 LSB. Typically use fast bipolar current switches, do not multiply.

amplifiers, and/or Zener reference-voltage sources. But the most accurate DACs are on separate chips to keep heat sources away. On small IC chips, amplifier heating not only affects dc drift but also can actually increase the settling time after large voltage excursions ("thermal tails"). Finally, care is required to separate analog and digital ground-current returns.

6.7 ANALOG-TO-DIGITAL CONVERSION

Analog-to-digital converters (*ADCs, encoders*) produce digital output proportional to an analog input voltage X_A or, more often, to the ratio of an analog input voltage X_A and an internally or externally supplied *reference voltage E*. As a rule, each conversion is started and timed by a *CONVERT command* in the form of a pulse or level change supplied by a switch, digital computer logic, timer, or sensing device. Each conversion requires a finite time, during which the input may have to be held constant by a sample-hold circuit (Sec. 5.8). End of conversion is normally indicated by a flip-flop level change (*DONE flag, CONVERSION COMPLETED flag*), since the digital result must not be used or transferred before the conversion is finished.

The basic ingredients of all electronic ADCs are *analog comparators* (Sec. 3.2), whose binary output indicates the sign of an analog input sum or difference.

6.8 PARALLEL-COMPARATOR AND CASCADE-ENCODER ADCs

The simplest analog-to-digital converters employ multiple parallel comparators for individual digital output levels (Fig. 6.15). This requires $2n$ comparators for n-bit conversion and is, therefore, not practical if high digital resolution is needed. But *parallel-comparator ADCs ("flash converters") are the fastest of all*, since all digits are produced simultaneously. 3- to 8-bit flash converters using bipolar or CMOS transistors-pair comparators are commercially available as monolithic integrated circuits with integral latches and decoders. Conversion times of 2 to 5 ns permit sampling rates as high as 500 MHz, for example, for digital television systems (Refs. 2 and 20 to 23). The built-in latching flip-flops following each comparator act as sample-holds, but very fast track-hold circuits are often used ahead of flash converters to permit better comparisons.

Cascade encoders (subranging encoders) reduce the number of comparators but *produce binary digits, or groups of binary digits, in succession*. Referring to Fig. 6.16*a* with $X_A > 0$, comparator 1 decides whether $X_A > E/2$. If $X_A > E/2$, then comparator 2 decides whether $X_A > E/2 + E/4$. If $X_A < E/2$, then comparator 2 decides whether $X_A > E/4$, etc.

The basic cascade converter of Fig. 6.16*a* requires one clock period per bit. But for periodic sampling, one can separate the converter stages with analog delay lines one clock period (say 5 to 100 ns) long, so that the earlier stages can begin to work on the next incoming sample before each conversion is completed ("pipelining"). Moreover, each comparator in Fig. 6.16*a* is replaced by a 4- to

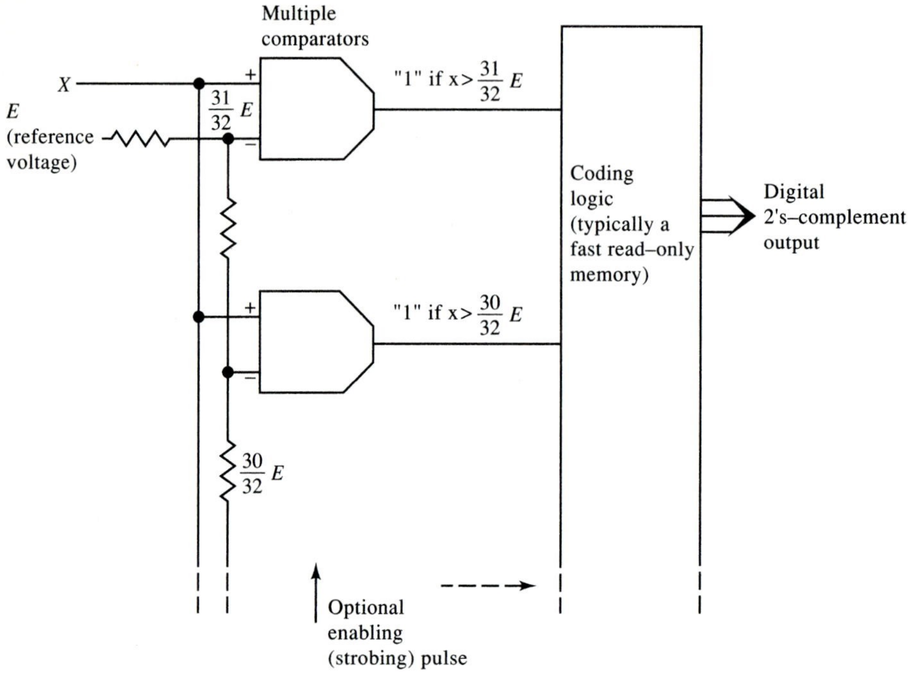

FIGURE 6.15
A "flash" converter using multiple parallel comparators. The integrated-circuit comparators are fast bipolar or CMOS differential-amplifier stages, each driving a latching flip-flop sampled by successive convert-command pulses. The latches are followed by bipolar or CMOS decoding logic. The comparators are fast (2 to 10 ns) but not very accurate (typically 100 to 300 mV).

8-bit flash converter delivering multiple bits during each clock period ("folded flash converter"). Figure 6.16*b* shows one of many possible circuit arrangements.

6.9 RAMP-COMPARATOR CONVERTERS

A simple *ramp-comparator converter or analog-to-time-to-digital* converter gates clock pulses into a counter while an accurate ramp voltage varies between zero and the analog input $X_A > 0$ (Fig. 6.17). The counter, which serves as the ADC output register, is then read and reset to zero. The second comparator response can initiate a DONE (conversion completed) signal.

Either binary or decimal counters can be used. To read positive or negative voltages between -10 and $+10$ V, we can set the first comparator at -10 V instead of 0 V and employ a binary (two's-complement) counter initially reset to -10.00 instead of 0. The converter will then read the correct two's-complement code.

The two comparators and the output logic can be a single-chip integrated circuit, so that comparator levels are closely matched and will drift together with temperature changes.

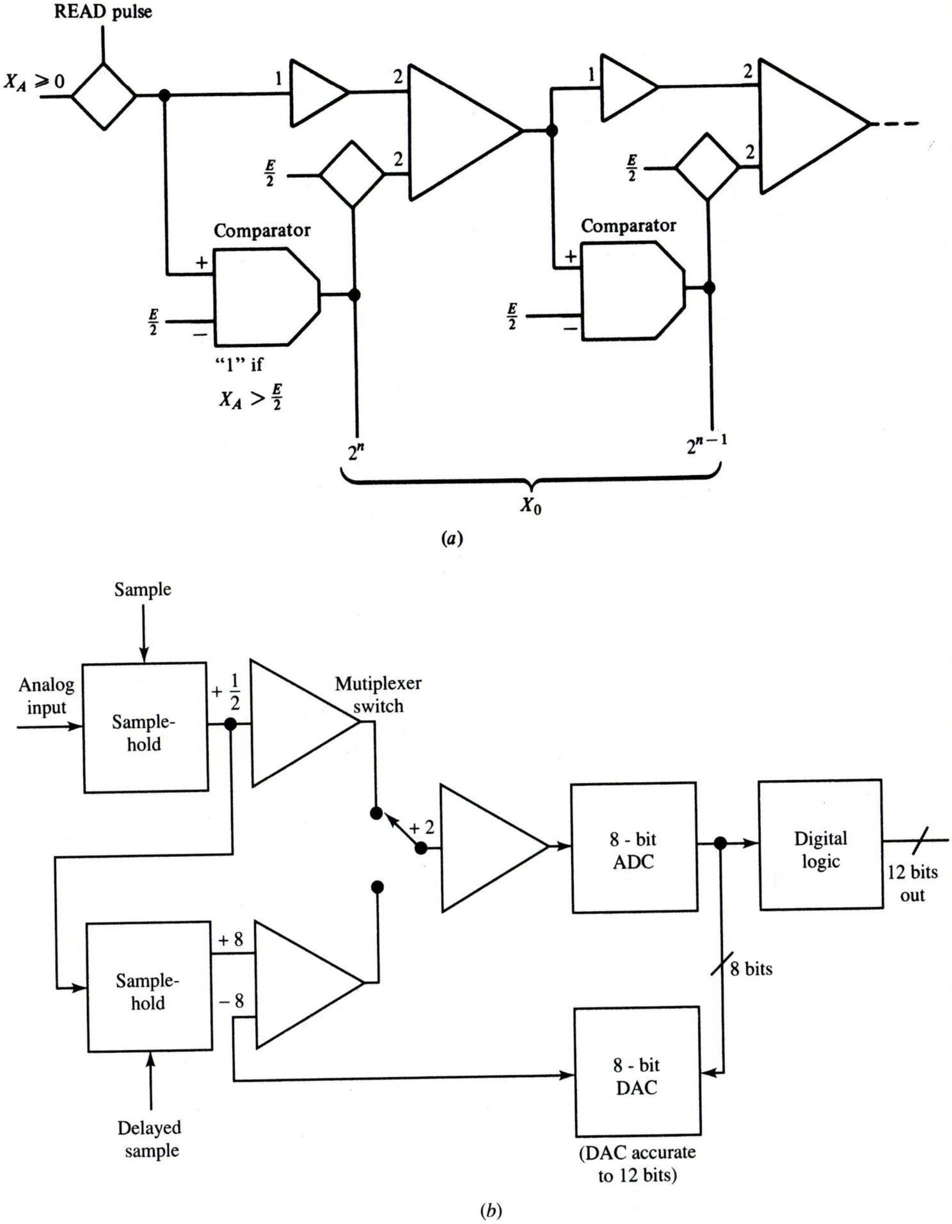

FIGURE 6.16
(*a*) Basic cascade encoder; (*b*) simplified circuit of a cascade encoder for a digital oscilloscope capable of sampling at 4 MHz (Ref. 25). A single fast ADC is multiplexed to convert first the most significant 8 bits and then 4 more bits of the analog error resulting from the 8-bit conversion. All 12 bits are combined by digital logic. The 8-bit DAC (but not the 8-bit ADC) must have 12-bit accuracy. Pipelining permits the DAC to settle while the ADC already works on the next sample.

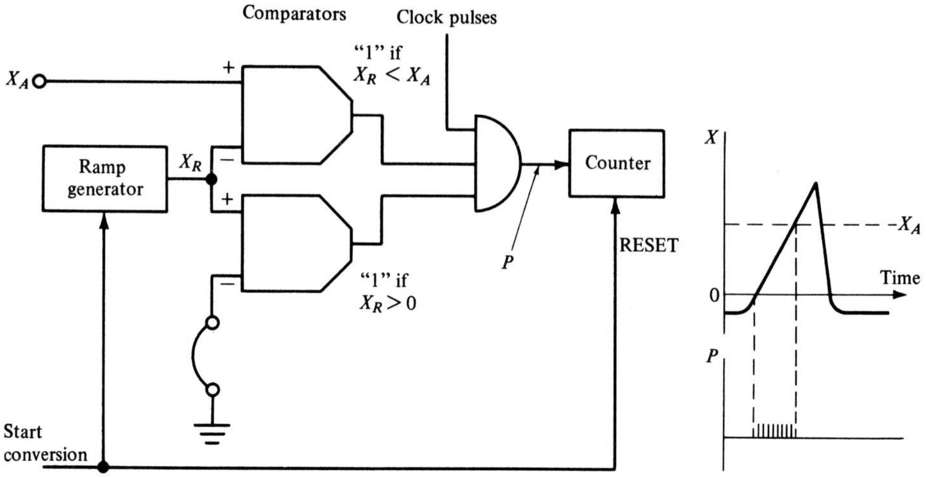

FIGURE 6.17
A simple ramp-comparator ADC.

6.10 INTEGRATING CONVERTERS

If a constant or variable voltage $X_A > 0$ (one could handle negative inputs by adding a known positive offset) is applied to the input of an electronic integrator initially reset to zero, the integrator output will take ERC/X_A s to reach the negative reference level $-E$. In *analog-to-frequency-converter ADCs* the integrator is then quickly reset to zero, and the process repeats (Fig. 6.18*a*). The average frequency of the reset pulses, measured by the pulse count during a specified *observation period* T, is approximately

$$\frac{N}{T} \approx \frac{1}{ERC}\overline{X}_A \qquad \overline{X}_A = \frac{1}{T}\int_0^T X_A(t)dt$$

where we have neglected the effect of the integrator-resetting time. This will be permissible if T_{RESET} is less than the desired measurement error percentage of the integrator time constant RC and $X_A \leq E$. In practice, T_{RESET} can be less than 10 μs, so that $RC = 10$ ms is large enough for 0.1 percent error (some of the reset-time error could be "calibrated out").

Reset-time errors could be reduced by employing two integrators, one integrating while the other one resets. A better voltage-to-frequency conversion scheme switches the input of a single integrator between X_A and $-X_A$ whenever the integrator output reaches $-E$ and $+E$. This method does away with reset switching and also compensates integrator dc offsets.

The *dual-slope integrating ADC* shown in Fig. 6.18*b* starts with its integrator output at 0 V and then integrates the analog input $-X_A < 0$ for N_R clock pulse periods, or $T = N_R\Delta t$/s. Digital logic now switches the integrator input to a $+E$ reference, and comparator/logic circuits count N pulses while the integrator

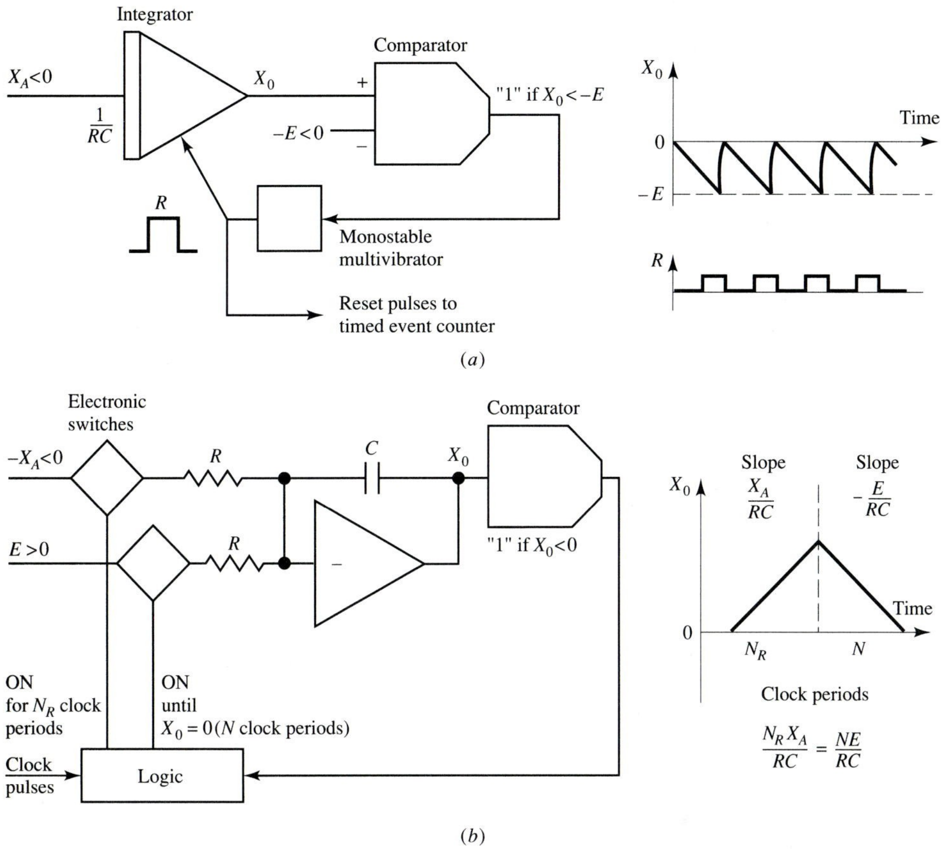

FIGURE 6.18
(*a*) Analog-to-frequency converter; (*b*) dual-slope converter.

output returns to 0 V. Then

$$N = \frac{N_R}{E}\overline{X}_A \qquad \overline{X}_A = \frac{1}{T}\int_0^T X_A(t)dt$$

which is independent of the integrator RC and also of the clock frequency, so that no precision crystal oscillator is needed. Dual-slope ADCs serve in both inexpensive and precision (0.01 percent of half-scale $+\frac{1}{2}$ LSB) digital voltmeters, with conversion times between 0.1 and 50 ms. Bipolar operation is obtained with an input voltage offset as in Sec. 6.8.

Integrating ADCs read the *time average* $\overline{X}_A$ of a possibly variable analog input X_A over a specified observation time T. *This is especially useful for rejecting periodic and nonperiodic high-frequency noise in low-level dc measurements*. In particular, line-frequency noise effects can be reduced by at least 50 dB if we make the observation interval T a multiple of the line-frequency period. It may also be possible to "synchronize" the observation time with other periodic noise. For *timed* measurements, integrating ADCs will require sample-hold circuits.

6.11 ANALOG-TO-DIGITAL CONVERTERS USING DIGITAL-TO-ANALOG FEEDBACK

For high accuracy and fast conversion independent of capacitor properties, one employs a *feedback digital-to-analog converter* which reconverts the digital output X_D to a voltage EX_D, where E is the *ADC reference voltage* (Fig. 6.19*a*). The feedback voltage EX_D is now compared with the analog input X_A. With $|X_A| \le E$ and $|X_D| \le 1$, we vary X_D during each conversion cycle until the *absolute analog error* $|e_A| = |EX_D - X_A|$ is sufficiently small, i.e., less than $\frac{1}{2}$ LSB:

$$X_D = \frac{X_A + e_A}{E} \qquad |e_A| < \frac{E}{2^n}$$

where n is the required number of bits, *including the sign bit*. Different schemes for varying X_D yield various compromises between circuit simplicity and maximum n-bit conversion time. We measure conversion time in clock periods, where each clock period must be sufficiently long to permit comparator, logic, and D/A converter switches to settle. Typical clock periods will be between 0.05 and 10 μs.

In the simplest scheme, the D/A converter register is a clock-driven 2^n-step counter which starts at the most negative value once per conversion cycle and counts up to produce a staircase D/A output analogous to the ramp voltage in Fig. 6.17*a*. X_D is read when the staircase voltage reaches X_A. *The maximum conversion time is* 2^n *clock* periods, since each staircase step will be just $2E/2^n$ V. Such converters are used in some inexpensive digital panel meters.

"Continuous" Converters

To reduce the average conversion time, the D/A converter register can be a reversible counter; we count up or down as required by the sign of the error $e_A = EX_D - X_A$ (*continuous* or *incremental converter, digital servo*, Fig. 6.19*b*). A small comparator dead space about $e_A = 0$ will prevent oscillation about the correct output. Incremental converters save an appreciable amount of time if X_A is not a random input (like a multiplexer output) but a time-variable analog voltage *known to increase and decrease no faster than* $R < 2E/2^n$ *V per clock period.* In this case, the first conversion can take up to $2^n R/2E$ clock periods; the converter will then *track the signal* in successive clock periods. We can track faster signals at the expense of circuit complexity if we add comparators which sense large analog errors and transfer the counter input to a higher digit position (variable-increment counter).

Successive-Approximation Converters

For faster conversion, we program the D/A converter register with digital logic to "weigh" X_A by successive binary-digit approximations. We start with $X_D = 0$

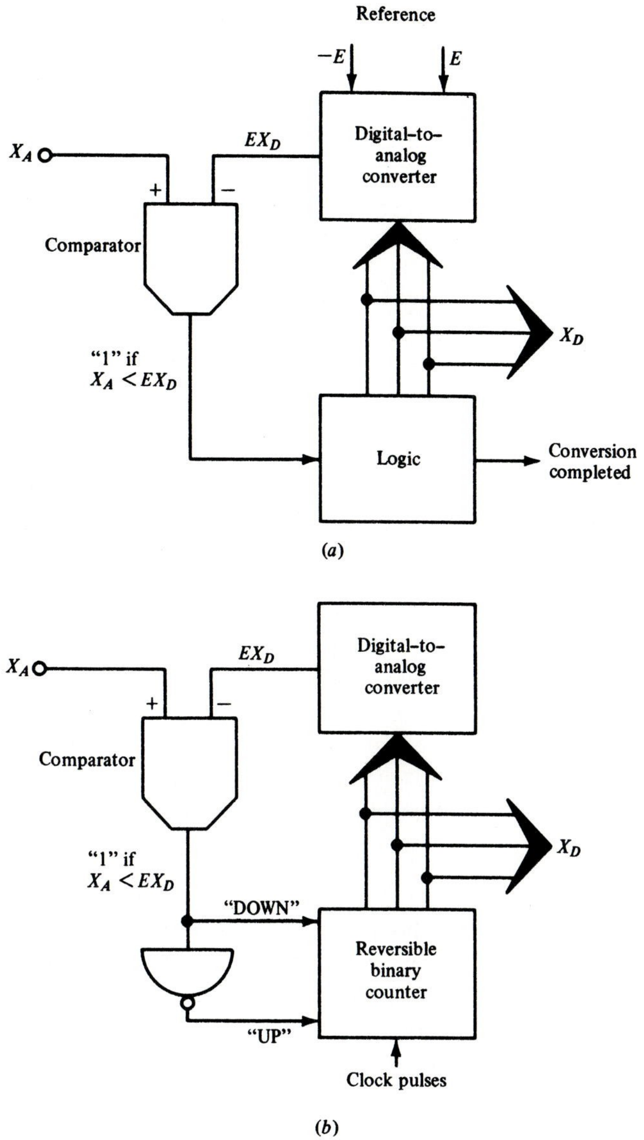

FIGURE 6.19
Principle of (*a*) a feedback-type ADC and (*b*) a "continuous" (reversible-counter-feedback) converter.

to find the sign of X_A. If $X_A > 0$, we next try $EX_D = E/2$ to see if the most significant binary digit is 0 or 1, and we continue to the last binary digit *(successive-approximation converter). The required conversion time will be n clock periods, one for each bit in* X_D.

Figure 6.20 illustrates the operation of a typical successive-approximation converter. The digital output X_D, and thus the DAC output voltage EX_D, is set

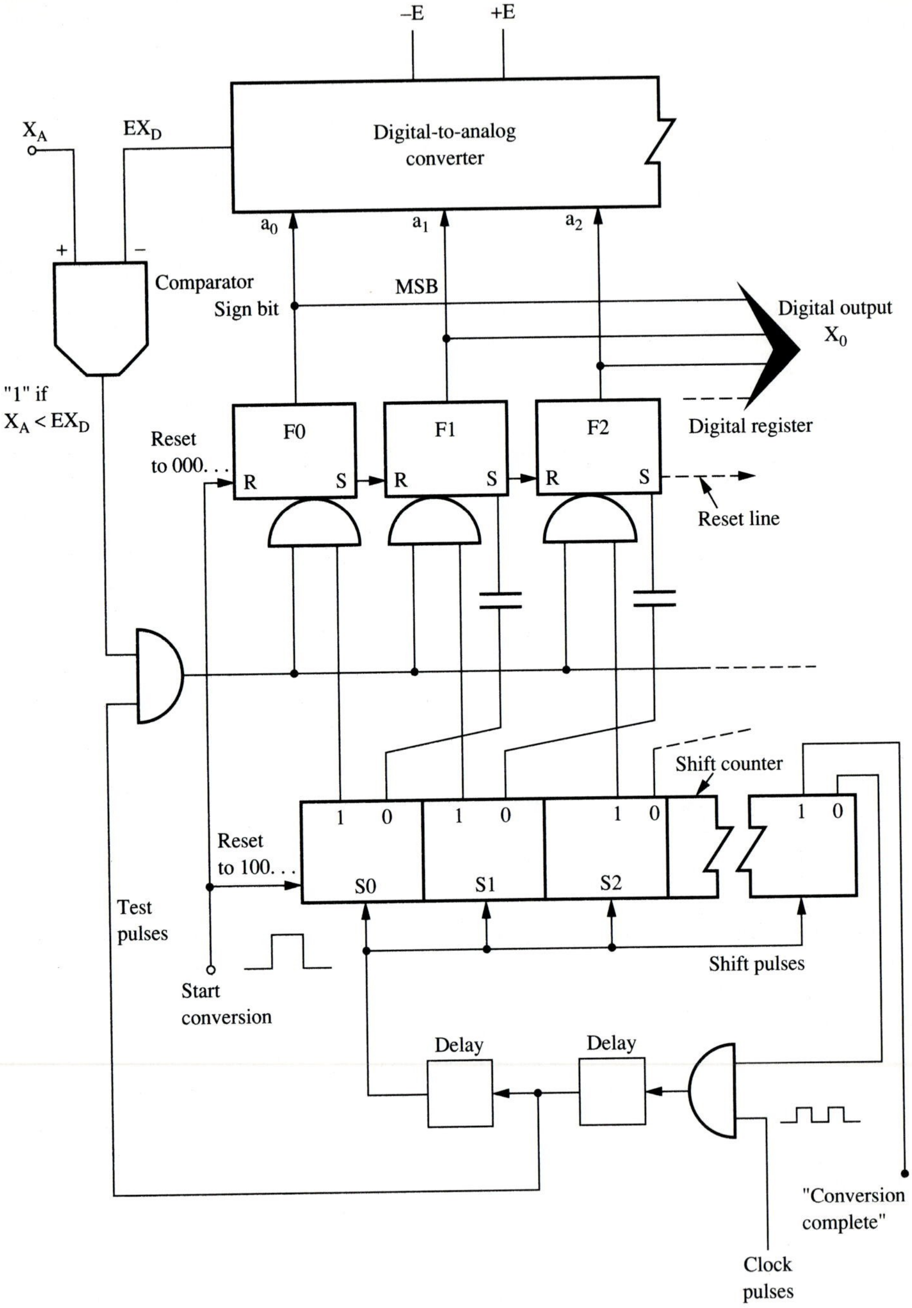

(*a*)

FIGURE 6.20
(*a*) Successive-approximation ADC. (*continued on next page*)

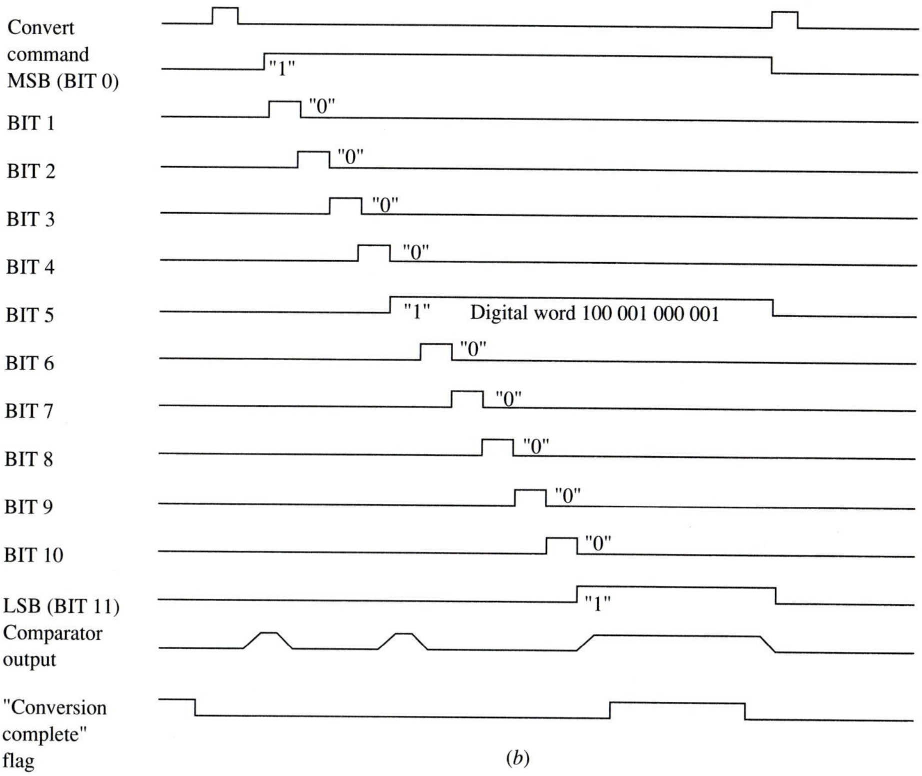

FIGURE 6.20 (continued)
(*b*) Timing diagram showing conversion of successive bits.

by the *ADC (output) register*, comprising *n* flip-flops F0, F1, F2, ... Successive X_D values will be programmed by a digital 1 passing through an $(n + 1)$-bit *shift counter* (flip-flops S0, S1, S2, ...).

The START CONVERSION pulse resets the shift counter to 1000... and the ADC register to 0000...; this also releases the test pulses and shift pulses produced by the ADC clock. Initially, $X_D = 0000\ldots$. If now $X < EX_D = 0$, the comparator output will be 1, and the first test pulse (suitably delayed to let flip-flops, DAC, and comparator settle) will set the sign-bit flip-flop F0 to 1. The first shift pulse follows and shifts the counter to 0100.... We then have

$$X_D = \begin{cases} 1100\ldots & \text{if } X_A < 0 \\ 0100\ldots & \text{if } X_A \geq 0 \end{cases}$$

and the corresponding DAC output ($-E/2$ or $E/2$) is compared with the analog input X_A. If (and only if) $X_A < -E/2$ or $0 \leq X_A < E/2$, the second test pulse will correctly reset the MSB flip-flop F1 to 0. Otherwise, the most significant bit remains, as it should, at 1. Successively following shift pulses and test pulses similarly determine the other bits of X_D in order of decreasing significance. The *n*th shift pulse shifts the 1 into the $(n + 1)$st shift-counter flip-flop, which serves as a DONE *flag* signaling completion of the conversion.

6.12 ADC SYSTEM DESIGN CONSIDERATIONS

Integrating ADCs can discriminate against periodic disturbances (such as line-frequency noise) and often incorporate autoranging and automatic offset and gain calibration circuits. They serve mainly for dc and low-frequency measurements, from low-cost digital panel meters and pocket multimeters to high-quality instruments; capacitor dielectric absorption typically limits accuracy to 0.01 percent of full scale.

A *data acquistion system* combines an ADC with a sample-hold circuit and possibly an input instrumentation amplifier, which may permit voltage-scale changes. Most data acquisition systems also include an input multiplexer to permit measurements on multiple input channels. This can cause large sample-hold or ADC input steps even when the individual input voltages vary continuously. Instead of a single sample-hold following the multiplexer, one can provide individual sample-holds at each multiplexer input to permit simultaneous sampling; this also reduces sample-hold input steps (but not ADC input steps). The best data acquisition systems have differential-input amplifiers and multiplexers to reduce common-mode errors.

Successive-approximation converters can convert continuous and/or multiplexed analog inputs at sampling rates up to several megahertz with resolutions to 14 bits. Successive-approximation converter *accuracy* and *conversion time* are essentially determined by the feedback DAC (Table 6.3). The comparator, reference supply (typically a built-in buried Zener diode), sample-hold, and multiplexer specifications are chosen so as not to compromise the DAC specifications. The bulk of data acquisition work is done with inexpensive monolithic 12-bit ADCs (which is conservative for 0.1 percent of full-scale instrumentation), at sampling rates below 100 kHz (which permits the use of personal computers). Such ADCs are mass-produced for the commercial 0 to 50°C temperature range. More expensive units accommodate larger temperature ranges and sampling rates.

ADCs for high sampling rates (16 bits at up to 350 kHz for high-fidelity audio, 4 to 8 bits at up to 250 MHz for digital television, 10 to 12 bits for digital oscilloscopes) employ *parallel comparators (flash conversion)*, with or without *subranging*.

Complete 8- to 12-bit ADC systems, including DAC, comparator, reference source, sample-hold, multiplexer, and digital logic can be fabricated on a single IC chip, e.g., using CMOS fabrication. Once again, the more accurate ADCs employ separate chips to separate DAC networks and switches from heat-generating amplifiers and logic. Proper *thermal design of IC chips* (and even printed-circuit boards) is, altogether, crucial for really accurate analog circuits. Careful layout must separate heat sources from heat-sensitive circuits as much as possible, but heat sources should also be arranged so as to heat different circuit components evenly. This can be checked by infrared imaging and with special computer programs for thermal IC design.

REFERENCES AND BIBLIOGRAPHY

1. G. A. Korn and T. M. Korn: *Electronic Analog and Hybrid Computers*, 2d ed., McGraw-Hill, New York, 1972.
2. D. H. Sheingold (Ed.): *Analog-Digital Conversion Handbook*, 3d ed., Prentice-Hall, Englewood Cliffs, N. J., 1986.
3. S. Franco: *Design with Operational Amplifiers and Analog Integrated Circuits*, McGraw-Hill, New York, 1988.
4. J. G. Graeme, G. E. Tobey, and L. P. Huelsman: *Operational Amplifiers*, McGraw-Hill, New York, 1971.
5. G. A. Korn: *Minicomputers for Engineers and Scientists*, McGraw-Hill, New York, 1973.
6. C. R. Pearman and A. E. Popodi: "How to Design High-Speed D/A Converters," *Electronics*, Feb. 21, 1964.
7. C. P. Pracht: "A New Digital-Attenuator System," *Simulation*, April 1967.
8. M. R. Aaron and S. K. Mitra: "Synthesis of Resistive D/A Conversion Ladders for Arbitrary Codes," *IEEETEC*, June 1967.
9. ———:"A Note on the Design of D/A Converters," *IEEETEC*, October 1967.
10. J. Freeman: "Ladder Networks Are Easy to Design," *Electron, Design*, July 5, 1967.
11. M. Walker: "Exploit Ladder-Network Design Potential," *EDN*, Feb. 1, 1969.
12. ———: "Expand Basic Ladders into Complex Networks," *EDN*, Mar. 1, 1969.
13. T. Henry: "Binary D/A Converters Can Provide BCD Conversion," *EDN*, August 1973.
14. O. H. Schade: "CMOS for Low-Power D/A Converters," *Computer Design*, April 1973.
15. J. Graeme: "Monolithic D/A Improves Conversion Times," *EDN*, Mar. 15, 1971.
16. W. C. Goeke: "An $8\frac{1}{2}$-Digit Integrating Analog-to-Digital Converter with 16-Bit, 100,000 Sample-per-Second Performance," *Hewlett-Packard Journal*, vol. 40, pp. 3–14, April 1989.
17. ———: "18-Bit Audio DAC," *Electronic Design*, p. 170. October 1988.
18. ———: "Dual 20-Bit DAC Packs Reference, Interface Circuits," *Electronic Design*, p. 171, January 1989.
19. M. Leonard: "Two-Step 10-Bit Converter," *Electronic Design*, p. 183, November 1988.
20. J. H. Mayer: "Flash A-D Converter Developments," *Computer Design*, pp. 74–79, Oct. 15, 1988.
21. J. J. Corcoran: "One-Gigasample-per-Second ADC," *Hewlett-Packard Journal*, vol. 39, pp. 59–66, June 1988.
22. S. Runyon: "One-Chip ADCs Reach 2 GHz," *Electronics*, p. 77, Feb. 18, 1988.
23. D. Cormier: "Analog-to-Digital Conversion in a Flash," *ESD*, p. 37, May 1988.
24. P. Burton: "CMOS DAC Application Guide," *Analog Devices, Inc.*, Norwood, Mass., 1984.
25. "Handbook of Linear IC Applications," Burr-Brown Corporation, Tucson, Ariz., 1988.
26. "Data Converters and Voltage References," Maxim Integrated Products, Inc., Sunnyvale, Calif., 1989.
27. "Data Acquisition Handbook," Intersil, Inc., Cupertino, Calif., 1988.
28. A. Gee and R. W. Young: "Signal Conditioning and Analog-to-Digital Conversion for a 4-MHz, 12-Bit Waveform Recorder," *Hewlett-Packard Journal*, pp. 15–22, February 1988.

PROBLEMS

6.1. Draw a voltage-ladder circuit with SPDT bit switches switching between $-X_1$ and $+X_1$ in the manner of Fig. 6.3*c*.

6.2. Prove, with the aid of suitable equivalent circuits, that the "inverted" ladder DAC circuit of Fig. 6.7 will indeed multiply correctly by

$$Y_D = -a_0 + \frac{1}{2}a_1 + \frac{1}{4}a_2 + \cdots + \frac{1}{2^{n-1}}a_{n-1}$$

6.3. Derive a similar proof for the two current-ladder DACs of Fig. 6.11.

6.4. The accuracy of DAC/operational amplifier circuits will depend on their feedback ratio β. Find β for each of the circuits of Figs. 6.1; 6.3*a*, *b*, and *c*; 6.5*c*; and 6.7.

6.5. Find the percentage error due to a 0.01 percent error in the input resistors R, $2R$, $4R$, etc., in Fig. 6.3*a*. This may be due to manufacturing tolerance, temperature effects, or switch on resistance.

6.6. For the (1, 2, 4, 8) BCD converter of Fig. 6.8, use the equivalent-circuit technique of Fig. 6.4 to prove

$$X_0 = -\alpha X_1 \left[a_1 + \frac{1}{2}a_2 + \frac{1}{4}a_3 + \frac{1}{8}a_4 + \frac{1}{10}\left(b_1 + \frac{1}{2}b_2 + \frac{1}{4}b_3 + \frac{1}{8}b_4\right) + \frac{1}{100}\left(c_1 + \frac{1}{2}c_2 + \frac{1}{4}c_3 + \frac{1}{8}c_4\right)\right]$$

What is the scaling constant?

INDEX

Absolute-value circuit, 171
Accuracy:
  multiplier dynamic, 204–205
  multiplier static, 204
ADC specifications, 374
Allen, P. E., 315
All-pass function, 243–244
Amplifier:
  AC, 70
  combined inverting and noninverting, 39
  differential, 32
  instrumentation, 52
  noninverting, 28
  operational (*see* Operational amplifier)
Analog comparator:
  dead-space, 231
  feedback, 150–159
Analog Dialog, 141, 222
Analog-digital converter, 365
Analog multiplexer (*see* Multiplexer)

Bandpass function, 242
Band-reject filter, 243
BCD code, in DAC, 364
Beta (*see* Voltage feedback ratio)
Binary codes, 344–347
  for voltage, 348
Binary digit, 344
Binary fraction code, 345
Biquadratic function, general form of, 274
Bistables, 187
Bit, 344
  switch, 353
Bode plots, 95
Boothroyd, A. R., 222
Bridge amplifier, 72
Bridge circuit. four diode, 158
Broderson, A. J., 141
Bruton, L. T., 289
Buffer register, 349
Burr-Brown Research Corp., 2, 275
Butterworth function, 250

Callahan, M. J., 222
Cascade encoder, 365–366
Cate, T., 222
Cauer function (*see* Elliptic function)
Chebychev function, 250
Chennette, E. R., 141
Chopper stabilization, 114
Clamps. level (*see* Feedback limiters)
Clayton, G. B., 58, 222
CMOS switch, 331, 332
  for DAC, 355
CMRR (*see* Common-mode rejection ratio)
Code, integer, 344
Common-mode effects, 103
Common-mode impedance, 76
Common-mode rejection ratio, 104
Common-mode voltage, 33
  maximum input, 128
Compact disk system, DAC, 364
Comparator:
  dead-space, 245
  feedback, 161
  noise-immunity, improving with
    hysteresis, 178
  Zener diode, 154
Compensation for stability, 101
Components:
  capacitors, 136
  resistors, 136
Converter, continuous, 370
Correction scheme, for DAC, 360, 364
Cosine function, 236
Current, maximum output, 7

Current sources, 55–57
  Howland, 56
Current switch, DAC. 357–360
Cutoff frequency:
  for high-pass second-order functions, 242
  for low-pass second-order functions, 241

DAC specifications, 360–365
Data acquisition, 374
DC blocking, 48
DC offset (*see* Offset)
Dead-space circuit:
  passive, 156
  precision, 184
Deliyannis, T., 269
Depletion mode, 328
Detector, peak, 187–190
Differential amplifier, 32
  effect of unbalance in source resistances, 66
  input impedance, 66
Differential nonlinearity, in DAC, 360–361
Differentiator, 49
  two-amplifier, 49
Diode bridge, 158
Diode model, 153
Diode switch, 319–321, 332
Division:
  implicit, 205
  current ratioing, 204
Dobkin, R. G., 222
Double-buffered DAC, 347
Drift, offset, 112
Dual-slope converter, 368–369

Electronic Design magazine, 58
Electronics magazine, 58
Elliptic filter, second-order, 262, 285
Elliptic function, 262
  third-and fourth-order, 285
Equal-ripple magnitude function (*see* Chebychev function)
Error:
  fractional, 82
  multiplier, 203
Exponentiation operator, 197

FDNR, 289
Feedthrough error, 362
Filter (*see* specific types)
Flash converter, 365–366
Fleischer, P. E., 274
Follower, unity gain, 28
Frequency, unity-gain, 75, 77
Full-power response, 126
Function generation, implicit, 185

Gain:
  closed-loop, 80
  open-loop, 79
    measurement of, 144
    polarity convention of, 76
Generalized impedance converter (*see* GIC)
Giacolletto, L. J., 58
GIC, 287
Gilbert, B., 222
Glitch, 363–364
Graeme, J. G., 141, 222
Ground systems, 131
Gupta, S. L., 141

High-pass filter, 242
High-pass function, 242, 255
High-Q filter, 272
Hold mode, 334
Huelsman. L. P., 58, 141, 263, 272, 315
Hysteresis, suppressing comparator noise with, 178

Impedance:
  common-mode, 75
  differential input, 4, 76
  input: differential amplifier, 33
    inverting amplifier, 9
    noninverting amplifier, 29
      output, closed-loop, 83
      open-loop, 76
  short-circuit transfer, 23
Initial conditions (*see* Mode control)
Instrumentation amplifier, 52
Integrating converter, 368–369
Integrator:
  block-diagram symbol, 16
  inverting, 12
  noninverting, 64
Inverted ladder network, 352
Inverted transistor, 321–323
Inverter:
  transistor, 157
    unity-gain, 9
Inverting amplifier, 7–12

Jaeger, R. C., 141
Johnson, D. E., 315
Johnson, J. R., 317
JFET switch, 325–327
Junction, summing, 11
Junction field-effect transistor (*see* JFET)

Kerwin, W. J., 263, 272
Key, E. L., 250
Kirchhoff summing, 36

Korn, G. A., 58, 141, 222
Kostanty, R. G., 58

Ladder circuit, ADC, 332
Ladder circuits, DAC, 347–356
Leakage:
  capacitance, 318
  current, 318
  resistance, 318
Least significant bit (*see* LSB)
Limitations, large-signal, 125
Limiter:
  feedback, 165
  precision bridge, 175
  soft diode, 165
  transistor, 174
  Zener diode, 163
  zero, 167
Linear phase function, 241
Logarithmic operator, 193, 195
Low-pass filter
  characteristics of, 241
  first-order, 277
Low-pass function, 240, 250
  characteristics of, 241
  *n*th order, 241
  transformation to bandpass, 283
Low-pass to bandpass transformation, 283
LSB, 344–346

Margin:
  gain, 96
  phase, 96
Maximally flat delay function (*see* Thompson function)
Maximum circuit, 187
MDAC, 362
Minimum circuit, 187
Mode control, integrator, 337–339
Model:
  bipolar transistor, 154
  general linear, 75
  ideal diode, 153
  dc offset, 107
  ideal, 5
  linear, 76
  noise, 116
  one-pole, 77
Module, multfunction programmable, 201
Moore, H. P., 315
Morrison, R., 141
Moschytz, G. S., 315
MOSFET switch, 328–332
  in DAC, 365
Most significant bit (*see* MSB)

MSB, 344–346
Multiple-feedback filter, 265
Multiplexing, 316
Multiplier:
  applications, 201
  four-quadrant, 201
  Gilbert, 201, 222
Multiplying digital-to-analog converter (*see* MDAC)

Network, tee, 25
Newcomb, R. W., 272
Noise,
  internal sources, 114
Noninverting amplifier, 28

Offset voltage, 318
Offset, dc, 107
One's complement code, 345, 346
Operational amplifier,
  advantages, 1, 2
  characteristics, 2, 3, 75
  packaging, 2
  symbol, 4
Orchard, H. J., 312
Oscillator:
  FM sine waves, 217
  FM square-wave, 227
  phase-shift, 218
  sine-wave, 215
  square-wave, 220, 221
  triangle-wave, 221
  Wien bridge, 218
Output, rate, 125
Output impedance, closed-loop, 85
Overload recovery, 128

Papoulis, A., 141
Peak detectors, 189
Peck, R. L., 222
Polarity convention, 5
Pole-zero pair circuit, 42
Power boosting, 129
Power spectral density, 120
Power supplies, 131
  single, 57
  current pulsations in, 132
Precision limiter, 175
PSPICE, 141

Ramp-comparator ADC, 366–367
Rate-of-closure, 96
Ratio:
  common-mode rejection, 101
  voltage feedback, 81

Reagan, J. A., 222
Rectifier:
  full-wave, 181
  half-wave, 171
Reference voltage, 344
  built-in, 358
Rejection, common-mode (*see* common-mode rejection)
RMS operator, 208

Sallen, R. P., 250
Sample-hold, 333–337
  specifications, 337
Saturation, 155, 322
Scale factor for multiplier, 202
Schwartz, M., 222
Searle, C. L., 222
Selectors, amplitude, 187
Sensitivity, 254
Series/shunt switch, 330, 332
Series switch, 317
Settling time, DAC, 362
  operational amplifier, 128
Shah, M. J., 73
Shehan, D. F., 312
Shielding, 134
Short-circuit stable, 100
Shunt switch, 317, 319
  transistor, 324
Sign-and-magnitude code, 345
Sine function, 236
Slewing rate, 126
Smith, John L., 58
Specification:
  noise, 115
  typical operational amplifier, 75
SPICE, 141
Spike (*see* Switching spike)
Square-rooting, 206
Stability:
  closed-loop, 96
  criteria for, 96
Static error, DAC, 360, 364–365
Step response, DAC, 362
Stewart, E. L., 141
Successive-approximation converter, 370–374
Summer-integrator, 12
Summer-inverter, 7
  block diagram symbol, 11
Summing, Kirchhoff, 37
Switch,
  BJT, 321
  diode, 319–321
  FET, 325
  ideal, 317
Switch, series, 318
Switch, shunt, 318
Switched amplifier, 340–342
Switching spike, 318
  MOSFET, 329
  in sample-hold, 335–336
  in DAC, 363

Tee network, linear, 26
Tee network, switch, 317, 319
  DAC, 351
    (*see also* series/shunt switch)
  of resistor network, 352
  MOSFET switch, 333
Thermal design, of ADC, 374
Timing error, in sample-hold, 336
Tobey, G. E., 222
TRACK mode, 334
Transfer function, first-order, 40
Transistor switch, 321–325, 332
  nonsaturating, 325, 357, 358
Tschebyshev function (*see* Chebychev function)
Tuning procedure, filter, 286
Turnoff time, diode, 321
Two's-complement code, 346, 351

Unity-gain follower, linear analysis of, 89
Unity-gain inverter, 9

Vanvalkenberg, M. E., 315
Varactor amplifier, 114
VCVS filters, 255
Vector resolution, 236
Voltage controlled voltage source (*see* VCVS)
Voltage differential input, 33
Voltage ladder, 352
Voltage regulators, 54
Voltage sources, reference, 54
Voltmeter:
  AC, full-wave, 181
    half-wave, 171
  RMS, 210

Wave form generation:
  sine-wave, 212
  square-wave, 217
  triangle wave, 217

Y parameter in active RC filter analysis, 248

Zener-voltage reference, 365